Radionuclide Technology

For Alma, Lâtife, Nuraya, Jenny, Jean, Linda, Mary, Madeleine, Yvonne and several Susans

Radionuclide Technology

an Introduction to Quantitative Nuclear Medicine

K.F. Chackett
Principal Physicist
Dudley Road Hospital, Birmingham

VNR VAN NOSTRAND REINHOLD COMPANY
New York—Cincinnati—Toronto—London—Melbourne

Published by Van Nostrand Reinhold Company Ltd., Molly Millars Lane, Wokingham, Berkshire, England

Published in 1981 by Van Nostrand Reinhold Company, 135 West 50th Street, New York, NY 10020, USA

Van Nostrand Reinhold Limited, 1410 Birchmount Road, Scarborough, Ontario, M1P 2E7, Canada

Van Nostrand Reinhold Australia Pty. Limited, 17 Queen Street, Mitcham, Victoria 3132, Australia

Library of Congress Cataloging in Publication Data

Chackett, K F

Radionuclide technology.

Bibliography: p. 391

Includes index.

1. Nuclear medicine. 2. Radioisotopes. 3. Radioisotopes in medical diagnosis. I. Title.

R895.C46 616.09′57 79-19258

ISBN 0-442-30170-7

ISBN 0-442-30171-5 pbk.

Printed and bound in Great Britain at The Pitman Press, Bath

Preface

There are very few people alive today who can remember the world before the discovery of radioactivity by Becquerel and the isolation of polonium and radium by the Curies. Ever since those early days, there has been interest in finding out whether radioactive materials could be used to the benefit of man, either diagnostically in the characterisation of disease states, or therapeutically in their direct treatment. At first, progress in these fields was slow, if sometimes spectacular; radioactive material was difficult to obtain and its use was bedevilled by lack of experience, lack of adequate facilities for measurement and monitoring, and ignorance of the inherent dangers both to the investigators and their patients. As far as supply problems are concerned, the discovery by Irene Curie and Joliot of 'artificial', i.e. induced, radioactivity in what are normally stable elements began to change the picture, and the development of the cyclotron by Lawrence and the means of generating neutrons by Chadwick paved the way to a vast expansion culminating in our own time in the proliferation of nuclear reactors and weapons. In all this time, a characteristic feature of the scene has been the continuous interplay between the roles of the chemist and the physicist, as well as of people of many nations—an interplay which has sometimes led to a clash of personalities. One of England's greatest physicists—I will not name names—is reputed to have disliked and mistrusted chemists; perhaps the sentiment was reciprocated; and one of the greatest French scientists—I will not be specific as to sex—is reputed to have said 'La radioactivité, c'est une science française!' But in Scandinavia the spectroscopic genius of Ångström flowered again in the persons of M. and K. Siegbahn, and Bohr initiated the drive towards the understanding of nuclear as well as atomic structure. It was the German chemist Otto Hahn who set the physicists' world at odds by the identification of the products of fission in uranium; the Italian physicist Segrè whose chemical tests proved the existence of the hitherto unknown element technetium—that element which, forty years on, is used far more than any other in hospitals the world over for diagnostic tests of nearly every major organ of the human body. The Austrian-born Fritz Paneth seems to have been the first properly to appreciate the potential use of radioactive nuclides as tracers, or indicators, to use his favourite term; in applying this idea in living creatures, Hevesy became the pioneer of what we now call nuclear medicine. At the time he was working at the Institute for Theoretical Physics in Copenhagen. The leader of the team which built the first nuclear reactor and made it work was the Italian Fermi; the basic design of the first Canadian reactor, which many people have considered the finest engineering achievement of all, was the Russian-born Kowarski. But all this is history, and to come closer to the present day would mean making an invidious choice of names, many of them of course American.

Yet in all of this exciting development, and perhaps even more so today, when the rate of change is greater than ever before, it seems to me that there are two factors without which the efforts of the famous men and women mentioned above would have been sadly hampered. One factor is the existence of the great army of students, research students, technicians and assistants to the team leaders, who have carried out the basic tests, made time-consuming calibrations and corrections, investigated the apparently bright ideas which in the event turned out to be impracticable, and carefully amassed the vast bulk of stubborn facts. These folk have always needed at least as clear a knowledge of what they were supposed to be doing, and why, as anyone else has; more; especially in these days of weltering paperwork and the insatiable urge to find something to publish in the hope that it marks a significant breakthrough, today's army needs a down-to-earth manual that sets out some fundamental principles and shows how these are relevant to a few practical problems. The other factor is less substantial, but perhaps even more effective; it is the assistance to clear thinking, in anyone's mind, that is provided by an essentially quantitative, i.e. mathematical, approach to the problems.

Accordingly, I have written the present volume, primarily about the techniques for the production and measurement of radioactive nuclides and their clinical use as tracers, and primarily addressed to students and technicians (in the broadest sense). I make no bones about whether the principles invoked belong to the realms of chemistry or physics, or indeed any particular discipline, but seek throughout to derive mathematical relationships so that the reader can see how experimental measurements give data that are meaningful in a variety of contexts.

My long teaching experience has shown that many extremely able people working in biologically oriented fields (and elsewhere, for that matter) do so without much mathematical preparation beyond a nodding acquaintance with statistics. Since the mathematical approach is so much a fundamental feature of the book, I have begun with a whole chapter, illustrated with worked examples and followed by a set of questions, which sets out what I feel to be the minimum requirement for the proper appreciation of the later chapters. Long as it is (and some recent authors have relegated the derivation of the radioactive decay law to a two-page appendix), it is by no means exhaustive and indeed is very far from rigorous in a truly mathematical sense. Yet no-one today who studies the transport of radioactivity in living material can get the best out of his quantitative experiments without differential equations and Laplace transforms, which are incomprehensible without an appreciation of what integration is all about. That appreciation must involve some familiarity with differential calculus, which is turn implies at least competence in ordinary algebra and geometry. This chapter will turn no-one into a professional mathematician, but it might prompt the perceptive reader to be properly sceptical when others, including his superiors, jump to conclusions on the basis of inadequate data.

My second chapter, on the basic physics of nuclides, begins with the fundamental properties of stable and radioactive nuclides, and then discusses how radioactive species are formed in nuclear reactions, separated by physicochemical processes, and finally treated in various ways to make them suitable for administration to human subjects in order to get information of clinical value. This chapter exemplifies what I have already said about the interpenetration of different disciplines, but I have been rather brief in my discussion of pharmaceutical problems as these are becoming more and more the province of specialists.

In the third chapter, the methods for the detection and estimation of radioactive materials are reviewed, and here I have taken as broad a view as possible, beginning with the elementary physics of collision processes between massive particles and electromagnetic quanta and ending with a brief survey of such topics as radiation chemistry. In between, I have treated the features of the interaction of particulate and electromagnetic radiation with matter as far as these are relevant to an understanding of the mode of action of radiation 'counters' and the absorption and scattering of beta and gamma rays.

The fourth chapter is concerned with the basic laws of radioactive growth and decay, and with statistical phenomena associated therewith. This leads to a discussion of the statistical errors inherent in radioactive measurements.

In the fifth chapter, I discuss the estimation of radioactive material *in vivo*, and, for the sake of completeness, of radioactive samples of body fluids *in vitro*. Some design features of scanners and gamma cameras (excluding detailed electronics) are described at some length. The clinical significance of all these measurements and displays is reviewed but only briefly and as a guide to the much more extensive treatments of these matters available in other books written by and for medical specialists.

Finally, the sixth chapter turns to dynamic studies, for which radioactive nuclide measurements *in vivo* provide information about material transport in living systems of unique value. This information is evaluated in its most succinct form by utilising the transform technique developed purely academically in Chapter 1. The book concludes with a discussion of how this same approach provides an extremely powerful way of assessing, in a physical way, the performance of radionuclide, and indeed other, imaging equipment.

My book will be of most value to those in, or contemplating entering, a hospital environment, as medical physics technicians, medical physicists, radiopharmacists, radiographers, or physicians specialising in nuclear medicine. The latter, particularly, may remark on the fact that in the body of the book I have studiously refrained from referring to 'the patient' and indeed from making any comment about patient care. This is not from any wish to appear coldly scientific, to the point of being inhumane, but rather that I would wish my readers to obtain instruction on the proper care of the sick from others much better qualified to give it than myself. But I believe that everyone who works in a hospital does so at least partly because of a love for his fellow beings and a desire to be of service. Let me be bold enough to assume that this belief is universally accepted, and that having been stated once explicitly on this page my readers will understand it to be implicit on all the others.

It is a pleasure to acknowledge the help and encouragement I have had from many colleagues in the preparation of the book. Especial thanks are due to Mr. R.F. Farr, Chief Physicist at the Queen Elizabeth Hospital, Birmingham, and to Dr. A.B.M.G. Mostafa, Senior Physicist at Dudley Road Hospital, Birmingham, who read all the frequently amended manuscripts and made valuable suggestions and corrections. Messrs. J.G. Cuninghame, M.F. Finlan, J.F.J. van der Grift, and D. Vonberg supplied the information on cyclotron characteristics tabulated in Chapter 2, and Messrs. P. Bradstock and H. Fletcher the information about the J and P tomographic scanner referred to in Chapter 5. There are many references in the text to radiopharmaceutical products available from The Radiochemical Centre, and I am grateful to the Director for permission to abstract from several leaflets. Some of the material presented in Chapter 6 was inspired by the Summer Schools in Dynamic Analysis and related topics, organised over several years by Dr. E.R. Carson at the City University, London. Mr. R.H. Holliday prepared all the line diagrams, and Dr. Mostafa and Miss J. McCulloch prepared all the photographs from scanner and gamma camera pictures obtained in this Department, using Ohio Nuclear and General Electric imaging equipment. Finally, I owe a deep debt of gratitude to my wife who has contributed much in the way of late-night coffee, many hours of critical reading, and a generous share of her own technical expertise. It would be pleasant to think that all these efforts have produced a perfect book, but I am well aware of some deficiencies and I suspect there are a whole host of others. I would like to thank in advance any readers who may care to suggest improvements.

Department of Physics and Nuclear Medicine
Dudley Road Hospital, Birmingham

Contents

1
Basic Mathematics

1.1 Introduction

The main aim of this book is to present the quantitative aspects of the production and measurement of radioactive materials and their utilisation for medical diagnosis. It is addressed primarily to the physicists and medical physics technicians concerned with what has come to be known as nuclear medicine, who will already have some familiarity with mathematical techniques including elementary calculus. Much of the book will, however, be of interest to many others, including technicians in other fields and, of course, biologists and medical staff, who may not have much mathematical preparation. To enable them to come to terms with what follows in this chapter, a brief introduction to imaginary and complex numbers is given in Section 1.23, and some elementary results in differential and integral calculus are summarised in Section 1.24. Proofs of these results will be found in standard mathematical texts.

The development of the use of radioactive materials in many fields, and particularly during the last few years, has necessitated the employment of mathematical methods which are still more sophisticated than those just mentioned. These are well known to mathematicians, but in the author's experience it is not easy for the practising physicist or technician to discover them for himself without protracted study of advanced texts, abstraction of relevant material from among a mass of what, to him, are only side-issues, and tedious transcription of symbols and notation. Accordingly, this chapter is devoted to a brief exposition of four topics of fundamental importance to the rest of the book. Preliminary study of these should facilitate the understanding of later discussions which can proceed untrammelled by the necessity of explaining purely mathematical manipulations. Relieved of this burden, the perceptive reader will perhaps more clearly realise some of the limitations of the mathematical techniques; by any standards, biological systems behave in extremely complicated ways and it is rarely that equations describing their behaviour can be found which are at once a true representation and of a kind simple enough to admit of a solution. We begin with a brief discussion (Sections 1.1–1.6) of Laplace transforms since these are of importance to the solution of differential equations (Sections 1.7–1.15) and, with Z transforms, to the evaluation of convolutions and deconvolutions (Sections 1.16–1.19). The chapter ends with a consideration of the statistics of randomly occurring events (Sections 1.20–1.22). The relevance of these three latter topics lies respectively in the description of the rates of transfer of radioactive materials within biological systems, in the interplay between radioactive material within the biological system and the instrumentation used to display it, and in the limitations of the precision to which any radioactive measurement is subject as a result of the random nature of radioactive decay. Disparate though these topics appear to be, they are in fact closely interwoven, both in mathematical formalism and in their practical significance.

TRANSFORMS

1.2 Laplace Transforms

It is well known that the area under the curve $y = f(x)$ is given by the general expression $\int f(x)\mathrm{d}x$, and in particular that the area under the curve between the ordinates for which $x = b$ and $x = a$ is $\int_a^b f(x)\mathrm{d}x$ (Fig.1.1). This expression is a *definite integral*; its value is calculated by finding $I(x)$, the integral of $f(x)$ with respect to x, and then evaluating $I(b) - I(a)$. For most curves met with in practice, the definite integral $\int_a^b f(x)\mathrm{d}x$ is finite when both a and b are also finite. However, this is not always the case when the integration limits are 0 and ∞.

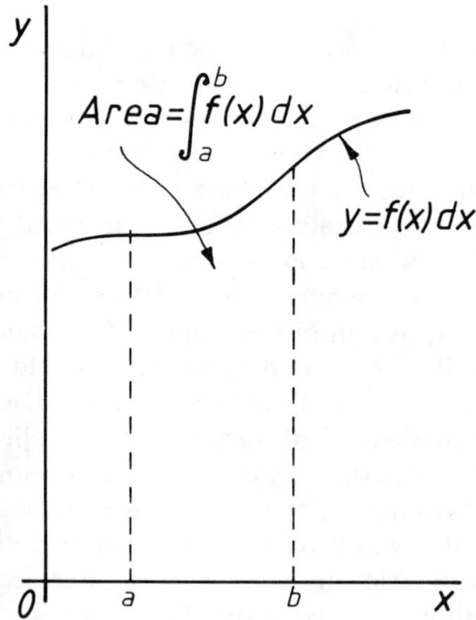

Fig. 1.1 Area under a curve expressed as a definite integral.

Example 1
Let

$$y = f(x) = k \qquad k \text{ any positive constant}$$

Then

$$\int_0^\infty k\,\mathrm{d}x = [kx]_0^\infty$$

$$= k.\infty - k.0$$

$$= \infty$$

Example 2
Let

$$y = f(x) = \exp(-sx) \qquad s \text{ any positive constant}$$

Then

$$\int_0^\infty \exp(-sx)\,\mathrm{d}x = \left[-\frac{\exp(-sx)}{s}\right]_0^\infty$$

$$= 0 - \left(\frac{-1}{s}\right)$$

$$= \frac{1}{s}$$

Let us consider a curve $y = f(x)$ and derive a new curve from it by multiplying the ordinates of all points on it by a factor $\exp(-sx)$ where again s is a constant greater than zero so that $\exp(-sx)$ is less than 1 when $x > 0$. The process is illustrated in Fig.1.2.

This figure shows how every value of y, the ordinate in curve (c), has a length $p.q$ which is the product of the corresponding ordinates in curves (a) and (b). The area under the derived curve to the right of the y axis is clearly $\int_0^\infty f(x) \exp(-sx)\mathrm{d}x$. It will be realised that when the integral is evaluated, it will no longer contain terms in x because the values $x = \infty$ and $x = 0$ will be inserted in the calculation. But it will in general contain the parameter s; in other words, it is a *function* of s. It is called the Laplace transform of $f(x)$ and we write formally

$$\text{Laplace transform of } y = \mathscr{L}(y) \equiv \mathscr{L}\{f(x)\} = \int_0^\infty f(x) \exp(-sx)\mathrm{d}x$$

Strictly speaking, the above definition is that of the 'one-sided' Laplace transform. An alternative approach to the definition is as follows. First, we define a function $u(x)$ which is zero for $x < 0$ and 1 for $x \geq 0$. Then for any given function $f(x)$ the product $f(x)u(x)$ represents a function which is zero

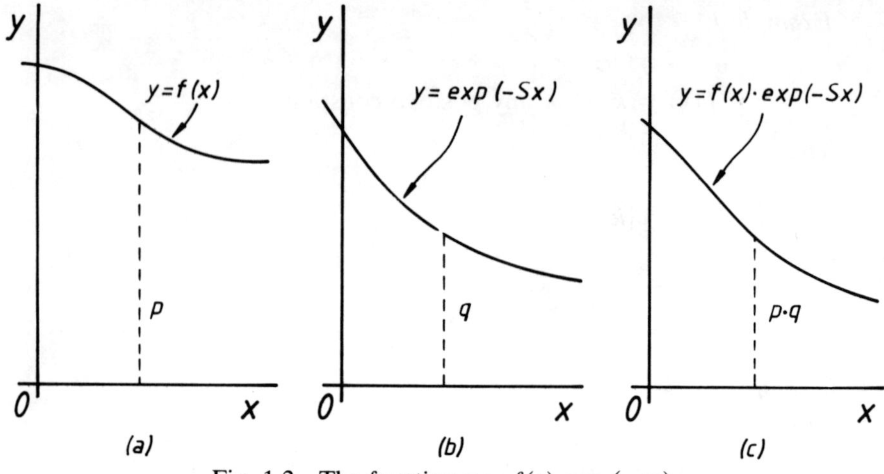

Fig. 1.2 The function $y = f(x) \exp(-sx)$.

for $x < 0$ and identically equal to y for $x \geq 0$. The Laplace transform is then $\int_{-\infty}^{+\infty} f(x).u(x) \exp(-sx)\mathrm{d}x$. For some functions, the two-sided Laplace transform $\int_{-\infty}^{+\infty} f(x) \exp(-sx)\mathrm{d}x$ is useful; an example occurs in Section 6.20. Both kinds have important applications but we here deal only with the one-sided form which we now evaluate for a few illustrative cases.

Example 3
Let

$$y = a \qquad \text{a constant}$$

Then

$$\begin{aligned}\mathscr{L}(y) &= \int_0^\infty a \exp(-sx)\mathrm{d}x \\ &= -\frac{a}{s}[\exp(-sx)]_0^\infty \\ &= -\frac{a}{s}(0-1) \qquad \text{if } s > 0 \\ &= \frac{a}{s}\end{aligned}$$

This result is illustrated in Fig. 1.3.

Example 4
Let

$$y = a \exp(bx) \qquad \text{where } a \text{ and } b \text{ are constants}$$

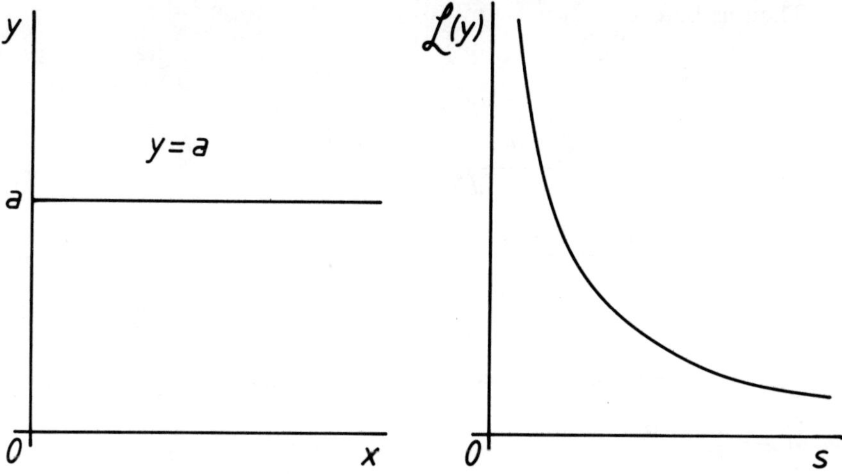

Fig. 1.3 Laplace transform of $y = a$ (a is a constant).

Then

$$\begin{aligned}
\mathscr{L}(y) &= \int_0^\infty a \exp(bx).\exp(-sx)\mathrm{d}x \\
&= \int_0^\infty a \exp\{(b - s)x\}\mathrm{d}x \\
&= \frac{a}{b - s}[\exp\{(b - s)x\}\mathrm{d}x]_0^\infty \\
&= \frac{a}{b - s}(0 - 1) \qquad \text{if } s > b \\
&= \frac{a}{s - b}
\end{aligned}$$

The curve for $\mathscr{L}(y)$ as a function of s is just the same as that in Example 3 except that it is shifted over a distance b to the right. (Notice that the transforms become infinite if s has values *less* than those given above, i.e. 0 and b in the respective examples.) We now find the Laplace transform of a function involving imaginary numbers (see Section 1.23).

Example 5
Let

$$y = a \exp(ibx) \qquad \text{where a and } b \text{ are constants}$$

Then as before

$$\mathscr{L}(y) = \frac{a}{s - ib}$$

$$= \frac{a(s + ib)}{(s - ib)(s + ib)}$$

$$= \frac{a(s - ib)}{s^2 + b^2}$$

But now

$$y = a \exp(ibx) = a\{\cos(bx) - i \sin(bx)\}$$

So we have proved that

$$\mathscr{L}[a\{\cos(bx) - i \sin(bx)\}] = a \frac{s - ib}{s^2 + b^2}$$

Therefore, cancelling a on both sides and equating the real parts of each side, we have

$$\mathscr{L}\{\cos(bx)\} = \frac{s}{s^2 + b^2}$$

Also equating the imaginary parts (coefficients of i), we have

$$\mathscr{L}\{\sin(bx)\} = \frac{b}{s^2 + b^2}$$

These results could of course have been proved directly.

In many applications, the variable y is a quantity which varies with time t rather than a spatial coordinate x, and has a Laplace transform expressed as $\int_0^\infty y(t) \exp(-st)\mathrm{d}t$. For example, y might represent the amount of radioactive material at a time t in some particular region of a processing plant or a biological system; in such a case, y might start at a value of zero, rise to a maximum, and then fall away to zero again in the course of time. The curve of y against t will therefore be of the form of a single 'hump'. We shall be particularly interested in the Laplace transforms of curves of this general shape.

1.3 Time Shifting

In this section we find how the Laplace transform of a function changes when the curve representing it is shifted along the x, or t, axis without change of its shape. To fix our ideas, it may be helpful to imagine the case of a fluid flowing steadily along a uniform pipe; a quantity of radioactive material is introduced into the fluid without appreciably altering the flow, and we take readings on some form of radiation detector placed outside the pipe at some distance downstream. Initially we shall detect no activity, but the readings will rise to a maximum while the material flows past the detector and will ultimately die away again as the material recedes. The readings plotted against t will therefore assume a shape something like the full line in Fig. 1.4. Now if the flow is steady and non-turbulent, we should expect to find the corresponding curve obtained from readings from an identical detector a little way downstream to have just the same shape but shifted along the time axis, say by an amount τ, which is the time taken for the material in the pipe to flow between the sites of the two detectors. We may not know an exact analytical form for the function $y = f(t)$, but it is evident that we can write for the Laplace transform of the first curve the function, say Y, which is

$$Y = \mathscr{L}(y) = \mathscr{L}(f(t)) = \int_0^\infty f(t) \exp(-st) \mathrm{d}t$$

If indeed the second curve has the same shape but is 'delayed' by a time τ, a little thought will show that we can write its functional form as $f(t - \tau)$, i.e. we get the delayed curve simply by replacing t by $(t - \tau)$. The Laplace transform for the delayed curve is then

$$Y_D = \int_0^\infty f(t - \tau) \exp(-st) \mathrm{d}t$$

To see what this means, it is convenient to define a new variable T, equal to $t - \tau$. (Remember that T and t are *variables*, whereas τ is a constant which just indicates the relative displacement of the two curves.) Then we can attach

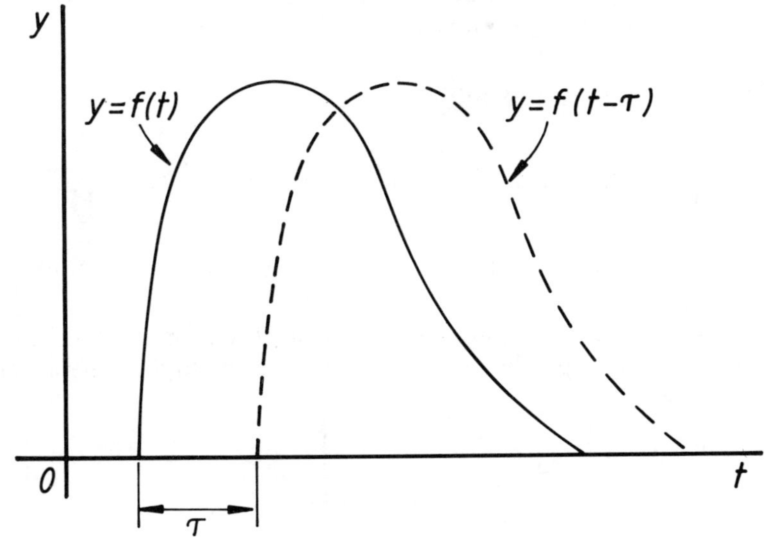

Fig. 1.4 The function $y = f(t)$ delayed by a time τ.

the same meaning to $\mathrm{d}T$ as to $\mathrm{d}t$, and rewrite the above equation in the form

$$Y_{\mathrm{D}} = \int_0^\infty f(T) \exp\{-s(T+\tau)\}\mathrm{d}T$$

$$= \int_0^\infty f(T) \exp(-sT) \exp(-s\tau)\mathrm{d}T$$

Now, as τ is constant, so is $\exp(-s\tau)$; therefore $\exp(-s\tau)$ is just a constant multiplier and can be brought outside the integration sign. We then have

$$Y_{\mathrm{D}} = \left(\int_0^\infty f(T) \exp(-sT)\mathrm{d}T\right) \exp(-s\tau)$$

The integral is now just the Laplace transform of $y(T)$, so

$$Y_{\mathrm{D}} = \mathscr{L}\{y(T)\}.\exp(-s\tau)$$

We see now that, for the special case of $\tau = 0$, $\exp(-s\tau) = 1$, Y_{D} becomes the Laplace transform of the *undelayed* curve. For any non-zero delay τ, we can calculate the Laplace transform of the delayed curve simply by taking that of the undelayed curve and multiplying it by $\exp(-s\tau)$.

1.4 Laplace Transforms for the Gate Function and the Unit Impulse

If we combine this last result with the Laplace transform for $y = a$, we can find the transform of the gate function shown in Fig.1.5a. Here we are defining a function $y = f(\Delta t)$ such that y has the value a during the time interval from $t = 0$ until $t = \Delta t$, after which it is zero. We can get this function by starting with the function $y = a$ (starting at $t = 0$), and subtracting from it the same function $y = a$ (but delayed by a time Δt). We get then, for the Laplace transform of the gate function, Y_{G},

$$Y_{\mathrm{G}} = \int_0^\infty f(\Delta t) \exp(-st)\mathrm{d}t$$

$$= \int_0^\infty a \exp(-st)\mathrm{d}t - \int_0^\infty a \exp(-st)\mathrm{d}t.\exp(-s\Delta t)$$

$$= \frac{a}{s}\{1 - \exp(-s\Delta t)\}$$

This result is correct for all positive values of Δt. If Δt is small enough, we can approximate to it by expanding $\exp(-s\Delta t)$ (according to the general rule for $\exp(x)$) and get

$$Y_{\mathrm{G}} = \frac{a}{s}\left\{1 - 1 + s\Delta t - \frac{(s\Delta t)^2}{2!} + \cdots\right\}$$

$$= a\Delta t - \frac{as\Delta t^2}{2!} + \cdots$$

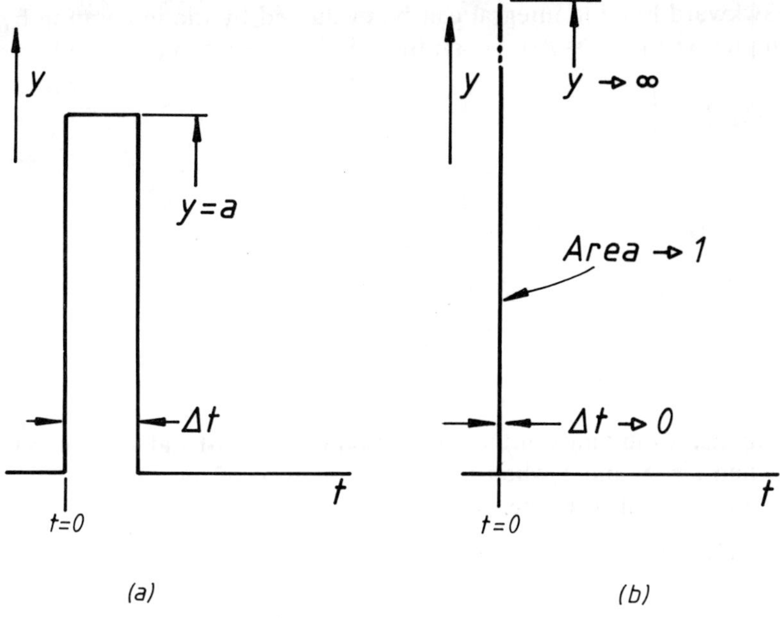

Fig. 1.5 (a) Gate and (b) unit functions.

Finally, it proves useful to define the *unit impulse* as a function which is an indefinitely narrow gate function, occurring at $t = 0$, and having unit area in the y–t plane. From the above equation, by putting $a\Delta t = 1$ and neglecting all later terms, which contain high powers of Δt, we get for the transform of the unit impulse, Y_U, that $Y_U = 1$. The unit impulse is depicted in Fig.1.5b.

1.5 Laplace Transforms of Differential Coefficients

Suppose we have some function $y = f(t)$, from which we can calculate the Laplace transform $\mathcal{L}(y)$. We shall also need to know the *initial* value of y, say y_0 or $y(0)$, i.e. the value of y when $t = 0$. We now consider the function y' or $\mathrm{d}y/\mathrm{d}t$, which of course is another function of t, and try to find an expression for its Laplace transform. By definition this is

$$\mathcal{L}(y') = \int_0^\infty (\mathrm{d}y/\mathrm{d}t) \exp(-st)\mathrm{d}t$$

which can be written

$$\mathcal{L}(y') = \int_0^\infty \exp(-st)\mathrm{d}y$$

This awkward-looking integral can be evaluated by the integration-by-parts technique, writing $u = \exp(-st)$; then $\mathrm{d}u/\mathrm{d}t = -s \exp(-st)$ and

$$\mathcal{L}(y') = \int_0^\infty u \,\mathrm{d}y$$

$$= [u.y]_0^\infty - \int_0^\infty y \,\mathrm{d}u$$

$$= [\exp(-st).y]_0^\infty - \int_0^\infty y(\mathrm{d}u/\mathrm{d}t)\mathrm{d}t$$

$$= [\exp(-st).y]_0^\infty + s\int_0^\infty y \exp(-st)\mathrm{d}t$$

Now we insert the limits of the integration in the left-hand term: $\exp(-st)$ is zero when $t = \infty$ and 1 when $t = 0$. The right-hand term is just s times the Laplace transform of y; therefore

$$\mathcal{L}(y') = 0 - y_0 + s\mathcal{L}(y)$$

i.e.

$$\mathcal{L}(y') = s\mathcal{L}(y) - y_0$$

Similarly, it may be shown that the Laplace transform of the second differential coefficient $y'' = \mathrm{d}^2y/\mathrm{d}t^2$ is

$$\mathcal{L}(y'') = s^2\,\mathcal{L}(y) - sy_0 - y_0'$$

where y_0' is the *initial* value of y', i.e. the initial *slope* of the curve for y plotted against t.

Expressions for the higher-order coefficients are also known. Some useful elementary Laplace transforms are collected in Table 1.1.

Example 6
Find the Laplace transform for $y = f(t)$ if

$$y = 0 \qquad 0 < t < \tau$$
$$y = m(t - \tau) \qquad t \geq \tau$$

This is a ramp function with slope m and delayed by a time τ. From the table, the transform is

$$Y(s) = \frac{m}{s^2}\exp(-s\tau)$$

A useful notation for expressing the relation between a function and its transform employs the double arrow. Thus

$$y(t) \leftrightarrow Y(s)$$

means that Y(s) is the transform of $y(t)$, and that $y(t)$ is the inverse transform of $Y(s)$. $Y(s)$ and $y(t)$ may be called a *transform pair*.

Table 1.1 Useful Laplace Transforms

	$y = f(t)$	$\mathcal{L}(y) = Y(s) = \int_0^\infty y \exp(-st)\mathrm{d}t$
Step $y = a$	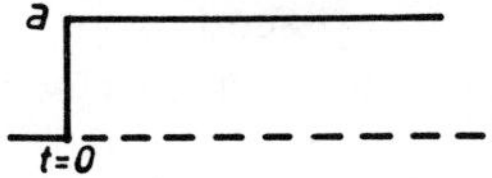	$\frac{a}{s}$
Gate	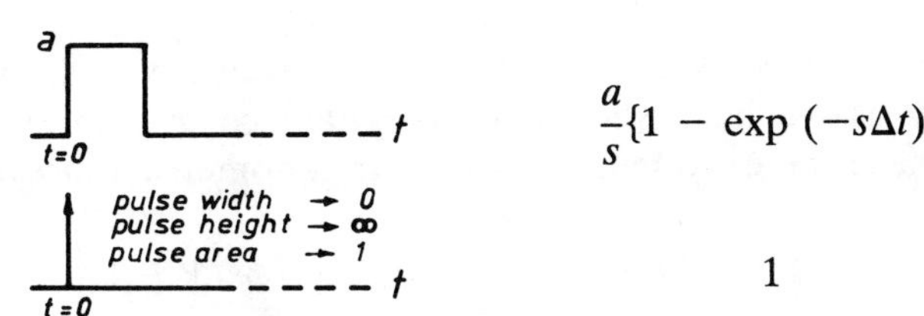	$\frac{a}{s}\{1 - \exp(-s\Delta t)\}$
Unit Impulse $y = \delta(t)$		1
Ramp $y = mt$	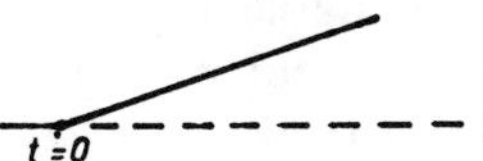	$\frac{m}{s^2}$
Exponential $y = a \exp(-kt)$		$\frac{a}{s + k}$
Cosine $y = a \cos(\omega t)$	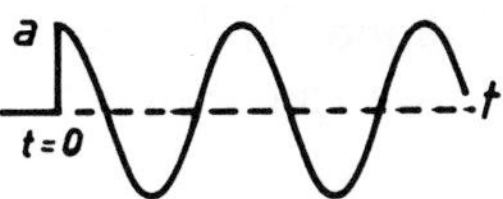	$\frac{as}{s^2 + \omega^2}$
Sine $y = a \sin(\omega t)$		$\frac{a\omega}{s^2 + \omega^2}$
(Time delay τ)		Multiply above by $\exp(-s\tau)$
$y' \equiv \frac{\mathrm{d}y}{\mathrm{d}t} \equiv \mathrm{D}y$		$sY(s) - y_0$
$y'' \equiv \frac{\mathrm{d}^2y}{\mathrm{d}t^2} \equiv \mathrm{D}^2y$		$s^2Y(s) - sy_0 - y_0'$
$y''' \equiv \frac{\mathrm{d}^3y}{\mathrm{d}t^3} \equiv \mathrm{D}^3y$		$s^3Y(s) - s^2y_0 - sy_0' - y_0''$
$I = \int_0^t y(\tau)\mathrm{d}t$		$\frac{Y(s)}{s}$

1.6 Fourier Transforms

Although Laplace transforms have several applications, as will be illustrated in later sections, it is found that for certain functions it is more convenient to use those introduced by Fourier. There are many ways of approaching these, but for our purposes it is simplest to regard them as an extension of Laplace transforms but with the parameter s replaced by an imaginary number, say $i\omega$. The two-sided Fourier transform of a function $y = f(t)$ is then written

$$\mathscr{F}(y) \equiv \mathscr{F}\{f(t)\} \int_{-\infty}^{+\infty} f(t) \exp(-i\omega t)dt$$

This form as it stands is often used to evaluate the transform, in just the same way as we have shown for the Laplace transforms. Otherwise, one may replace the exponential term by trigonometric functions and obtain

$$\mathscr{F}(y) \equiv \mathscr{F}\{f(t)\} = \int_{-\infty}^{+\infty} f(t) \cos(\omega t)dt - i \int_{-\infty}^{+\infty} f(t) \sin(\omega t)dt$$

The transform has a real and an imaginary part; in Section 6.18 a Fourier transform occurs in which only the real part appears. One-sided Fourier transforms are defined in the same way as one-sided Laplace transforms, and compilations of Fourier transforms of many elementary functions are available. The introduction of imaginary numbers is something of a complication, and this is the main reason for preferring Laplace transforms wherever possible in this book. The applications of the two transforms are however so similar that we could have used Fourier transforms throughout.

In some applications Fourier transforms have an important advantage in that their *inverses* are easily calculated. The Fourier transform of a function $f(t)$ is clearly a function of ω and we can write

$$\mathscr{F}\{f(t)\} = \int_{-\infty}^{+\infty} f(t) \exp(-i\omega t)dt = F(\omega)$$

Now it can be shown that

$$\frac{1}{2\pi} \int_{-\infty}^{+\infty} F(\omega) \exp(i\omega t)d\omega = f(t)$$

There is thus a certain symmetry in the relation between $F(\omega)$ and $f(t)$ and indeed some authors emphasise this by redefining the Fourier transform such that

$$F(\omega) = \frac{1}{\sqrt{(2\pi)}} \int_{-\infty}^{+\infty} f(t) \exp(-i\omega t)dt$$

and then

$$f(t) = \frac{1}{\sqrt{(2\pi)}} \int_{-\infty}^{+\infty} F(\omega) \exp(i\omega t)d\omega$$

These equations imply that the problem of finding $F(\omega)$ given $f(t)$ is just the same as that of finding $f(t)$ given $F(\omega)$, apart from appropriate changes in sign and introducing a factor 2π if necessary, depending on which definitions are used. This close symmetry property is not possessed by Laplace transforms in quite the same way. Inverse Laplace transformations (i.e. the identification of the function $f(t)$ whose Laplace transform is $Y(s)$) are however easily found in most cases by reference to published tables of transform pairs.

In this very superficial discussion of Fourier transforms we have space only to state the existence of some restrictions on the form of the function $f(t)$. The transform does not exist unless $f(t)$ obeys the so-called Dirichlet conditions, viz.

(a) it must be absolutely integrable, i.e.

$$\int_{-\infty}^{+\infty} |f(t)|\mathrm{d}t < \infty$$

(b) it must have a finite number of maxima, minima and discontinuities in any finite interval.

If these conditions are met, a unique transform $F(\omega)$ can be found; also this transform has the unique inverse $f(t)$. All the functions we shall use are of this kind.

DIFFERENTIAL EQUATIONS

1.7 Differential Equations: General

In many experimental situations, we may find that we can measure some quantity y in terms of a variable x or t, and we express our results in the form of an equation $y = f(x)$ or $y = f(t)$, where the function f is purely empirical (i.e. based on experimental data alone) or is at least partly based on some theoretical considerations, in which case it will generally have some definite algebraic form, e.g. $y = ax + bx^2$, or $y = a \cos(\omega t)$. Even when we have no theoretical basis, we find it most useful to find some such analytical relation, perhaps by plotting the pairs of values of y and x (or t), and trying to fit them to a straight line or other recognisable curve. What we are saying is that an essential part of trying to understand a natural phenomenon is the expression of relations between two measured quantities in a definite way, ideally in the form of an algebraic equation.

But all too often we find ourselves measuring not precisely a quantity we are really concerned with, but the rate at which it is changing; for instance, we may measure the velocity of a particle but may really want to have information about its position, or we may measure the rate at which a radioactive substance is decaying but want to know the amount present in a sample at a given instant of time. Mathematically expressed, this means that we are given some function $y' = \mathrm{d}y/\mathrm{d}t$ as a function of t, say $f'(t)$ and we want to find y

itself. In this case, we express the problem as:

Given

$$y' = f'(t) = \frac{\mathrm{d}}{\mathrm{d}t}(f(t))$$

Find

$$y = \int (f'(t))\mathrm{d}t = f(t) + c$$

—and the solution is found by a single process of integration. The initial equation

$$y' = \frac{\mathrm{d}}{\mathrm{d}t}(f(t))$$

is in fact the simplest possible example of a *differential equation*; its solution, obtained in this case by a single integration, is $y = f(t) + c$, an equation which is characterised by involving *no* differential coefficients.

We note that the solution as written contains an arbitrary constant. In practical applications, the value of this constant can be found not through the properties of the equation itself, but because it represents a physical problem for which we generally know the value of y for one particular value of t. These known values are called the *initial conditions*, $y = y_0$ when $t = t_0$, where y_0 and t_0 are actual numbers. It follows immediately from the solution equation that $y_0 = f(t_0) + c$, i.e. that c can be evaluated as $y_0 - f(t_0)$.

Example 7

A body starts from rest at time zero at a point 3 m along the y axis and travels along it with a velocity 7 m s^{-1}. Find its position at any subsequent time.

We know that velocity is given by $\mathrm{d}y/\mathrm{d}t$, so the problem can be expressed as the differential equation

$$\mathrm{d}y/\mathrm{d}t = 7$$

and the initial conditions are $y_0 = 3$, $t_0 = 0$.

The general solution is then, by integration,

$$y = 7t + c$$

and the values of the initial conditions give

$$c = y_0 - 7.(t_0) = 3$$

so that the complete solution is

$$y = 7t + 3$$

which gives the displacement y explicitly in terms of the time t.

Many physical problems, however, cannot be reduced to such simple terms. A complicated differential equation might be expressed as a function ϕ in the form

$$\phi \equiv \phi[y, t, \mathrm{d}y/\mathrm{d}t, \mathrm{d}^2y/\mathrm{d}t^2, \ldots] = 0$$

e.g.

$$\phi = ky^2 + \exp(\lambda t)\,\mathrm{d}y/\mathrm{d}t + \ln(y)\,\mathrm{d}^2y/\mathrm{d}t^2 = 0$$

and the solution

$$y = f(t) + c_1 + c_2 + \cdots$$

will be found—if indeed it can be found at all—only by quite sophisticated techniques, of which integration is only one. Integration will be necessary once if ϕ includes terms only as far as $\mathrm{d}y/\mathrm{d}t$, twice if it contains terms in $\mathrm{d}^2y/\mathrm{d}t^2$, and so on.

Equations implying one, two, . . . integrations are called 'first-order', 'second-order' and so on. Corresponding to each integration, integration constants $c_1, c_2, \ldots$ will appear in the solution and their values will have to be found from initial conditions such as the values of $y = y_0$, $\mathrm{d}y/\mathrm{d}t = y_0'$, $\mathrm{d}^2y/\mathrm{d}t^2 = y_0''$, . . . for a particular value of $t = t_0$.

There are several techniques for solving differential equations, some of which we review in the next sections. It is characteristic of the methods that each will work only for certain kinds of equation, and that sometimes when two methods work with the same equation they do not necessarily give the same result. Thus one may find *two* solutions, say $y = f_1(t) + c_1$ and $y = f_2(t) + c_2$, where the two functions f_1 and f_2 are not the same, and neither are the integration constants, for the matter.

In these cases, and when in the original differential equation the factors multiplying $\mathrm{d}y/\mathrm{d}t$, $\mathrm{d}^2y/\mathrm{d}t^2$ etc., are constants, it can be shown that the two solutions can be combined to read $y = f_1(t) + f_2(t) + c$ so that *both* solutions and their sum are correct (with suitable evaluation of the integration constants).

We shall begin with some simple forms of first-order equations, and with the insight gained go on to examples of higher orders. It should be noted that even some first-order equations are known which have no known solutions. Higher-order equations are tractable in general only if the quantities multiplying the differential coefficients are constants (i.e. are not functions of the variables). Very few *nonlinear* equations (i.e. those having differentials raised to more than the first power, e.g. $(\mathrm{d}^2y/\mathrm{d}t^2)^3$) have known solutions. In what follows, we shall regard the solution as complete when we have found an explicit expression for y in terms of the independent variable and one or more integration constants, it being understood that in practice the solution to be of real use will have to have the values of the integration constants found by insertion of the relevant initial conditions.

1.8 Differential Equations: Variables Separable

If a differential equation can be written in the form

$$\mathrm{d}y/\mathrm{d}t = \frac{\phi_1(t)}{\phi_2(y)}$$

where ϕ_1 is a function of t but not of y, and ϕ_2 is a function of y but not of t, we say that the variables are separable and write it in the equivalent form

$$\int \phi_2(y)\mathrm{d}y = \int \phi_1(t)\mathrm{d}t$$

Both sides of this equation can be integrated straight away and the result can if necessary be solved for y.

Example 8
Let $\mathrm{d}y/\mathrm{d}t = at \exp(-by) = at/\exp(by)$. We can identify ϕ_1 as at and ϕ_2 as $\exp(by)$. Then

$$\int \exp(by)\mathrm{d}y = \int at\mathrm{d}t$$

i.e.

$$\frac{\exp(by)}{b} = \frac{at^2}{2} + c$$

which leads to

$$y = \frac{1}{b} \ln \left[b\left(\frac{at^2}{2} + C\right)\right]$$

1.9 Differential Equations: Homogeneous Forms

If a differential equation can be written in the form

$$\mathrm{d}y/\mathrm{d}t = \phi\left(\frac{y}{t}\right)$$

where ϕ is any function, it is said to be homogeneous, and can always be solved by putting $y = ut$, u being a new variable. For, if we differentiate the last equation, regarding it as a product, we have by the rule for product differentiation that

$$\mathrm{d}y/\mathrm{d}t = u + t\,\mathrm{d}u/\mathrm{d}t$$

Therefore

$$\phi(u) = u + t\,\mathrm{d}u/\mathrm{d}t$$

and

$$\frac{du}{dt} = \frac{\phi(u) - u}{t}$$

This is now a variables-separable equation and gives

$$\int \frac{du}{\phi(u) - u} = \int \frac{dt}{t}$$

Example 9
Let

$$dy/dt = 1 + \frac{y}{t}$$

We put $y = ut$ and identify $\phi(u)$ as $1 + u$. The equation for du/dt then reads

$$du/dt = \frac{1 + u - u}{t} = \frac{1}{t}$$

so

$$\int du = \int \frac{1}{t} dt$$

i.e.

$$u = \ln (t) + c$$

and

$$y = ut = t[\ln (t) + c]$$

1.10 Differential Equations: Exact Forms

Neither of the above methods will work for the equation

$$dy/dt + 2ty/(t^2 + y^2) = 0$$

However, if we multiply the equation throughout by $(t^2 + y^2)$ we get

$$(t^2 + y^2)\, dy/dt + 2ty = 0$$

If now we consider the term $(t^2 + y^2)$ and work out its *partial* differential

$$\frac{\partial}{\partial t}(t^2 + y^2)$$

with y being temporarily regarded as a constant, we get $2t$; again if we consider the term $2ty$ and find its partial differential

$$\frac{\partial}{\partial y}(2ty)$$

with t being regarded temporarily as a constant, we also get $2t$. Any such equation, of form

$$A\ \mathrm{d}y/\mathrm{d}t + B = 0$$

fulfilling the criterion that

$$\partial A/\partial t = \partial B/\partial y$$

is known as an *exact form*. For all these equations there exists a function ϕ, involving y and t, for which

$$\phi(y, t) - c = 0$$

c being a constant.

We find the function ϕ in the following way. Since ϕ is a function of y and t, we have

$$\mathrm{d}y/\mathrm{d}t = -(\partial\phi/\mathrm{d}t)/(\partial\phi/\mathrm{d}y)$$

which can be rewritten to read

$$(\partial\phi/\partial y).\mathrm{d}y/\mathrm{d}t + \partial\phi/\partial t = 0$$

Comparing this with our equation

$$A\ \mathrm{d}y/\mathrm{d}t + B = 0$$

we see that A corresponds to $\partial\phi/\partial y$ and B to $\partial\phi/\mathrm{d}t$. These relations imply that ϕ can be evaluated as either $\phi = \int A\mathrm{d}y$ (with t constant) or $\phi = \int B\mathrm{d}t$ (with y constant). Performing the integrations in our special case with $A = t^2 + y^2$ and $B = 2ty$ we find

$$\phi = \int (t^2 + y^2)\mathrm{d}y = t^2y + \frac{y^3}{3} + \text{(any function of } t\text{)}$$

$$\phi = \int (2ty)\mathrm{d}t = t^2y + \text{(any function of } y\text{)}$$

where the usual integration constants are functions of t and y respectively, since these are regarded temporarily as constants during the two integration operations. However in general terms both t and y must be variable; nevertheless it is clear that if we write

$$\phi = t^2y + \frac{y^3}{3} = c$$

we have a relation which satisfies both the above conditions (since in this case the first includes the second) and this can be regarded as the solution of the original differential equation.

Example 10

Let $\mathrm{d}y/\mathrm{d}t = y \tan(t)$. This can be solved as a variables-separable equation, an exercise we leave to the reader. Alternatively, recalling that

$\tan(t) = \sin(t)/\cos(t)$ we may write the equation as

$$\cos(t)\, dy/dt - y \sin(t) = 0$$

This is an exact form because

$$\frac{\partial}{\partial t}\{\cos(t)\} = -\sin(t) = \frac{\partial}{\partial y}\{-y \sin(t)\}$$

Therefore the required function ϕ is given by

$$\begin{cases} \phi = \int \cos(t) dy = y \cos(t) + c \\ \phi = \int -y \sin(t) dt = y \cos(t) + c \end{cases}$$

In this case both the conditions read the same and the solution is

$$y \cos(t) + c = 0$$

or

$$y = c' \sec(t) \qquad c' \text{ being a constant}$$

1.11 Differential Equations: Linear Equations of the First Order

These equations are of the form

$$dy/dt + Py = Q$$

where P and Q may be any functions of t alone, but not of y. These can always be solved by finding an *integrating factor*, which is $\exp(\int P dt)$. In other words, we must first find $u = \int P dt$ and then obtain $\exp(u)$. Multiplying the equation throughout by this factor and rearranging, we get

$$\exp\left(\int P dt\right).dy/dt + Py \exp\left(\int P dt\right) - Q \exp\left(\int P dt\right) = 0$$

This is an exact form because

$$\frac{\partial}{\partial t}\left\{\exp\left(\int P dt\right)\right\} = \frac{\partial}{\partial t}\{\exp(u)\} = \frac{\partial}{\partial u}\{\exp(u)\}\frac{\partial u}{\partial t} = \exp(u).P$$

and

$$\frac{\partial}{\partial y}(Py - Q)\exp\left(\int P dt\right) = P \exp\left(\int P dt\right) = P \exp(u)$$

The function ϕ can therefore be found from

$$\phi = \int \exp\left(\int P dt\right) dy = y \exp\left(\int P dt\right) = f_1(t)$$

and

$$\phi = \int \left\{ Py \exp\left(\int P\mathrm{d}t\right) - Q \exp\left(\int P\mathrm{d}t\right)\right\} \mathrm{d}t = f_2 y$$

where $f_1(t)$ is a function of t alone and $f_2(y)$ is a function of y alone. The first condition gives

$$y = \exp\left(-\int P\mathrm{d}t\right) . f_1(t)$$

The second gives

$$\int \left\{ Py \exp\left(\int P\mathrm{d}t\right)\right\} \mathrm{d}t - \int \left\{ Q \exp\left(\int P\mathrm{d}t\right)\right\} \mathrm{d}t = f_2(y)$$

or

$$y \exp\left(\int P\mathrm{d}t\right) = \int \left\{ Q \exp\left(\int P\mathrm{d}t\right)\right\} \mathrm{d}t + f_2(y)$$

i.e.

$$y = \exp\left(-\int P\mathrm{d}t\right)\left[\int \left\{ Q \exp\left(\int P\mathrm{d}t\right) \mathrm{d}t + f_2(y)\right\}\right]$$

For this condition to be compatible with the first, it is evident that $f_1(t)$ in the first must be identified with $\int Q \exp(\int P\mathrm{d}t)\mathrm{d}t$ in the second. Also, in the second condition, if $f_2(y)$ actually contained such a term as ky (k any constant), the expression on multiplying out would give us $ky \exp(-\int P\mathrm{d}t)$; this is inconsistent with the first condition, which allows terms like $\exp(-\int P\mathrm{d}t)$ to be multiplied only by a function of t. We are forced to conclude that $f_2(y)$ can only be a constant, i.e. $f_2(y) = c$. The final solution must therefore be

$$y = \exp\left(-\int P\mathrm{d}t\right)\left\{\int Q \exp\left(\int P\mathrm{d}t\right)\mathrm{d}t + c\right\}$$

The solution still contains integrals, but since they involve only known functions of t they can in general be evaluated to give the final solution.

Example 11

Let $\mathrm{d}y/\mathrm{d}t + 2ty = 10t$. This means that $P = 2t$ and $Q = 10t$. The integrating factor is $\exp(\int P\mathrm{d}t) = \exp(t^2)$ and the solution is $\exp(-t^2).(\int 10t \exp(t^2)\mathrm{d}t + c)$.

To evaluate the integral we put $u = t^2$, so that $\mathrm{d}u/\mathrm{d}t = 2t$ and we can replace $2t\mathrm{d}t$ by $\mathrm{d}u$. Then

$$\begin{aligned} y &= \exp(-t^2)\left(5 \int \exp(u)\mathrm{d}u + c\right) \\ &= \exp(-t^2)(5 \exp(u) + c) \\ &= 5 + c \exp(-t^2) \end{aligned}$$

1.12 Differential Equations: Linear Equations of Higher Orders, with Constant Coefficients

These equations are particularly important as they can be used to represent a great variety of physical problems. Fortunately their solutions are easy to obtain and we now explain the two well established methods for doing this.

It is convenient first of all to introduce the shorthand notation in which we replace d/dt (or d/dx) by the single symbol D. (It is generally clear from the nature of the physical problem whether differentiation with respect to time t or some distance coordinate x is implied, and the principle is the same in either case.) On this understanding, we may rewrite the first-order equation, e.g.

$$dy/dt + ay = 0$$

as

$$Dy + ay = 0$$

In this second form the operator D looks like a simple multiplier—which it is not—but in fact it does no harm if we treat it as such for the moment and write instead

$$(D + a)y = 0$$

as long as it is clearly understood what the new equation means; it tells us that the slope of a curve at any point plus a times the value of the ordinate at that point is just zero. It must not be taken to mean that either $D + a = 0$ or that $y = 0$, which it would do if D were just a number, not an operator equal to d/dt. In the same way we are going to use D^2 to mean 'differentiate twice', but not 'multiply D by D'.

With this notation we can write the differential equation we wish to solve as

$$\begin{aligned} d^n y/dt^n + a\, d^{n-1}y/dt^{n-1} + b\, d^{n-2}y/dt^{n-2} \cdots \\ = (D^n + aD^{n-1} + bD^{n-2} \cdots)y = f(t) \end{aligned}$$

To begin with, however, we examine a very simple example.

Application of the methods we have already met shows that the solution of

$$(D + a)y = 0$$

is

$$y = c_1 \exp(-at)$$

and also that the solution of

$$(D + a)y = \exp(-bt)$$

is

$$y = c_1 \exp(-at) + c_2 \exp(-bt)$$

where the c's are arbitrary constants.

Clearly the solution of the second of these differential equations is the sum of two parts:

(a) $\exp(-at)$, which is in fact the solution we would get if the $\exp(-bt)$ term on the right-hand side had been replaced by zero—the term $c_1 \exp(-at)$ is called the *complementary function*;

(b) $c_2 \exp(-bt)$, which only exists because the differential equation actually has the term $\exp(-bt)$ on its right-hand side—the term $c_2 \exp(-bt)$ is called the *particular integral*.

It can be shown that this situation is a special case of a general rule, that any equation of the form

$$(\mathrm{D} + a)y = f(t)$$

has a complementary function $c_1 \exp(-at)$ and a particular integral which depends on the function $f(t)$; the whole solution is the sum of these expressions. Clearly we can always write down the value of the complementary function straight away, but we need a method for finding particular integrals in the general case when $f(t)$ is other than $\exp(-bt)$. One such method is suggested by the following procedure.

If we consider again $(\mathrm{D} + a)y = \exp(-bt)$ we can differentiate it without in any way altering what its solutions will be. Using D to mean $\mathrm{d}/\mathrm{d}t$ we get

$$\begin{aligned} \mathrm{D}(\mathrm{D} + a)y &= \mathrm{D} \exp(-bt) \\ &= \frac{\mathrm{d}}{\mathrm{d}t} \exp(-bt) \\ &= -b \exp(-bt) \\ &= -b(\mathrm{D} + a)y \end{aligned}$$

which by rearrangement gives

$$[\mathrm{D}^2 + (a + b)\mathrm{D} + ab]y = 0$$

or

$$(\mathrm{D} + a)(\mathrm{D} + b)y = 0$$

Here we have been treating D as if it were an ordinary algebraic symbol; which is justifiable only if we interpret the final result in the right way. The interpretation we postulate as being correct is that the original differential equation is equivalent to the *pair* of equations

$$\left.\begin{aligned} (\mathrm{D} + a)y &= 0 \\ (\mathrm{D} + b)y &= 0 \end{aligned}\right\}$$

whose solutions are obviously $y = c_1 \exp(-at)$, $y = c_2 \exp(-bt)$; so that, in view of the general rule mentioned in Section 1.7, the whole solution is $y = c_1 \exp(-at) + c_2 \exp(-bt)$. (We already have enough integration constants.) The real justification for this apparently arbitrary procedure is of course that it works in practice; it clearly does in the present case and mathematicians have shown indeed that it is of general validity. So the routine to be followed in solving any equation of the kind we are reviewing, viz.

$$(\mathrm{D}^n + a\mathrm{D}^{n-1} + \cdots)y = f(t)$$

is to differentiate it (perhaps more than once) and rearrange the result to form a higher-order equation with zero on the right-hand side; factorisation of the left-hand side will give the solution in the form

$$y = c_1 \exp(\lambda_1 t) + c_2 \exp(\lambda_2 t) + \cdots$$

where the λ's are the roots of the auxiliary equation

$$r^n + ar^{n-1} + br^{n-2} + \cdots = 0$$

Example 12
Let

$$\mathrm{d}y/\mathrm{d}t + ay = bk \exp(-bt) \qquad (a,\ b \text{ and } k \text{ constants})$$

Write this as

$$(\mathrm{D} + a)y = bk \exp(-bt)$$

Differentiate:

$$\begin{aligned} \mathrm{D}(\mathrm{D} + a)y &= -b^2k \exp(-bt) \\ &= -b(\mathrm{D} + a)y \end{aligned}$$

Rearrange:

$$(\mathrm{D}^2 + (a + b)\mathrm{D} + ab)y = 0$$

Factorise:

$$(\mathrm{D} + a)(\mathrm{D} + b)y = 0$$

Solution:

$$y = c_1 \exp(-at) + c_2 \exp(-bt)$$

Example 13
Let

$$\mathrm{d}^2y/\mathrm{d}t^2 + 5\ \mathrm{d}y/\mathrm{d}t + by = 7$$

i.e.

$$(\mathrm{D}^2 + 5\mathrm{D} + 6)y = 7.$$

Differentiate:

$$\mathrm{D}(\mathrm{D}^2 + 5\mathrm{D} + 6)y = 0$$

Factorise:

$$\mathrm{D}(\mathrm{D} + 2)(\mathrm{D} + 3)y = 0$$

Solution:

$$y = c_1 + c_2 \exp(-2t) + c_3 \exp(-3t)$$

Example 14
Let

$$d^2y/dt^2 - y = k \exp(3t)$$

i.e.

$$(D^2 - 1)y = k \exp(3t)$$

Differentiate:

$$\begin{aligned} D(D^2 - 1)y &= 3k \exp(3t) \\ &= 3(D^2 - 1)y \end{aligned}$$

Rearrange:

$$(D - 3)(D^2 - 1)y = 0$$

Factorise:

$$(D - 3)(D - 1)(D + 1)y = 0$$

Solution:

$$y = c_1 \exp(3t) + c_2 \exp(t) + c_3 \exp(-t)$$

1.13 Extension to Trigonometric Functions: The *RCL* Circuit

The differential equation

$$d^2y/dt^2 + y = 0$$

or

$$(D^2 + 1)y = 0$$

factorises to give $(D + i)(D - i)y = 0$ and the solution is

$$\begin{aligned} y &= c_1 \exp(-ix) + c_2 \exp(ix) \\ &= c_1[\cos(ix) - i \sin(ix)] + c_2[\cos(ix) - i \sin(ix)] \\ &= c_3 \cos(ix) + ic_4 \sin(ix) \end{aligned}$$

where the constants c_3 and c_4 are respectively equal to $c_2 + c_1$ and $c_2 - c_1$.

A concrete example taken from electrical circuit theory will show an extension of this idea, which is of great practical importance. The circuit shown in Fig. 1.6 consists of a resistor R, a capacitor C and an inductor L connected in series. We imagine the capacitor to be charged initially to a potential V volts, so that it carries a charge CV coulombs. Then at time zero, or $t = t_0$, the points marked A and B are short-circuited by closing a switch. We desire to know the charge remaining on C, $Q(t)$, as a function of time. This charge will obviously decrease, at a rate which constitutes the current i

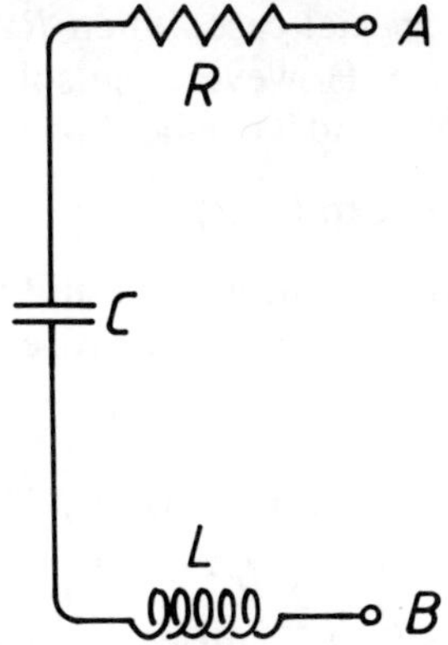

Fig. 1.6 Series *RCL* circuit.

flowing in the circuit. According to Ohm's law there will be a voltage drop across R of magnitude $Ri = R\,\mathrm{d}Q/\mathrm{d}t$. Also, according to Faraday's law of induction, a voltage will be generated in the inductor of magnitude $L\,\mathrm{d}i/\mathrm{d}t = L\,\mathrm{d}^2Q/\mathrm{d}t^2$. Adding these to the voltage $V = Q/C$ remaining on the capacitor we have (since A and B are at the same potential after the short-circuit) that

$$L\,\mathrm{d}^2Q/\mathrm{d}t^2 + R\,\mathrm{d}Q/\mathrm{d}t + Q/C = 0$$

i.e.

$$\mathrm{d}^2Q/\mathrm{d}t^2 + (R/L).\mathrm{d}Q/\mathrm{d}t + (1/LC).Q = 0$$

For convenience we will denote R/L by $2k$ and $1/LC$ by n^2 and rewrite this equation as

$$(\mathrm{D}^2 + 2k\mathrm{D} + n^2)Q = 0$$

It can be factorised in the form

$$(\mathrm{D} - r_1)(\mathrm{D} - r_2)Q = 0$$

where r_1 and r_2 are the roots of the equation $r^2 + 2kr + n^2 = 0$, i.e. $r_1 = -k + \surd(k^2 - n^2)$ and $r_2 = -k - \surd(k^2 - n^2)$. The solution for Q must then be

$$Q = c_1 \exp\{[-k - \surd(k^2 - n^2)]t\} + c_2 \exp\{[-k + \surd(k^2 - n^2)]t\}$$

and this solution will be purely exponential if $k^2 \geq n^2$ but will involve trigonometric forms if $k^2 < n^2$. In more detail:

(a) $k^2 > n^2$. This means that $(R/2L)^2 > (1/LC)$ or $R^2 > 4L/C$. Then $k^2 - n^2$ is real; its square root $\surd(k^2 - n^2)$ is less than k so that both coefficients $-k - \surd(k^2 - n^2)$ and $-k + \surd(k^2 - n^2)$ are negative. Calling these $-p$ and $-q$ (where p and q are positive) we get

$$Q = c_1 \exp(-pt) + c_2 \exp(-qt)$$

— a pair of 'exponential decays'.

(b) $k^2 = n^2$. This is a special case, when $R^2 = 4L/C$. Here the square root terms disappear. However, the solution must still contain two integration constants, and it can be shown to be

$$Q = (c_1 + c_2 t) \exp(-kt)$$

That this is indeed a solution is easily verified by differentiating it twice. Clearly it is an exceptional case, even in a purely physical system, and although a theoretical possibility is hardly likely ever to occur in a biological system.

(c) $k^2 < n^2$. This means that $R^2 < 4L/C$. The solution is

$$\begin{aligned} Q &= c_1 \exp\{[-k + \mathrm{i}\sqrt{(n^2 - k^2)}]t\} \\ &\quad + c_2 \exp\{[-k - i\sqrt{(n^2 - k^2)}]t\} \\ &= \exp(-kt)\{c_1 \exp[\mathrm{i}\sqrt{(n^2 - k^2)}t] + c_2 \exp[-\mathrm{i}\sqrt{(n^2 - k^2)}t]\} \\ &= \exp(-kt)\{c_3 \cos[\sqrt{(n^2 - k^2)}t] + \mathrm{i}c_4 \sin[\sqrt{(n^2 - k^2)}t]\} \end{aligned}$$

The real part of Q, $\mathcal{R}(Q)$, (that is to say the part of Q which is experimentally observable) is thus

$$Q = c_3 \exp(-kt) \cos[\sqrt{(n^2 - k^2)}t]$$

— a product of an exponential and a cosine term.

The function $Q(t)$ for the three cases is shown in Fig. 1.7. In case (c) the curve has an 'envelope' which is just the curve for $Q = c_3 \exp(-kt)$. The circuit is said to 'ring' with frequency $\sqrt{(n^2 - k^2)}/2\pi$ because the value of the cosine term repeats itself whenever t is a whole number times this quantity.

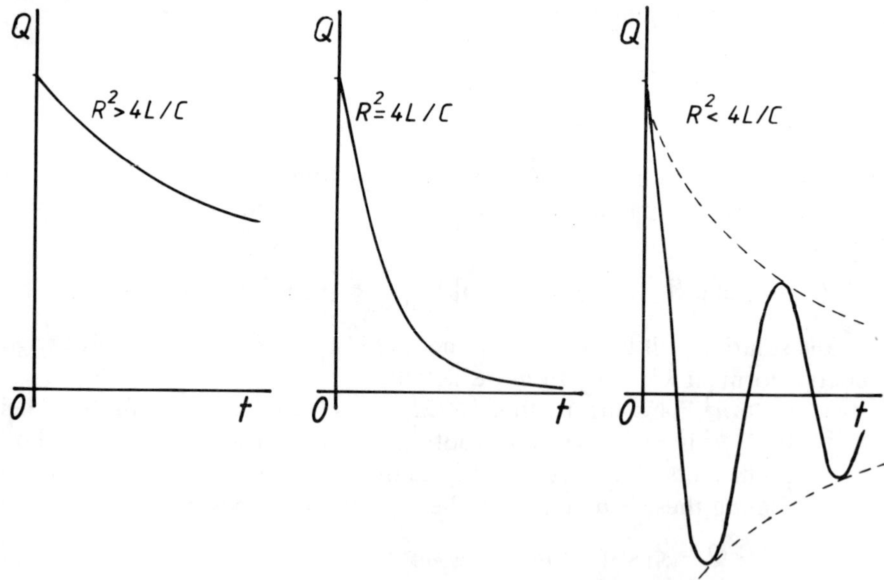

Fig. 1.7 Discharge of capacitor in RCL circuit.

It is instructive to examine the behaviour of an *RCL* circuit in the laboratory, using a pulse generator to charge a capacitor repetitively, while displaying the voltage across it on a sweep oscilloscope. Adjustment of R by using a variable resistor will then demonstrate each kind of curve illustrated.

The motion of some mechanical systems (e.g. pointers on voltmeters and ammeters) is exactly analogous to this electrical case. When the motion corresponds to the curves (c), (b) and (a) respectively, we refer to it as being under-damped, critically damped, or over-damped. Instruments which are very slightly under-damped are generally regarded as being the best, since the pointer comes to rest at the correct reading in the shortest possible time, and approaches its rest position from the side opposite to its former position, a situation which is believed to minimise the effect of 'sticking' in the movement.

1.14 Differential Equations Solved by the Laplace Transform

The use of Laplace transforms in solving differential equations constitutes a quite different approach from the methods of Sections 1.7–1.13. The essential step is to *transform* the given differential equation by writing the Laplace transform of each of its terms. The resulting equation is then a purely algebraic equation in one unknown, viz. the Laplace transform of y, $\mathscr{L}(y)$. Solution of the equation for $\mathscr{L}(y)$ is generally easy, and the problem is completed if we can find its inverse transform, i.e. y itself. An example will show how the method works.

Example 15
Let

$$\mathrm{d}y/\mathrm{d}t + a \exp(-kt) = 0$$

In accordance with Section 1.5, in which we discussed the transforms of differential coefficients, we will denote $\mathrm{d}y/\mathrm{d}t$ by y', instead of using the D notation. So we have

$$y' + a \exp(-kt) = 0$$

as our given equation; now using the results collected in Table 1.1 and transforming both terms we get immediately

$$s\mathscr{L}(y) - y_0 + \frac{a}{s + k} = 0$$

where y_0 is the initial value of y. Solving for $\mathscr{L}(y)$ we get

$$\mathscr{L}(y) = \frac{y_0}{s} - \frac{a}{s(s + k)}$$

and to find y we have to find the inverse transform of the right-hand side. To do this we express it as a sum of partial fractions (see

Section 1.25) and get

$$\mathscr{L}(y) = \frac{y_0}{s} - \frac{1}{k}\left(\frac{a}{s} - \frac{a}{s + k}\right)$$

We can now recognise that y_0/s, a/s, and $a/(s + k)$ are the transforms of y_0, a and $a \exp(-kt)$ respectively. Therefore

$$y = y_0 - \frac{1}{k}\{\mathrm{a} - a \exp(-kt)\}$$

or

$$y = \frac{a}{k} \exp(-kt) + \left(y_0 - \frac{a}{k}\right)$$

This result is interesting in that it does *not* contain an arbitrary constant; instead the constant term $y_0 - (a/k)$ occurs and this is indeed the value of the arbitrary constant which would have been found by employing the previous methods. The fact that the transform method 'automatically' inserts the correct values of the arbitrary constants is one reason for its great popularity.

Of course the above is a particularly simple example, in that the process of finding the inverse transform involved only the recognition of transforms of functions we have already studied. To use the technique in a general way, one needs access to a compilation of Laplace transforms for many more functions. (Some 120 transforms are quoted in the *Handbook of Chemistry and Physics* (Chemical Rubber Company), taken from *Modern Operational Mathematics in Engineering* by R. V. Churchill (McGraw-Hill).) But another example using only the functions in Table 1.1 we insert here as we shall use the result directly in Section 6.10.

Example 16
Let

$$\mathrm{d}y/\mathrm{d}t + k_1 y = k_2 \exp(-k_3 t)$$

where k_1, k_2 and k_3 are constants, and the initial value of y, $y_0 = y(0)$ is 0. Then using Y to mean $\mathscr{L}(y)$ we have

$$sY - y(0) + k_1 Y = \frac{k_2}{s + k_3}$$

Therefore

$$Y = \frac{k_2}{(s + k_1)(s + k_3)}$$

$$= \frac{k_2}{k_3 - k_1}\left\{\frac{1}{s + k_1} - \frac{1}{s + k_3}\right\}$$

and, by transforming back,

$$y = \frac{k_2}{k_3 - k_1} \{\exp(-k_1 t) + \exp(-k_2 t)\}$$

The elegance of the transform method is well shown in this example.

1.15 A General Solution for Linear Differential Equations with Constant Coefficients

In many circumstances it turns out that two variables, x and y, will be related by a differential equation in which both occur as time derivatives. (One may think of x as representing some sort of 'cause' which exists through time, and y as representing an 'effect' which will in general be something which varies as time goes by.) So let us consider the equation

$$a_0 y + a_1\, \mathrm{d}y/\mathrm{d}t + a_2\, \mathrm{d}^2y/\mathrm{d}t^2 + \cdots$$
$$= b_0 x + b_1\, \mathrm{d}x/\mathrm{d}t + b_2 \mathrm{d}^2x/\mathrm{d}t^2 + \cdots$$

where the a's and b's are constants. Taking Laplace transforms term by term, and writing Y for $\mathscr{L}(y)$, X for $\mathscr{L}(x)$, we have:

$$a_0 Y + a_1\{sY - y(0)\} + a_2\{s^2 Y - sy(0) - y(0)\} + \cdots$$
$$= b_0 X + b_1\{sX - x(0)\} + \cdots$$

This equation leads to a very simple and important result if x and y are such that their initial values and those of all their differential coefficients are zero, i.e.

$$y(0) = y'(0) = \cdots = x(0) = x'(0) = \cdots = 0$$

Then we get

$$Y(a_0 + a_1 s + a_2 s^2 + \cdots) = X(b_0 + b_1 s + b_2 s^2 + \cdots)$$

or

$$Y = \frac{b_0 + b_1 s + b_2 s^2 + \cdots}{a_0 + a_1 s + a_2 s^2 + \cdots} X$$

It is now always possible to factorise the polynomial in the denominator and express it as a product $(s + k_1)(s + k_2)(s + k_3) \cdots$ using a sufficient number of terms. Assuming that the numerator $b_0 + b_1 s + b_2 s^2 + \cdots$ is of no larger order than this, the technique of partial fractions enables the equation to be rewritten as

$$Y = \left(\frac{K_1}{s + k_1} + \frac{K_2}{s + k_2} + \frac{K_3}{s + k_3} + \cdots\right) X$$

with proper choice of the constants $k_1, k_2, k_3 \cdots$ and $K_1, K_2, K_3 \cdots$. If finally x can be made to have the form of the *unit impulse* (Section 1.4) we have that $X = 1$ and can see immediately by taking the inverse transform of Y

that

$$y = K_1 \exp(-k_1 t) + K_2 \exp(-k_2 t) \ldots$$

Application of this result to any mechanical, electrical or biological system in which changes occur in time as the result of some stimulus means that (if the system obeys the differential equation above) the application of a sudden stimulus of very short time duration will give a response described by a set of exponential functions. (Of course, some of the exponential terms may be imaginary as in Section 1.13 depending on how the denominator factorises, so that sine and cosine terms may appear in the solution.)

It is clear that the expression

$$\frac{b_0 + b_1 s + b_2 s^2 + \cdots}{a_0 + a_1 s + a_2 s^2 + \cdots} \equiv \frac{K_1}{s + k_1} + \frac{K_2}{s + k_2} + \cdots$$

just depends on the form of the original equation and is the same whatever the values given to x or the resulting values of y. It is called the *transfer function*, H, and we can say that the original equation is equivalent to

$$Y = H.X$$

H can be regarded as the ratio of the Laplace transforms of $y(t)$ and $x(t)$ when the stimulus $x(t)$ is allowed to have any form; alternatively it is the Laplace transform of the particular function $y(t)$ which arises when the stimulus $x(t)$ is just a unit impulse. This particular function is called the *impulse response* and is denoted by $h(t)$;

$$\mathcal{L}\{h(t)\} = H$$

CONVOLUTION

1.16 Convolution: The Discrete Case

Whenever a scientific investigation is made, the experimental results are obviously a function of the physical phenomena being investigated, and perhaps less obviously a function of the instrumentation being used. Ideally one would like to have instruments that are 100% reliable and unbiased in any way; an analogy for a perfect instrument might be a perfectly plane mirror with 100% reflectivity, so that the image seen in it would be completely indistinguishable from a given object (except for left-to-right inversion). But, in practice, no mirror is perfect and to the extent that imperfections are present the image will be distorted. We can express these ideas semi-mathematically by the conceptual equation.

(Phenomena) $*$ (Instrumental characteristics) $=$ Experimental results

where the $*$ sign is analogous to a multiplication sign. We shall shortly give it a precise mathematical definition and refer to it as a 'convolution'. The basic problem in physical science is to solve the equation explicitly for the quantity we have labelled (Phenomena). To get an accurate image of natural phenomena, we have to take our experimental results and remove from them the blurring effects of our imperfect instrumentation; this inverse operation is a 'deconvolution'.

Very often a physical quantity we want to measure cannot be represented merely by one number, but by a series of numbers we shall call a *sequence*. Examples would be the voltage generated by an electrical circuit or a physiological system which we would like to know at time instants 0, Δt, $2\Delta t$, . . . ; or the strength of a radioactive preparation at distances 0, Δx, $2\Delta x$, . . . from some origin along a line imagined within the body of a patient in hospital. Let us represent such sequences generally by the quantities

$$\ldots 0, 0, 0, u_1, u_2, u_3, u_4, \ldots, 0, 0, \ldots$$

where the zeros are inserted to remind us that our measurements might extend over an indefinitely long time interval, or space interval. It is assumed here that the measurements are taken at equal intervals of time, or space, as appropriate.

Similarly we may express the performance of the instrumentation we are using by a sequence

$$\ldots 0, 0, 0, v_1, v_2, v_3, \ldots, 0, 0, \ldots$$

which would represent the instrumental readings we would obtain if the physical quantity being measured were represented by the special sequence

$$\ldots 0, 0, 0, 0, 0, 1, 0, 0, 0, \ldots$$

This special sequence will not correspond to a 'real' situation, e.g. a measurement of radioactivity distributed in a region of space, for example within living tissue, but to a measurement on a standard object, such as a 'point' source of radioactivity. In a sense, the artificial test object provides a 'test pulse'. Further, we can regard an ideal instrument as one for which the 'performance' sequence 0, 0, v_1, v_2, . . . , 0, 0, . . . is in fact 0, 0, 0, 1, 0, 0, 0, . . . This implies that when an ideal instrument is used to measure an object whose sequence is 0, 0, u_1, u_2, u_3, . . . , 0, 0, . . . the experimental results will be the identical sequence 0, 0, u_1, u_2, u_3, . . . , 0, 0, . . . , that is to say the results are an exact replica of the object.

For any practical instrument, however, several of the v's in its performance sequence will be non-zero. For instance, if a voltmeter is used to indicate a fluctuating voltage, a time lag in the movement of the pointer will mean that a reading at any instant is somewhat influenced by the true voltage at a previous instant. Radiation detectors are almost never perfectly collimated, so that, if a 'point' source of radiation is placed at a distance X from the origin, there will be a relatively large response in the detector when it is aligned exactly with this point but smaller responses when it is directed at points at distances $X - 2\Delta x$, $X - \Delta x$, $X + \Delta x$, $X + 2\Delta x$, for example.

If we are to represent instrumental performance by a sequence of numbers, we have to decide what *scale* to use. For the present purpose, the relative values of these bear more significance than their absolute magnitudes, and we find it most convenient to express them as a *normalised* sequence, i.e. one in which the sum of the terms is unity. Each term represents the relative efficiency of the instrument at different instants in time or positions in space as the case may be, these efficiencies being so scaled that the total efficiency over all time or all space is 100%. We will find it convenient to denote this sequence by $\{v\}$; then for a perfect instrument we would say

$$\{v\} = 0, 0, 0, 1, 0, 0, \ldots$$

whereas in practice we might have an instrument whose sequence is

$$\{v\} = 0, 0, 0.1, 0.2, 0.4, 0.2, 0.1, 0, 0, \ldots$$

Also the quantity we are measuring will be denoted by the sequence $\{u\}$, which will not be a normalised sequence, since it is a set of voltages or the strengths of a number of radioactive sources or some such entity which we need to know in absolute as well as in relative terms.

We now want to know exactly how to form the convolution of the two sequences $\{u\} * \{v\}$, which will be the sequence $\{w\}$ say, and will represent the results obtained when an imperfect instrument is used to carry out measurements on the physical system being investigated.

To fix our ideas let us consider a radiation detector being moved over a set of radioactive sources distributed along the line of movement, as shown in Fig. 1.8. This set of sources might approximate to the actual distribution of activity within a biological system (e.g. a patient). As the detector moves from left to right along the source sequence, it is clear that the extreme, left-hand source will first be measured with an efficiency given by the extreme right-hand term of the $\{v\}$ sequence; the response of the detector will therefore be a measure of u_0v_0. This is then the first term in the $\{w\}$ sequence and is plotted as the extreme left-hand point in the $\{w\}$ sequence shown on the right of the diagram. As the detector moves to the next position we find that the source u_0 is now being measured with efficiency v_1 and the source u_1 is being measured with efficiency v_0. It seems reasonable to suppose that the total effect on the detector given by the two sources is just the sum of the effects of the two separately; this means that the second term in $\{w\}$ is $u_0v_1 + u_1v_0$. Further terms in $\{w\}$ are formed in the same way as the detector moves, until it has passed completely over the set of sources.

We conclude that the convolution of two sequences can be obtained in the general case by the following steps.

(a) Write down the two sequences on two lines, the second $\{v\}$, being reversed end to end so that the last term in $\{u\}$ overlaps the first term in $\{v\}$. Write down the product of overlapping terms; this number is the first term in $\{w\}$.

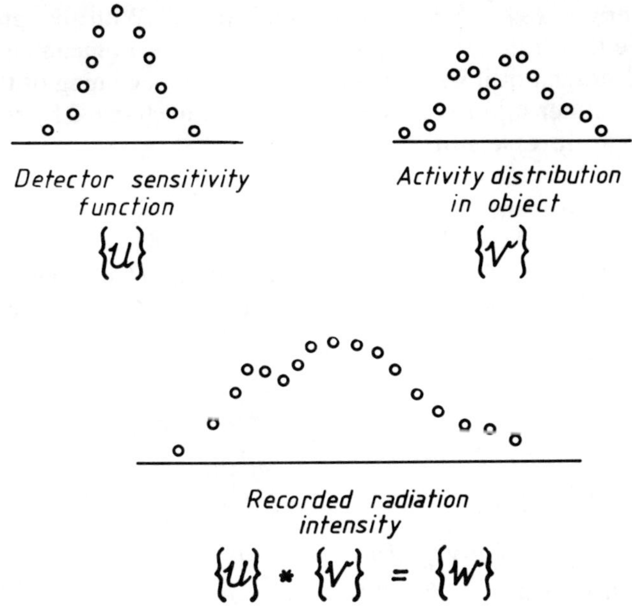

Fig. 1.8 Convolution in radiation intensity measurement.

(b) Shift the second sequence along so that two terms overlap. Form the sum of the products of overlapping terms: this is the second term in $\{w\}$.

(c) Shift again and from the next term in $\{w\}$ in the same way.

(d) Continue until only zero terms overlap; then $\{w\}$ is complete.

To illustrate, we here work out the convolution of two sequences which each have three non-zero terms; zeros are omitted although they are of course implied.

$$
\begin{array}{lccccccccl}
\text{(a)} & & & u_1 & u_2 & u_3 & & & \rightarrow & u_1v_1 \\
 & v_3 & v_2 & v_1 & & & & & & \\
\text{(b)} & & & u_1 & u_2 & u_3 & & & \rightarrow & u_1v_2 + u_2v_1 \\
 & & v_3 & v_2 & v_1 & & & & & \\
\text{(c)} & & & u_1 & u_2 & u_3 & & & \rightarrow & u_1v_3 + u_2v_2 + u_3v_1 \\
 & & & v_3 & v_2 & v_1 & & & & \\
\text{(d)} & & & u_1 & u_2 & u_3 & & & \rightarrow & u_2v_3 + v_3u_2 \\
 & & & & v_3 & v_2 & v_1 & & & \\
\text{(e)} & & & u_1 & u_2 & u_3 & & & \rightarrow & u_3v_3 \\
 & & & & & v_3 & v_2 & v_1 & &
\end{array}
$$

The convolution therefore produces the new sequence $\{w\}$ which is displayed (except for zeros) vertically in the third column.

It is clear from the above that it does not in fact matter *which* of the original sequences is reversed end to end so long as the first step overlaps u_1

and v_1—otherwise $\{w\}$ will be reversed end to end. With this provision we may conclude that $\{u\} * \{v\} = \{w\} = \{v\} * \{u\}$, the $*$ sign meaning 'convolute with' in the mathematical way foreshadowed at the beginning of this section. Clearly also if either $\{u\}$ or $\{v\}$ is *symmetrical*, i.e. unchanged by reversal, it is not necessary to reverse either sequence.

1.17 The *Z* Transform

Another simple way of working out convolutions is by means of the Z transform. We define the Z transform of a sequence $\{u\}$ by the polynomial

$$Z(\{u\}) = u_1 + u_2 z + u_3 z^2 + u_4 z^3 + \cdots$$

and similarly

$$Z(\{v\}) = v_1 + v_2 z + v_3 z^2 + v_4 z^3 + \cdots$$

whence we easily find that

$$Z(\{u\}) \times Z(\{v\}) = u_1 v_1 + (u_2 v_1 + u_1 v_2)z \\ + (u_1 v_3 + u_2 v_2 + u_3 v_1)z^2 + \cdots$$

and the right-hand side is clearly the Z transform of the convoluted sequence $\{w\}$. That is to say, if

$$\{u\} * \{v\} = \{w\}$$

then

$$Z(\{u\}) \times Z(\{v\}) = Z(\{w\})$$

Notice that the sign between the Z transforms is just an ordinary multiplication sign, so that we are perfectly justified in rewriting this equation to read

$$Z(\{u\}) = Z(\{w\})/Z(\{v\})$$

The implication is that we now have a simple way of performing the *deconvolution* envisaged at the beginning of Section 1.16. In outline, the process of deconvolution requires the following steps:

(a) obtain experimental results on the object being studied, writing them as a sequence $\{w\}$ and forming the Z transform $Z(\{w\})$;
(b) measure the instrumental performance on a 'test' object, giving the sequence $\{v\}$ whose Z transform is $Z(\{v\})$;
(c) find $Z(\{u\}) = Z(\{w\})/Z(\{v\})$;
(d) write the sequence $\{u\}$ which is a perfect image of the object being studied, the effect of instrumental imperfection having been eliminated.

It will be noticed that the numerical value assigned to the parameter z (except for zero) in all the above is of no consequence, and in this respect z plays much the same part as the parameter s in Laplace transforms or $i\omega$ in Fourier transforms. Indeed there are many parallels between all three transforms, as will become clear as we proceed, and in some sense each can be regarded as a special case or an extension of the others.

Two points have been rather glossed over in the above treatment. One is that we are justified in adding together the terms forming each summation in the convolution; this is a question of what we call the *linearity* of the measuring system, and we take up this point again specifically in Section 6.2. The other point is that we have assumed all measurements as being taken at isolated, and equally spaced, intervals of time or increments in space—hence the term 'discrete case'. While we do indeed sometimes move an instrument from point to point in space and keep it stationary during a measurement, it is often more convenient to drive it in a continuous manner. We therefore turn now to a discussion of the convolution of continuous functions.

1.18 Convolution: The Continuous Case

If we are investigating the activity of a radioactive source which is distributed in a region of space by taking measurements from point to point along a line within it, or the change in the activity of a source by taking measurements in successive intervals of time, it is natural in the interests of accuracy to take many 'readings' with small interval between them rather than only a few at great intervals. In the limit we might imagine getting so many experimental points that when plotted they would look almost indistinguishable from a continuous curve. Likewise, of course, we would try to express our instrumental characteristics on as finely divided a scale as we could manage. Let us therefore consider the convolution of continuous functions $u(x_0)$ and $v(x_0)$ instead of the sequences $\{u\}$ and $\{v\}$, where x_0 is the space (or, for that matter, time) coordinate. Again u represents the actual distribution we want to measure and v represents the response of the instrument to a 'test' object. Fig. 1.9 gives a plot of u and of v (Fig 1.9a and b). As will be seen, we are

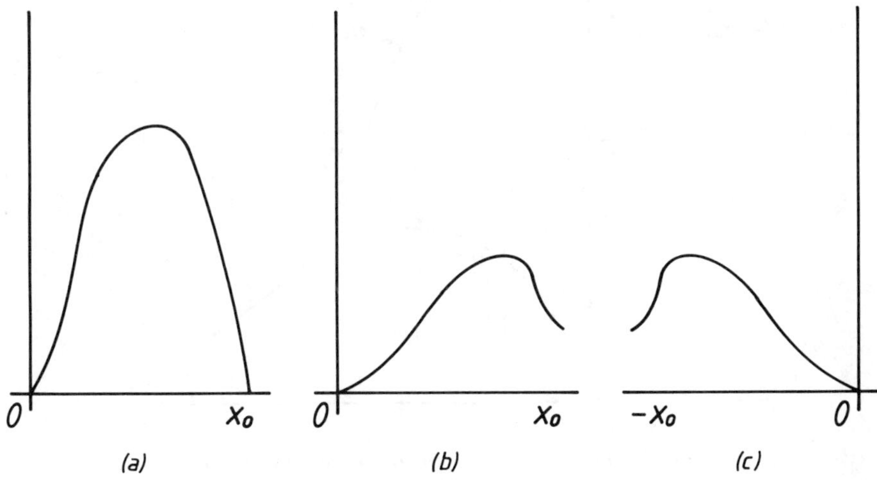

Fig. 1.9 The functions $u(x_0)$, $v(x_0)$ and $v(-x_0)$.

going to consider functions that start at zero, rise to a maximum, and decrease again, just as the $\{u\}$ and $\{v\}$ sequences did. As with the $\{v\}$ sequence, it is useful to consider v as being normalised, that is to say

$$\int_0^\infty v(x_0)\mathrm{d}x_0 = 1$$

Now if we think of the function $v(x_0)$ and replace x_0 by $(-x_0)$, we get the function $v(-x_0)$ which is the mirror image of $v(x_0)$, the mirror being along the line of the vertical axis (Fig. 1.9c). Further, the function $v(x - x_0)$, where x is given a set of values starting at zero and gradually increasing, will be represented by the curve for $v(-x_0)$, which shifts steadily along the x_0 axis as x increases. The movement is exactly analogous to the shifting of the reversed sequence $\{v\}$ discussed in the previous section.

Fig. 1.10 illustrates the situation when for some particular value of x there is some overlap between $u(x_0)$ and $v(x - x_0)$. In the region of the overlap, we now form, for all possible pairs of ordinates, the product $u(x_0).v(x - x_0)$. This new function is plotted as a dashed curve in the figure. The area under the new curve has its value plotted as the point P in the lower graph. It is clearly given by $\int_0^\infty u(x_0)v(x - x_0)\mathrm{d}x_0$ and is called the *convolution integral*. If the functions u and v had been functions of time, the integral would be $\int_0^\infty u(t_0)v(t - t_0)\mathrm{d}t_0$. It may be difficult at this stage to see what can be meant by the two times t and t_0 in this expression, but the point is taken up again in Section 6.14 in connection with dynamic studies in which time is the independent variable rather than a space coordinate.

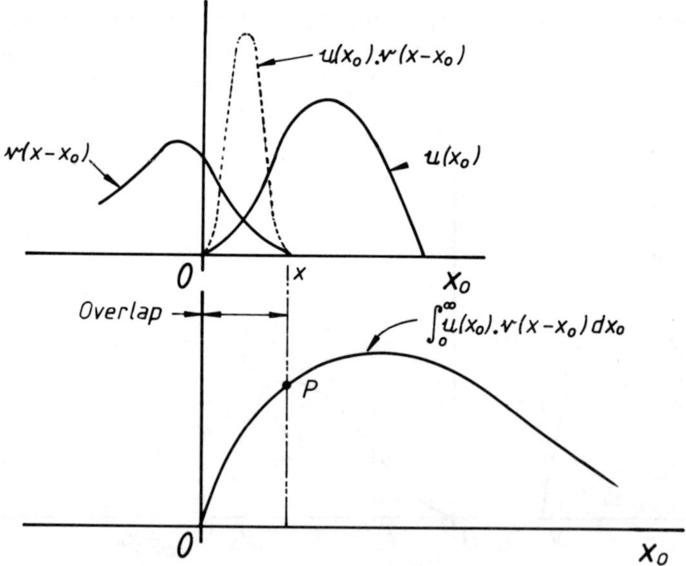

Fig. 1.10 Convolution of $u(x_0)$ and $v(x - x_0)$.

Three points should be immediately made about the convolution integral; they are:

(a) It does not really matter what we take to be the integration limits, so long as these include the interval from 0 to x, because the area is zero elsewhere. We have restricted the discussion to cases in which u and v are zero for negative x_0 (or t_0); more generally when u and v can have non-zero values to the left of the vertical axis the convolution integral must be written

$$\int_{-\infty}^{+\infty} u(x_0)v(x - x_0)\mathrm{d}x_0 \qquad \text{or} \qquad \int_{-\infty}^{+\infty} u(t_0)v(t - t_0)\mathrm{d}t_0 .$$

(b) Whatever the limits, when the integral is evaluated x_0 (or t_0) disappears from the expression, as it is replaced by the limits. The function is then one of x (or t) only. Clearly it will give a 'hump-shaped' curve like u and v but rather more 'spread-out' if plotted on the same scale (i.e. x on the same scale as x_0). It is of course convenient to have the same scale; thus x and x_0 could both be in millimetres, say.

(c) Just as with the discrete case when the convolution was obtained by shifting $\{u\}$ with respect to $\{v\}$, or the other way about, we can write the integral as $\int_0^\infty u(x - x_0)v(x_0)\mathrm{d}x_0$. As with the discrete case, the physical reality behind this interchangeability is simply that we get the same result, for example, by moving a detector from left to right across an activity distribution, as we would if we kept the detector stationary and moved the activity across it in the reverse direction.

1.19 Convolution in Terms of the Laplace Transform

We have seen how the Z transform provides a particularly simple way of carrying out the convolution of two sequences $\{u\}$ and $\{v\}$; the general term in $Z(\{u\})$ is $u_n z^{n-1}$ and the transform itself is the sum of all such terms. If we plot the $\{u\}$ sequence as a histogram, we can also plot the value of $u_n z^{n-1}$ for each value of u_n, making any arbitrary assumption we like about the magnitude of z. The Z transform itself is then represented by the area beneath this second histogram, as is shown in Fig. 1.11. Now a continuous function, say $u(t)$, can be approximated to by a sequence $u(0)$, $u(\Delta t)$, $u(2\Delta t)$, . . . , $u(N\Delta t)$, . . . , where Δt is a small increment in t; the approximation can be made as close as we please by making the interval Δt indefinitely small and increasing the number of terms suitably. Let us suppose that the Nth term, $u(N\Delta t)$, always corresponds to a point $u(t)$ on the continuous function, i.e. $t = N\Delta t$.

As we are free to choose any value of z we like, let us put $z = \exp(-s\Delta t)$ where s is some other number. The Nth strip in the z histogram therefore has a height

$$u(N\Delta t).\exp\{-s(N-1)\Delta t\} \approx u(N\Delta t).\exp(-sN\Delta t)$$

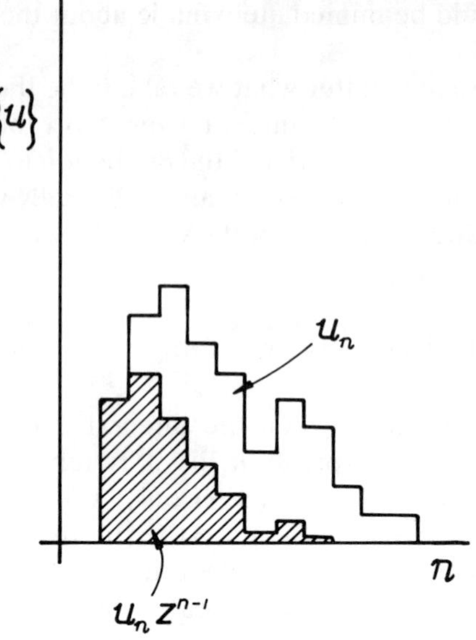

Fig. 1.11 The Z transform.

since we may neglect 1 in comparison with the large number N. The transform is then the sum of the areas of all the strips, and passing to the limit when $\Delta t \to 0$ but still with $N\Delta t = t$ the sum becomes the integral

$$A = \int_0^\infty u(t) \exp(-st)\mathrm{d}t$$

which is identical with the Laplace transform for $u(t)$. We conclude that *continuous* functions of t (or of x) may be convoluted in exactly the same way as the corresponding discrete functions, with Laplace transforms replacing Z transforms. In particular, if u and v are continuous functions and have a convolution w given by

$$w = u * v$$

we can take Laplace transforms of each side and obtain

$$\begin{aligned} W &= \mathcal{L}(w) \\ &= \mathcal{L}(u).\mathcal{L}(v) \\ &= U.V \end{aligned}$$

This means we have an alternative method for finding the function w given u and v; instead of having to evaluate the integral $\int_0^\infty u(t_0)v(t_0 - t)\mathrm{d}t_0$ we may

(a) find the Laplace transforms of u and v,
(b) multiply them together, and
(c) find the inverse transform of the product.

A statement of the equivalence of these alternatives is known as the *convolution theorem*.

Furthermore, we showed in Section 1.15 that the solution of the equation

$$a_0 y + a_1 dy/dt + a_2\, d^2y/dt^2 + \cdots$$
$$= b_0 x + b_1\, dx/dt + b_2\, d^2x/dt^2 + \cdots$$

can be written, if the initial conditions are suitable, as

$$Y = H.X$$

where Y and X are the Laplace transforms of y and x respectively. Comparing this result with the above it is clear that the differential equation is just another way of expressing the convolution

$$y(t) = h(t) * x(t) = \int_0^\infty h(t_0)x(t_0 - t)dt_0 = \int_0^\infty h(t - t_0)x(t_0)dt_0$$

with y, h and x replacing w, u and v respectively. It is perhaps worth emphasising that while the impulse response function $h(t)$ does not appear explicitly in the differential equation, its Laplace transform H depends only on the constants $a_0, a_1, \ldots$ and $b_0, b_1, \ldots$. The equivalence of linear differential equations and convolution forms the theoretical basis for much of the content of Chapter 6.

STATISTICS

1.20 Elementary Statistics

Any single measurement of a physical quantity, giving a numerical value say m, can hardly be regarded as reliable until confirmed by repeated measurement. Yet, however carefully measurements are taken, we generally do not find exact agreement between the successive values, say $m_1, m_2, m_3, \ldots, m_n$ (n being the number of observations). The best we can do is to accept the *average* or mean value $\bar{m} = (m_1 + m_2 + \cdots + m_n)/n$ because we feel intuitively that it is less likely to be in error than any of the extreme values, whether high or low. Using a convenient notation, we write

$$\bar{m} = \frac{\sum_1^n m_n}{n}$$

the symbol $\sum$ denoting the sum of the n experimental values.

If the number of observations is very large, we may introduce a useful modification of this formula. For we may discover that each of the experimental values is one of the discrete numbers M_1, M_2, M_3, . . . , there being no instances of an intermediate value. (For example, individual measurements might be values like 10.3, 10.4, 10.5, 10.6, 10.7, but we will never record a result as 10.45 because we know that we are not justified in using an extra place of decimals but would put such a result into either the category 10.4 or the category 10.5.) In general of course we will not find merely *one* result M_1, one result M_2, etc.; rather we shall find a_1 occasions in which the result is M_1, a_2 occasions in which it is M_2, and so on, where $a_1, a_2, \ldots$ may be quite large numbers.

The mean $\bar{m}$ is then given by

$$\bar{m} = \frac{a_1M_1 + a_2M_2 + a_3M_3 + \cdots}{n}$$

$$= \frac{a_1}{n}M_1 + \frac{a_2}{n}M_2 + \frac{a_3}{n}M_3 + \cdots$$

It is clear that the coefficients a_1/n, a_2/n, a_3/n, . . . are fractions whose sum is just 1, because $a_1 + a_2 + a_3 + \cdots = n$. Since these coefficients represent the relative frequencies with which the results $M_1, M_2, M_3, \ldots$ are obtained, we call them the 'frequency distribution' of the results. They are often denoted by the symbols $W(M_1)$, $W(M_2)$, $W(M_3)$, etc., and we can write our formula for the mean as

$$\bar{m} = W(M_1).M_1 + W(M_2).M_2 + \cdots = \sum W(M).M$$

where $W(M)$ is the frequency for the general result M and the summation sign $\sum$ implies summation over all the terms.

When a very large number of measurements are made, it is generally found that they tend to cluster around the mean value; put another way, we find the chance of any particular value being very far from $\bar{m}$ is much less than that of being very close to $\bar{m}$. A plot of a frequency distribution $W(M)$ against M is therefore bell-shaped as in Fig. 1.12. $W(M)$ reaches a peak value when M is very close to $\bar{m}$—or even identical with $\bar{m}$ as nearly as we can tell. The shape

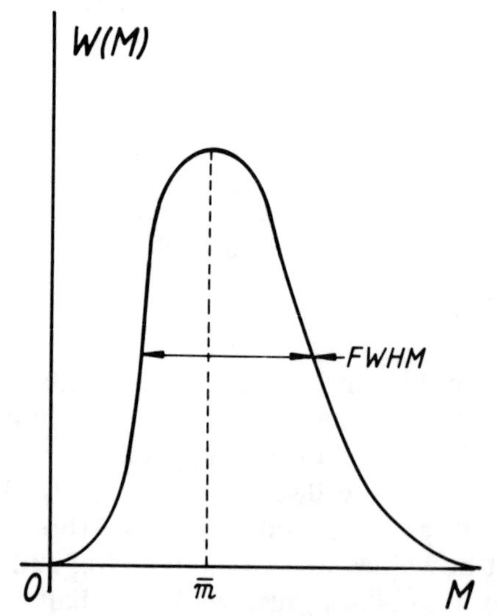

Fig. 1.12 A typical frequency distribution.

of the curve is clearly related to the 'accuracy' of the set of observations, in the sense that observations of poor quality made with crude apparatus and rough methods can be expected to give a broad peak whereas careful work with good apparatus and technique will give a sharp peak. The peak sharpness is measured by the full width at half maximum (FWHM) as shown by the horizontal line in the figure.

Another measure of the way a whole set of measurements is related to the mean value of the set is given by the *standard deviation*, or root mean square (RMS) deviation. If we have a set of values $m_1, m_2, \ldots, m_r, \ldots, m_n$, we can define for any one reading m_r, a *deviation* from the mean $\bar{m}$ of $(m_r - \bar{m})$. Squaring this we get $(m_r - \bar{m})^2$, and the sum of all such terms as r takes on all values from 1 to n is

$$(m_1 - \bar{m})^2 + (m_2 - \bar{m})^2 + \cdots (m_n - \bar{m})^2 = \sum_0^N (m_r - \bar{m})^2$$

We then define the *variance* of the set of values as

$$\nu = \sum_{r=1}^{n} (m_r - \bar{m})^2 \Big/ n$$

and the standard deviation as the square root of the variance as

$$\sigma = \sqrt{\left[\sum (m_r - \bar{m})^2 \Big/ n\right]}$$

Clearly the standard deviation σ will be large if there are many values of m_r which are much greater than, or much smaller than, the mean value $\bar{m}$; whereas, if all m_r values are close to $\bar{m}$, σ will be relatively small.

The expression defining σ can be used as it stands for calculation purposes, but alternatively we have

$$\begin{aligned}
\nu &= \{(m_1 - \bar{m})^2 + (m_2 - \bar{m})^2 \cdots (m_n - \bar{m})^2\}/n \\
&= \{m_1^2 + m_2^2 + \cdots + m_n^2 - 2\bar{m}(m_1 + m_2 + \cdots + m_n) \\
&\quad + (\bar{m}^2 + \bar{m}^2 + \cdots)\}/n \\
&= \left\{\sum_{r=1}^{n} (m_r^2) - 2\bar{m} \sum_{r=1}^{n} (m_r) + n\bar{m}^2\right\} \Big/ n \\
&= \left\{\sum_{r=1}^{n} (m_r^2) - 2\bar{m}.n\bar{m} + n\bar{m}^2\right\} \Big/ n \\
&= \left\{\sum_{r=1}^{n} (m_r^2) - n\bar{m}^2\right\} \Big/ n \\
&= \sum_{r=1}^{n} (m_r^2) \Big/ n - \bar{m}^2
\end{aligned}$$

This means that the variance is equal to the mean of the squares of all the m values, *minus* the square of the mean value of the m's. So for the standard deviation, using the bar sign as usual to denote the mean value, we can write

$$\sigma = \sqrt{[\overline{m_r^2} - (\bar{m})^2]}$$

An advantage of the standard deviation as the measure of the self-consistency of a set of results is that it can be worked out for small values of n (it is not very significant for less than four or five results), whereas the FWHM for a distribution of results can obviously not be found unless there are a great many. However, in a good many cases in which distribution curves have been investigated it turns out that they fit a certain equation, variously called the *normal law of error* and the *Gaussian distribution*, which reads

$$W(M) = \frac{1}{\sqrt{(2\pi)}\sigma} \exp\{-(M - \bar{m})^2/2\sigma^2\}$$

In this equation the exponential term is important in expressing the way in which $W(M)$ changes with M; the factor $1/\sqrt{(2\pi)}\sigma$ is just a constant multiplier which ensures that the sum of all the $W(M)$ terms is unity. $\bar{m}$ is the mean of all the m values in the distribution and σ is the standard deviation of all these values. The values of $\bar{m}$ and σ derived from a great many results are clearly more reliable than those derived from only a few, and are probably slightly different too. For this reason the value of σ obtained from an indefinitely large number of results is sometimes denoted by the letter s, and it is generally assumed that, when only a few results are available, it can be estimated from the relation

$$s = \sigma\sqrt{\left[\frac{n}{(n - 1)}\right]}$$

However, for n as small as 5, s and σ differ by only about 10%, and for 10 observations only 5%. Since it is rarely necessary to estimate standard deviations with high precision the factor $\sqrt{[n/(n - 1)]}$ (Bessel's correction) can usually be ignored. The equation above for $W(M)$ therefore gives a rough estimate of the distribution of a rather small number of observations.

For any result M (or m_r), it has been found very convenient to define a new parameter t where $t = (M - \bar{m})/\sigma$. This is equivalent to measuring the deviations not in absolute terms but as multiples of σ. If then we work in units of σ, we can replace the σ in the multiplying term $1/\sqrt{(2\pi)}\sigma$ by unity and write

$$W(M) = \frac{1}{\sqrt{(2\pi)}} \exp(t^2/2)$$

—an expression which has been extensively tabulated (see, for example, *Handbook of Chemistry and Physics*, CRC). In particular, for $M = \bar{m}$, $t = 0$ and

$$W(M) = \frac{1}{\sqrt{(2\pi)}} = 0.3989$$

while for $M - \bar{m} = \pm\sigma$

$$W(M) = \frac{1}{\sqrt{(2\pi)}} \exp(-0.5) = 0.2420$$

and for $M - \bar{m} = \pm 2\sigma$

$$W(M) = \frac{1}{\sqrt{(2\pi)}} \exp(-2) = 0.054$$

Conversely for

$$W(M) = \frac{0.3989}{2} = 0.1994$$

i.e. at half the peak height we find

$$t^2 = -2 \ln(\sqrt{(2\pi)}.0.1994)$$

$$\therefore t = \pm 1.18$$

so that

$$(M - \bar{m}) = \pm 1.18\sigma$$

The FWHM is therefore twice this value, or 2.36σ. These relationships are illustrated in Fig. 1.13.

Tables have also beeen worked out for the areas beneath various parts of the normal curve. (They have to be found by numerical methods as there is no way of carrying out the integration analytically.) Thus the area under the curve from $+\sigma$ to $-\sigma$ is 0.704; under the FWHM it is 0.76; under the curve from 2σ to -2σ it is 0.955. These figures mean, for instance, that if one takes a fairly large number of measurements one can expect some 70.4% of them to

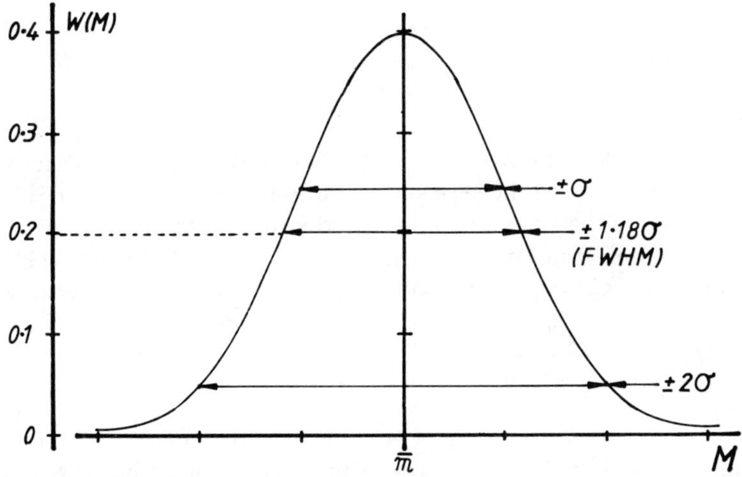

Fig. 1.13 Relationship between standard deviation and FWHM for a normal curve.

lie within one standard deviation of the mean and 95.5% of them to lie within two standard deviations of the mean. If the results do not conform to these rules, one should pick out the very discrepant ones and discard them, working out revised values of the mean and standard deviation of the remaining terms.

While the normal, or Gaussian, distribution often seems to be satisfactory in practice, it does not appear to have a good theoretical justification. However, there are others which have a logical basis and it turns out that the Gaussian distribution is, within some limits, a fair approximation to these. One important example is derived from the binomial theorem, as shown in the next section.

1.21 The Binomial Distribution

Direct multiplication shows that for any pair of numbers p and q

$$(p + q)^2 = p^2 + 2pq + q^2$$
$$(p + q)^3 = p^3 + 3p^2q + 3pq^2 + q^3$$
$$(p + q)^4 = p^4 + 4p^3q + 6p^2q^2 + 4pq^3 + q^4$$
$$\vdots \qquad\qquad \vdots$$

and, in general, if $(p + q)$ is raised to any positive integer power N,

$$(p + q)^N = {}^N_0\mathrm{C}p^0q^N + {}^N_1\mathrm{C}pq^{N-1} + {}^N_2\mathrm{C}p^2q^{N-2} + \cdots$$

where the coefficients

$${}^N_0\mathrm{C} = \frac{N!}{0!\,N!}$$

$${}^N_1\mathrm{C} = \frac{N!}{1!\,(N-1)!}$$

$${}^N_2\mathrm{C} = \frac{N!}{2!\,(N-2)!}$$

$$\vdots \qquad \vdots$$

and $N!$ means $N(N-1)(N-2)\cdots 1$. (Note: $0! = 1$.) This is the *binomial theorem*. We may regard the series of terms in the expansion of $(p + q)^N$ as a possible form of a distribution as discussed in the previous section; using the same notation we write $W(0)$ for the first term $(= {}^N_0\mathrm{C}p^0q^N)$, $W(1)$ for the second term $(= {}^N_1\mathrm{C}p^1q^{N-1})$ and so on. On this basis the Mth term will be

$$W(M) = \frac{N!}{M!\,(N-M)!}\,p^Mq^{N-M}$$

For example, for the special case of $N = 10$, $p = 0.2$ and $q = 1 - p = 0.8$, a histogram of $W(M)$ values is shown in Fig. 1.14 (dotted line). Notice that, because we have chosen $p + q$ to have the value 1, the sum of all the $W(M)$ values is also 1. Obviously for very large values of N the histogram will

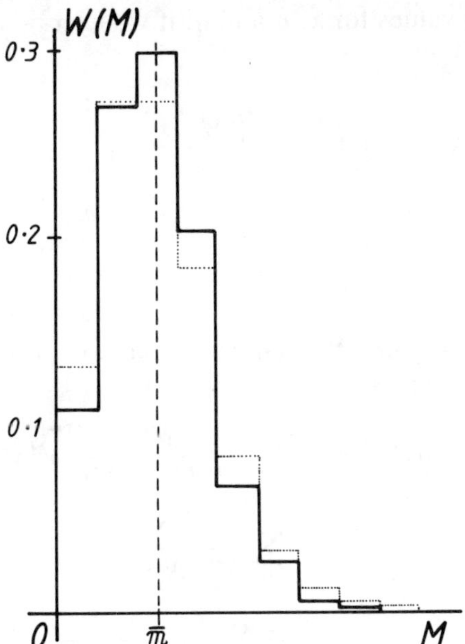

Fig. 1.14 Histogram of a Binomial distribution (full line) and Poisson distribution (dotted line).

approximate to a smooth bell-shaped curve — just in fact the kind of curve we were considering in the sections on convolution, as well as the normal distribution of Section 1.20, although the binomial distribution curve is not symmetrical over its whole extent unless $p = q = 0.5$.

The binomial distribution applies to many physical measurements, and in particular, as will be shown in Chapter 4, to the number of radioactive atoms that are observed to decay in a given fixed time. It is therefore of interest to calculate $\bar{m}$, the mean value of M in the distribution; this will correspond, for example, to the *average* number of radioactive atoms decaying in a fixed time interval, the *actual* number being measured in a large number of trial intervals. Since an actual value M occurs with a relative frequency $W(M)$, we must have that $\bar{m}$ is the sum of all terms like $W(M) \times M$, i.e.

$$\bar{m} = \sum W(M).M$$

This can be found as follows. Let us replace p in the binomial distribution by px, where x is going to be considered as a variable while all the other quantities are constants. We have then

$$(px + q)^N = \sum_0^N \frac{N!}{M!\,(N - M)!}\, p^M x^M q^{N-M}$$

Differentiating both sides with respect to x we get

$$N(px + q)^{N-1}.p = \sum_0^N \frac{N!}{M!\,(N - M)!}\, p^M M x^{M-1}.q^{N-M}$$

Now we can choose values for x, p and q; if we let $x = 1$ and $p + q = 1$ we get

$$Np = \sum_0^N \frac{N!}{M!\,(N-M)!} p^M q^{N-M}.M$$
$$= W(M).M$$

so that

$$\bar{m} = Np$$

Moreover, we can easily find the standard deviation of the distribution; for a second differentiation gives

$$N(N-1)(px+q)^{N-1}.p^2 = \sum_0^N \frac{N!}{M!\,(N-M)!} p^M M(M-1)x^{M-2}.q^{N-M}$$
$$= \sum_0^N \frac{N!}{M!\,(N-M)!} p^M M^2 q^{N-M}$$
$$- \sum_0^N \frac{N!}{M!\,(N-M)!} p^M.Mq^{N-M}$$

and again putting $x = 1$ and $p + q = 1$ we get

$$N(N-1)p^2 = \sum_0^N W(M)M^2 - \sum_0^N W(M)M$$
$$= \sum_0^N W(M)M^2 - Np$$

by the previous result. We recognise $\sum_0^N W(M)M^2$ as the mean of the *squares* of the M values, i.e. $\overline{m^2}$. Recalling that, for any distribution of experimental values, the variance ν or σ^2 is given by $\overline{m^2} - (\bar{m})^2$ (see Section 1.20) we have

$$\nu = \sigma^2 = \overline{m^2} - (\bar{m})^2$$
$$= N(N-1)p^2 + Np - N^2p^2$$
$$= Np - Np^2$$
$$= Np(1-p)$$
$$= Npq$$

so that

$$\sigma = \surd(Npq)$$

In summary we have the important results that $\bar{m} = Np$ and that $\sigma = \surd(Npq)$. It is also worth noting that if p is very small compared with 1, $q = 1 - p$ is very nearly 1 and σ is approximately $\surd(Np)$ or $\surd\bar{m}$. As already implied, we shall use these results directly in the discussion of radioactive decay in Chapter 4.

1.22 Approximations to the Binomial Distribution: The Poisson and Gaussian Distributions

The expression for the binomial distribution

$$W(M) = \frac{N!}{M!(N - M)!} p^M q^{N-M}$$

is mathematically rather clumsy and in practice we may make some approximations which result in a more tractable expression without detracting from its range of usefulness. Let us see what approximations can be made if we assume N to be a very large number but p to be very small. (For example if we take $N = 10^6$ and $p = 10^{-3}$ we find $\bar{m} = Np = 10^3$. Then if we examine the first 10^4 terms in the expansion we find that although they are only 1% of all the terms they easily encompass all of the peak.) Now when we consider the term $N!/(N - M)!$ in the binomial expression, which is the product $N(N - 1)(N - 2) \cdots$ carried on for M terms, we realise that each term after the first is only slightly smaller than N itself, and we will make little error if we replace it by N^M. Next we may replace p^M by $(\bar{m}/N)^M$. Finally when we consider $q^{N-M} = (1 - p)^{N-M}$ we can replace $(1 - p)$ by $\exp(-p)$. This term is then approximately $\exp\{(-p).(N - M)\} \approx \exp(-pN)$, since we may neglect M in comparison with N, and this is now equal to $\exp(-\bar{m})$. The whole expression for $W(M)$ then becomes

$$W(M) = \frac{N^M}{M!}\frac{(\bar{m})^M}{N^M} \exp(-\bar{m})$$

$$= \frac{(\bar{m})^M}{M!} \exp(-\bar{m})$$

which is the well known Poisson distribution. The terms $W(0)$, $W(1)$, $W(2)$, . . . are respectively $\exp(-\bar{m}).1$, $\exp(-\bar{m}).\bar{m}$, $\exp(-\bar{m}).\bar{m}^2/2$, $\exp(-\bar{m}).m^3/3!$, etc.

Even for N as small as 10, the Poisson distribution can be a fair approximation to the binomial; this is shown in Fig. 1.14, where the short horizontal lines show the value of the Poisson terms for $\bar{m} = Np = 2$.

A further approximation may be made if, in addition to restricting ourselves to large N and small p values, we consider only that part of the distribution for which M is nearly equal to $\bar{m}$. This further approximation will be acceptable near the peak of the histogram but not so good on the 'wings'. We first of all put in an approximation due to Stirling which reads

$$M! = \sqrt{(2\pi M)}.M^M \exp(-M)$$

and get

$$W(M) = \frac{(\bar{m})^M}{\sqrt{(2\pi M)}\, M^M \exp(-m)} \exp(-\bar{m})$$

$$= \frac{(\bar{m}/M)^M}{\sqrt{(2\pi M)}} \mathit{exp}\ (M - \bar{m})$$

We then deal with the term $(\bar{m}/M)^M$; we have

$$\ln\left(\frac{\bar{m}}{M}\right)^M = M\ln\left(\frac{\bar{m}}{M}\right)$$

$$= M\ln\left(1 + \frac{\bar{m} - M}{M}\right)$$

$$= M\left(\frac{\bar{m} - M}{M}\right) - \frac{(\bar{m} = M)^2}{2M^2} + \dots$$

by substitution of $(\bar{m} - M)/M$ for x in the series expansion for $\ln(1 + x)$. Then

$$\ln\left(\frac{\bar{m}}{M}\right)^M \approx \bar{m} - M - \frac{(\bar{m} - M)^2}{2M}$$

assuming that later terms can be neglected. Therefore

$$\left(\frac{\bar{m}}{M}\right)^M = \exp\left\{\bar{m} - M - \frac{(\bar{m} - M)^2}{2M}\right\}$$

and the whole expression for $W(M)$ becomes

$$W(M) = \frac{\exp\{-(\bar{m} - M)^2/2M\}}{\sqrt{(2\pi M)}}$$

Also having regard to the fact that M is assumed to be close to $\bar{m}$, and $\bar{m} = \sigma^2$, we get

$$\mathrm{W}(M) = \frac{\exp\{-(\bar{m} - M)^2/2\sigma^2\}}{\sqrt{(2\pi)}\sigma}$$

—the gaussian distribution.

It will be noticed that an essential step in this argument is the replacement of $\bar{m}$ by σ^2—a relation which we derived from the binomial distribution (and which can incidentially be derived also independently from the Poisson distribution). But it is important to realise that there are many sets of data that fit a gaussian distribution for which it is *not* true to say that $\bar{m} = \sigma^2$; the reason is that these data do not fit a binomial distribution. As we shall see in Chapter 5, the number of radiations detectable from a radioactive source in a fixed time interval fluctuates about a mean value according to a binomial law and therefore approximately to a gaussian. But the distribution of weights among adult human beings does not obey a binomial law although the distribution is nearly gaussian and has a certain standard deviation. This standard deviation can only be determined experimentally and is *not* given by the square root of the mean weight. It is worth remembering that the binomial distribution can apply only to quantities that are *dimensionless*.

1.23 Appendix 1 Imaginary and Complex Numbers

If a and b are any real numbers, we know from elementary algebra that the expression $a^2 - b^2$ may be written as the product of two factors

$$a^2 - b^2 = (a + b)(a - b)$$

Complex numbers arise naturally from the attempt to express $a^2 + b^2$ as a similar product. This can only be done by introducing the imaginary number i, whose square is defined to be -1. It is assumed that the ordinary operations of addition, subtraction, multiplication and division can be applied to i and any expressions involving it. Now if we replace b in the above equation by ib we get

$$a^2 - \mathrm{i}^2b^2 = (a + \mathrm{i}b)(a - \mathrm{i}b)$$

or

$$a^2 + b^2 = (a + \mathrm{i}b)(a - \mathrm{i}b)$$

The factorisation of $a^2 + b^2$ has therefore been found, but only in terms of the so-called complex numbers $a + \mathrm{i}b$ and $a - \mathrm{i}b$; each of these is partly a real number a and partly an imaginary number ib or $-\mathrm{i}b$.

It is sometimes helpful to regard complex numbers as points on a plot constructed by taking the values of the real part along the direction of the x-axis and the values of the imaginary part along the y-axis. This plot is called the Argand diagram (Fig. 1.15). The diagram makes it clear that any point P is *uniquely* defined by its coordinates a and b; that is to say that if P has to represent a complex number $a + \mathrm{i}b$ and at the same time a complex number

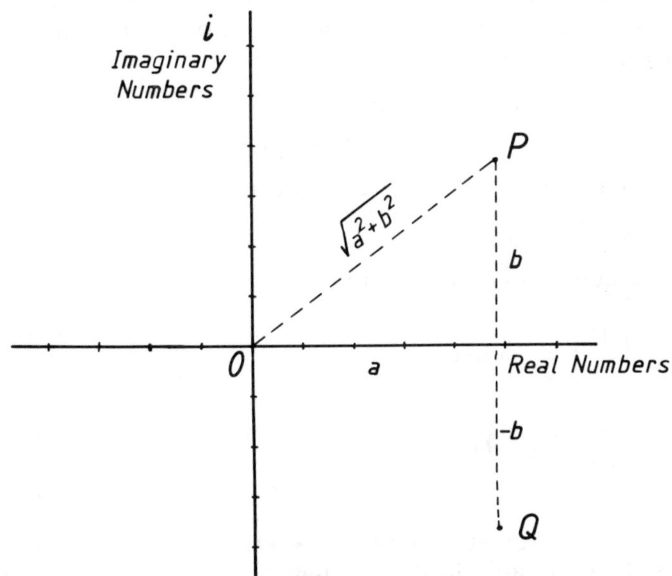

Fig. 1.15 The Argand diagram, showing $(a + \mathrm{i}b)$ and $(a - \mathrm{i}b)$ at P and Q.

$c + id$, it can only be that $a = c$ and $b = d$. An algebraic proof is that if

$$a + ib = c + id$$

then

$$(a - c) = i(d - b)$$

Now the real number $a - c$ cannot equal the imaginary number $i(d - b)$ unless both are zero; hence $a = c$ and $b = d$.

The Argand diagram shows that $a^2 + b^2$ is to be represented as the square drawn on OP (or OQ) as one side. The length OP, which is equal to $\sqrt{(a^2 + b^2)}$, is called the *modulus* of the complex number $a + ib$, and of course is also equal to the modulus of $a - ib$. Also the slope of OP is called the *amplitude* of $a + ib$; it is clearly $\tan^{-1}(b/a)$. The position of a point P is fixed if we know the values of a and b; we could alternatively specify where P is by quoting the values of its modulus $\sqrt{(a^2 + b^2)}$ and its amplitude $\tan^{-1}(b/a)$. These quantities are much used in the theory of alternating currents, when a modulus is used to represent the complex impedance of a circuit while the amplitude gives the phase difference between the current and the voltage.

1.24 Appendix 2 Elementary Calculus

(a) If y is given as a function of x, $y = f(x)$, the differential coefficient of y with respect to x is written $dy/dx \equiv f'(x)$ and is defined as the limit of the ratio of small increments Δy, Δx, as Δx tends to zero. If y is plotted against x, dy/dx is the slope of the curve at any point with coordinates x and y. For a curve possessing a maximum or a minimum value for y, dy/dx is zero at that point.

(b) If $y = ax^n$, $\quad dy/dx = nax^{n-1}$

$y = \exp(ax)$, $\quad dy/dx = a\exp(ax)$

$y = \sin(ax)$, $\quad dy/dx = a\cos(ax)$

(c) $y = u.v$, $\quad dy/dx = u\, dv/dx + v\, du/dx$

(d) Higher differentials may be written

$$\frac{d}{dx}\left(\frac{dy}{dx}\right) = d^2y/dx^2$$

$$\frac{d}{dx}\left[\frac{d}{dx}(dy/dx)\right] = d^3y/dx^3$$

etc.

Differentials are sometimes written, when there is no ambiguity as to the independent variable, in the shortened form $dy/dx \equiv Dy$, $d^2y/dx^2 \equiv D^2y$, etc.

(e) Integration is the inverse process to differentiation; thus, if $dy/dx = z$, $\int z\,dx = y + c$, where the integration constant c can have any value.

(f) The definite integral $\int_a^b z\,dx = y_{x=b} - y_{x=a}$, i.e. the value of y when $x = b$, *less* the value of y when $x = a$. The geometrical interpretation of $\int_a^b z dx$ is the area beneath the curve $z = f(x)$ between the ordinates $x = b$, $x = a$.

(g) If

$$z = ax^m, \quad \int z dx = ax^{m+1}/(m + 1) + c$$

$$z = \exp(ax), \quad \int z dx = \frac{1}{a}\exp(ax) + c$$

$$z = \sin(ax), \quad \int z dx = -\frac{1}{a}\cos(ax) + c$$

$$z = a/x, \quad \int z dx = a \ln(x) + c$$

(h) Areas in a plane may be written as double integrals, volumes by triple integrals, and so on.

(i) Integration by parts makes use of the fact that if u and v are functions of x,

$$\int u dv = uv - \int v du$$

(j) There are theorems (Taylor's and Maclaurin's theorems) which use successive differentiation to evaluate the constants $a_0, a_1, a_2, \ldots$, etc., in infinite series expansions such as

$$\cos(x) = a_0 + a_1x + a_2x^2 + a_3x^3 + \cdots$$

Useful expansions are

$$\begin{aligned}\exp(x) &= 1 + x + x^2/2! + x^3/3! + \cdots \\ \ln(1 + x) &= x - x^2/2 + x^3/3 - \cdots \qquad (-1 < x < 1) \\ \cos(x) &= 1 - x^2/2! + x^4/4! - \cdots \\ \sin(x) &= x - x^3/3! + x^5/5! - \cdots\end{aligned}$$

Using the expansions for $\cos(x)$, $\sin(x)$ and $\exp(ix)$ and with i defined as $\sqrt{-1}$, one may show that

$$\cos(x) + i\sin(x) = \exp(ix)$$

and, more generally, that

$$[\cos(x) + i\sin(x)]^n = \exp(nix)$$

These results are known as de Moivre's theorem.

(k) If ϕ is any function of x and y, the *partial* differential $(\partial\phi/\partial x)$ means the differential coefficient of ϕ with respect to x, with y assumed constant.

Similarly for $(\partial\phi/\partial y)$. A small change in ϕ is then related to small changes in x and y by

$$\Delta\phi = \left(\frac{\partial\phi}{\partial x}\right)\Delta x + \left(\frac{\partial\phi}{\partial y}\right)\Delta y$$

If in addition $\phi = c$, where c is any constant,

$$\frac{\mathrm{d}y}{\mathrm{d}x} = -\left(\frac{\partial\phi}{\partial x}\right)\Big/\left(\frac{\partial\phi}{\partial y}\right)$$

1.25 Appendix 3 Partial Fractions

It is often useful, particularly when working with Laplace transforms, to be able to express a polynomial fraction in terms of partial fractions. The problem resolves itself into evaluating the constants A, B, C, . . . in the identity

$$\frac{1}{(s + a)(s + b)(s + c) \ldots} \equiv \frac{A}{s + a} + \frac{B}{s + b} + \frac{C}{s + c} + \cdots$$

where there are as many terms multiplied together in the denominator on the left-hand side as there are fractional terms on the right-hand side. In our application s is a real quantity but any of the letters a, b, c, . . . might be imaginary. There is a simple rule by which A, B, C, . . . can be found; it is called the 'cover-up rule'. To find A, we

(1) write down the expression on the left-hand side but omitting the term $(s + a)$; (thus $(s + a)$ is covered up);
(2) substitute $-a$ for s in the remaining terms.

So

$$A = \left\{\frac{1}{(s + b)(s + c) \cdots}\right\}_{s=-a} = \frac{1}{(b - a)(c - a) \cdots}$$

Similarly for B, we get by covering up $(s + b)$

$$B = \frac{1}{(a - b)(c - b) \cdots}$$

and so on for the rest of the terms.

If repeated factors occur the problem is more difficult. Thus for

$$\frac{1}{(s + a)(s + b)^2} \equiv \frac{A}{s + a} + \frac{B}{s + b} + \frac{C}{(s + b)^2}$$

we find by covering up $(s + a)$ and $(s + b)^2$ in turn that

$$A = \frac{1}{(a - b)^2} \quad \text{and} \quad C = \frac{1}{a - b}$$

In order to get B we have to put these values in the original identity and solve for B, getting

$$B = -\frac{1}{(a - b)^2}$$

1.26 Exercises

1. Find from first principles the Laplace transforms of

   (a) $y = mt$, m a constant,
   (b) $y = \cos(bt)$, b a constant,
   (c) $y = \sinh(bt) = \{\exp(bt) - \exp(-bt)\}/2$,
   (d) $y = \exp\{(a + ib)t\}$, a and b constants,
   (e) $y = \exp(at).\cos(bt)$,
   (f) $y = \exp(at).\sin(bt)$.

2. Show that the Fourier transform of the gate function

   $$y = 0, \quad t < -\tau$$
   $$y = a, \quad -\tau \leq t \leq \tau$$
   $$y = 0, \quad t > \tau$$

   is

   $$\mathcal{F}(y) = 2a\tau \frac{\sin(\omega\tau)}{\omega\tau} \equiv 2a\tau \sin c(\omega\tau)$$

3. Solve the differential equations

   (a) $$\frac{d^2y}{dt^2} + \frac{dy}{dt} - 2y = 4 \exp(3t);\ y_0 = 0 \text{ and } y_0' = 1$$

   (b) $$\frac{d^2y}{dt^2} - 2\frac{dy}{dt} + 9y = 6 \exp(2t).\cos(3t);\ y_0 = 0 \text{ and } y_0' = 3$$

4. Consider the simultaneous differential equations

   $$\frac{dx}{dt} = k_1x + k_2y \qquad \frac{dy}{dt} = k_3x + k_4y$$

   for which the initial conditions are

   $$x(0) = x'(0) = 0 \qquad y(0) = y'(0) = 0$$

   Write the Laplace transform for each equation and show that

   $$Y = \mathcal{L}(y) = (s - k_1)/\{s^2 - s(k_1 + k_4) + k_1k_4 - k_2k_3\}$$

   Solve for y explicitly if $k_1 = 2$, $k_2 = 2$, $k_3 = 1$, $k_4 = 3$.

5. Find the convolution of the sequences $\{u\} = 0.1, 0.2, 0.5, 0.3$ and $\{v\} = 2.5, 6.5, 10, 4$. (All other terms are zero.) Use the method of

Table 1.2

Time interval (s)	0–1	1–2	2–3	3–4	4–5	5–6	6–7	Over 7
t = starting time of each interval	0	1	2	3	4	5	6	7
No. of cases	12	27	28	19	9	4	1	0

Section 1.16. Write the Z transforms for all the sequences and prove the correctness of your result by deconvolution.

6. A radiation detector equipped with an electronic timer was used to measure the length of the time intervals between successive gamma rays detected from a weak source. The number, n, of instances in which a time interval of 0–1, 1–2, 2–3, . . . seconds was recorded is given in Table 1.2. Find $\bar{t} = \sum nt/\sum n$, the mean starting time of the intervals. Calculate the terms $\exp(-\bar{t})$, $\exp(-\bar{t}).\bar{t}$, $\exp(-\bar{t}).t^2/2$, . . . and hence show that the values of n closely follow a Poisson distribution.

1.27 Bibliography

BOAS, M.L. *Mathematics in the Physical Sciences* (John Wiley)

GABEL, R.A., and ROBERTS R.A. *Signals and Linear Systems* (John Wiley)

DE SAPIRO *Calculus for the Life Sciences* (Freeman)

SPENCER *et al.* *Engineering Mathematics Volume 1* (Van Nostrand Reinhold)

KENNEDY, J.B. and NEVILLE A.M. *Basic Statistical Methods for Engineers and Scientists* (Harper International)

2
Basic Physics of Nuclides

NUCLEAR PHYSICS

2.1 Atomic Structure

A typical atom consists of a nucleus composed of particles called *nucleons* of which there are two kinds, *protons* and *neutrons*. Outside the nucleus there is a set of extranuclear *electrons*. Protons each carry a positive electric charge, neutrons are neutral and electrons each carry a negative charge. A proton and an electron together have a mass of 1.6734×10^{-27} kg, of which the electron contributes only 9.109×10^{-31} kg. The neutron is slightly more massive, at 1.6748×10^{-27} kg. In any sample of ordinary hydrogen, whether the atoms are chemically combined or not, the vast majority (some 99.98%) of the atoms have nuclei composed of just one proton. The remainder have nuclei containing one neutron as well as one proton; hence these nuclei have almost exactly twice the mass of the others. The two kinds of atom (nucleus plus one extranuclear electron) are referred to as *isotopes of hydrogen*, the word isotope meaning that they occupy the same place in the periodic table of the elements and have virtually the same chemical properties. All the other elements also exhibit this property of isotopy. Thus carbon is a mixture of two isotopes: 98.893% of them have 6 protons and 6 neutrons in their nuclei. The atomic number of an element is defined to be the number of protons present and is generally given the symbol Z. The neutron number N is the number of neutrons. The mass number A is just the sum of these, $A = N + Z$. Thus, for carbon, Z is 6 for both isotopes, whereas A is either 12 or 13. Any particular atom for which we specify the Z and A values is referred to as a *nuclide*. For example, the lighter isotope of carbon would be referred to as consisting of the nuclide $^{12}_{6}C$ (or, for orthographic reasons, carbon-12). A group of nuclides all of the same A are called *isobars*; a group all of the same N are called *isotones*. The masses of nuclides are so small that it is convenient also to define a new unit of mass, the unified mass unit, symbol u, which is defined to be one-twelfth of that of $^{12}_{6}C$. Alternatively we may say that the nuclide $^{12}_{6}C$ is assigned a mass of exactly 12.0000 unified mass units. On this scale the proton mass (strictly, including an electron) is 1.007 825 u, and the neutron mass is 1.008 665 u.

Whereas the particles (protons and neutrons) which make up a typical nucleus occupy a space whose linear dimensions are only a few times 10^{-15} m, the atom as a whole has a diameter of a few times 10^{-10} m, a factor 10^5 larger. Within this space, the extranuclear electrons are dispersed. Many years ago it was thought that these must move in orbits around the nucleus in much the same way that planets are in orbit around the sun. Although the idea of a strict localisation of the electrons has been discarded, it is still clear that one must picture each of them as having a definite total energy. A measure of this energy is to be obtained conceptually as the amount of work necessary to just remove the electron in question from the Coulomb field of its nucleus and set it free with zero kinetic energy. It is referred to as the *binding energy*.

Binding energies are not continuously variable, that is to say the electrons in any given atom must have energies which are restricted to certain fixed values, and it is impossible for an electron to remain in the atom unless its energy conforms exactly to one or other of the available values. We speak of electrons occupying fixed energy levels. For any particular configuration of electrons within the allowed levels, the atom is said to be in a certain quantum state. In particular, if the electrons are all in the lowest energy level possible (i.e. in the most highly bound states) the atom is said to be in its *ground* state.

The electron energy levels form a well defined series, which fall naturally into groups. In a multi-electron atom there will be two electrons in particular, called the K electrons, whose binding energy is almost the same as each other, and at the same time much larger than that of the others. There will be a group of up to eight of the L electrons, which have somewhat smaller binding energy. The M group are still more loosely bound, and there may be up to 18 of these. The numbers, 2, 8, 18, . . . form a sequence of form $2n^2$ ($n = 1, 2, 3, \ldots$) corresponding to the K, L, M, . . . groups. The groups are often referred to as the K, L, M, . . . *shells*; this word is a carry-over from the time when it was thought that the K electrons orbited in a region entirely within the orbit of the L electrons, and so on. The unit of energy in which the binding energies are commonly expressed is the *electron volt* or its multiples kilo-electron volt, and mega-electron volt. The electron volt is the energy gained by an electron in moving through a potential of 1 V. Since the electronic charge is 1.6×10^{-19} C, this is $e \times V = 1.6 \times 10^{-19} \times 1 = 1.6 \times 10^{-19}$ J. The K electron in a hydrogen atom has a binding energy of 13.5 eV. The binding energies of K electrons in a heavy element, say tungsten, ($Z = 74$) are about 70 keV. The least strongly bound electrons, however, need only a few electron volts to remove them from the tungsten atom. In general, the one, two, or three least strongly bound levels in atoms are responsible for the formation of chemical bonds, either by donation of electrons to, or receipt of electrons from, levels in neighbouring atoms.

Exceptionally, the so-called inert gases, He, Ne, Ar, Kr, Xe, Rn, do not have electrons which can easily be transferred to other atoms, or available levels which could receive electrons from others. They therefore form very few chemical compounds. We refer to them as having 'closed shells'.

2.2 Production of X-rays

It was a great triumph of early quantum theory that, at any rate for simple cases, the binding energy of electrons in atoms could be precisely calculated. For this calculation to be made, one needs only two assumptions: (*a*) The electron moves under an inverse square law of force with respect to the nucleus—the Coulomb field exactly paralleling the gravitational field in planetary motion—and (*b*) the electron possesses an angular momentum of form $nh/2\pi$ where n is restricted to exact whole numbers, h is Planck's constant, 6.66×10^{-34} J s. For a hydrogen-like atom, with only one electron, the calculation is simple and exact. For multi-electron atoms the effect of electrons upon one another can be allowed for but only with some degree of approximation.

Under certain conditions, an atom in its ground state may receive energy as a result of which an electron makes a transition from one possible energy level to another; for this to happen the energy absorbed by the atom must theoretically be of a certain well defined magnitude, as is confirmed by experiment. Subsequently an electron may make a transition from a higher level to the level previously vacated. In doing so, it releases a quantity of energy equal to the energy difference between the levels, and this appears as a *quantum* of electromagnetic radiation, or photon. According to a relation proposed by Einstein, the energy available, E, is related to the wavelength λ of the radiation by $E = hc/\lambda$, c being the velocity of light. In SI units with E in kilo-electronvolts and λ in nanometres,

$$E = 1.24/\lambda \qquad (\lambda \text{ in nm})$$

(In the older literature λ was expressed in Ångström units; 1 Å $= 10^{-10}$ m, whence

$$E = 12.4/\lambda \qquad (\lambda \text{ in Å})$$

In a conventional X-ray set, tungsten (or sometimes molybdenum) atoms forming the target of the X-ray tube are bombarded with electrons which have been accelerated by a potential of say 100 kV. These fast electrons on hitting the target are capable of ejecting K electrons from the target; sometimes L or M electrons are ejected instead. The process is an example of ionisation. When a vacancy is created in the K shell, it will be almost immediately filled by an electron from one of the upper levels, L, M, N, ... Since the K electron has a binding energy of 70 keV and an L electron has a binding energy of only 14 keV, the transition of an L electron into the K shell should bring about the production of a quantum of radiation of wavelength $1.24/56 = 0.022$ nm, which is in the X-ray region. The fact that X-rays of this predicted wavelength are indeed observed is only one of hundreds of examples which could be quoted, so that the quantum theory on which the predictions are based is amply justified.

It is true that these so-called *characteristic* X-rays are not the only ones observed. The electrons used to excite the target are themselves subject to sudden energy losses whenever they pass close to the nucleus of a target atom. Such encounters although not causing ionisation will generate an X-ray of wavelength $1.24/\Delta E$ nm where ΔE is the relevant energy loss. ΔE can have a value anywhere between zero to the whole energy carried by the electron (100 keV in the case quoted). A plot of the relative intensities I of X-rays within small ranges of wavelength, against the wavelength itself, has the shape shown in Fig. 2.1.

The absolute intensity of an X-ray beam may be defined as the total radiant energy passing through a unit area at right angles to the beam in unit time. Since we are here considering the energy distribution of the X-ray photons in a given beam, we can usefully define a relative intensity as the number of photons in the beam with energies between E and $E + \Delta E$ which are generated in a fixed time, multiplied by the energy E itself. The variation of this quantity with wavelength is all that matters in this discussion and we are not concerned with absolute values.

There are no X-rays with wavelength less than a limiting value λ_0 corresponding to the operating kilovoltage. At greater wavelengths, there is a continuous distribution as the bombarding electrons pass close to atomic nuclei, each encounter giving rise to the generation of an X-ray of randomly determined energy as the electrons are slowed down. The German name for this radiation is bremsstrahlung—i.e. slowing-down radiation—which aptly describes it, although it must be remembered that some energy is lost in electron–electron collisions which may lead to ionisation and perhaps other

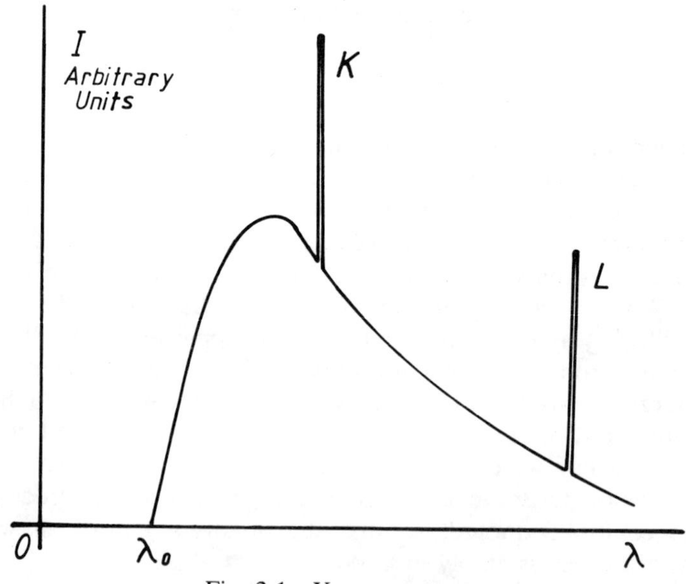

Fig. 2.1 X-ray spectrum.

processes. Superimposed on the continuous spectrum are sharp 'spikes', the characteristic radiation, originating from the ejection of K, L, M, . . . electrons as discussed in the previous paragraph. Accurate experiments show that these spikes are not single, but multiple, showing that all the electrons in each group (K,L,M, . . .) do not have exactly the same energy.

2.3 Review of Stable Nuclides

With minor exceptions (e.g. the elements technetium, $Z = 43$, and promethium, $Z = 61$) the elements from hydrogen, $Z = 1$, to bismuth $Z = 83$, are composed of nuclides which are said to be *stable*. This means that as far as we have been able to show experimentally, these nuclides have no tendency to change their compositions, as regards to numbers of protons and neutrons in their nuclei. The elements do not all possess the same number of isotopes, thus hydrogen has two ($Z = 1, N = 0$ or 1), carbon has two ($Z = 6, N = 6$ or 7), oxygen has three ($Z = 8, N = 8$, 9 or 10), tin has 10, ($Z = 50, N = 112$, 114, 115, 116, 117, 118, 119, 120, 122, 124). On the other hand, iodine is monoisotopic ($Z = 53, N = 74$,) and so is gold ($Z = 79, N = 118$). Technetium and promethium have no stable isotopes at all, and there are at least 16 elements beyond bismuth which are known to us only in the form of unstable isotopes.

The unusually large number of stable isotopes for $Z = 50$ indicates unusual stability for nuclei containing this number of protons, suggesting an analogy with the inert gases which have unusual chemical stability and closed shells of electrons. Thus without knowing anything about nuclear structure we can suspect a quantised regime of some sort, leading to closed shells of nuclear particles. Further examination shows that there is an unusually large number of stable nuclides with $N = 50$, and $Z = 36$, 37, 38, 39, 40, 42, covering the elements from krypton to molybdenum. (One of these, an isotope of rubidium, is actually very slightly unstable.) Other, but less well marked examples of unusually stable configurations occur, and the list now comprises N, or Z, = 2, 8, 20, 50, 82, and 126. The existence of these numbers and their actual values are cornerstones in the development of theories of nuclear structure; they are known as 'magic' numbers.

2.4 Masses of Nuclides 1

Although the atomic nuclei are composed of protons and neutrons, the mass of any given nuclide in its ground state is not equal to the sum of the masses of the constituents. Thus, to take an example at random, the mass in unified mass units of the nuclide $^{197}_{79}Au$, is 196.966 55, whereas we find for $Z = 79$, $N = 118$, the mass $79 \times 1.008\,825 + 118 \times 1.008\,665 = 79.620\,15 + 119.020\,70 = 198.642\,62$ u.

Thus if we could assemble 79 protons and 118 neutrons to form a single gold nucleus in its ground state, there would be a loss of mass of 1.676 07 u, or about 0.85% of the mass involved. This means that if we consider the synthesis of say a whole kilogram of gold, the mass loss would be about 8.5 g. According to relativity theory in which mass and energy are interconvertible, this loss would have to be accompanied by a release of energy. The energy E is given by the product mc^2, i.e. $0.0085 \times (3 \times 10^8)^2 = 7.65 \times 10^{14}$ J, or approximately 2×10^8 kW h. Yet one might have thought that a great deal of work would have to be *done* to force the protons together against their mutual electrostatic repulsion. This calculation therefore indicates the immensity of the forces which must operate within nuclei. Correspondingly, while the energies associated with ordinary chemical bonds are of the order of a few electronvolts, the energies needed to produce noticeable effects in nuclei are often measured in kilo-electron volts or mega-electron volts (keV and MeV). The mass defect, as calculated above, for any nuclide can be expressed in energy units by Einstein's relation. It is called the binding energy of the nucleus. A useful quantity derived from this is the binding energy *per nucleon*, i.e. the binding energy divided by A, or $(N + Z)$. This parameter plotted as a function of A is shown in Fig. 2.2.

The points, one for each stable nuclide, do not lie quite on a smooth curve, partly because the binding energy per nucleon is slightly greater for some nuclei than that for their immediate neighbours. Thus noticeable peaks occur in the curve where $A = 4$ ($Z = 2$, $N = 2$), $A = 16$ ($Z = 8$, $N = 8$), $A \approx 90$ ($N = 50$), $A \approx 140$ ($N = 82$) and $A \approx 205$ ($N = 126$).

Here again we have evidence for the shell closure especially at the 'magic' numbers in N.

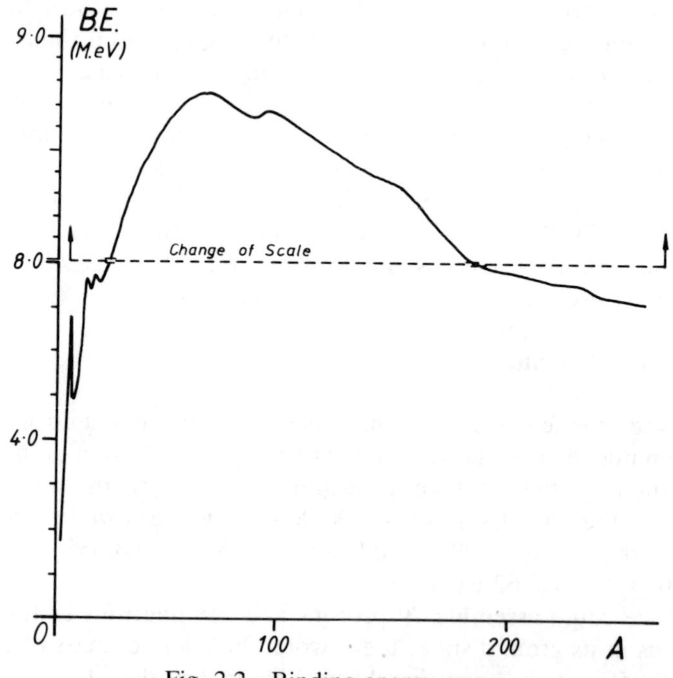

Fig. 2.2 Binding energy curve.

(Note that on this plot the ordinate scale has been expanded between 8 and 9 MeV to show the shape of the curve for medium mass nuclei more clearly.)

2.5 Unstable Nuclides

In addition to the naturally occurring stable nuclides so far discussed, there are two other groups.

(*a*) *Naturally occurring unstable nuclides.* These are the isotopes of the elements from $Z = 84$ (polonium) to $Z = 92$ (uranium), some of the isotopes of thallium ($Z = 81$), lead ($Z = 82$), and bismuth ($Z = 83$), and the nuclides ${}^{147}_{62}Sm$, ${}^{87}_{37}Rb$, ${}^{40}_{19}K$, and possibly a few others. In contrast to the stable nuclides, these tend to change spontaneously, with release of energy, in a variety of ways. Ultimately they will all be converted into stable nuclides, in the natural course of events.

(*b*) *Artificially produced unstable nuclides.* These cover the whole range of elements. They have either a smaller, or a greater, number of neutrons than the naturally occurring nuclides. Isotopes of technetium ($Z = 43$) and promethium ($Z = 61$) and at least eight elements with Z greater than 92 do not occur at all in nature except possibly in very minute amounts but most are readily obtained artificially. They also tend to change spontaneously into stable nuclides; in this respect and in the mechanisms of the changes there is no difference between the two groups. The spontaneous changes are called radioactive decays and the nuclides undergoing the changes are called radioactive nuclides, or more shortly radionuclides.

2.6 Units of Radioactivity

The obvious unit of radioactivity would seem to be that amount of radionuclide which is changing at the rate of one nucleus per second. This unit, formally s^{-1}, has been named the becquerel (symbol Bq) after the discovery of radioactivity by Henri Becquerel in 1896, and was adopted in 1975 as a special SI unit for radioactivity by the Fifteenth Conférence Generale des Poids et Mésures. From a practical point of view the unit is inconveniently small, even a megabecquerel (MBq) corresponding to only 27 μg of radium. A theoretical objection is that radionuclides do not decay at an exactly uniform rate, so that a 'becquerel of radioactive material' would be meaningful only in terms of an *average* decay rate measured over a time very long compared with one second. Nevertheless for the sake of conformity with other SI units we will use the becquerel throughout but with occasional reference to the older unit, the curie (Ci), whenever it might be convenient for the reader to consult earlier published work. The curie is that amount of radioactive material whose decay rate is 3.7×10^{10} Bq, which is the same to within about 0.5% as the activity of one gram of radium (${}^{226}Ra$).

2.7 Radioactive Decay Constant and Half-Life

Suppose we have a sample of a single pure radioactive nuclide containing N nuclei, N being a function of time and decreasing as time goes by. Over successive equal time intervals Δt we may measure the number of nuclei decaying in each interval; this number we denote by $-\Delta N$ since it represents a decrease in N. Experiment shows that successive $-\Delta N$ values show random variations, called fluctuations, which are superimposed on a steady decrease. The decrease is however not constant, but slows down in the same way as N itself decreases. It is postulated that, ignoring the fluctuations, the (negative) rate of change of N is just proportional to N, i.e.

$$-\Delta N/\Delta t \approx \lambda N$$

and in the limit when Δt tends to zero

$$-(\mathrm{d}N/\mathrm{d}t) = \lambda N$$

where λ is a constant characteristic of the particular nuclide being studied, it is called the *decay constant.* Closely related to the decay constant is the half-life $t_{1/2}$ for the nuclide, which may be defined as the time taken for the number of nuclei in a sample to be reduced by decay to half its original value. Since the decay *rate* is proportional to N at all times, the half-life may be alternatively defined by the time taken for the rate itself to decrease to half its initial value. The precise relation between λ and $t_{1/2}$ is

$$t_{1/2} = (\ln 2)/\lambda = 0.6932/\lambda$$

The derivation of this equation and an explanation of the fluctuations mentioned above are deferred to Chapter 4.

For the moment all we shall do is to comment on the extremely wide range of half-lives encountered, from times as short as 10^{-11} s up to 10^{11} yr or more. Radium (^{226}Ra) has a half-life of 1650 yr; many radionuclides in common use have half-lives in the range of a few seconds up to a few years. A list of important ones is included later in this Chapter.

2.8 Masses of Nuclides 2

Well over a thousand nuclides are now well known and their masses have been tabulated. It is interesting to try, for any given nuclide, to find a general formula for its mass in terms of the mass number A and the atomic number Z. As already noted, the mass is only approximately equal to the sum of N neutron masses and Z proton masses, so cannot be exactly proportional to A $(=N + Z)$. In an explicit formula for the mass, of the form $M = f(A, Z)$, there will be terms which express deviations from a simple proportionality to A on account of the binding energies of the nuclear particles (nucleons) and the fact that they are not all equally strongly bound. For example, in a roughly spherical collection of nucleons, the peripheral ones will be subject to forces directed towards the centre of the group whereas the innermost nucleons will be bound in all directions. The total binding energy, and therefore the mass, will therefore be affected by the extent of the surface area of the nucleus. Since the nucleus contains A nucleons, we can expect the function $f(A,Z)$ to include a term proportional to $A^{2/3}$.

Another consideration is that there are Z positively charged particles each one in an electric field created by the remaining $(Z - 1)$ particles. Now the energy needed to bring a charge Q_1 from infinity to a point distant r from a charge Q_2, is proportional to Q_1Q_2/r. Since the radius of a spherical nucleus will be proportional to $A^{1/3}$, the average spacing between the protons might also be proportional to $A^{1/3}$. The energy of the nucleus should therefore contain a term of the form $Z(Z - 1)/A^{1/3}$ and a quantity proportional to this is to be expected in the mass formula. Again, there is some evidence that neutrons and protons are 'paired'; since N is generally greater than Z, there will be $N - Z = A - 2Z$ unpaired neutrons (or alternatively there will be unpaired protons if $N < Z$). The unpaired nucleons are not so strongly bound as the paired ones and so contribute differently to the final mass. This effect brings in a term $(A - 2Z)^2/A$. Finally there is the question of whether A is even or not. If it is *odd* there must be an odd neutron, or an odd proton, which cannot be paired either with a like or an unlike particle. If A is even, it could be that both N and Z are even; alternatively, both must be odd. The three alternatives (no odd particles, one odd particle, or two odd particles) lead to a term δ/A where δ is a positive quantity, zero, or a negative quantity respectively. The numerical factors to be associated with these various terms, can only be found by essentially trial and error, hence the term *semi-empirical mass equation*. One reasonably accurate expression is

$$M = 925.55A - 0.78Z + 13.1A^{2/3} + 0.585Z(Z - 1)A^{-1/3} + 18.1(A - 2Z)^2A^{-1} - \delta A^{-1} \quad \text{MeV}$$

with $\delta = 132$, 0, or -132, as explained above. This equation may be simplified to read

$$M = aZ^2 + bZ + c - \delta A^{-1} \quad \text{MeV}$$

with

$$a = 0.585A^{-1/3} + 72.4A^{-1}$$

$$b = -0.585A^{-1/3} - 73.18$$

$$c = 943.65A + 13.1A^{2/3}$$

Thus for any given value of A, M varies quadratically with Z. This means that if we plot a graph of the masses of all nuclides with a certain mass number but with different atomic numbers (i.e. a set of isobars), the points should lie on a parabola. Since the coefficient of Z^2 in the expression for M is always positive, the parabola is concave upwards.

Some possibilities are shown in Figs. 2.3 and 2.4. In Fig. 2.3 (A odd) we have a single parabola because δ is zero. The parabola joins points which

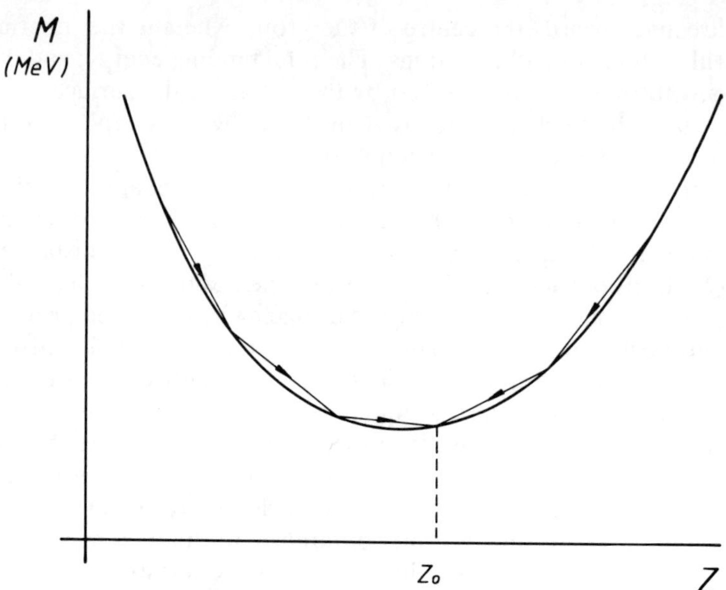

Fig. 2.3 Masses of isobars with A odd (i.e. N or Z odd).

correspond to *discrete* Z values, and has no physical significance between the points. The Z values are alternately even and odd, while the N values are alternately odd and even. It is clear then that there must be a particular Z value, denoted Z_0, for which the mass is a minimum. This will correspond to a stable isobar, all others with larger or smaller Z being fundamentally unstable. They can lose mass by making transitions towards the minimum value on the curve. Note that such transitions will involve protons being exchanged for neutrons, or vice versa. The curve explains why, for any given *odd* value of A, there is only one possible stable isobar.

The situation is different for the cases when Z is even, because of the non-zero δ term. In Fig. 2.4a there are *two* possible cases of minimum mass. An actual example is the pair $^{40}_{18}Ar$ and $^{40}_{20}Ca$, the intermediate isobar $^{40}_{19}K$ being unstable in both directions. The triplet case of $^{124}_{50}Sn$, $^{124}_{52}Te$ and $^{124}_{54}Xe$ shows three minimum masses with intervening unstable isobars. Fig. 2.4b shows the case, common among the light elements, where there is only one stable isobar.

2.9 Other Nuclear Properties: Spin and Parity

We have space to consider these properties only very briefly. In basic units of angular momentum, both neutrons and protons have angular momentum, or spin, of one-half. In a nucleus of even A it is possible for the total nuclear spin to be zero, as many nucleons spinning in one direction as in the other. As a rule, this would happen when the nucleus is in its ground state, but higher energy states are possible (excited states), in which the nucleons are not all 'paired'. These will have a total spin of 1, 2, 3, . . . units. (Very rarely a

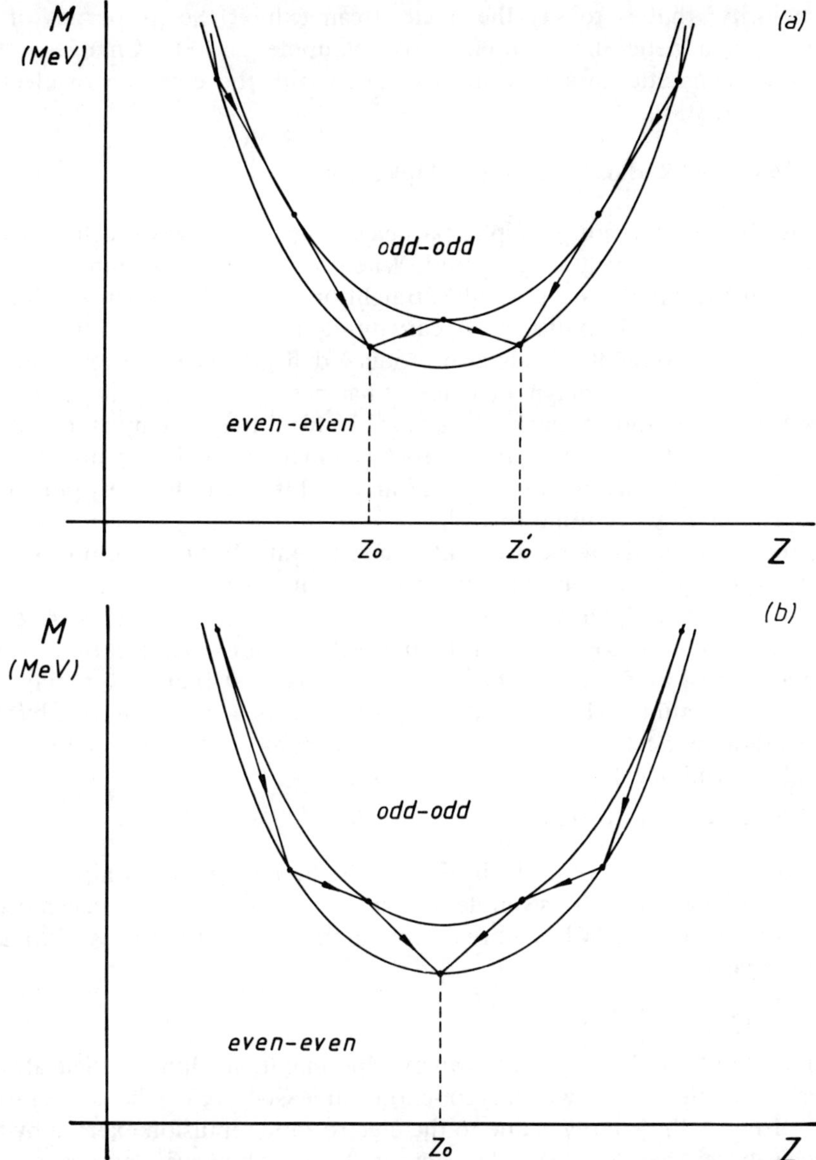

Fig. 2.4 Masses of isobars with *A* even (i.e. both *Z* and *N* odd, or both even).

high-spin state is the ground state, e.g. $^{90}_{41}$Nb, spin 8.) Even-numbered spin states, (2,4,6, . . .) are very common. In a nucleus of *odd A* the total spins must be half-integral, e.g. $\frac{1}{2}$, $\frac{3}{2}$, $\frac{5}{2}$, . . . , and often, but not always, the spin state $\frac{1}{2}$ has the lowest energy (ground state). Parity is a property which is akin to left- or right-handed symmetry in the nucleus, and is denoted by the symbols + and − (positive and negative parity). The motions, including spin, of the nucleons may result in a nucleus possessing electric, and magnetic multipolarity; that is to say the nucleus can exhibit the properties of an electric, or magnetic, dipole, quadrupole, octupole, . . . etc. Changes in the electric or magnetic moments are associated with the emission of electromagnetic radiation.

2.10 Modes of Radioactive Decay: Alpha Emission

Historically the emission of alpha particles from nuclei was the first decay process to be extensively investigated. When travelling in vacuum, it can be shown that their paths are normally straight lines, but that when an electric field is present, their paths show curvature. This proves that they carry electric charge. Magnetic fields also produce deflections. From experiments using both electric and magnetic fields, it has been proved that the charge is positive, and the ratio of charge in units of electron charge, to mass in units of u, is 0.5. The particles do not travel very far in matter, and if they are allowed to enter a vessel filled with a gas such as argon, it is found that they penetrate no more than a few centimetres.

Moreover, analysis of the argon after a sufficient number of particles have been stopped in it reveals the presence of helium gas. These experiments make it clear that alpha particles are in fact helium nuclei of mass 4, or in other words helium atoms without their usual pair of extranuclear electrons. The emission of an alpha particle from a nucleus must therefore reduce the latter's atomic number by 2 and its mass number by 4. For example, by this process radium nuclei ($Z = 88$, $A = 226$), become radon nuclei ($Z = 86$, $A = 222$). Symbolically

$$^{226}_{88}\mathrm{Ra} \rightarrow {}^{222}_{86}\mathrm{Rn} + {}^{4}_{2}\mathrm{He}$$

The three-dimensional plot of Fig. 2.5 shows the mass relationships.

It would appear from this simple description that alpha emission can occur whenever a nuclide $^{A+4}_{Z+2}X$ has more mass than the sum of the masses of $^{A}_{Z}Y$ and an alpha particle

$$^{A+4}_{Z+2}X \rightarrow {}^{A}_{Z}Y + {}^{4}_{2}\mathrm{He}$$

This is not quite so, because as shown by experiment, an alpha particle always possesses kinetic energy, which is commonly at least 4 MeV. The acceleration which provides the energy is due to the electrostatic repulsion exerted by the Y nucleus after the alpha particle leaves it. There is however, no *new* source of energy here; conservation of (mass + energy) means that the $^{A+4}_{Z+2}X$ nucleus has to provide the mass of $^{A}_{Z}Y$ and $^{4}_{2}\mathrm{He}$ *and* the kinetic energy of the alpha particle, and in addition the small but not negligible recoil energy of the $^{A}_{Z}Y$ nucleus.

Since the $^{A+4}_{Z+2}X$ and $^{A}_{Z}Y$ nuclei are in definite quantum energy states, the alpha particles observed must be monoenergetic. A few exceptional cases are known in which more than one quantum state of either nucleus is involved, and then there are two or more *groups* of alpha particles, each group being monoenergetic.

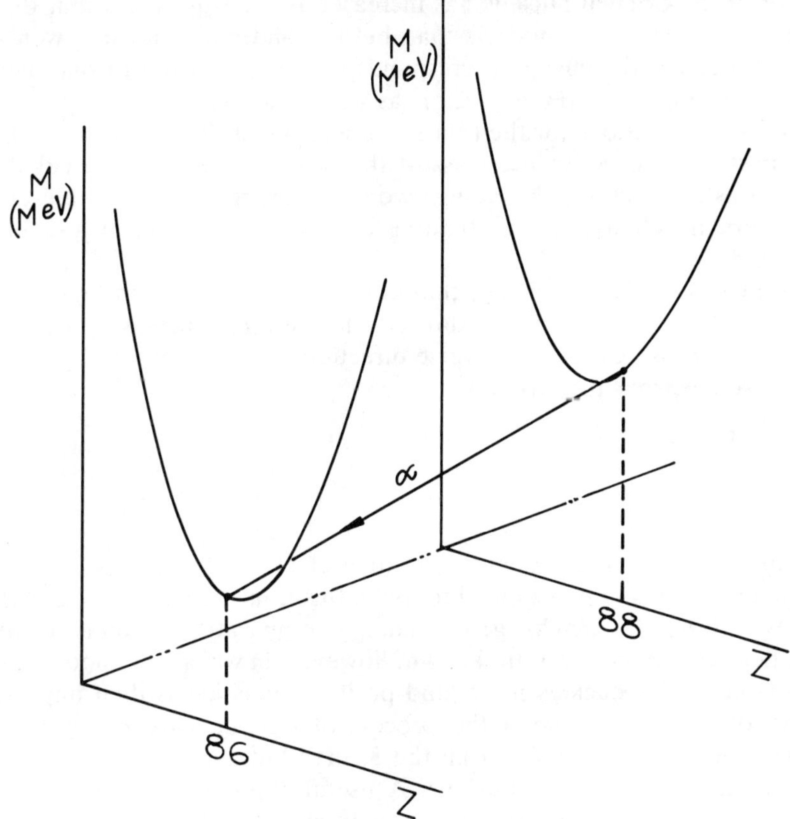

Fig. 2.5 Three-dimensional plot of mass (A,Z).

However, 'ground state to ground state' transitions are by far the most common, giving a single group.

2.11 Modes of Radioactive Decay: Beta Emission

Referring again to Figs. 2.4 and 2.5, it will be remembered that among a group of isobars there will be one, or more rarely two or even three, whose masses are lower than that of their neighbours. With respect to these, the neighbours are therefore unstable. The directions of the expected transitions are shown by arrows in the diagrams. As these transitions are between neighbouring isobars, no heavy particles (nucleons) are emitted, but conservation of charge dictates that an electronic charge must be given to, or supplied from, the region outside the nucleus. Ordinary *beta* emission in fact means the emission of a (negative) electron,

$${}_{Z-1}^{A}X \rightarrow {}_{Z}^{A}Y + {}_{-1}^{0}\mathrm{e}$$

the superscript zero on the electron symbol indicating that the mass number of the nucleus has not changed, the subscript -1 showing that the atomic number of the original nucleus has increased by 1. This means that there is one more proton in the nucleus than before, and one neutron fewer. Formally, therefore, the change corresponds to the conversion of one neutron into one proton, ${}^{1}_{0}\mathrm{n} \rightarrow {}^{1}_{1}\mathrm{H} + {}^{\;0}_{-1}\mathrm{e}$, or, as sometimes written, $\mathrm{n} \rightarrow \mathrm{p} + \mathrm{e}^{-}$.

The same conditions for the (mass + energy) conservation apply as in the alpha particle case. Since the mass of the neutron is greater than that of a proton plus an electron, the change would be energetically possible even if there were no change in the binding energy of the neutron and proton concerned.

The process of beta emission thus corresponds to cases, in Figs. 2.4 and 2.5, where the arrows are in the direction left to right (increasing Z). Correspondingly, transitions in the reverse direction are accompanied by emission of positive electrons (positrons):

$${}^{\;\;\;A}_{Z+1}X \rightarrow {}^{A}_{Z}Y + {}^{0}_{1}\mathrm{e}$$

i.e.

$$\mathrm{p} \rightarrow \mathrm{n} + \mathrm{e}^{+}$$

Positrons have the same mass as negative electrons but a positive charge. Clearly energy has to be supplied to make this transition possible, and it can be provided only by the change in binding energy as the proton is converted into a neutron. Many cases are known, however, in which not enough energy is available in the nucleus itself, and positron emission is then impossible. Instead of positron emission the process of *electron capture* takes place, whereby an electron, usually from the K shell but more rarely from higher shells, is absorbed into the nucleus. A useful approximate formula for the ratio of L to K electron capture resulting in an atom of atomic number Z is

$$\frac{\mathrm{L}}{\mathrm{K}} = (0.06 + 0.0011Z)\left(\frac{E_{\mathrm{D}} - E_{\mathrm{B}}^{\mathrm{L}}}{E_{\mathrm{D}} - E_{\mathrm{B}}^{\mathrm{K}}}\right)^{2}$$

where E_{D} is the energy available in the decay and $E_{\mathrm{B}}^{\mathrm{L}}$, $E_{\mathrm{B}}^{\mathrm{K}}$ are the L and K binding energies; we must have that $E_{\mathrm{D}} > E_{\mathrm{B}}^{\mathrm{K}}$. Electron capture is always energetically possible, although among very heavy elements it may involve only L or even M electrons. It may be symbolised:

$${}^{\;\;\;A}_{Z+1}X + {}^{\;0}_{-1}\mathrm{e} \rightarrow {}^{A}_{Z}Y$$

i.e.

$$\mathrm{p} + \mathrm{e}^{-} \rightarrow \mathrm{n}$$

It produces an atom in which there is a vacancy in an extranuclear shell; it is therefore followed by X-ray emission or some equivalent process.

In some atoms in which there is only just enough energy for positron emission, there is competition between this mode of decay and electron capture; that is to say, some nuclei transform by one mechanism and some by the other. In general, the more energy available, the more likely positron emission is to occur.

A very rough equation for the ratio R of K-electron capture to positron emission as a function of the maximum energy E_{max} of the positron and the Z of the radioactive nucleus is

$$R = \text{antilog}\,\{(0.0447R - 1.6) + (0.005Z + 3.5)\log E_{max}\}$$

Thus for example for $Z = 40$ and $E_{max} = 0.5$ MeV we find $R = 15$; for $E_{max} = 2$ MeV the value of R is 0.15. These are typical figures and show how very strongly favoured the electron capture process is when the energy available is low. The formula applies strictly when the change of spin of the decaying nucleus is 0 or 1 unit; if the spin change is 2 units the R value as calculated should be multiplied by $(2 + 0.51/E_{max})$. The formula agrees well with the theoretical curves given in *Table of Isotopes* (Lederer, Hollander and Perlman) when R lies between 10 and 0.1; there are however discrepancies to the extent of some 30% between the calculated values and those determined by experiment for a few radionuclides, viz. those for which there is a change of parity during the transition. The formula is accurate enough for the purpose for which we need it in Chapter 3 (Calculation of radiation exposure rates, Section 3.10).

It should be noted that when a positron is slowed down by making collisions in matter, until it reaches only thermal energy (approx. 0.025 eV), it unites with an atomic electron in a process of mutual annihilation. The mass of the pair of particles is converted into electromagnetic energy, in the form of a pair of photons of 0.51 MeV. These leave the site of the annihilation at 180° to each other, so conserving both energy (1.02 MeV) and momentum (almost zero). The two photons are known as annihilation radiation.

2.12 Beta Particle Spectra

As the transitions just discussed are between nuclides each of a definite fixed mass, the energy available is single valued. We would expect the beta particles (or positrons) emitted from any one species of nuclide all to have the same kinetic energy—as is indeed the case with alpha particles. But experiment contradicts this expectation. In fact, the relative numbers N of beta particles within a small range ΔE of an energy E varies with E in a complex way, as illustrated in Fig. 2.6. This energy spectrum shows that most of the electrons have energies less than half that of the most energetic. There is a definite upper limit of energy, denoted E_{max} on the diagram. This upper limit corresponds to the total energy available in the transition. In the vast majority of transitions, the electron is given considerably less than the calculated value.

In 1934 Fermi proposed that a second particle called the *neutrino* was emitted simultaneously with the electron, so that our earlier equation

$$\mathrm{n} \rightarrow \mathrm{p} + \mathrm{e}^-$$

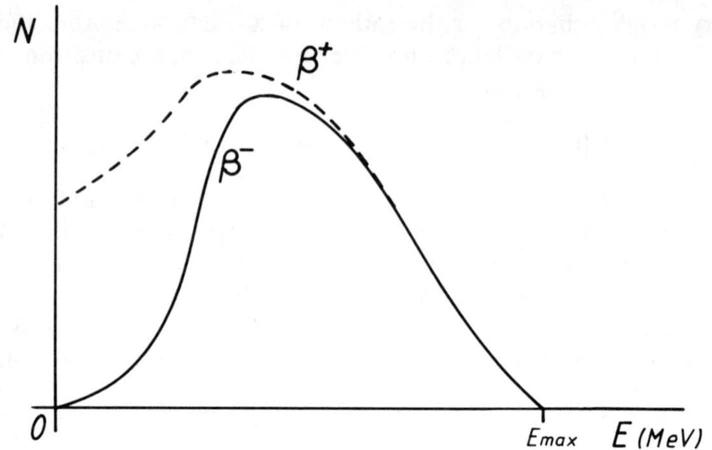

Fig. 2.6 Relative numbers of beta particles (+ and −) at different energies.

should be written

$$n \rightarrow p + e^- + \nu$$

ν being used to symbolise the neutrino. The properties of neutrinos are extraordinary; they have no mass and almost no effect on ordinary matter; we are all continuously bathed in a sea of these particles which have no apparent effect on us. Nevertheless, they carry away the missing energy in beta-emission processes.

While beta-particle energy spectra are always of the same form as that shown in Fig. 2.6, different nuclides give spectra which (besides having different end-points) are of slightly different shape. This has to do with the fact that the transitions may involve different changes of angular momentum (spin) of the nucleus. From the exact details of the spectral shape, it can be deduced whether the nucleus has changed its electric dipole, quadrupole, or octupole moment, or the corresponding magnetic properties. This is a large topic in nuclear physics, but can be allowed only a brief mention here. However, for all observed spectra it is the case that the *average* beta particle energy is roughly one-third of the maximum in the spectrum, E_{max}. The latter value is the one generally quoted in nuclear physics tables. It must be divided by three in order to get an estimate of the energy actually deposited by the beta particle, e.g. in tissue.

2.13 The Origin of Gamma-Rays

We have so far implied that when a nucleus decays by beta particle emission, the product nucleus is in its ground state, with all its protons and neutrons in the lowest possible energy state. It has however been found that in most cases the newly formed nucleus will emit electromagnetic radiation, physically indistinguishable from X-rays. Generally this radiation is of shorter wavelength than X-rays, but not invariably so. Its important characteristic is that it is monoenergetic; that is to say, a given nuclide will emit a group, or a few groups, of rays, each group having exactly the same wavelength. This is evidence that newly formed nuclei can exist (although usually only for very short times) in excited states. These are important in giving evidence about the law of force between nuclear particles (which is certainly not that of the inverse square). Just as atoms emit X-rays when electrons make transitions from one shell to another, so nuclei making transitions from one energy level to another emit these characteristic radiations. They are called gamma-rays. A typical simple gamma-ray spectrum, i.e. a plot of the number of gamma-ray photons $N(E)\Delta E$ observed with energies between E and $(E + \Delta E)$ is given in Fig. 2.7 which shows results obtained when the gamma radiation from ^{60}Co is examined by germanium-lithium detector (Section 3.17). The smooth part of the curve is not due to the emission of gamma-rays of a continuous range in energy, but to the mechanism of interaction of the main group of gamma-rays with the material of the detector (Section 3.6). Clearly there are two main groups of gamma-rays; accurate measurements show that their energies are very close to 1.173 MeV and 1.3325 MeV respectively. The detector is not equally sensitive to the two groups, and when this is taken into account it is found that the 1.3325 MeV group is slightly the more numerous (by 0.12%). These results are explained by assuming the existence of two excited states in the nucleus ^{60}Ni with energies 2.5057 and 1.3325 MeV above the ground state. The emission of beta particles from ^{60}Co nuclei produces ^{60}Ni nuclei in the upper state in 99.88% of cases. These ^{60}Ni nuclei then emit gamma-rays of (2.5057 − 1.3325) MeV and of 1.3325 MeV, as they decay to the ground state. In the remaining 0.12% of ^{60}Co beta decays, the 1.3325 MeV state in

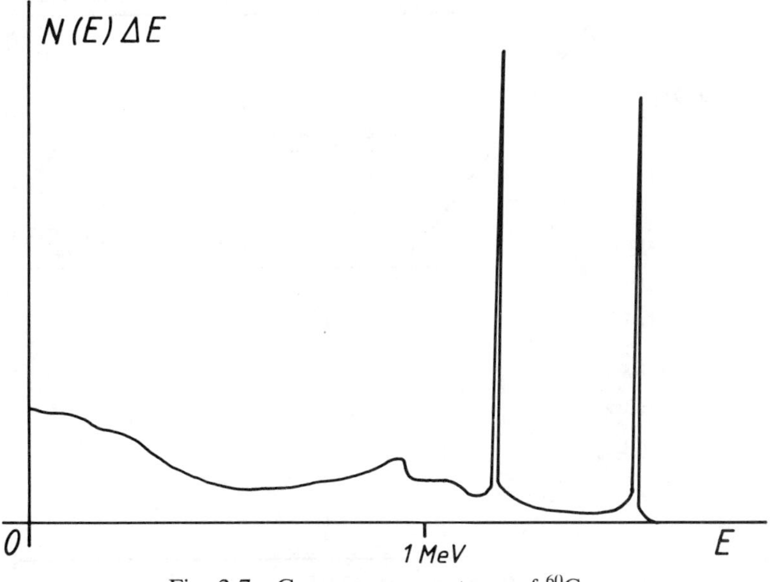

Fig. 2.7 Gamma-ray spectrum of ^{60}Co.

^{60}Ni is populated directly. The whole process is depicted in Fig. 2.8 and is called the *decay scheme* for ^{60}Co.

In the figure, the numbers in brackets give the units of spin and the parity associated with each of the energy states. The half-lives of the upper states in ^{60}Ni have been measured and are less than 10^{-12} s (1 ps), so that in this example it appears that the gamma-ray emission can readily 'remove' the two units of nuclear spin as implied in the diagram. (A gamma-ray possesses momentum $h\nu/c$ and its emission must change the angular momentum of the parent nucleus by at least one unit in order to conserve momentum). The fact that gamma-rays of 2.5057 MeV are not observed means that gamma emission resulting in a decrease of 4 units of spin is a much less favoured process. (Such 'cross-over' transitions are in fact observed in other nuclei, and even in this case, the detector may sometimes appear to be responding to gamma-rays of 2.5057 MeV. The explanation of these rather rare events is that two of lower energy gamma-rays are affecting the detector simultaneously, so that the signal generated has the same amplitude as it would have for a single full-energy gamma-ray. A 'spurious' peak at 2.5057 MeV therefore appears in the spectrum; this is called a 'sum-peak'.) The elucidation of decay schemes has been a major effort in low-energy nuclear physics over many years, the main experimental methods being the measurement of beta and gamma-ray spectra, and of the half-lives of the intermediate states. Some decay schemes

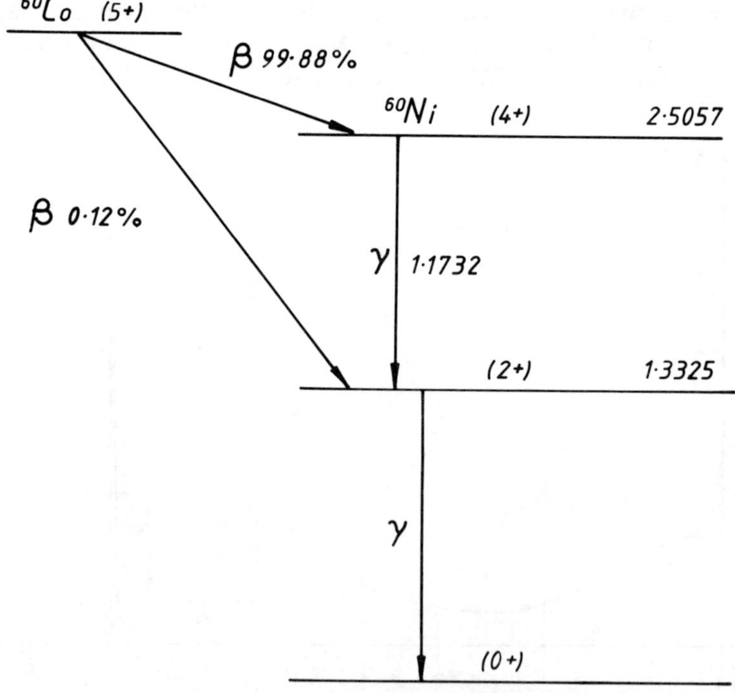

Fig. 2.8 Decay scheme for ^{60}Co.

are extremely complicated with up to a hundred possible transitions between the energy states, and there is still much to do to characterise them in terms of the spins and parities of the states. But among the simpler decay schemes, it seems that we can ascribe the gamma-ray emission to the behaviour of a single nucleon, namely the one concerned in the preceding beta emission.

Remembering that a beta emission can in general be symbolised

$$n \rightarrow p + e^- + \nu$$

we can well appreciate that the neutron (originally at a fixed energy level) when transformed into a proton may find itself energetically some way above a vacant proton level. On descending to this level, it emits its surplus energy as a gamma-ray in precisely the same way that an extranuclear electron emits an X-ray. We refer to this process as a 'single-particle model of the nucleus'. Mathematical development of the model shows that transitions which involve only a small change in spin of the whole system can follow the beta emission with no appreciable delay, but that when a large spin change is necessary the probability (per unit time) of the transition occurring is much reduced. Spin changes in general will alter the electric, or magnetic, multipolarity of the nucleus and so give rise to electromagnetic radiation in the form of gamma rays. It has been found possible to relate the half-life of the transition and the energy change to the change in multipolarity and it turns out that for large spin changes and small energy radiated the half-life can be very long indeed. A nuclear state with half-life longer than 10^{-6} s is conventionally known as a *nuclear isomer*. Some 50 examples of these are known, with half-lives in a range extending to many years.

In more detail, it has been recognised that when the spin I changes in the transition by only one unit ($\Delta I = 1$) and there is no parity change, the decay simply changes the magnetic dipole moment of the nucleus; this is an 'M1'-type transition. A transition for which $\Delta I = 1$, but having a parity change, gives a change in electric dipole moment and is denoted E1. When the spin change is 2 ($\Delta I = 2$), we get changes in quadrupole moment; the transition is electric quadrupole E2 for *no* parity change, M2 if the parity does change. The effect of parity change reverses again for the E3 and M3 transitions, and so on. Calculated half-lives for the transitions E1 to E5, for medium-mass nuclei emitting gamma rays of 0.1 MeV, range from 10^{-13} to 10^{15} s, in steps of a factor about 10^7. For the magnetic transitions M1 to M5, the calculated half-lives are about a factor 100 longer than for the corresponding electric transitions. Many of the isomers with half-lives of a few minutes up to a few days are of the M4 type. A very important nuclear isomer occurs as one of the upper states in technetium-99. The decay scheme is shown in Fig. 2.9.

The upper state in technetium produced by beta decay of ^{99}Mo is denoted ^{99m}Tc, the letter m meaning 'metastable'—to distinguish it from the ground state ^{99g}Tc. Since the transition ^{99m}Tc–^{99g}Tc involves a spin change of 4 units and a parity change, it can be designated M4. However, only about 1.4% of the transitions actually occur directly, as there exists another state with spin $\frac{7}{2}$

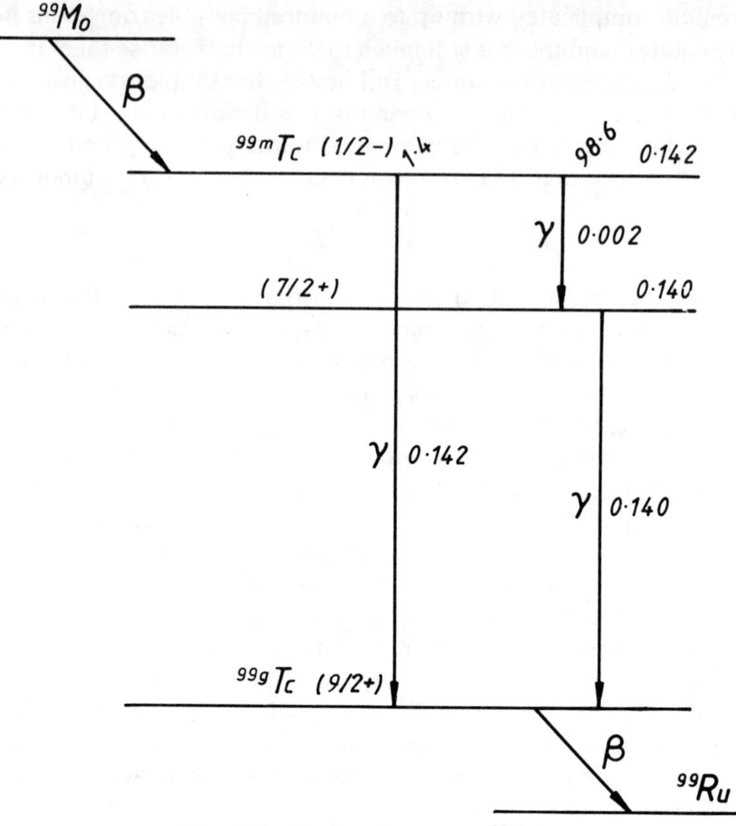

Fig. 2.9 Decay scheme for ^{99m}Tc.

and positive parity only 0.002 MeV below the isomeric state; 98.6% of the ^{99m}Tc nuclei decay via this intermediate state, emitting first of all a gamma-ray of 0.002 MeV in an E3 transition and then, after a very short delay, a gamma-ray of 0.140 MeV in a M1 transition. The M4 transition is an example of a cross-over, as mentioned earlier.

The half-life of the isomeric state ^{99m}Tc is 6.05 h and of the ground state ^{99g}Tc, 2.15 × 10^5 yr.

2.14 Fluorescence Yields and Internal Conversion

Sometimes when an atom loses an electron from one of the inner shells, the filling of the vacancy by an outer electron does not give rise to an X-ray as would be expected. Instead an electron is emitted from an outer shell, carrying the appropriate energy. The process suggests that an X-ray generated by a vacancy in say the K shell can be absorbed in the *same* atom, using its energy to eject an L or a M electron. Out of a large number of such events, the fraction in which X-rays are observed is called the fluorescence yield (sometimes alternatively defined as the ratio of the number of X-rays to the number of electrons emitted). In a rather similar way, a process called *internal conversion* occurs in gamma-ray emission. Although called a 'conversion' process for historical reasons, it is in fact an additional mode of decay. For any transition in which the emission of a gamma-ray is expected, this process will indeed occur, some nuclei emitting gamma-rays in accordance with the calculated half-life, and *in addition* some nuclei will pass on their excess energy to their own extranuclear electrons, from which a K, L, or more rarely M, electron will be ejected. The ratio of electrons emitted to gamma-rays emitted is called the internal conversion coefficient α. The effect is energy dependent; clearly a K electron cannot be emitted if the transition energy is less than the K binding energy. It is also very strongly dependent on the multipolarity of the transition. Values of α for different multipolarities are plotted as a function of transition energy and shown in Fig. 2.10. Since electron capture provides an additional mode of decay, the half-life as calculated in the previous section must be corrected by multiplying by a factor $(1 + \alpha)$.

The curves are calculated for $Z = 40$; for $Z = 30$ the values of α are smaller by a factor of about two, and for $Z = 50$ they are larger by about the same factor. They are for K shell conversion only, as this is usually the most common. From the graph we can estimate that the 0.140 and 0.142 MeV gamma rays from ^{99m}Tc ($Z = 43$) are accompanied by conversion electrons to

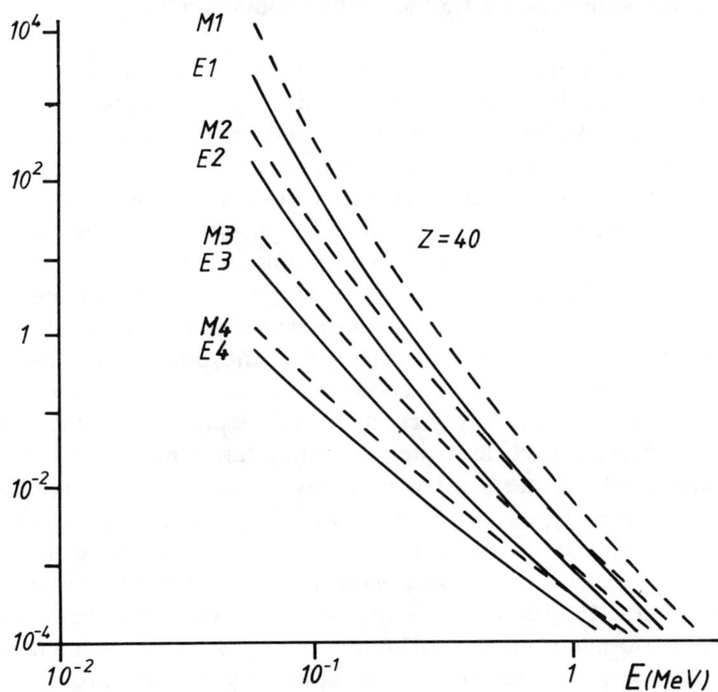

Fig. 2.10 Conversion coefficient for K shell α_K. (Reproduced from C.M. Lederer, J.M. Hollander and I. Perlman, *Table of Isotopes*, 6th ed., by permission of John Wiley & Sons, Inc.)

quite different extents. The 0.140 MeV gamma is a M1 transition with $\alpha_K \approx 0.1$ only. The 0.142 MeV gamma is a M4 transition with $\alpha_K \approx 80$; conversion in the other shells will give about another 30 conversion electrons per gamma-ray emitted. (The 0.002 MeV gamma is a E3 transition with a very high α value, although the energy is so low that conversion occurs only in the M shell.) As it happens, the relative frequency of the M4 transition is very low (1.4%) so that for every 100 gamma-rays of ^{99m}Tc nuclei we get $1.4 \times 110 \approx 15$ conversions from this branch, against $98.6 \times 0.1 \approx 9.9$ conversion electrons from the other. The rather low total of conversion electrons per 100 gamma-rays emitted is of practical importance because it enables rather high activities of the radionuclide to be administered for diagnostic tests while still ensuring acceptable radiation hazard to the patient. It is of interest to see what would have happened if the ($\frac{1}{2}$+) intermediate state had not existed, for then ^{99m}Tc would have decayed with a half-life 100/1.4 times its actual half-life of 6 h, i.e. about 18 d; also it would emit about 110 times as many conversion electrons as gamma rays. These properties would have rendered it almost useless in the field of diagnostic nuclear medicine.

NUCLEAR REACTIONS

2.15 Nuclear Reactions for Radionuclide Production

Radionuclides are produced by (a) artificial transmutation of stable nuclides and (b) radioactive decay of some of the already existing radionuclides. We review the transmutation processes in this and the following Sections; examples of method (b) are discussed in Section 2.21.

Nuclei undergo various changes when bombarded by particles such as neutrons, protons, deuterons ($^2_1H^+$), $^3_2He^{2+}$ and $^4_2He^{2+}$ particles. Occasionally nuclei of atoms heavier than these have been used, but they are not generally available. The nuclei undergoing bombardment we refer to as the 'target'.

Whatever the course of the transmutation reaction, a fundamental consideration is that the total (mass + energy) of the bombarding particle and target nucleus must be at least equal to the corresponding total for the reaction products, i.e. the newly created nucleus plus any particles that may be emitted. Thus if a particle p interacts with a target nucleus T to produce a new nucleus P plus a particle (or particles) p′ we can write the course of the reaction as $p + T \rightarrow P + p'$, or, more briefly, T(p,p′)P. If the masses of the entities concerned are m_p, m_T, m_P, $m_{p'}$ we can define a 'Q value', $Q = m_P + m_{p'} - (m_p + m_T)$, analogous to a heat of reaction in elementary thermodynamics. If Q is negative the reaction cannot proceed, unless the total mass of target nuclide and bombarding particle is augmented by an energy term—viz. the kinetic energy brought in by the bombarding particle p. (*Excess* energy can be used up in creating gamma-rays or providing kinetic energy for the reaction products.) For *exothermic* reactions, for example the absorption of a neutron into a nucleus, the neutron does not have to contribute any kinetic energy, and the reaction will proceed with so-called 'slow' neutrons. However, for most nuclear reactions Q is negative, and there is a corresponding threshold energy which the bombarding particle must have to make the reaction energetically possible.

Bombarding particles carrying a positive charge (e.g. protons, ${}^{4}_{2}He^{2+}$ ions) can effect no changes in a nucleus unless they have kinetic energies sufficient to carry them reasonably close to the nuclear surface against the repulsive Coulomb force acting on them. The energy needed to enable a particle to reach the nuclear surface, sometimes referred to as the *barrier height*, B, is proportional to zZ/r; z, Z, being the atomic numbers of particle and target nuclide and r the latter's radius. A rough estimate of the barrier height for a target of mass number A is given by

$$B = \frac{zZ}{A^{1/3}} \quad \text{MeV}$$

Strictly this expression should be multiplied by a numerical factor but in the units chosen the value of the factor is nearly enough unity.

If bombarding particles of sufficient energy are available, at least two kinds of reaction are possible. These are as follows:

(*a*) *Direct Interaction.* In this process, the bombarding particle interacts with only one or two nucleons within the target nucleus. Thus a bombarding proton may collide with, and eject, a neutron as in the reaction ${}^{63}_{29}Cu + {}^{1}_{1}H \rightarrow {}^{62}_{24}Cu + {}^{1}_{1}H + {}^{1}_{0}n$, or ${}^{63}Cu(p,pn){}^{62}_{24}Cu$. If the bombarding proton happens to collide 'head-on' with the neutron it will lose almost all its own kinetic energy, and might suffer capture by the rest of the nucleus, so the reaction will be ${}^{62}_{29}Cu(p,n){}^{63}_{28}Ni$. High speed ('fast') neutrons used as bombarding particles probably interact similarly, for example in the reactions ${}^{32}_{16}S(n,p){}^{32}_{15}P$, ${}^{35}_{17}Cl(n,p){}^{35}_{16}S$ and ${}^{58}_{28}Ni(n,p){}^{58}_{27}Co$. These reactions are technically important as they are used in the preparation of ${}^{32}_{15}P$, ${}^{35}_{16}S$ and ${}^{58}_{27}Co$ on a large scale. Another kind of direct interaction occurs in bombardments with deuterons (${}^{2}_{1}H^{+}$) when the neutron part of the deuteron is absorbed but the proton part is not, giving e.g. ${}^{63}_{29}Cu(d,p){}^{64}_{29}Cu$. This is called *stripping* reaction; its overall effect is the same as a bombardment with neutrons, the target nucleus merely absorbing an extra neutron.

(*b*) *Compound-Nucleus Formation.* Here the bombarding particle is absorbed into the target nucleus and takes its place in one of the available unoccupied energy levels. Since it is now a bound particle, its binding energy as well as its kinetic energy become available to the nucleus as a whole. Numerous secondary effects then become possible as the excited nucleus makes transitions towards a ground state, and one or more of the following may occur:

(i) *Gamma emission.* This may be the sole process following absorption of neutrons of low energy. We may therefore refer to the reaction by the notation (n,γ), e.g. ${}^{12}_{6}C(n,\gamma){}^{13}_{6}C$.

(ii) *Neutron emission.* Depending on the energy available one or more neutrons may leave the compound nucleus, each carrying some kinetic energy. The process has been likened to the evaporation of molecules from a drop of hot liquid.

(iii) *Charged particle emission.* Occasionally protons are evaporated from the compound nucleus, though less frequently than are neutrons. The potential barrier which hinders the approach of protons to a nucleus is also effective in hindering their escape.

(iv) *Fission.* In this process the compound nucleus breaks into two more or less equal halves, together with a small number of neutrons. When fission is caused by very high energy particles (e.g. protons of 1000 MeV) the fission fragments are most commonly of nearly equal mass, although some asymmetric fission does occur. Some heavy nuclides such as $^{235}_{92}U$ and $^{239}_{94}Pu$ undergo fission after slow neutron capture, when the most commonly occurring fission fragments have masses with a ratio 2:1. On average about 3.5 neutrons are also produced per fission making a nett gain of 2.5. It is this fact which makes possible the construction of nuclear reactors and fission bombs.

The direct-interaction and compound-nucleus mechanisms just outlined must not be thought of as mutually exclusive, but rather as extreme cases of a spectrum of reaction mechanisms. If the bombardment energy is high, there is evidence that in one and the same struck nucleus some particles are ejected by the direct action process and others escape by a subsequent evaporation mechanism. This kind of nuclear reaction is called 'spallation'.

2.16 Reaction Cross-Sections

Some idea of the yield of radionuclides available from nuclear reactions may be gained from considerations of nuclear cross-sections. Various experiments in nuclear physics suggest that nuclei are approximately spherical, with a radius given by $r = 1.3 \times 10^{-15} A^{1/3}$ m, A being the mass number. This expression is clearly unreliable for nuclei in which the number of nucleons is small. For large nuclei the numerical factor has been quoted in a range of values up to 1.8×10^{-15}, depending on the method of measurement. But for medium sized nuclei the radius will be on the scale of 10^{-14} m, and a cross-section of a nucleus through the midplane will have an area in the region of 10^{-28} m^2. An area of this size was chosen as the unit and humorously named, by Fermi in 1942, a 'barn'. (He was referring to uranium nuclei and said their cross-sections must be as big as a barn.) The cross-section of any nucleus is usually denoted by σ, measured in barns or sometimes millibarns. Now suppose we have a target containing N_T nuclei. Their total cross-sectional area will be $N_T\sigma \times 10^{-28}$ m^2. As a rule this quantity will be small compared with 1 m^2 so that we are justified in neglecting 'overlap'; from whatever direction in bombarding particle approaches the target, the chances of one nucleus obscuring another is negligible. While the target is under bombardment we can visualise a stream of bombarding particles moving through it, some penetrating without making collisions, others hitting a target nucleus. Let the flux density of the particles be f m^{-2} s^{-1}. We may disregard the cross-section of the particles themselves as they are generally much smaller than the target nuclei; in a sense, the choice of the numerical factor in the expression for r allows for their finite size. The probability of a bombarding particle making a collision is $N_T\sigma \times 10^{-28}$, since this is the fraction of 1 square metre occupied by target nuclei. Then the total number of collisions in a time t is simply $N_T\sigma ft \times 10^{-28}$. (Some confusion might arise in using this formula, because hitherto the units of area used were square centimetres, not square metres. Earlier textbooks therefore give the formula as $N_T\sigma ft \times 10^{-24}$. Also the quantity f is often mistakenly called 'flux' instead of 'flux density'.) This simple result, it must be emphasised, only gives the number of *collisions* occurring within the target, and this is often very far from being the number of new nuclei, of the desired nature, produced. Many collisions result merely in scattering of the bombarding particles, particularly when the energy given to the struck nucleus is high. Some bombarding particles will react in one way, and some in another, as outlined in the previous section. An empirical approach to this problem is to ascribe a partial cross-section to each of the possible reactions, including scattering. What we have so far called just the cross-section, we rename the geometrical cross-section σ_g. This area is the sum of quantities such as σ_s, σ_n, σ_p corresponding to scattering processes, emission of a neutron, or a proton, respectively. The ratio $\sigma_n/(\sigma_s + \sigma_n + \sigma_p + \cdots)$ thus represents the fraction of all interactions in which just one neutron is emitted.

The partial reaction cross-sections depend very strongly on the bombardment energy, and their calculation from first principles presents tremendous difficulties. However, much information has been obtained experimentally and may be briefly summarised thus:

(1) Reactions in which the bombarding particles are slow neutrons usually have very high cross-sections, amounting to hundreds or even thousands of times the geometrical cross-section. This apparent anomaly can be explained only on a wave mechanical basis, the neutron being associated with a very long wavelength corresponding to its low energy. It is, therefore, always possible for a neutron to be absorbed into a target nucleus, even from a great distance, although a geometrical point moving on the same trajectory would miss the nucleus by many nuclear diameters. The existence of such very large cross-sections has the effect that the assumption made above (i.e. that the nuclei do not obscure each other) will be invalid for targets containing more than a certain number of target nuclei. The yield of radionuclide will be less than the calculated value and the target can be described as showing 'self-shielding'.

(2) Reactions which are energetically possible, but in which the bombarding particle has an energy a little less than the potential barrier (Section 2.15) are sometimes observed, but with low cross-section (rising with energy). It is clear that a 'barrier-penetration' process is possible and an explanation is

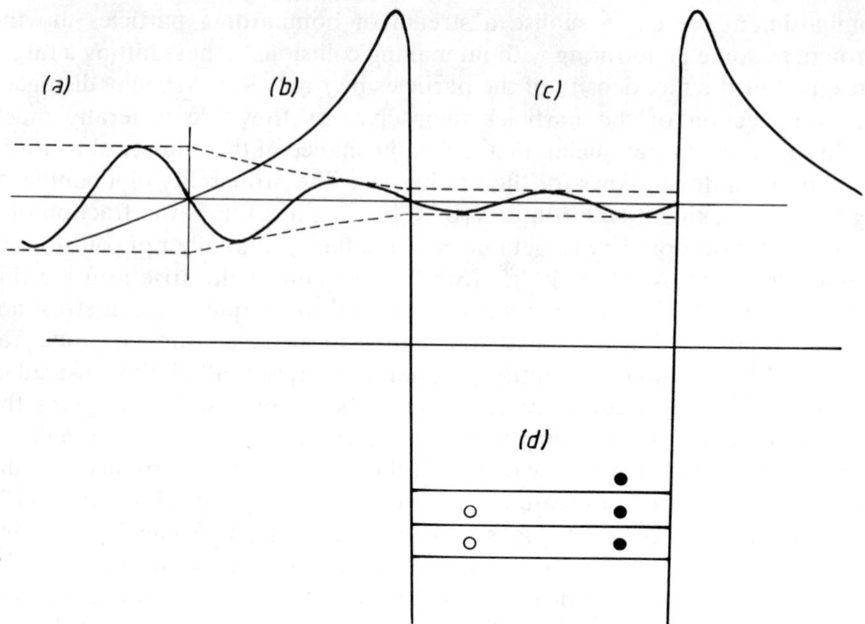

Fig. 2.11 Barrier penetration as a wave-mechanical process.

found in terms of wave-mechanics (Fig. 2.11) which shows the energy of particles or waves, at distances x from the centre of a nucleus.

In the diagram we show three regions of space, in which we have:

(a) a wave corresponding to a bombarding particle of energy E,

(b) the wave being attenuated within the potential barrier, just as light is attenuated in an absorbing medium,

(c) a wave representing the particle which has emerged into the nucleus.

The probability of penetration then depends on the ratio of the squares of the wave amplitudes at (c) and (a). Also at (d) the horizontal lines represent energy levels within the nucleus; these levels will be occupied by the bound nucleons.

(3) When bombarding particles have energies above the potential barrier, many reactions may compete with one another. In general, the greater the energy available, the greater the number of particles which can escape from the compound nucleus. This means that as the bombarding particle energy is progressively raised, reactions in which 1, 2, 3, . . . particles are emitted predominate in turn. A plot of reaction yield, or partial cross-section, versus particle energy is called an *excitation function*. Fig. 2.12 depicts typical excitation functions. The curves (1), (2), (3), depict the yields, as a function of energy, for products which arise through emission of 1, 2, 3, neutrons. Points A, B, C, . . . represent the 'threshold' energies below which emission of 1, 2, 3, . . . neutrons is not possible energetically. These represent the Q values for the respective reactions. Each curve goes through a maximum at approximately the same energy at which a competing reaction begins to appear.

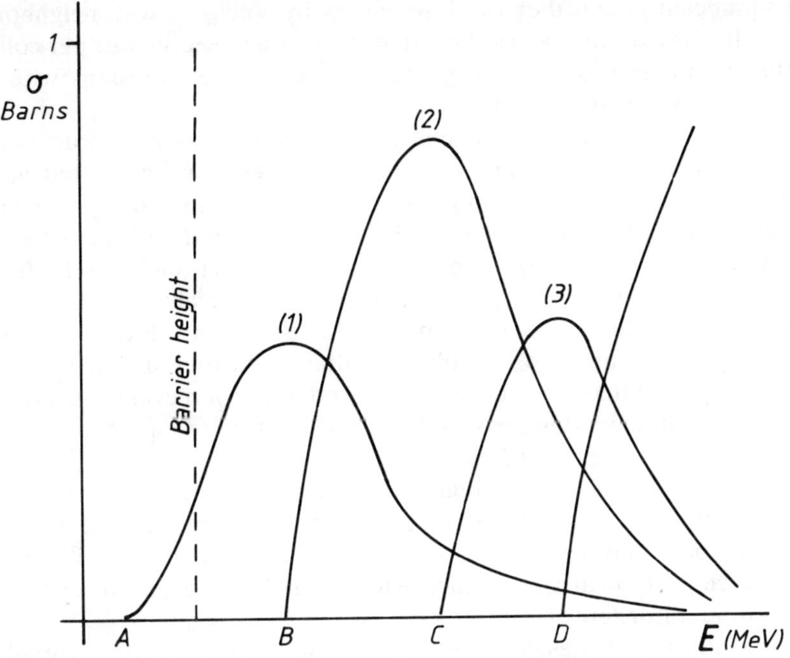

Fig. 2.12 Schematic excitation functions.

The maxima are by no means all the same height—there often being 'odd–even' effects. Low yields from reactions in which one neutron is emitted may persist to high energies. Commonly the spacing between peaks is 7–10 MeV. The situation is complicated by the existence of partial cross-sections for proton emission in competition with neutron emission, these cross-sections often reaching maxima which are smaller (by a factor between 2 and 10) than those for neutron emission. A particular practical difficulty occurs with charged-particle reactions because the bombarding particles are rapidly slowed down within the target by atomic collisions, and thereby lose energy in causing dense ionisation. Thus they have effective ranges in the target material which are quite small—fractions of a millimetre as against several centimetres for slow neutrons. This effect limits the thickness of material which can be usefully bombarded, a problem which is taken up again in Section 2.20.

2.17 Sources of Bombarding Particles: Neutrons

Neutrons are generated most prolifically in nuclear reactors by fission in uranium or plutonium. Neutrons created in a reactor fuel element have initially some 2 MeV of kinetic energy. (Neutrons have no charge; what this statement means is that they have nearly the same velocity as that of protons which have been accelerated by a potential of 2 MV). Such 'fast' neutrons are known to produce nuclear reactions such as (n,p) and (n,α), but these are relatively unimportant except for a few targets (Section 2.15). What is far more significant is that they can lose energy by collisions with neighbouring nuclei. The laws of mechanics show that the average energy loss per collision with heavy nuclei is small, but conversely is high for collisions with light nuclei, especially with hydrogen.

Direct energy loss by *ionising* collisions in atoms does not occur, but in a hydrogenous material the struck hydrogen nuclei may be ejected at high speeds from bound states in molecules (so-called 'knock-on' protons) and these will cause ionisation. The slowing-down, or *moderation*, of fast neutrons is achieved in some reactors by collision with ${}^{2}_{1}H$ nuclei in the form of heavy water (D_2O) or with ${}^{12}_{6}C$ nuclei in graphite. Eventually the neutrons become 'slow', with an energy of about 0.025 eV and a velocity 2200 m s^{-1}, and at this stage they are capable of initiating further fission in the fuel element. As more than one neutron is emitted per fission event, the process is multiplicative and might increase indefinitely. However, in practice, there are several limiting factors, as follows:

(a) If the volume of the reactor is too small, many neutrons will be created near its surface and some will escape before causing further fissions. In other words a reactor is self-sustaining only if it is above a certain 'critical size', which will depend on the nature of the fuel, moderator, etc.

(b) Materials such as cadmium which absorb neutrons very strongly are incorporated into reactors as a design feature. Withdrawal of these 'control rods' allows the multiplication process to build up; reinsertion tends to shut the reactor off. At an intermediate position the reactor will run at a constant power output level.

(c) Some neutrons are lost adventitiously by absorption in impurities, structural materials, etc.

(d) Some neutrons are utilised in the preparation of radionuclides by bombardment of 'target' materials inserted in special channels within the reactor. Neutron flux densities available at these sites in a large reactor may be as high as 10^{19} m^{-2} s^{-1}.

Smaller neutron flux densities are produced by 'neutron generators' which use the following nuclear reaction:

$$ {}^{2}_{1}H + {}^{2}_{1}H \rightarrow {}^{3}_{2}He + {}^{1}_{0}n \qquad \text{or} \qquad {}^{2}_{1}H(d,n){}^{3}_{2}He $$

In these machines a beam of deuterons (nuclei of ${}^{2}_{1}H$) are accelerated to 500 KeV or more, and the beam is directed on to a target consisting of deuterium gas absorbed in a plate of zirconium. The neutrons produced have energies of a little over 2 MeV, the exact figure depending on the bombarding energy and the angle of emission. A similar reaction with a tritium (${}^{3}_{1}H$) target gives 14 MeV neutrons. These may be moderated by surrounding the target with hydrogenous materials such as paraffin wax. A commercially available machine will give a flux density of slow neutrons of up to 10^{15} m^{-2} s^{-1} at a distance of a few centimetres from the target.

On a still smaller scale, but one still large enough for simple experiments, one may obtain neutrons from nuclear reactions in which alpha particles, or gamma-rays, from long-lived nuclides are used to bombard beryllium. Thus 'radium-beryllium sources' consist of an intimate mixture of a radium salt ($^{226}_{83}Ra$) and powdered beryllium, the whole enclosed in a steel capsule. These sources have the practical disadvantage that the high gamma ray emission from the radium causes a radiation hazard. Recently americium–beryllium sources ($^{241}_{33}Am$) have become available instead, which are much safer in this respect. Sources containing 400 GBq of alpha-active material, i.e. approximately 1 g of radium or 0.3 g americium, will produce flux densities of moderated neutrons of 10^9 m^{-2} s^{-1}. The nuclear reaction here is

$$^{9}_{4}Be + ^{4}_{2}He \rightarrow ^{12}_{6}C + ^{1}_{0}n \qquad \text{or} \qquad ^{9}_{4}Be(\alpha,n)^{12}_{6}C$$

Still another laboratory source of neutrons is the nuclide ^{252}Cf (californium), which decays partly by spontaneous fission, giving about 3.5 neutrons from each fission event.

A very popular method of moderating the neutrons from americium–beryllium or californium sources is to support the capsule containing the source in the middle of a large tank of water (say 1 m^3). Targets to be irradiated with the neutrons can then be suspended in the tank in polythene or soft glass tubes; the highest yields are obtained with the target material as close as possible to the capsule where the slow neutron flux is greatest. The mean free path for neutrons in water is about 60 mm, and about 20 collisions are needed to bring the neutrons to thermal energies. The flux density close to the capsule falls off with distance much less slowly than the inverse square law.

2.18 Sources of Bombarding Particles: Charged Particles

By far the most commonly used charged particles are the nuclei of hydrogen, deuterium and helium, that is to say protons $^{1}_{1}H$, deuterons $^{2}_{1}H$, and the helium ions $^{3}_{2}He^{2+}$ and $^{4}_{2}He^{2+}$. The ions are easily obtained by bombarding hydrogen or helium gas at low pressure (10^{-3} Pa) by electrons from an electrically heated filament after these have been accelerated by a potential of a few hundred volts. The positive ions can be drawn out of the ionising region (plasma) by electrostatic fields; a well designed ion source will deliver many milliamps of positive-ion current. Before the ions can produce nuclear reactions they must be accelerated until their energies are at least comparable with the potential barrier of the target nucleus. This amounts to a few million electronvolts per mass unit. Acceleration is achieved in two main types of machine. These are 'linear' accelerators, in which the ions move essentially along the axis of a straight evacuated tube, and 'Circular' machines in which their path is constrained by magnetic fields to be approximately spiral, or approximately circular. In either case acceleration is brought about by electrostatic fields applied to the ions in flight. In Van de Graaff machines this is done by generating a high potential—commonly 4, 6 or 8 million volts—and applying it to grids across opposite ends of the evacuated tube through which the ions pass. For extra energy two or more Van de Graaff generators are used 'in tandem'. In small 'linac' machines, e.g. deuteron accelerators for neutron generators, a number of hollow cylindrical electrodes are placed axially along the vacuum tube. Alternate ones, i.e. nos. 1, 3, 5, 7, . . . and nos. 2, 4, 6, 8, . . . are connected together and an alternating potential at a constant frequency is applied between the two groups.

Ions from an ion source at one end of the tube are drawn into the first electrode when this is at a negative potential, and continue to move along the axis in a 'bunch'. Conditions are arranged so that the potentials are reversed by the time the bunch arrives at the gap between the first and second electrodes. Acceleration away from the ion source and towards the target occurs and this is repeated at each crossing of electrode gaps throughout the whole sequence. Clearly, as the ions accelerate the electrodes must be made longer so that the same time is spent in each. As the speed of the ions approaches that of light, their energy is increased more and more by the relativistic increase of mass rather than an increase in speed. Therefore, the electrode length does not increase linearly as the highest energies are achieved, but tends to become constant. Another important effect is that the bunch of ions tends to disperse partly through their mutual repulsion and partly because of the effect of the curvature of the electric lines of force between the electrodes. These defocusing effects have to be counteracted by additional electrodes or magnetic lenses. In machines accelerating electrons to very high energies, the particles are always relativistic and the accelerating tube consists of a series of resonant cavities obtained by inserting equally spaced discs, with a central hole for the beam, along the tube axis. Radiofrequency power from a high-powered oscillator is then fed into the tube producing a travelling wave. In this way, for instance, the Stanford University accelerator has produced electrons at 1 GeV (10^3 MeV). It has not however, been found practical to accelerate positive particles (protons) to such high energies in linear machines.

Accelerating machines in which the charged particles move along curved paths under the influence of magnetic fields take many forms. The simplest of these is the cyclotron. In the earlier designs of these machines—some of which are still operating—a nearly uniform field is maintained across a shallow cylindrical vacuum chamber. Within the chamber are two hollow electrodes, called dees because of their semicircular shape. The ion-source is near the centre of the system, so that ions emerging from it enter one dee. While within the dee, their paths curve round through 180°, and they then cross the gap into the other dee. The process repeats many times. A high-frequency alternating electric field is maintained between the dees so that at each crossing, ions which happen to start out at the right instant and in the right direction are progressively accelerated. Their paths are therefore a series of semi-circles with steadily increasing radius of curvature. Ions with a charge e moving at velocity v constitute a current element ev so that the magnetic force acting in the flux B is Bev. If the ions are moving on a path with a radius of curvature r the centripetal acceleration is v^2/r so that

$$Bev = mv^2/r$$

or

$$mv = Ber$$

Therefore, the momentum of the ions is proportional to the flux B and to the radius of curvature. From this it is easy to show that the kinetic energy $\frac{1}{2}mv^2$ is just $B^2e^2r^2/2m$. (This calculation ignores relativity effects but is useful up to energies of about 10 Mev per unit mass number.) Clearly the maximum energy the ions can achieve is $B^2e^2R^2/2m$ where R is the radius of curvature of the dees themselves, because if the energy rises higher the ions must collide with the dees and be absorbed in them. Notice that this final energy does not depend on the voltage across the dees; in principle a low voltage (with the ions crossing the gap many times) is just as effective as a high voltage (with the ions crossing only a few times). In practice one uses as high a voltage as is available, viz. in the region of 100 kV to 1 MV. The reasons for this are:

(1) The higher the voltage, the shorter the total path length out to full radius, so the smaller the losses of accelerated ions due to collision with gas molecules in the vacuum tank. (It is easy to show that even at a pressure of 10^{-5} Pa there are about 3×10^{17} gas molecules present per m^3.)

(2) The higher the voltage the greater the separation between successive orbits of the ions. Thus a machine with $R = 1.0$ m, accelerating protons to 10 MeV and using a dee-to-dee voltage of 100 kV, will have an orbit-to-orbit separation of about 10 mm, as is easily shown from the expression given above for the energy. With such a separation it becomes practicable to use a dee having a slot at the centre of the curved face to allow fully accelerated ions to escape into the space *behind* the dee. Once in this new path they can be deflected by a suitable electrode and be caused to move into an *exit port* and thus out of the machine altogether, giving an 'extracted beam' (see Fig. 2.13).

It is, of course, possible in principle to use the accelerated ions *within* a cyclotron to bombard targets for radionuclide production, simply by inserting

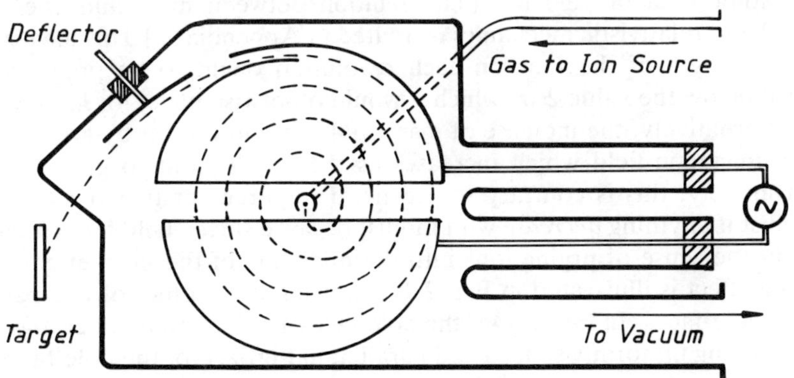

Fig. 2.13 Schematic diagram of showing exit channel (plain view).

targets into the narrow gap between the dees. This technique enables a choice of bombarding energy to made by specifying the radius at which the target is placed. The practical difficulty however which severely limits the technique is that of efficiently cooling the target. Target materials are generally confined to metals with good thermal conductivity in the form of hollow blocks through which cooling water passes. Some metals can be bombarded 'internally' after soldering or welding them to copper blocks which can be water cooled during bombardment. Although there are inevitably some losses of ions during the extraction process mentioned above, an 'external' beam of ions is often of greater utility because it allows the bombardment of powdery materials which would be scattered in a vacuum chamber, and also because the face of the target can be cooled by a blast of air or even an inert gas such as helium. An important feature here is that cyclotrons are usually designed in such a way that the internal beam of ions is well focused, i.e. the ions are bunched as closely together as possible, in order to achieve high efficiency of extraction. Good focusing depends on a number of design factors such as slight non-uniformity in the magnetic field and the shape of the dee-edges, which are not quite parallel. This aim of good focusing is, of course, the very antithesis of what would be required for internal bombardment of targets because the heat is thereby concentrated on as *small* an area as possible. However, the *external* beam may easily be defocused using electric or magnetic fields, and targets may be presented to it in the form of long strips of material at a glancing angle, a method impossible within the machine because of the narrowness of the gap between the dees.

A simple cyclotron of the kind described above would have a beam-extraction efficiency of perhaps 10%. Some 90% of the accelerated particles are therefore wasted, all of them contributing to undesirable heating effects and many of them creating undesirable radioactivity in the dees and parts of the exit channel. Another limitation of the simple cyclotron is in the maximum energy (about 10 MeV per mass unit) that can be achieved. The limit arises because at higher energies the velocity of the ions becomes an appreciable fraction of that of light, and increments of energy produce an increase of mass rather than of velocity. (The relation between mass and energy in classical and relativistic mechanics is treated in Appendix 2.) This means that the path length of the ions on each revolution ought to be progressively reduced below the value $2\pi r$, which it would otherwise have, for large values of r. Alternatively, the increase of mass could be allowed for by arranging to have a magnetic field which increases as the ions go out to greater radii. Unfortunately, this is contrary to a general requirement that the magnetic field must if anything *decrease* with radius, because such a field has a focusing effect in the sense of urging ions into the midplane of the magnet gap. The focusing effect is illustrated in Fig. 2.14, showing the profiles of two pairs of magnet pole-faces. In Fig. 2.14a, the pole-faces are flat and parallel, the lines of force being uniformly spaced and parallel; in Fig. 2.14b, the pole-faces are convexly curved, giving curved lines of force and a field at the midplane which

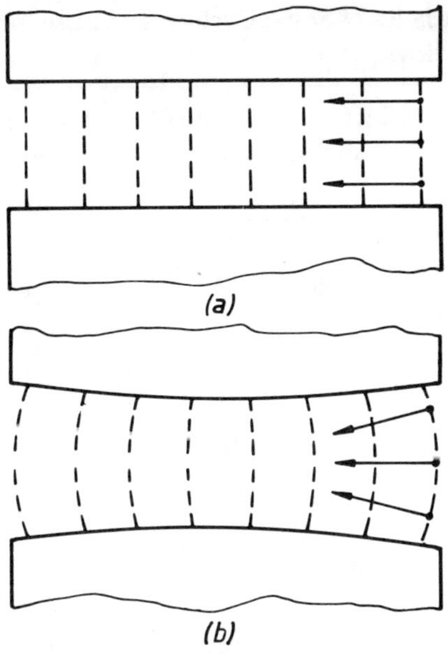

Fig. 2.14 Focusing effect of non-uniform magnetic field.

becomes weaker at large radii. ('Fringing' fields are omitted in both diagrams.)

The figure shows how ions which are either above or below the midplane and moving at right angles to the plane of the paper and also at right angles to *parallel* lines of force (uniform field) are acted on by forces towards the centre of the magnet, but parallel to one another. In Fig. 2.14b, where the outer lines of force are curved as the field decreases radially, the centripetal forces are not quite parallel but have a component directed towards the midplane. Obviously the curvature in the opposite sense caused by a radially increasing field would have a defocusing effect and the beam of particles would be dispersed.

This difficulty is overcome in the so called *radial-ridge* and *spiral-ridge* machines. These have magnetic fields which are extremely non-uniform azimuthally; as ions travel in their roughly circular paths they traverse alternately 'strong' and 'weak' magnetic fields. The alternation of field strength is achieved by having 60° segments of magnetic material fixed to the otherwise nearly flat pole faces. The field alternation has a strongly focusing action, in much the same way that a beam of light is focused by an alternation of converging and diverging lenses. An overall average increase of field with radius becomes possible, enabling ions to be accelerated to 50 MeV or more per mass unit, while still retaining focusing into the midplane. The focusing is in fact so strong that only one dee is necessary, whereas two are always required to maintain focusing in a 'flat pole' machine. The separation between orbits near the maximum radius is of the order 1 mm, but the extreme

Table 2.1 Characteristics of Some Cyclotrons in Use for Production of Radionuclides

Machine	TRC, Amersham, UK	VEC, AERE, UK	Hammersmith Hospital, UK	Philips, Petten, Netherlands
Particles in main use, energy (meV)	$^{1}_{1}H$ 25 $^{2}_{1}D$ 16	$^{1}_{1}H$ 50 $^{4}_{2}He$ 84	$^{2}_{1}D$ 16 $^{4}_{2}He$ 32	$^{1}_{1}H$ 30 $^{2}_{1}D$ 34 $^{3}_{2}He$ 44
Other particles available, maximum energy (meV)	—	B,C,N,O,Ne (charge Z) 84 Z^2/M	$^{1}_{1}H$ 8	—
Current, internal (μA)	750 max 200–250 routinely	Not used	1000	100
Current, external (μA)	Not available	$^{1}_{1}H$ 20 $^{4}_{2}He$ 20 others 1	$^{2}_{1}D$ 320 $^{4}_{2}He$ 250	Not available
Radius (m)	0.70	0.88	0.63	0.70
Type Magnet B_{max} (T)	Spiral ridge 1.7	Spiral ridge 1.35	Flat pole 1.5	Spiral ridge 1.35
Frequency (MHz)	10–21	7.5–23	11.275 (fixed)	10–21
Dee voltage (kV)	27–30 to ground	80 to ground	120 dee-to-dee	50 to ground
Main radionuclide production	^{57}Co ^{67}Ga ^{111}In	^{123}I	^{11}C, ^{13}N, ^{15}O, ^{18}F, ^{81}Rb	^{67}Ga, ^{111}In, ^{201}Tl
Special features	Computer control for continuous running	8 external beam lines. Particle energy infinitely variable	Hospital environment allows use of very short-lived radionuclides. Machine also use for neutron therapy	—

precision with which radial-ridge machines are built nevertheless permits construction of beam-extraction systems which are at least as efficient as in the earlier designs. In England the MRC (Medical Research Council) cyclotron at the Hammersmith Hospital is a conventional flat-pole machine whereas the newer VEC (Variable Energy Cyclotron) at AERE, Harwell, is of the radial-ridge design.

The latter machine was built for very general purposes, but also has been used recently for specialised radionuclide production. Some parameters of these and other machines are collected in Table 2.1. It will be noticed that in these cyclotrons several bunches of ions are being accelerated simultaneously. Since the period of revolution of the ions remains constant, the machines are termed fixed-frequency or isochronous. The problem of relativistic mass increase is dealt with in other machines called synchrocyclotrons, not by field shaping, but by progressively lengthening the oscillation time. Synchrocyclotrons can therefore accelerate only one bunch of ions at a time. After the ions are fully accelerated and sent to the exit port the machine reverts to a faster oscillation frequency and starts with a new bunch. These machines are said to have a low *duty cycle*. Nevertheless, high effective beam currents are possible and this in combination with high energies (e.g. the CERN S.C. gives 600 MeV protons) quite large amounts of activity can be produced. The other main kind of circular accelerator is the *synchrotron*. In these machines charged particles—usually protons—are injected from a linear machine into an approximately circular orbit which remains nearly the same throughout the acceleration cycle. In a large machine the orbit has alternate straight sections. with no magnetic field, and curved sections in each of which the beam is turned through only a few degrees. For example, in the CERN synchrotron there are 24 straight sections and curved sections, the whole orbit having a diameter of 200 m. Large machines such as this can accelerate protons up to many GeV but are hardly suitable for radionuclide production because of the relatively low beam current and the extreme complexity of the mixture of radionuclides produced.

It has recently been proposed to use high-beam-current synchrocyclotrons for radionuclide production as a sort of by-product of their main function in nuclear physics research, viz. the production of the subnuclear particles such as π mesons (pions). These could be used to provide strong sources of neutron-deficient radionuclides from spallation reactions.

2.19 Bombardment Conditions: Neutron Bombardments

As noted in Section 2.15, neutron capture reactions with suitable targets give high yields of products isotopic with the target and one mass unit heavier. Target material for irradiation in reactors may weigh up to several grams, and if chemically unreactive the material can be contained in a screw-top aluminium can or capsule. Double encapsulation may be required for bombardment at the highest flux density available and for additional safety it may be necessary to seal the sample in a quartz tube. The sample must be capable of withstanding a moderate temperature rise without decomposition; the temperature rise is due to the heat generated in the reactor itself more than to energy dissipated within the target, but is rarely as much as 100 °C.

Radionuclide yields are usually quite high. For example, suppose one gram of gold ($^{197}_{74}$Au) is bombarded with slow neutrons at a flux density 10^{16} m^{-2} s^{-1} for 10^4 s. During this time the number of ^{198}Au nuclei formed will be

$$(\text{number of target atoms}) \times (\text{cross section}) \times (\text{flux density}) \times (\text{time})$$

$$= \frac{6.02 \times 10^{23}}{197} \times 96 \times 10^{-28} \times 10^{16} \times 10^{4} = 2.93 \times 10^{15}$$

There will be some decay during this time but this may be neglected in a rough calculation. Assuming a decay constant $\lambda = 2.97 \times 10^{-6}$ s^{-1} (calculated from the half-life of 2.7d), the decay rate is then $2.93 \times 10^{15} \times 2.97 \times 10^{-6}$ s^{-1} or 8.7×10^9 Bq (about 200 mCi).

When it is recalled that nuclear reactors can produce fluxes of at least this magnitude, that bombardments of some weeks' duration are possible, and that hundreds of samples can be irradiated simultaneously, it is clear that total production can run into many terabecquerels. Indeed The Radiochemical Centre at Amersham, Bucks, is currently processing, at this level, supplies of radionuclides of all kinds every week, the vast majority of which are reactor-produced. On a laboratory scale (flux densities in the range 10^9 m^{-2} s^{-1} from Am–Be sources to 10^{15} m^{-2} s^{-1} from neutron generators) yields are of course correspondingly lower, but by no means insignificant for some research purposes.

One disadvantage of neutron capture reactions is that the radioactive product is an isotope of the target element, so there is a practical limit to the *specific activity* of the product. For example, in the example cited the specific acitivity could be expressed as 8.7 GBq per gram of gold. The actual weight of $^{198}_{79}$Au atoms (3×10^{15} atoms) would be only 10^{-6} g. The specific activity of the product is, therefore, a million times smaller than the theoretical figure for pure $^{198}_{79}$Au. In other words, the technique has the disadvantage that the product is not 'carrier-free', being inevitably mixed with residual stable atoms. Extremely long bombardment times and/or increases of flux density would be necessary in practice to approximate to a carrier-free product. Fortunately for the purposes we consider in this book, specific activities from reactor-produced isotopes are generally high enough.

The specific activities in targets bombarded in neutron generators and by Am–Be sources are of course correspondingly lower. For some radionuclides useful special techniques have been developed, such as the so-called Szilard–Chalmers process, named after its discoverers. They found that when an aliphatic or aromatic halogen compound, e.g. ethyl iodide, was exposed to slow neutrons, the $^{128}_{53}$I atoms formed by neutron-capture in the stable $^{127}_{53}$I could escape from their chemical bonds and become isolated unbound atoms, dispersed within the bulk of the organic medium. After terminating the bombardment, they could be removed into an aqueous medium by agitating the organic liquid with water, or preferably an aqueous solution of a mild reducing agent such as sulphite ion. The process is very similar to classical 'solvent extraction'. Efficiencies of extraction are commonly 20–50% overall, and it seems clear that not all the activated iodine atoms are released from their parent organic molecules. Possibly some newly formed $^{128}_{53}I$ atoms encounter fragments of decomposed organic molecules and combine with them. However, the important result is that the final aqueous solution is virtually 'carrier-free' as regards inactive iodine atoms, being contaminated only by free iodine if this was initially present as an impurity, and by free iodine produced as a result of radiation damage to the organic target during irradiation. (For a general discussion of radiation-chemical effects, see Section 3.22.)

In summary, the technique allows the use of perhaps hundreds of grams of iodine in the form of an organic compound as a target, but enables the product radioiodine to be isolated together with only milligram amounts of stable iodine.

2.20 Bombardment Conditions: Charged Particle Bombardments

It is convenient to measure the *flux* of bombarding particles in an accelerator beam in terms of the ion current. Thus a flux of protons of 6×10^{14} s^{-1}, each carrying a charge of 1.6×10^{-19} C, will constitute a current of $6 \times 10^{14} \times 1.6 \times 10^{-19}$ C s^{-1} or 96 μA, and indeed 'internal beam' currents of this magnitude are typical of many cyclotrons. Equally typical is the fact that the beam available for bombardment of targets has a cross-section of the order of 10 mm^2 or 10^{-5} m^2, so that the *flux density* is 6×10^{19} m^{-2} s^{-1}, or much the same as the highest flux density of slow neutrons in a large reactor. There are, however, several reasons why radionuclide production using charged particle bombardments is inevitably on a smaller scale than that which can be achieved with a reactor.

The most obvious factor is the small lateral spread of the beam. External beams are often 'defocused' electromagnetically before being allowed to strike the target, but of course only at the cost of reducing the flux density; alternatively they may be directed, by using a variable magnetic field, to each of several targets in turn, but then each target receives only a fraction of the flux. It is, therefore, scarcely practical to bombard more than two or three targets concurrently, in contrast to the many hundreds which can be irradiated simultaneously in a large reactor.

Secondly, there is the question of the range of the bombarding particles in target material. The ranges in copper of protons with energies up to 100 MeV and of $^{4}_{2}He^{2+}$ with energies up to 400 MeV are shown graphically in Fig. 2.15.

Clearly the ranges of interest are of the order of only 1 mm (in solid materials). Now in its passage through matter, an accelerated particle will dissipate 99.9% of its kinetic energy in collisions which ultimately degrade the energy to heat; less than 0.1% of the collisions are with atomic nuclei. Thus a

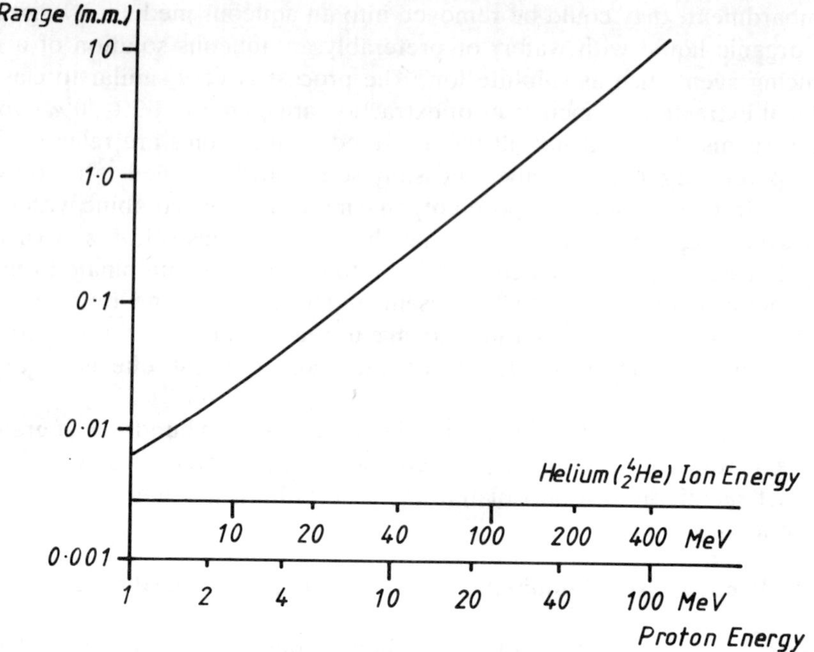

Fig. 2.15 Range of protons of energy 1–100 MeV in copper. These are the same as helium ions $^4_2He^{2+}$ of energy 4–400 MeV.

beam current of 96 μA of protons at, say, 20 MeV will produce a heating effect equivalent to $96 \times 10^{-6} \times 20 \times 10^6$ J s^{-1} or 1.92 kW, and for a beam cross-section of 10 mm^2 incident on a target in which the range of the protons is 1 mm, all this thermal power is concentrated into a volume of only 10 mm^3. The target must therefore be able to withstand very high temperatures or be provided with extremely effective cooling. Radionuclide production in bombardments with charged particles is much more restricted by target heating than by the beam current available at the machine.

Finally, there is the consideration that the cross-section σ for a particular nuclear reaction with a given target nucleus is a function of the energy of the bombarding particle (Section 2.15). The yield of useful radionuclide will depend on many other factors, but aside from these we may write

$$Y \propto \int_{E_1}^{E_2} \sigma(E)\mathrm{d}E$$

In this integral, the upper limit E_2 may be set as the maximum particle energy provided by the machine, but sometimes it is necessary to *reduce* this energy in order to avoid the production of some unwanted competing radionuclide. The lower limit E_1 is the threshold energy for the nuclear reaction. The production of a particular required radionuclide, therefore, occurs in a layer of target material which may be much thinner than the range of the particles (which in turn may be less than the actual thickness of the target). The thickness of target material in which useful reactions occurs can be calculated, given the values of E_1 and E_2, from range/energy equations which have been obtained empirically from experimental data.

For protons and helium ions ($^4_2He^{2+}$) in standard air, these are—for protons with $1 < E < 100$ Mev

$$R_p = \text{antilog}\,[-1.4839 + 1.3717 \log E + 0.31355 (\log E)^2 - 0.0714 (\log E)^3] \quad \text{kg m}^{-2}$$

for helium ions with $4 < E < 400$ MeV

$$R_{He} = \text{antilog}\,[-2.1805 + 0.9166 \log E + 0.4425 (\log E)^2 - 0.0714 (\log E)^3] \quad \text{kg m}^{-2}$$

(It is useful to remember that $^4_2He^{2+}$ particles of energy E have the same range as protons of energy $E/4$; $^3_2He^{2+}$ particles of energy E have three-quarters of this range.)

The corresponding ranges in materials other than air are given by the following formula, where R_Z is the range in material of atomic number Z, R_{air} is the range the particle would have in air, and M is the particle's mass number:

$$R_Z = R_{air}\{0.90 + 0.0275Z + (0.06 - 0.0086Z) \log (E/M)\} \quad \text{(valid for } Z > 10)$$

The design of a cyclotron for radionuclide production is inevitably a compromise between economy and the desirability of high energy, high beam current, and flexibility in the choice of bombarding particle to be available. But, in general, radionuclide production using charged particle beams has a unique advantage in that in most cases the product nuclide has a different atomic number Z from that of the target. This consideration is very important because it means that if accelerator targets are suitably treated chemically, one may obtain very high specific activities, i.e. virtually carrier-free radionuclides. The limit here depends on the purity of the initial target material and the elegance of the chemical processing. The following sections, 2.21–2.30, review the topic of nuclear chemistry, i.e. the application of chemical methods to the separation and characterisation of radionuclides.

NUCLEAR CHEMISTRY

2.21 Chemical Processing of Reactor and Accelerator Targets

We turn now to the chemical problems of preparing useful amounts of radionuclides in a 'radiochemically' pure form, that is to say the nuclide being sought is to be separated from undesirable radioactive contaminants. The additional problems of treating the radionuclide so as to be suitable for medical use—that is to say the preparation of radiopharmaceuticals—are deferred to Section 2.32.

From the previous discussions, it will be clear that reactor targets in which radionuclides are made by the neutron capture process may very well need no chemical treatment at all. Often a pure element can be bombarded and the product is a single isotope chemically identical with the target material. Sometimes, when the target element is very reactive chemically, a relatively inert compound is chosen for irradiation, with the disadvantage that the other elements in the compound are also activated. Even here, however, chemical processing is often avoidable. For example, if sodium chloride is irradiated to prepare $^{24}_{11}Na$ by the reaction $^{23}_{11}Na(n,\gamma)^{24}_{11}Na$, the product $^{24}_{11}Na$ ($t_{1/2}$ = 15.0 h) is initially contaminated, from a radioactive point of view, by $^{38}_{17}Cl$ arising from the $^{37}_{17}Cl(n,\gamma)^{38}_{17}Cl$ reaction. But $^{38}_{17}Cl$ has a half-life of only 38 min, so that a 'cooling' period of say 15 h after the end of the bombardment will result in the decay of only half the $^{24}_{11}Na$ but of more than 99.9% of the $^{38}_{17}Cl$. In this case, therefore, chemical separation of the reaction products is scarcely worthwhile. Consider, on the other hand, the preparation of $^{82}_{35}Br$ (half-life 34 h) by the neutron irradiation of sodium bromide. A delay period after bombardment of 150 h would be necessary to allow the $^{24}_{11}Na$ to decay by a factor 2^{10}, or approximately 1000. Such a long delay period would be very inconvenient and, of course, would allow substantial decay of the desired bromine activity. Chemical separation of the bromine soon after bombardment would, therefore, be almost indispensable. Alternatively, one should consider the bombardment of a different bromine compound, such as lithium bromide, which produces no such contaminating activity.

An important, and rather exceptional, example of slow neutron activation is that of tellurium. This element is a mixture of isotopes of which $^{130}_{52}Te$ gives the 1.2 day $^{131}_{52}Te$. The latter is of no particular interest except as the parent of $^{131}_{53}I$, which until recently was the radionuclide most widely used in medicine. The ^{131}I can easily be chemically separated from its parent and other Te radionuclides simultaneously formed by distillation (see Section 2.24).

At the other end of the scale, radionuclides formed by the fission of heavy elements in a reactor are such a complicated mixture, even after some delay period to allow short-lived products to decay away, that chemical processing is rarely economic. Among fission products which are easily accessible, mention should be made of $^{137}_{55}Cs$ (half-life 30 yr) which is much used as a gamma ray standard and in terabecquerel units for radiotherapy. The fission product $^{90}_{38}Sr$ (half-life 28 yr) constitutes an important environmental hazard, as after ingestion it is deposited in bone. Isotopes of the inert gases, such as $^{85}_{36}Kr$ and $^{133}_{54}Xe$, are easily separated by taking advantage of their ready volatility.

Fig. 2.16 shows the yield of nuclides of mass number A per 100 fissions initiated by slow neutrons in ^{235}U. It indicates, for instance, that nuclides of masses in the ranges 85 to 105 and 129 to 149 are produced to the extent of 1–5 nuclei per 100 fissions. Each fission product in the 'light' range is accompanied by one in the 'heavy' range, so that the total number of fission product

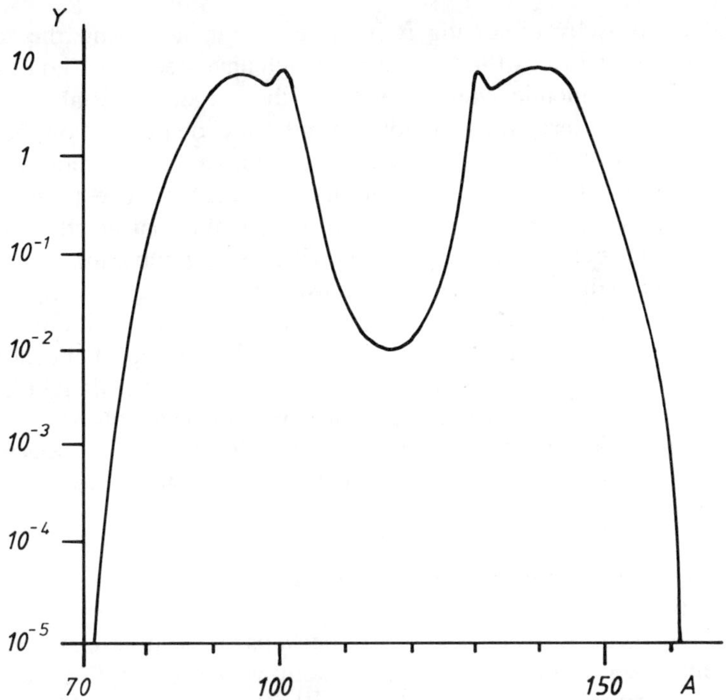

Fig. 2.16 Fission-yield curve (yield of atoms of mass number A, per 100 fissions).

nuclei is 200. At each mass number A there is a range of possible Z values. For instance, at mass number 131, for which the fission yield is 2.9%, all the following nuclides are formed: ^{131}Sn, ^{131}Sb, ^{131}Te and ^{131}I. The figure of 2.9 refers to the total numbers of these, on average, per 100 fissions. (Among the group, the individual yield of ^{131}Sb is the greatest.) All members of the group decay forming a chain which terminates in stable ^{131}Xe:

$$^{131}Sn \rightarrow {}^{131}Sb \rightarrow {}^{131}Te \rightarrow {}^{131}I \rightarrow {}^{131}Xe$$

A chain length of four radioactive members, like this, is fairly common. This means that the number of different nuclear species formed with yields of more than 1%, either directly or via decay of a precursor, is about 150. They extend over the elements selenium to rhodium and tin to promethium.

In between the extremes, wherein chemical separation is unnecessary or, alternatively, nearly impractical, there are a vast number of cases in which, standard processes of analytical chemistry have been modified to deal with the special problems entailed by the radioactive nature of the desired product. Most reactor or cyclotron targets from which radionuclides need to be separated weigh something in the region of milligrams to grams, whereas the radioactive product itself may weigh 10^{-12} g or less, and its presence is detectable only by the radiation emitted. It is usually easy to convert the target into a form (e.g. by solution in an aqueous medium) suitable for chemical analysis by dissolving it in water or an acid, and the following paragraphs review briefly the techniques which have been evolved to separate the desired radionuclide chemically from the target material.

A particularly interesting technique of radionuclide production, because of its wide application and practical use, is applicable when a radionuclide, as produced in the target, has a useful radioactive daughter. The 'parent' nuclide may not need to be separated from the target material at all, although it generally is so separated in the interests of achieving maximum purity of the daughter, and reduction of radiation hazard to personnel. Such a parent–daughter system is colloquially called a 'cow' or 'isotope generator' and the process of obtaining daughter from it, by a suitable physicochemical process, is called 'milking'. If the parent is relatively long-lived, the daughter may be 'milked' from it on several separate occasions. In what follows, we discuss separation methods many of which are suitable in either connection, i.e. separation of desired radionuclide directly from target materials, or from parent radioactive species.

2.22 Separation Technique Using Precipitation

Precipitation from aqueous solution is perhaps the simplest and most well known technique in inorganic chemical analysis. An example of its use as a milking procedure is in the separation of $^{137}_{56}$Ba from $^{137}_{35}$Cs. The latter is a 30 yr nuclide produced in high yield in uranium fission. Aqueous solutions of this nuclide in the form of its soluble chloride or nitrate are easily obtained. Decay of the $^{137}_{55}$Cs nuclei gives rise to $^{137}_{56}$Ba nuclei whose half-life is 2.5 min. The barium atoms formed by the decay remain in the aqueous solution as barium ions Ba^{2+}, and at first sight it might appear that they could form a precipitate of the 'insoluble' barium sulphate by the addition of sulphate ion. But, in fact, no such precipitate could form because the radioactive barium atoms would in practice be far too few to enable the solubility product of barium sulphate to be exceeded. However, if a few milligrams of ordinary inactive barium chloride are added to the solution, subsequent addition of a soluble sulphate will produce a precipitate in which the active barium atoms are incorporated. The precipitate may be separated from the remaining 'mother' solution by filtration or by centrifugation. It will be found to emit gamma rays of 0.66 MeV and half-life 2.5 min, while an identical activity grows again in the solution, in accordance with the law given in Section 4.9, as new barium nuclei ^{137}Ba are formed. In cases like this, the material added to the solution before carrying out the precipitation—in this case, an inactive barium salt—is called a 'carrier'. In the present example, a technical improvement can be effected by also adding inactive caesium, as a soluble salt, before precipitation. This step inhibits the contamination of the barium sulphate precipitate by adsorbed or occluded radioactive caesium atoms.

Since the caesium carrier has the effect of reducing the appearance of 'mother' activity in the precipitate it is referred to as a 'hold-back carrier'. Hold-back carriers are also very frequently used in order to reduce adsorption of otherwise 'carrier-free' activities on the walls of containing vessels, particularly of glass. Some carrier-free activities are very strongly adsorbed on to glass surfaces from aqueous solutions; the process is only partially reversible and the alkali metals in particular seem to be able to diffuse well inside the physical surface of the wall and to resist removal by all known cleaning agents.

2.23 Separation Technique Using Solvent Extraction

Many elements form compounds which tend to become distributed very unequally between pairs of immiscible solvents. For example, if gallium chloride is dissolved in dilute hydrochloric acid and the solution is put into contact with di-isopropyl ether, the gallium migrates preferentially into the ether. The process is slow if the two solvents are not agitated, but may be made quite rapid if they are vigorously shaken so that a mixture of small droplets, with large total surface area, is formed. After shaking, the liquids may be left to settle out again into two distinct layers; better still the mixture can be centrifuged. In general, the tendency is for the ratio of the concentrations of the *solute* (e.g. gallium) in the two phases to approach a fixed value, called the distribution constant. Distribution constant values are commonly in the range 100–10^8 in many practical cases. Often the distribution is very dependent on pH. For example, in the case of gallium the distribution between *water* and ether is such that at equilibrium more than 99% of the gallium would pass into the water layer.

Radioactive gallium ($^{68}_{31}$Ga, half-life 68 min) can easily be milked from its parent $^{68}_{32}$Ge (half-life 275 d) by the application of this technique. Bombardment of zinc with $^{3}_{2}$He or $^{4}_{2}$He particles yields, *inter alia*, $^{68}_{31}$Ge atoms and these are obtainable in acid solution simply by dissolving the target in hydrochloric acid. The daughter $^{68}_{31}$Ga atoms can then be removed to the extent of at least 99% from this solution by carrying out a 'solvent extraction' step with ether. After mechanical separation of the two liquids, the gallium atoms are easily 'back-extracted' into water by shaking the ether solution with water. The process may of course be repeated as often as required while sufficient of the parent activity remains in the hydrochloric acid solution.

2.24 Separation Technique Using Distillation

Here a good example is the preparation of the iodine isotope $^{131}_{53}$I ($t_{1/2}$ = 8.0 d) from the parent ^{131}Te ($t_{1/2}$ = 1.2 d) produced by neutron irradiation of tellurium. The target may be dissolved in nitric acid forming a solution of telluric acid. The daughter $^{131}_{53}$I atoms which grow in the solution are so volatile that gentle boiling of the solution enables them to pass into the gas phase—possibly as elemental iodine. On cooling the evolved gases and collecting them in water containing a mild reducing agent such as bisulphite ion the isotope is trapped as the involatile iodide ion I^-.

2.25 Separation Technique Using Gas Evolution

A somewhat similar process to the above is available for the inert gases. For instance, radium ($^{226}_{88}Ra$) solutions constantly evolve the inert gas $^{222}_{86}Rn$; the half-lives are 1650 yr and 3.85 d respectively. It is found that if a fine stream of bubbles of air—or any other fairly insoluble gas—is passed through a radium solution, the radon is quantitatively removed. If necessary, radon can be separated from the 'carrier' gas by condensing it in a liquid-air-cooled U-tube, and pumping away the more volatile carrier.

The U-tube can then be isolated from the pump and heated, when the radon will be released again. It may then be compressed into thin-walled glass or gold capillary tubes which can be sealed off at both ends. Such 'radon needles' or 'radon seeds' were used therapeutically for many years until superseded by radioactive gold ($^{198}_{79}Au$, $t_{1/2}$ = 2.7 d) and other artificial radionuclides. Similar techniques have been used for other radioactive inert gases, including the lighter fission-product gases krypton and xenon which are easily condensed in U-tubes containing activated charcoal at liquid-air temperature. A recent development is the production of ^{81m}Kr (half-life 13 s) which is generated by the decay of ^{81}Rb (half-life 4.7 h). The latter is produced by bombardment of bromine, in the form of sodium bromide targets, with helium ions.

2.26 Separation Technique Using Ion Exchangers

Important as the previously discussed methods are, the use of ion exchangers has, in recent years, proved exceptionally convenient and applicable to almost all known examples of genetically related radionuclides with half-lives greater than a few minutes. While some qualitative accounts of exchanger techniques are available in several textbooks, their application to radioactive materials does not seem to have been systematically covered. Yet the extremely low concentrations so often experienced in solutions of radioactive materials enables some simplifying assumptions to be made. We review, first, some of the elementary physical chemistry underlying the technique.

2.27 Structure of Cation and Anion Exchangers

The earliest ion exchangers known were the naturally occurring silicates called zeolites. The molecular structure of these is based on large rings of alternate silicon and oxygen atoms, the rings being superposed to form tubes so that the bulk material has a sponge-like form and is porous to water and aqueous solutions. While in pure silica the ratio of silicon to oxygen atoms is 1:2, in zeolites about one atom in six of the silicons is replaced by aluminium. Now silicon atoms have four valence electrons, but aluminium atoms have three. This means that in order to maintain electrical neutrality, additional positive ions must be present. Some at least of these are not very strongly bound in the alumino-silicate lattice, and can easily be replaced by others. For example, if a sample of zeolite in which the loosely bound ions are sodium, Na^+, is immersed in an aqueous solution of potassium chloride, a certain degree of interchange will occur, sodium ions leaving their sites near the walls of the pores, and entering the aqueous solution, the vacated sites being occupied by K^+ ions. Within any pore in the bulk solid it appears that a dynamic equilibrium is rapidly set up, which results in the ratio of sodium–potassium ions in the liquid phase (in the pores) being linearly dependent on the ratio of sodium to potassium ions in the available sites in the lattice. If the zeolite sample is a single large lump, of course it will take a very long time for the same state of equilibrium to be attained throughout, and if there is a large volume of aqueous solution surrounding the lump it will take a long time for its composition to change appreciably. If, however, the zeolite is in the form of a fine powder and it is stirred vigorously in the solution, it is possible for an equilibrium state to be attained over the whole system in a matter of minutes or even seconds. After equilibration, the solid and liquid can be separated, e.g. by using a filter, and it becomes possible in theory at least, to analyse each phase for its content of the two ions involved.

In general, calling the ions X_1 and X_2, and using the conventional square brackets to mean concentrations, we can express the equilibrium relationship by

$$\frac{[X_1]_E}{[X_1]_S} \div \frac{[X_2]_E}{[X_2]_S} = \text{constant}$$

or

$$\frac{[X_1]_E . [X_2]_S}{[X_1]_S . [X_2]_E} = k$$

the subscripts E and S referring to the solid exchanger and the solution respectively. This equation, as it stands, applies when both X_1 and X_2 are singly charged (or both doubly charged) ions but needs obvious modification if, say, a single and a doubly charged ion are exchanging.

Nowadays, many exchangers of this type are made synthetically, the products having much improved reliability in operation and reproducibility from one sample to another. In addition, exchangers with an organic base (e.g. the Amberlites) have been produced. These have a structure based on polymerised styrene, $C_6H_5—CH{=}CH_2$ (polystyrene) and related compounds, the molecular architecture resembling a very open basket work in three dimensions. By various chemical processes applied to the polymer, exchangeable atoms are introduced in more or less random positions in this lattice. For instance, the acidic organic groupings $-COOH$ and $-SO_2OH$ provide replaceable hydrogen atoms which can enter aqueous solutions as H^+ and be replaced by other positive ions. Alternatively, basic amino groups $-NH_2$ enable the exchanger to function anionically. Thus, after soaking the exchanger material with dilute hydrochloric acid, the $-NH_2$ groups from the loose combination $NH_2.HCl$, in which the chlorine atom is replaceable by negative ions, e.g. $B\gamma^-$, I^-, TcO_4^-, to name three in very common use.

Anion exchangers are in fact usually available in the 'chloride form', a term which expresses the fact that chlorine atoms are present in all available exchanging sites. With anion exchangers the same equilibrium considerations apply as to the cation exchangers.

An important development in the use of anion exchangers has been their use of ions of the electropositive elements (i.e. those which, in solution, normally give cations and are therefore unaffected by anion exchangers). However, many positive ions undergo 'complexing' reactions which yield negative ions. For example the reaction

$$Co^{2+} + 6Cl^- \rightleftharpoons CoCl_6^{4-}$$

implies that in aqueous solutions of a cobalt salt containing high concentrations of chloride ion, a fraction of the cobalt ions will be positively charged and the remainder will be 'complexed' and carry negative charges as the ion $CoCl_6^{4-}$. The latter will be very strongly exchanged by anion exchangers. We, thus, have the situation where the extent of exchange can be controlled by alteration in the concentration of chloride ion. A more complicated example of a complexing agent is lactic acid (2-hydroxypropanoic acid), a weak acid ionising in aqueous solution.

$$CH_3CH(OH)C(=O)OH \rightleftharpoons CH_3CH(OH)C(=O)O^- + H^+$$

The concentration of the lactate ion itself in any given solution can be controlled by adjustment of the pH (or hydrogen ion concentration). Thus, addition of the strong acid HCl increases the hydrogen ion concentration and depresses the lactate ion concentration. Now lactate ion complexes very strongly with, for example, the trivalent rare-earth ions. For example, the equlibrium gives negatively charged complex ions which also strongly interact with anion exchangers. Again the equilibria, and therefore the extent of exchange, are pH controlled, and slight differences in the equilibrium con-

$$3\left(CH_3CH(OH)C(=O)O^-\right) + La^{+++} \rightleftharpoons \left[\left(CH_3CH(O)C(=O)O\right)_3 La\right]^{---} + 3H^+$$

stants among the rate-earth elements La–Lu have been exploited to separate these otherwise very similar elements from each other.

2.28 Relative Affinities of Ions in Exchangers

If a solution containing a pair of ions X_1 and X_2 is allowed to come to equilibrium with an exchanger—either anionic or cationic—we may think of the two species X_1 and X_2 competing with each other for the available exchange sites. Which of the two wins the competition, i.e. is more *strongly absorbed* to use a common term, depends on a number of factors which are not completely understood. However, there are some simple guidelines. Between ions both of the same charge, the one with the larger radius is often the more strongly absorbed. This rule is however often misleading because some ions (e.g. Li^+) are strongly hydrated in aqueous solution, so in this case it is the effective radius of a complex $(Li.4H_2O)^+$ ion which is operative, not that of the bare Li^+ ion. Secondly, of a pair of ions of the same size, but different charges, the one with the higher charge is invariably the more strongly absorbed. Thus Ca^{2+} is much more strongly absorbed than K^+ although the latter ion has the larger radius.

A noteworthy fact is the very large range in k values possessed by different ions. Thus ions of the same charge and very nearly the same radius may have closely similar K values, but differences in charge and radius between a pair of ions may easily give rise to k values different by two or three orders of magnitude. Chloride ion Cl^- and iodide ion I^- which are closely similar in many chemical senses have very different affinities for many anion exchangers and it is found that iodide ion can be removed almost completely, from a solution containing both chloride and iodide, by equilibrating with a sufficient amount of almost any anion exchanger (in the chloride form).

The number of exchangeable atoms per kilogram of exchanger, of course, depends on the method of manufacture, but is generally of the order 10^{24} per kilogram so that at least one mole of ions can be exchanged per kilogram of exchanger. On the laboratory scale we often work with only a few grams of exchanger, which still will exchange up to say 10^{21} atoms. In terms of radioactive atoms 10^{21} is a very large number indeed, and the fact that the actual weights of radioactive materials used in practice are so extremely small makes for considerable simplification. Suppose that the ion X_1 referred to in Section 2.27 is a radioactive tracer—i.e. present in only submicroscopic amount, the other being present in macroscopic amount. We can rearrange the basic equation to read

$$\frac{[X_2]_E}{[X_1]_S} = k\frac{[X_2]_E}{[X_2]_S}$$

Now all the terms on the RHS are virtually constant. The actual amount of the X_2 ion present in either exchanger or solution will be virtually the same whether the tracer ion X_1 is completely exchanged or remains entirely in the solution. To take a concrete example, consider a solution of technetium, as TcO_4^- ions, in saline—i.e. the solution containing also macroscopic amounts of Cl^-. If, now, to the solution we add an anion exchanger in the chloride form, the system will tend towards an equilibrium in which a certain fraction of the TcO_4^- ions will exchange. But the change in the number of Cl^- ions in the solution and in the exchanger will be extremely small compared with the original numbers. We are therefore justified in writing

$$\frac{[Tc]_E}{[Tc]_S} = k'$$

and this equation will apply whatever the amount of technetium present in the system so long as it remains in the 'tracer' range. It should be noted however that the numerical value of k' can be changed at will over a large range, by changing the concentration of chloride ion in the solution (the $[X_2]_S$ term) either by adding more soldium chloride to the solution or else by diluting it. In general for a 'tracer' solute X_1 we write

$$[X_1]_E = k'[X_1]_S$$

We have not so far specified what are the units of concentration to be used in the above equations, and in the past several different conventions have been employed. If the equilibrium were that of a solute distributed between two immiscible solvents, it would be natural to measure the concentrations in moles per litre of the solutions. Although exchangers as such as solids, they are invariably used in a finely divided form, and it appears that exchanging solute ions can penetrate into them almost as freely as if they were liquids. We will assume then that we can express concentrations in the exchanger in moles of solute per litre, the real justification for the assumption of course being that its consequences are borne out in practice. Let us then consider a certain volume of solution of a tracer into which we introduce a fixed amount of an exchanger. In order to attain the equilibrium, in which part of the solute remains in the solution phase and the remainder undergoes exchange, it will be necessary to stir the solution and the exchanger together but we now imagine the exchanger to have settled to the bottom of the containing vessel where it occupies a volume V_B say. This space will contain a certain quantity of liquid identical in composition with the supernatant solution. The deposited exchanger is termed the 'bed'; fraction of it which is truly liquid, or the 'fractional void space' is denoted by i. The volume of entrained solution is therefore iV_B and the true volume of the exchanger is $(1 - i)V_B$. Let us denote the volume of the supernatant solution by V_S, the amount of solute in the supernatant m_S moles, and the amount of solute bound to the exchanger m_E moles. Then substituting in the equation governing the equilibrium we have

$$\frac{m_E}{(1 - i)V_B} = k' \frac{m_S}{V_S}$$

Therefore

$$\frac{m_E}{V_B} = k'(1 - i)\frac{m_S}{V_S}$$

The term $k'(1 - i)$ is clearly a constant for a particular exchanger and conditions of operation. It has been denoted by the symbol D_v and called the *distribution constant* for the exchange. (The subscript v is used as a reminder that the concentration terms are expressed as moles per unit volume, not per unit mass, as was the convention at one time.)

The total amount of solute present in the bed, m_B, is the sum of the amount truly exchanged and the amount in solution in the interstitial liquid; therefore

$$m_B = m_E + \left(\frac{m_S}{V_S}\right) iV_B$$

The *apparent* concentration of solute in the bed is then given by

$$\frac{m_B}{V_B} = \frac{m_E}{V_B} + \frac{m_S}{V_S}i$$

$$= \{k'(1 - i) + i\}\frac{m_S}{V_S}$$

$$= (D_v + i)\frac{m_S}{V_S}$$

Therefore

$$D_v = \left(\frac{m_B}{V_B}\Big/\frac{m_S}{V_S}\right) - i$$

Now for a radioactive solute, the activities A_B and A_S in the two phases (bed and supernatant respectively) are just proportional to the amounts m_B and m_S. The ratio in the bracket in this last equation is therefore equal to the ratio of count-rates obtained, under identical geometrical conditions, from equal-volume samples of bed and supernatant. The estimation of i may be made by taking activity measurements on supernatant and bed for a solute whose D_v value is known (preferably one whose D_v is zero). A typical value is 0.6, a value which shows incidentally that a considerable fraction of the entrained liquid is *within* the particles of exchanger, since these are usually approximately spherical and pack together closely. Extensive compilations of D_v values have been made; see, for example, '*Anion Exchange Studies of the Fission Products*', K.A. Kraus, F. Nelson, in *Proceedings of the International Conference on the Peaceful Uses of Atomic Energy*, Geneva, 1955, Vol. 7, 113 (1956).

It is instructive to consider the equilibrium relation as applied to the contents of the bed itself. For if the amount of solute in the solution in the interstitial volume is m'_S, we must have

$$\frac{m_E}{(1 - i)V_B} = k' \frac{m_S'}{iV_B}$$

Therefore

$$\frac{m_E}{m_S'} = \frac{k'(1 - i)}{i} = \frac{D_v}{i}$$

Now if we consider a portion of the bed which contains a unit amount of solute, of which a fraction a is truly exchanged, leaving the remainder $(1 - a)$ in the interstitial volume, it is evident that the ratios m_E/m_S' and $a(1 - a)$ are the same, and

$$\frac{a}{1 - a} = \frac{D_v}{i}$$

This form of the equilibrium relation, as applying to exchanging material *within* the bed, is used in the next section. There we shall make use of the fact that the equilibrium is *dynamic*, that is to say each solute ion frequently changes its state, sometimes being bound to a particular site on the exchanger and at other times existing in the 'free' state in the body of the solution.

2.29 Use of Exchangers in Ion-Exchange Columns

Mention has been made of the fact that the behaviour of some pairs of ions, e.g. Cl^- and I^-, are so different that analytical separation is possible simply by equilibrating a suitable exchanger with the mixed solution. Here the tracer iodide in a solution containing tracer iodide + macroscopic chloride can be almost entirely removed from the solution into the exchanger. (Notice that removal of tracer chloride from macroscopic iodide cannot be done in the same way.) Such 'one-stage' processes are rather rare and in most practical cases what is effectively a multistage process must be used because the D_v values for the pairs of ions considered are not very different.

Multistage processes are most conveniently carried out by passing a solution containing the mixture of ions whose separation is required, through a bed of exchanger held in a vertical tube (see Fig. 2.17). Care must be taken to ensure uniform packing of the exchanger, as the performance will be impaired if channels are available, allowing the liquid phase to bypass the bed. Differential exchange occurs as the solution encounters fresh exchanger in its passage through the bed. Eventually both species of ion pass through, the one whose distribution coefficient is the smaller percolating through first. The whole process is called *elution*; the liquid emerging at the base of the column is the *eluate*; a plot of concentration of either ion appearing in successive small samples of eluate, as a function of volume of eluate, is called an *elution curve*. The form of these curves obviously conveys much information about the performance of the column and it is very instructive to carry out the measure-

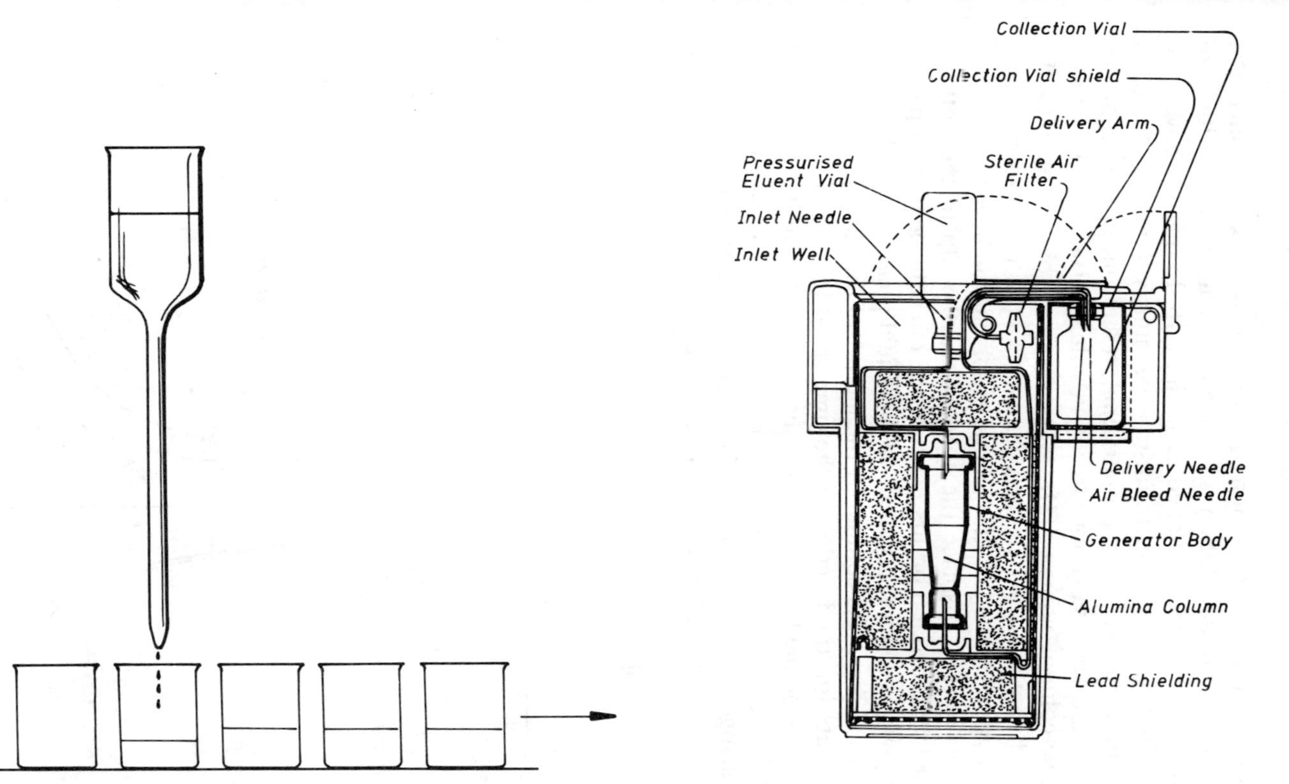

Fig. 2.17 Ion-exchange columns. (a) Exchange is supported above is a glass-wool plug in a column; the eluate is being caught in a series of fraction collector tubes. (b) The arrangement of a sterile generator for ^{99m}Tc (reproduced by permission of The Radiochemical Centre, Amersham.)

ments needed to be able to plot them and compare them with theoretical models. We cannot review this topic in detail but some important points are as follows:

(a) If a very small amount of solution containing a dissolved radionuclide is delivered onto a column and elution is carried out by following it with pure *solvent*, the elution curve is bell-shaped as in Fig. 2.18. In this curve we have plotted activity in Bq ml^{-1}, or in count-rate per unit volume of eluate, as ordinate since these quantities are proportional to the concentration. Notice that the scale on the abscissa is not directly in volumes such as ml, but as a number of 'free-column volumes' (f.c.v.), one f.c.v. being equal to the volume of liquid contained within the column. Clearly, if the solute is not absorbed at all by the exchanger, this volume of solvent will have to be passed down the column just to displace the solute from the top to the bottom. However if partial absorption occurs, each solute ion will spend only part of the time during which elution proceeds in the liquid phase, and only in this time will it be swept down the column by the motion of the eluting liquid. For the 'average' solute ion the fraction of the time spent in motion will just be $(1 - a)$, where a is defined as in the previous section; during the fraction of the time a it will be combined in one or other of the many available reactive sites in the exchanger. The ion will therefore reach the bottom of the column when the volume of eluting liquid is F_p say, where

$$\frac{1(\text{f.c.v.})}{F_p(\text{f.c.v.})} = 1 - a$$

i.e.

$$F_p = \frac{1}{1 - a} = \frac{D_v}{i} + 1$$

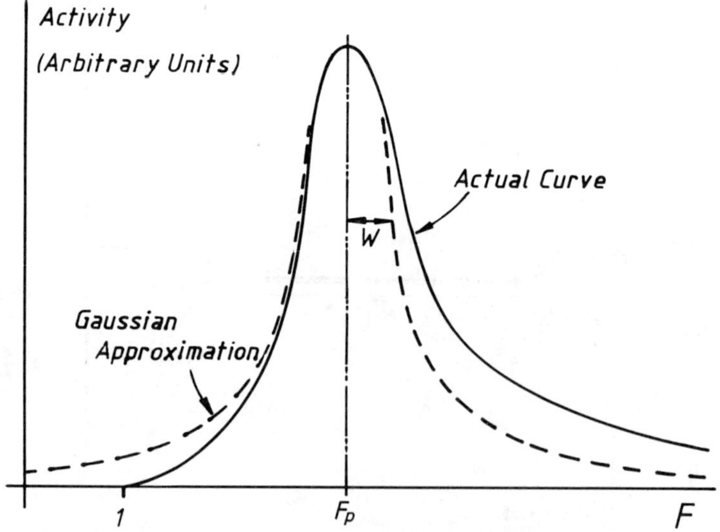

Fig. 2.18 Elution curve for a small sample of solute initially at the top of a column.

F_p, therefore, represents the volume of eluting liquid which is necessary to bring the elution curve to its peak value, as the behaviour of the solute ions will tend to cluster about that of the average.

(b) The 'spread' in the elution curve peak is due to a number of effects such as statistical fluctuations in the times spent by individual solute ions in the absorbed state, and in the dissolved state in the column, and the fact that they are subject to diffusion effects while being carried along in the solution. A detailed analysis of the processes involved would be very tedious, but an empirical treatment based on the idea of 'theoretical plates' is quite useful. We consider the column to be modelled by a collection of thin individual discs through which the solute percolates in steps, one disc at a time. Between each step, equilibration occurs in each disc. Thus, to begin with, the solute will be in the first disc and will be distributed as a fraction a in the absorbed state, and $(1 - a)$ in the solution. An element of solvent is now added, displacing the fraction $(1 - a)$ downwards into the second disc (where it re-equilibrates, giving a fraction $a(1 - a)$ in the absorbed state and $(1 - a) \times (1 - a)$ in the solution).

The new solvent in the first disc will produce a distribution in which a solute fraction a^2 is absorbed and $a(1 - a)$ is produced in the solution. Continuation of the process leads to the result that when the eluting liquid has gone down to the pth plate, the nth plate $(n < p)$ will contain an amount of solute roughly equal to the nth term in the binomial expansion of $[a + (1 - a)]^p$. (More exactly the terms in the middle of this expansion give the sum of the amount of solute combined with exchanger in each plate, plus the amount in the liquid portion of the previous plate.) If the column contains just p theoretical plates and we continue to elute beyond the stage when solute reaches the last plate, we will get an elution curve whose shape is given approximately by the $(p + 1)$th term in the expansions of

$$[a + (1 - a)]^p, \qquad [a + (1 - a)]^{p+1}, \qquad [a + (1 - a)]^{p+2}, \ldots \text{ etc.}$$

For large values of p these successive terms are quite closely approximated by a gaussian. Near the peak the activity corresponding to the passage of F free-column volumes is then proportional to $\exp\{-(F_p - F)^2/2W\}$, where W is the 'standard deviation' of the gaussian, and given by

$$W^2 = a/\{p(1 - a)^2\} = \frac{D_v}{i}\left(\frac{D_v}{i} + 1\right)\bigg/ p$$

Fig. 2.18 shows a typical elution curve (full line) and the approximating gaussian, normalised to the same peak height. The FWHM is, of course, 2.36 W, and

$$p = 0.18\,\frac{D_v}{i}\left(\frac{D_v}{i} + 1\right)\bigg/ (\text{FWHM})^2$$

Fig. 2.19 shows in a diagrammatic form the contents of the first six theoretical plates in a column, as the solute is developed downwards. Initially the

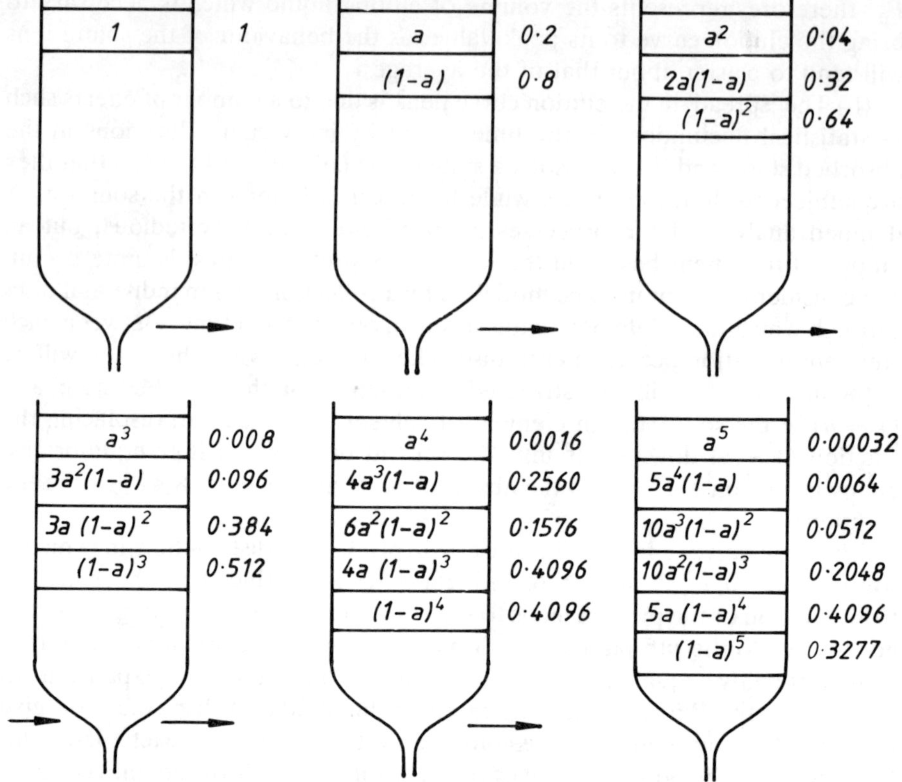

Fig. 2.19 Contents of theoretical plates as elution proceeds.

solute is all in the first plate at the top of the column. The algebraic expressions within the columns give the binomial expansions, each being evaluated for $a = 0.2$ in the accompanying figures. The value of p for any column can be found experimentally by determining an elution curve for a solute with known D_v and fitting a normal curve to this. It is usually sufficient to evaluate the FWHM for the curve and substitute in the above equation. For a gaussian curve, we find on reference to tables that the area enclosed between the ordinates at $(F_p - W)$ and $(F_p + W)$ is about 68% of the total, while that between $(F_p - 2W)$ and $(F_p + 2W)$ is about 95%. For laboratory size columns with fine-grain exchangers (say 400 mesh) a theoretical plate is of the order 1 mm deep, so that a column 100 mm long will give about 100 theoretical plates. For a solute having $D_v = 9$, and for $i = 0.6$ one finds $D_v/i = 15$, so that 16 free-colume volumes are needed to reach the peak in the elution curve.

The value of W is $\sqrt{(15 \times 16/100)} \approx 1.5$ free-column volumes. This means that some 68% of the solute will be eluted after about 14.5 free-column volumes have been passed through the column and before 17.5 have been passed through. More than 95% of the solute will be eluted in the volume between about 13 and 19 free-column volumes (plus or minus two standard deviations). In practice, one might allow for the approximate nature of the calculation by including one or two more free-column volumes on either side if one wanted to collect such a large fraction of the solute.

(c) On the tracer scale (i.e. very low solute concentrations), it is a fair assumption that *two* solutes will behave independently of one another, so that a *composite* elution curve for the pair will have the form shown in Fig. 2.20. Clearly in such cases there will be a particular value for F, the volume of eluate used, which will give an optimum analytical separation, shown by the vertical dashed ordinate in the diagram. Once the D_v values for the two solutes are known, the calculation of the elution volume for optimum separation is not difficult (although requiring numerical methods). In practice, one rarely has to make this calculation because for the vast majority of practical cases exchangers are available for which the D_v values (for a certain pH of eluate) are different by about one order of magnitude, and the elution curve for the more strongly absorbed component of the mixture is so 'spread out' that essentially pure samples of the less strongly absorbed component can easily be eluted nearly quantitatively. Nevertheless, some contamination of component A with B is inevitable, and it becomes a question of convention as to when 'break-through' of B, producing significant contamination of A, occurs. A practical case is that of the solutes MoO_4^{2-} and TcO_4^-, in which a small amount of molybdate ion as a carrier for the isotope ^{99}Mo, is 'loaded' onto a column of alumina. The D_v value for MoO_4^{2-} is very high, for the eluate used (perhaps at least 10^3), but as decay proceeds pertechnetate ions are formed within the narrow 'band' of absorbed molybdate. For pertechnetate ion in the same eluting liquid, the D_v value is quite low (near unity). On eluting the column with enough liquid to remove virtually all the pertechnetate (say 5–10 free-column volumes), the position of the molybdenum is

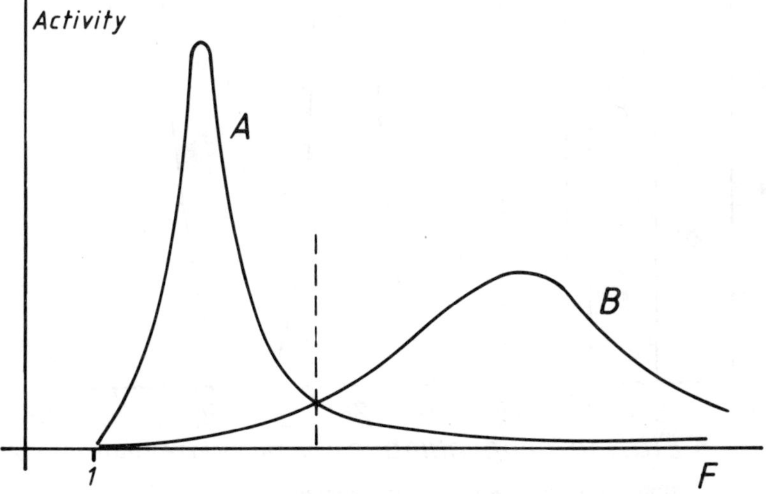

Fig. 2.20 Elution curves for two solutes, A and B.

moved only very slightly. One may then wait for a few hours to allow more technetium to grow, and elute again with virtually the same result as before (except of course that the molybdenum has itself decayed slightly so that the potential yield of technetium will be smaller). The column may in fact be stripped of technetium many times before the shape of the technetium elution curve is appreciably changed. It is, of course, good practice to check after a column has been in use for a few days whether significant molybdenum breakthrough is occurring.

(d) We have assumed in the above that the composition of the eluting liquid remains constant throughout the process; for a complex mixture of many components this might lead to a requirement of extremely large volumes due to the excessive spreading of the more strongly absorbed components. A way out of this difficulty is to progressively change the composition of the eluting liquid in such a way that the D_v values for the solutes change during the elution (although still remaining in the same relative order). Such a change might be, for instance, a gradual change of pH.

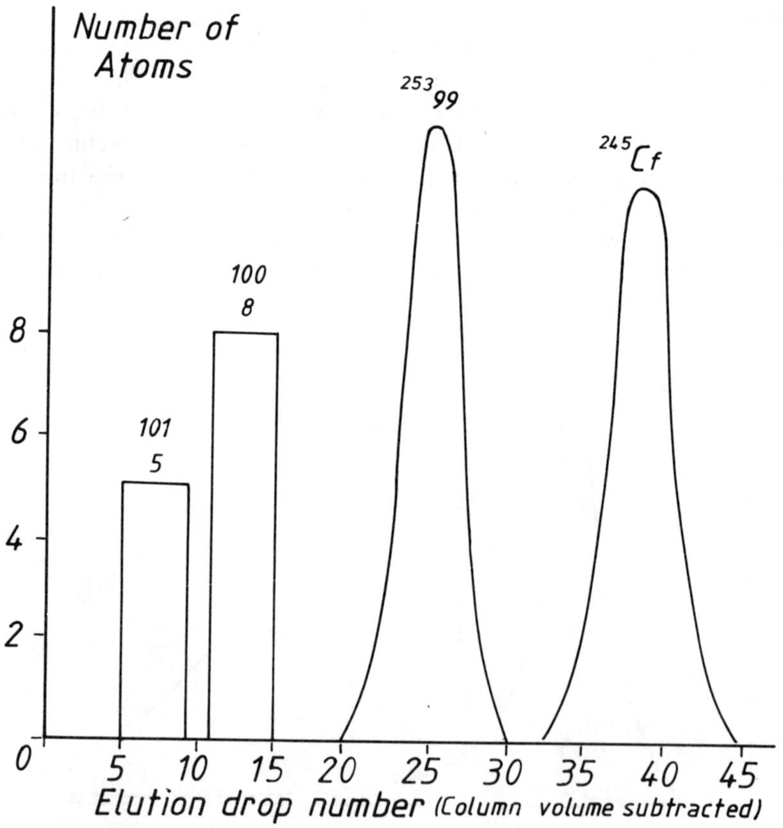

Fig. 2.21 Elution curves for transuranium elements. (Reproduced from A. Ghiorso *et al.*, *Physical Review*, **98**, 1518 (1955) by permission.)

The separation of some complex mixtures, such as the rare-earths (lanthanides) and transuranium elements (actinides) is facilitated by working at elevated temperature (87 °). The ultimate in ion exchange of actinide elements was possibly the characterisation of five atoms of lawrencium (Lw, element 101) (Fig. 2.21). Heavier elements are probably too short-lived to be handled by conventional chemical techniques.

2.30 Other Columnar Separation Techniques

The theoretical-plate conception of the mode of action of ion-exchange columns has an exact parallel in the theory of distillation columns—and in some industrial systems distillation columns have been built which incorporate individual 'plates' in which liquid-gas equilibration occurs with the gas passing upwards and the liquid downwards—hence the term theoretical plate.' On a laboratory scale many substances including biological materials can be separated by differential transport in cellulosic materials such as paper. Even single strands of cotton have been used on a very small scale. A popular system is that of 'ascending paper chromatography' in which a suitable solvent is made to ascend a strip of filter paper upon which a mixture of substances to be separated has been loaded as a single spot at the bottom of the paper.

In this kind of analytical separation, it is not convenient to have succeeding components of the mixture arriving at the top of the paper strip, but rather to stop the process after sufficient time has elapsed for the components to have separated out into individual 'bands' part-way up. In suitable cases, the positions of the separated bands can be made visible (if the solutes are not already strongly coloured) by spraying the paper with a developing agent which gives a colour reaction. Scanning with a narrowly collimated radiation detector is another possibility in the case of radioactive materials. In these methods a useful experimental parameter is the so-called R_f value, which is the ratio of the distance travelled by the solute and that travelled by the transporting solvent (the latter being indicated by the leading edge of the ascending liquid). It is formally equivalent to the reciprocal of the F_p value of an ion-exchange column, since this measures the number of free-column volumes of solvent which displaces the solute through a space of 1 free-column volume.

In still other systems transport of the solute mixture is achieved, not through movement of a solvent, but through the action of an electrical potential applied across the ends of the strip. This is the electrophoretic strip method, particularly useful for proteins.

Gases are easily separated by differential absorption on charcoal and other materials in the dry state, or with a liquid film on a finely divided substrate. Here, temperature control of the absorbing material is an important controlling factor. Hydrogen or helium gas can be used to maintain the flow in a tube containing the absorbing material. All known gases including the inert gases (He, Ne, Ar, Kr, Xe, Rn) are easily and completely separable using this gas-chromatography technique.

2.31 Summary of Available Radionuclides

We give here in tabular form (Table 2.2) a list of the most commonly encountered radionuclides, with their main radiation characteristics. For other nuclides the manufacturers' catalogues should be consulted. For each nuclide a production reaction is suggested; many radionuclides can be produced in several alternative ways, but the one given may be preferred on grounds of convenience or of giving a carrier-free product. Half-lives and λ values are as commonly accepted and are accurate enough for most purposes. Only the main radiations are mentioned, and in particular the X-rays which follow electron capture are not included (although they may add significantly to tissue doses). Energies of beta particles (E_{max}) and gamma-rays are in MeV and are followed by an abundance figure in per cent. An entry (ann) signifies annihilation radiation; if a disintegration is accompanied by the production of just one positron, giving two annihilation quanta, the abundance entry is 200. Multiple entries are separated by a solidus (/).

2.32 Radiopharmaceuticals

The production of radionuclides having suitable radiation characteristics and half-life and in a state of radiochemical purity is, of course, only the first step along the road to obtaining useful clinical data. If a radioactive preparation is to be administered to a human subject, it becomes a radiopharmaceutical and, as with all pharmaceuticals, must comply with certain criteria before being acceptable. One obvious criterion is that the preparation must not be chemically toxic. Fortunately, this requirement is easily met as the radionuclide is generally of very high specific acitivity and the actual mass administered, whatever its chemical nature, is well below toxic levels. For instance 100 MBq of ^{99m}Tc as sodium pertechnetate weighs only about 1×10^{-9} g. As far as the toxicity of the radionuclide itself is concerned, there is, therefore, no objection to incorporating it into food or drink, or into isotonic saline for injection. But there are other criteria which are by no means so easily met; these are the requirements of sterility and of apyrogenicity. There is a wide variety of living organisms—many of them harmless—which can survive in media otherwise suitable for injection, but which should be excluded whenever possible. Sterilisation of injection material can be achieved either by heat treatment (e.g. a temperature of 120 °C maintained for 15 min) or by filtration through filters with very fine pores (e.g. 0.22 μm which are commercially available). Unfortunately a medium cannot be tested for its sterility in under a day or so after such treatment, as this time is necessary for the incubation of the organisms. There is also the difficulty that some preparations (e.g. albumin labelled with ^{99m}Tc) cannot be heat-treated because this causes aggregation with the formation of unwanted colloidal particles. Since many radionuclides in current use have short half-lives sterility testing cannot be done until *after* the radiopharmaceutical has had to be used—a situation which runs counter to good pharmaceutical practice. The only way out of the difficulty, and admittedly a not wholly satisfactory one, is to establish procedures for radiopharmaceuticals which include the most rigorous quality checks, use of sterility tested reagents and equipment, and testing of samples reserved from batch production at regular intervals. The hope is that if a laboratory can pass such retrospective tests over a period and does not relax its standards of cleanliness and 'good housekeeping', it will continue to produce acceptable materials. The presence of pyrogens, which are probably largely the metabolic products of living organisms, in pharmaceutical preparations creates a similar difficulty as these can be detected only some hours post-injection into experimental animals. They cannot be removed by filtration or destroyed by heating except at inconveniently high temperatures. It is, of course, necessary to keep these dangers in a true perspective. It is certain that the rate of occurrence of adverse reactions in hospital patients following administration of radiopharmaceuticals has been extremely low over at least the past 20 years, although there is no statistical evidence on which to base numerical estimates. On the other hand, human beings are more than mere statistics and risks must be acknowledged and reduced to a minimum, having regard to the positive value which a diagnostic test may bring to the patient.

With this background, one can appreciate that radiopharmaceutical preparation has tended to develop along two lines which have been determined by the timescale imposed by the half-life of the radionuclide used. We can distinguish radiopharmaceuticals of long half-life, such as the vast range of products containing ^{14}C or ^{3}H which are commercially available. Their production has involved the syntheses of extremely complicated molecules containing these radionuclides, starting with such simple compounds as carbon dioxide and tritiated water. On a shorter timescale ^{57}Co (half-life 270 d) and ^{58}Co (half-life 72 d) have been incorporated into vitamin B_{12} (although here the sterility conditions are not so stringent since the vitamin is generally administered orally). Among shorter-lived radionuclides mention should be made of ^{67}Ga, marketed as the citrate, and many isotopes of iodine, available as iodide and also as compounds with various proteins, hippuran, etc. For a full list, the manufacturers' catalogues must be consulted. For our present purposes we put them in a single category of radiopharmaceuticals which are essentially 'off the shelf'.

The other main category is the radiopharmaceuticals which belong to the short timescale and must be prepared essentially 'on-site'. In the past this has meant that each institute such as a hospital using radiopharmaceuticals has set up its own production laboratory, although the present tendency in the UK is to organise supplies of radiopharmaceuticals from one centre to a group of hospitals in the near neighbourhood. A noticeable development of the preparative methods themselves has been the progressive replacement of lengthy procedures, which were sometimes unreliable unless carried out by very skilled hands, by very much quicker and simpler ones. The newer methods are of course based on the older ones but in the light of experience it has been possible to identify the essential physical conditions, such as the range of pH

Table 2.2 Characteristics of Some Commonly Used Radionuclides

Radio-nuclide	Production reaction	$t_{1/2}$	λ	Type of decay	β^- or β^+ emission E_{max} and abundance	γ emission E_γ and abundance
^{3}H	$^{6}Li(n,\alpha)$	12.26 yr	0.0565 yr^{-1}	β^-	0.0186; 100	
^{11}C	$^{11}B(p,n)$	20.34 min	0.0341 min^{-1}	β^+,EC	0.097; 99+	0.511 (ann); 200
^{14}C	$^{14}N(n,p)$	5730 yr	121 × 10^{-6} yr^{-1}	β^-	0.156; 100	
^{13}N	$^{10}B(d,n)$	9.96 min	0.0696 min^{-1}	β^+	1.20; 100	0.511 (ann); 200
^{15}O	$^{14}N(d,n)$ $^{16}O(^{3}He,d)$	2.07 min	0.335 min^{-1}	β^+	1.74; 100	0.511 (ann); 200
^{18}F	$^{18}O(p,n)$	109.7 min	0.00632 min^{-1}	β^+,EC	0.635; 97	0.511 (ann); 194
^{22}Na	$^{24}Mg(d,\alpha)$	2.62 yr	0.265 yr^{-1}	β^+,EC,γ	0.545; 99.95	0.511 (ann); 180 / 1.275; 100
^{24}Na	$^{23}Na(n,\gamma)$	14.96 h	0.0463 h^{-1}	β^-,γ	1.389; 99.997	1.369; 100 / 2.754; 100
^{32}P	$^{31}P(n,\gamma)$ $^{35}Cl(N,\alpha)$	14.3 d	0.0485 d^{-1}	β^-	1.71; 100	
^{35}S	$^{35}Cl(n,p)$	87.1 d	0.00796 d^{-1}	β^-	0.167; 100	
^{42}K	$^{41}K(n,\gamma)$	12.36 h	0.0561 h^{-1}	β^-,γ	3.52; 100	1.524; 18 / others
^{47}Ca	$^{46}Ca(n,\gamma)$	4.535 d	0.153 d^{-1}	β^-,γ	0.67; 82 / 1.98; 18	1.31; 76 / 0.49; 6 / 0.81; 6
^{47}Sc	^{47}Ca decay	3.43 d	0.202 d^{-1}	β^-,γ	0.44; 73 / 0.62; 27	0.16; 73
^{51}Cr	$^{51}V(d,2n)$	27.8 d	0.0249 d^{-1}	EC,γ		0.320; 9
^{55}Fe	$^{55}Mn(d,2n)$	3.0 yr	0.231 yr^{-1}	EC		
^{59}Fe	$^{59}Co(d,2p)$	45.6 d	0.0152 d^{-1}	EC,γ	0.475; 53 / 0.455; 45	1.292; 43 / 1.095; 56
^{57}Co	$^{56}Fe(d,n)$	270 d	0.00257 d^{-1}	EC,γ		0.122; 87 / 0.136; 11 / others
^{58}Co	$^{55}Mn(\alpha,n)$	71.3 d	0.00972 d^{-1}	EC,β^+	0.474; 15	0.511 (ann); 30 / 0.810; 99 / others
^{60}Co	$^{59}Co(n,\gamma)$	5.263 yr	0.132 yr^{-1}	β^-,γ	0.314; 99+	1.173; 100 / 1.322; 100
^{65}Zn	$^{65}Cu(d,2n)$	245 d	0.00283 d^{-1}	EC,β^+,γ	0.327; 1.7	0.511 (ann); 3.4 / 1.115; 49
^{67}Ga	$^{67}Zn(d,2n)$	77.9 h	0.00890 h^{-1}	EC,γ		0.093; 40 / 0.184; 24 / 0.296; 22
^{75}Se	$^{75}As(d,2n)$	120.4 d	0.00576 d^{-1}	EC,γ		0.136; 57 / 0.265; 60 / others

^{81m}Kr	^{81}Rb decay	13 s	0.0533 s^{-1}	LT		0.191; 65
^{85}Kr	Fission	10.76 yr	0.0644 yr^{-1}	β^-,γ	0.67;	0.514; 41
^{81}Rb	$^{79}Br(\alpha,2n)$	4.7 h	0.1475 h^{-1}	EC,β^+	1.03; 13	0.511 (ann); 26 /
^{86}Rb	$^{85}Rb(n,\gamma)$	18.66 d	0.0371 d^{-1}	β^-,γ	1.78;	1.078; 8.8
^{87m}Sr	^{87}Y decay	2.83 h	0.245 h^{-1}	IT,EC		0.388; 80
^{87}Y	$^{86}Sr(d,n)$	80 h	0.0087 h^{-1}	EC,β^+	0.7; 0.3	0.483;
^{99}Mo	$^{98}Mo(n,\gamma)$	66.7 h	0.0104 h^{-1}	β^-,γ	1.23;	0.740; 12 / 0.181; 7 / others
^{99m}Tc	^{99}Mo decay	6.05 h	0.1146 h^{-1}	IT		0.140; 98.4 / others
^{113}Sn	$^{112}Cd(\alpha,3n)$	115 d	0.00603 d^{-1}	EC		0.255; 1.8
^{111}In	$^{10°}Ag(\alpha,2n)$	2.84 d	0.244 d^{-1}	EC,γ		0 247; 94 / 0.173; 89
^{113}In	^{113}Sn decay	99.8 min	0.00695 min^{-1}	IT		0.393; 64
^{123}I	$^{121}Sb(\alpha,2n)$	13.3 h	0.0521 h^{-1}	EC		0.159; 83
^{125}I	$^{123}Sb(\alpha,2n)$	60.2 d	0.0115 d^{-1}	EC		0.035; 7
^{131}I	Fission, ^{131}Te decay	8.05 d	0.0861 d^{-1}	β^-,γ	0.806; 6 / 0.606;	0.364; 82 / 0.637; 6.8 / 0.284; 5.4 / others
^{125}Ke	$^{122}Te(\alpha,n)$	16.8 h	0.0413 h^{-1}	EC,γ		0.118; / 0.242;
^{127}Xe	$^{127}I(p,n)$	36.4 d	0.0190 d^{-1}	EC,γ		0.203; 65 / 0.172; 22 / 0.375; 20 / others
^{133}Xe	Fission	5.27 d	0.132 d^{-1}	β^-,γ	0.346;	0.081; 37 / 0.031; 47(X-rays)
^{137}Cs	Fission	30.0 yr	0.231 yr^{-1}	β	0.52;	
^{137m}Ba	^{137}Cs decay	2.60 min	0.267 min^{-1}	IT,γ		0.662;
^{144}Ce	Fission	284 d	0.00244 d^{-1}	β^-,γ	0.31;	0.080; 2 / 0.134;
^{198}Au	$^{197}Au(n,\gamma)$	2.70 d	0.257 d^{-1}	β^-,γ	0.962;	0.412; 95 / 0.676; 1 / others
^{203}Pb	$^{203}Tl(d,2n)$	52.1 h	0.0133 h^{-1}	EC,γ		0.279; 81 / 0.401; 5 / others

values, temperature of reaction, reaction times, etc., and to find reagents which can be pre-sterilised and prepacked in sealed ampoules containing the exact amount necessary for an individual preparation. In this way the pharmacist can be provided with a kit of reagents which need only to be mixed with a solution of the required radionuclide in a specified order and treated by the most elementary processes (e.g. stirring, dipping into boiling water for a specified time) to ensure a satisfactory end-product. The technique, of course, presupposes a radionuclide supply which is also sterile and pyrogen-free and here again the manufacture of sterile generators and their proper use in local laboratories are matters for strict control. Incidentally, it has been found that many commercially available ^{99}Mo–^{99m}Tc generators are themselves quite good filters for suspended organisms; thus a population of, for example, *E. coli* in a solution used as an eluting agent may be reduced by a factor of up to 10^6 on passage through an exchange column. It would be fair to say that the development of radiopharmaceutical techniques requires much expertise but that once a satisfactory kit is available as a result, the procedure

Table 2.3 Radiopharmaceuticals

Acronyms: DTPA = diethylenetriamine pentaacetic acid; DMSA = dimercaptosuccinic acid; MDP = methylenediphosphonic acid; HSA = human serum albumin.

Name or acronym	Half-life of radionuclide	Clinical use
^{75}Se selenomethionine	127 d	Pancreas and liver imaging
^{59}Fe citrate	45.1 d	Haematological studies
^{51}Cr RBC*	27.8 d	Haematological studies
^{131}I iodide	8.0 d	Thyroid imaging
^{131}I hippuran	8.0 d	Renography
^{111}In bleomycin	2.84 d	Tumour localisation
^{111}In DTPA complex (and others)	2.84 d	Cerebrospinal fluid study (after lumbar injection)
^{67}Ga citrate	78 h	Tumour localisation
^{99m}Tc, pertechnetate	6.05 h	Brain, thyroid imaging
^{99m}Tc denatured RBC*	6.05 h	Spleen imaging
^{99m}Tc DTPA complex*	6.05 h	Dynamic renal function
^{99m}Tc DMSA complex*	6.05 h	Renal imaging
^{99m}Tc Fe-ascrobate*	6.05 h	Renal imaging
^{99m}Tc MDP complex*	6.05 h	Skeletal (bone) imaging
^{99m}Tc HSA*	6.05 h	Blood-pool imaging
^{99m}Tc MAA*	6.05 h	Lung imaging
^{99m}Tc sulphur colloid*	6.05 h	Liver imaging
^{113m}In chloride	104 min	Brain, placenta imaging

itself should be technically extremely simple, although of course it will require the exact following of a set of precise instructions.

We give here in tabular form (Table 2.3) a list of what seem to be the most frequently used radiopharmaceuticals today. It is by no means exhaustive. The table omits the inert gases, ^{81m}Kr, ^{125}Xe, ^{127}Xe and ^{133}Xe in particular which are used for lung ventilation studies (half-lives 13 s, 18 h, 36.4 d and 5.27 d respectively) since they have to be handled by special methods. The compounds distinguished in the table by asterisks have been the subjects of detailed investigations by the Hospital Physicists' Association. Their preparation is outlined in the following pages.

4. RELIABLE METHODS OF PREPARING ^{99m}Tc-RADIOPHARMACEUTICALS AND ^{51}Cr-RED CELLS*

4.1 THE TESTING OF METHODS

Methods of preparing some common ^{99m}Tc-radiopharmaceuticals are set out in detail in this section. These methods have been selected for reliability at a number of U.K. laboratories; each individual method being tested in at least five laboratories. The methods listed are the ones most highly recommended by the testing laboratories from an original list of over 100. The present list is by no means complete and the authors are well aware that there are other well tried methods for each radiopharmaceutical. The reason for the absence of particular methods is either that they were not found to be reproducible at all the testing laboratories or were not included in the original list which was compiled from information on routine methods of preparation used by radiopharmaceutical laboratories throughout the UK.

The design aims for a good method of preparation have been set out [in section 2.3.1] earlier. In practice, it is often impossible to achieve all these aims, but one should attempt to achieve as many as possible. Therefore, whilst the authors strongly recommend that preparations should be made in closed sterile vials wherever possible, some methods are included because they were preferred by the testing laboratories on the basis of good quality and reproducibility of the product.

4.2 GENERAL PROCEDURES

A small stock of very clean glassware should be set aside for preparing radiopharmaceuticals. This and the followers for magnetic stirrers should be sterilised by heating. This can be done by the hospital pharmacy or by a central sterile supplies department.

Where possible the sterile pyrogen-free reagents required for a preparation should be available in ampoules for single use. This can be done in the hospital pharmacy and each ampoule should contain sufficient material for a single injection or the small number of injections based on one generator elution.

pH measurements should not be made by inserting the measuring system into the preparation. Instead a sample of the solution should be withdrawn from the container with a sterile disposable syringe and the pH measured with a pH meter or calibrated pH paper.

Sometimes it is necessary to use a resin column to remove free TcO_4^- from a preparation. Columns which have a height of 100 mm and a diameter of 10 mm are suitable. These should be filled with a resin such as Dowex 1 × 8–50–100 mesh or Amberlite IRA 400 Cl^-. The column should be sterilised by heat and the resin sterilised by autoclaving. Before autoclaving, the resin should be thoroughly washed with pyrogen-free water to remove any pyrogenic substances present. The column should be filled with the resin under aseptic conditions and at least 50 ml water for injection BP poured down the column to remove any further pyrogenic substances and to settle the resin bed before use.

All ingredients in the methods which follow must be sterile and pyrogen-free.

* Reproduced from *The Hospital Preparation of Radiopharmaceuticals*, Scientific Report Series, No. 16, by permission of the Hospital Physicists Association.

4.3 ^{99m}Tc-DENATURED RED CELLS

METHOD I

Materials:

Physiological saline.
Sodium citrate solution (2.8%) in water for injection.
Stannous chloride ($SnCl_2.2H_2O$) (4.4 mmol/l) (freshly prepared).
Closed sterile container (20 ml).

Method:

1. Put 2 ml sodium citrate solution into container.
2. Take 8 ml whole blood from the patient and add to the container immediately after venipuncture.
3. Mix thoroughly, but gently.
4. Centrifuge at 1000g for 5 mins.
5. Remove plasma and replace with an equal volume of saline.
6. Mix thoroughly, but gently.
7. Centrifuge at 1000g for 5 mins.
8. Remove supernatant.
9. Add 0.5 ml stannous chloride solution.
10. Incubate in water bath at 37 °C for 5 mins.
11. Add 1–3 ml $^{99m}TcO_4^-$ solution of required activity in saline.
12. Incubate in water bath at 37 °C for 5 mins.
13. Incubate in water bath at 55–57 °C for 10 mins (to denature the red cells).
14. Add saline to re-suspend rell cells.
15. Inject cells intravenously into the same patient.

4.3a ^{99m}Tc-DENATURED RED CELLS

METHOD II

Materials:

Physiological saline.
Acid citrate dextrose (ACD) solution.
(30 g trisodium citrate, 0.15 g sodium dihydrogen phosphate and 2 g dextrose in 1 l water for injection).
Stannous chloride ($SnCl_2.2H_2O$) sodium (4.4 mmol/l ACD solution) (freshly prepared).
Closed sterile container (20 ml).

Method:

1. Put 4 ml ACD solution into container.
2. Add 16 ml whole blood from the patient immediately after venipuncture.
3. Mix thoroughly.
4. Centrifuge at 1000g for 5 mins.
5. Remove plasma supernatant.
6. Add 0.1 ml $^{99m}TcO_4^-$ solution of required radioactivity in saline.
7. Incubate in water bath at 37 °C for 30 mins.
8. Add 0.7 ml stannous chloride solution.
9. Agitate gently for 15 mins.
10. Add 10 ml saline and mix gently.
11. Centrifuge at 1000g for 5 mins.
12. Remove supernatant.
13. Add 5 ml saline and mix gently.
14. Repeat steps 11 and 12 until little radioactivity is present in the supernatant.
15. Re-suspend red cells with 5 ml saline.
16. Inject red cells into the same patient.

4.4 ^{99m}Tc-DMSA (2,3-DIMERCAPTOSUCCINIC ACID)

Materials:

DMSA* solution (0.55 g/l water) (freshly prepared).
Stannous chloride ($SnCl_2.2H_2O$) solution (0.9 mmol/l) (freshly prepared).
Two closed sterile vials.

* Sigma Labs. Ltd.

Method:
1. Add equal volumes of the DMSA solution and the stannous chloride solution to a vial, immediately both solutions are ready.
2. Take required $^{99m}TcO_4^-$ activity in saline and add to the second vial, noting the volume.
3. Add the mixture prepared in step 1 to the radioactivity. The volume added should be half the volume of the $^{99m}TcO_4^-$ solution present.
4. Allow to stand for 10 mins.
5. Inject into the patient within 20 mins.

4.5 ^{99m}Tc-DTPA COMPLEX

METHOD

Materials:
Molar sterile sodium hydroxide.
Stannous chloride solution ($SnCl_2.2H_2O$) (4.4 mmol/l M HCL) (freshly prepared).
DTPA Penta Sodium salt.
Sterile Dowex resin column.
Millipore filter 0.22 μm.
Sterile beaker with magnetic stirrer.
Closed sterile vial.

Method:
1. Add to beaker:
   (*a*) 4 ml $^{99m}TcO_4^-$ solution in saline.
   (*b*) 1 ml stannous chloride solution.
2. Stir for 5 mins at 750 rev/min.
3. Add 400 mg DTPA.
4. Stir for 2 mins at 750 rev/min.
5. Adjust pH to 6.0–7.5 with sodium hydroxide.
6. Pass solution through the resin column.
7. Pass solution through Millipore filter into vial.
8. Assay total radioactivity.

4.5a ^{99m}Tc-DTPA COMPLEX

METHOD II

Materials:
10M sterile hydrochloric acid.
Molar sterile sodium hydroxide.
Ferrous sulphate ($FeSO_4.7H_2O$).
DTPA Penta Sodium salt.
Shaker.
Closed sterile vial.

Method:
1. Add to vial:
   (*a*) 4 ml $^{99m}TcO_4^-$ solution in saline.
   (*b*) 20 mg ferrous sulphate.
   (*c*) 1 ml hydrochloric acid.
   (*d*) 30 mg DTPA.
2. Shake vigorously for 2 mins.
3. Adjust pH to 3.0 with sodium hydroxide.
4. Shake for 2–3 mins.
5. Adjust pH to 5.5–6.0 with hydrochloric acid.
6. Assay total radioactivity.

4.6 ^{99m}Tc-HUMAN SERUM ALBUMIN (HSA)

METHOD I

Materials:

Physiological saline.
Molar sterile hydrochloric acid.
Molar sterile sodium hydroxide.
10% human serum albumin solution.*
Ferric chloride ($FeCl_3.6H_2O$) (37 mmol/l water for injection) (freshly prepared).
Ascorbic acid (114 mmol/l water for injection) (freshly prepared).
Sterile Dowex resin column.
Millipore filter 0.22 μm.
Two closed sterile vials.
Sterile beaker.

Method:

1. Add to sterile vial:
   (*a*) 2 ml $^{99m}TcO_4^-$ solution in saline.
   (*b*) 0.5 ml ferric chloride solution (5 mg).
   (*c*) 5 ml ascorbic acid solution (10 mg).
2. Adjust to pH 7.5 by adding sodium hydroxide drop by drop until dark violet-brown colour is seen. (Green coloration denotes excess sodium hydroxide has been added).
3. Add 0.1 ml albumin solution (10 mg).
4. Adjust pH to 1.5 with hydrochloric acid.
5. Adjust pH to 9.5 with sodium hydroxide.
6. Adjust pH to 1.5 with hydrochloric acid.
7. Adjust pH to 5.0 with sodium hydroxide.
8. Pass solution through resin column and wash through with 8 ml saline into beaker.
9. Pass solution through Millipore filter into vial.
10. Assay total radioactivity.

4.6a ^{99m}Tc-HUMAN SERUM ALBUMIN (HSA)

METHOD II

Materials:

Physiological saline.
Molar sterile hydrochloric acid.
Molar sterile sodium hydroxide.
10% human serum albumin solution.*
Ferric chloride ($FeCl_3.6H_2O$) (74 mmol/l saline) (freshly prepared).
Ascorbic acid (114 mmol/l saline) (freshly prepared).
Sodium acetate ($NaC_2H_3O_2$) (332 mmol/l saline) (freshly prepared).
Glacial acetic acid (190 mmol/l saline) (freshly prepared).
Sterile Dowex resin column.
Millipore filter 0.22 μm.
Closed sterile vial.
Sterile beaker.
Two sterile universal bottles with magnetic stirrers.

Method:

1. Prepare acetate buffer pH 5.6 by adding together 45 ml sodium acetate solution and 5 ml glacial acetic acid.

* Lister Institute, Elstree, Herts.

2. Add 4 ml $^{99m}TcO_4^-$ solution in saline (of known radioactivity) to universal bottle with stirrer.
3. Stir at speed 750 rev/min and add:
   (*a*) two drops of hydrochloric acid.
   (*b*) 0.5 ml ferric chloride solution (10 mg).
   (*c*) 0.5 ml ascorbic acid solution (10 mg).
4. Adjust pH to 5.0–5.6 with sodium hydroxide.
5. Add 0.5 ml acetate buffer pH 5.6.
6. Take contents of bottle up into a syringe.
7. Add to other universal bottle containing a magnetic stirrer:
   (*a*) 2 ml albumin solution (200 mg).
   (*b*) 0.5 ml acetate buffer pH 5.6,
   and stir at speed 750 rev/min.
8. Add ^{99m}Tc-solution drop by drop from the syringe.
9. Adjust pH to 2.5 by adding hydrochloric acid.
10. Pass solution through resin column and collect in beaker.
11. Adjust pH to 5.6 with sodium hydroxide.
12. Pass solution through Millipore filter into vial.
13. Assay total radioactivity.

4.7 ^{99m}Tc-IRON ASCORBATE COMPLEX

Materials:

Molar sterile sodium hydroxide.
Ferric chloride ($FeCl_3.6H_2O$) (18.5 mmol/l water for injection) (freshly prepared).
Ascorbic acid (114 mmol/l water for injection) (freshly prepared).
Two closed sterile vials.
Shaker.
Millipore filter 0.22 μm.

Method:

1. Add to vial:
   (*a*) 2 ml $^{99m}TcO_4^-$ solution in saline.
   (*b*) 1 ml ascorbic acid (20 mg).
   (*c*) 1 ml ferric chloride solution (5 mg).
2. Adjust to pH 7.5 with sodium hydroxide.
3. Shake for 15 mins—a very deep purple colour is obtained.
4. Pass solution through Millipore filter into vial.
5. Assay total radioactivity.

4.8 ^{99m}Tc-MACROAGGREGATED ALBUMIN (MAA)

METHOD I

Materials:

Physiological saline.
4.3% human serum albumin solution*.
Ferric chloride ($FeCl_3.6H_2O$))74 mmol/l saline) (freshly prepared).
Ascorbic acid (114 mmol/l saline) (freshly prepared).
Hydrochloric acid 0.15N (freshly prepared).
Sodium hydroxide 0.15N (pH 7) (freshly prepared).
Millipore filter 0.22 μm.
Closed sterile container (20 ml).
Closed sterile vial.
Sterile beaker with magnetic stirrer.

* Prepared from 10% human serum albumin solution obtained from Lister Institute, Elstree, Herts.

Method:

1. Add 4 ml $^{99m}TcO_4^-$ solution in saline to beaker with stirrer.
2. Stir at speed 750 rev/min and add:
   (*a*) 0.5 ml ascorbic acid (10 mg).
   (*b*) 0.5 ml ferric chloride (10 mg).
   (*c*) 0.85 ml sodium hydroxide.
   (*d*) 0.5 ml albumin solution.
   (*e*) 3.0 ml hydrochloric acid.
3. Leave for 15 mins.
4. Adjust pH to 5.0 with 2.3 ml sodium hydroxide.
5. Pass solution through 0.22 μm Millipore filter into container.
6. Make up to 20 ml with saline.
7. Shake in water bath at 75 °C for 20 mins.
8. Allow to cool.
9. Centrifuge at 2000 rpm for 5 mins.
10. Remove supernatant and re-suspend in 10 ml saline.
11. Pass through 25G needle into vial.
12. Assay total radioactivity.

4.8a ^{99m}Tc-MACROAGGREGATED ALBUMIN (MAA)

METHOD II

Materials:

Sterile 2M hydrochloric acid.
Sodium thiosulphate (255 mmol/l water for injection).
Zirconium sulphate (160 mmol/l water for injection).
Glacial acetic acid.
Sterile 10M sodium hydroxide.
Water for injection.
10% human serum albumin solution.*
Two closed sterile vials.

Method:

1. Prepare sodium acetate buffer pH 7.0 by adding together glacial acetic acid (9% by volume), sodium hydroxide (14.5% by volume) and water (76.5% by volume).
2. Add to vial:
   (*a*) 2 ml $^{99m}TcO_4^-$ solution in saline.
   (*b*) 0.2 ml sodium thiosulphate solution.
   (*c*) 0.2 ml hydrochloric acid.
3. Heat in boiling water for 3–5 mins.
4. Add 0.2 ml zirconium sulphate solution.
5. Add 0.2 ml sodium acetate buffer.
6. Add 0.1 ml albumin solution, injecting it in below the level of the liquid.
7. Autoclave for 5 mins.
8. Remove and shake well.
9. Draw up into a syringe and deliver into vial using a 25G needle.
10. Assay total radioactivity.

4.8b ^{99m}Tc-MACROAGGREGATED ALBUMIN (MAA)

METHOD III

Materials:

Stannous chloride ($SnCl_2.2H_2O$) solution (27 mmol/l M HCL) (freshly prepared).
Sodium acetate anhydride solution (255 mmol/l water for injection).

* Lister Institute, Elstree, Herts.

10% human serum albumin solution.*
Sterile 250 ml container with magnetic stirrer.
Closed sterile vials.
Large beaker.

Method:
1. Add 94 ml sodium acetate anhydride solution to the large container and autoclave it.
2. Allow to cool.
3. Add 4 ml stannous chloride solution and 2 ml albumin solution.
4. Heat 600 ml water in a large beaker to 95 °C.
5. Place the 250 ml container in the beaker of hot water and allow it to warm up for 5 mins. At the end of this period, the water temperature will have fallen to about 80 °C.
6. Stir for 12 mins.
7. Remove the container from the beaker and continue to stir for 8 mins during cooling.
8. Transfer 1 ml aliquots of the solution to the vials using a syringe and 25G needle.
9. Freeze dry or keep the vials frozen until required.

   To prepare individual doses:
10. Add required $^{99m}TcO_4^-$ activity in saline solution (1–5 ml) to a vial and shake the vial.

4.9 ^{99m}Tc-SULPHUR COLLOID

Materials:
Sterile M/5 Hydrochloric acid.
Sodium thiosulphate (50 mmol/l water for injection).
Sodium perrhennate (18 mmol/l water for injection).
6% Dextran solution.
Phosphate buffer pH 7.4 (152g disodium phosphate ($Na_2HPO_4.7H_2O$) and 9 g sodium phosphate ($NaH_2PO_4.H_2O$) in 1 l water for injection).
Closed sterile container (20 ml).

Method:
1. Add to container:
   (*a*) 6 ml $^{99m}TcO_4^-$ solution in saline.
   (*b*) 1.5 ml sodium thiosulphate solution.
   (*c*) 0.5 ml sodium perrhennate solution.
   (*d*) 2 ml Dextran solution.
   (*e*) 1 ml hydrochloric acid.
2. Heat for 3 mins in a boiling water bath (100 °C).
3. Add 2 ml phosphate buffer to container and shake.
4. Assay total radioactivity.

4.10 ^{51}Cr-RED CELLS

(*Recommended Method of the International Committee for Standardisation in Haematology*)

Materials:
Acid citrate dextrose (ACD) solution (22g trisodium citrate dihydride, 8g citric acid and 25g dextrose in 1 l water for injection).
Physiological saline.
Closed sterile container (20 ml).

Method:
1. Put 1 ml ACD solution into container.
2. Add 10 ml whole blood from the patient immediately after venipuncture.
3. Mix thoroughly.
4. Centrifuge at 1000g for 5 mins.

* Lister Institute, Elstree, Herts.

5. Remove supernatant plasma, noting the volume.
6. Add sodium (^{51}Cr) chromate of required activity in 1 ml saline and mix gently.
7. Incubate for 15 mins at a temperature between 15 °C and 37 °C.
8. Re-suspend the red cells by the addition of saline whose volume equals that of the removed supernatant.
9. Centrifuge at 1000g for 5 mins and remove the supernatant.
10. Repeat steps 8 and 9 once again.
11. Re-suspend red cells in saline.
12. Inject red cells into the same patient.

4.11 ACKNOWLEDGEMENTS

The authors gratefully acknowledge the help given by the following Medical Physics Departments who evaluated the efficiency and reproducibility of the methods.

Aberdeen.
Bristol.
Cardiff.
Coventry.
Edinburgh.
Glasgow—Department of Clinical Physics and Bio-Engineering.
Guy's Hospital, London.
Leeds—Nuclear Medicine Department.
Manchester.
Newcastle.
Northwick Park Hospital, Middlesex.
Plymouth.
Southampton.
St. Bathlomew's Hospital, London.
University College Hospital, London.

So many departments and so many individuals helped in the evaluation that some may have been overlooked and the authors apologise and offer their grateful thanks to them also if this should be the case.

2.33 Exercises

1. The mass of the nuclide ^4_2He is 4.002 604 unified mass units. From this figure and those given for protons and neutrons in Section 2.1, calculate the energy which would be released if a kilogram of helium were synthesised from hydrogen and neutrons.
2. In an atom of mass M, possessing a single electron of mass m, the binding energy of the electron is given by the formula

$$E = -2\pi^2 \frac{Mm}{M+m} \cdot \frac{e^4Z^2}{h^3} \cdot \left(\frac{1}{n^2}\right)$$

where $n = 1$ for the K shell, $n = 2$ for the L shell, and so on. (Note that the $-$ sign means that a positive amount of energy must be given to the electron to enable it to leave the atom with zero kinetic energy.) Assume that the formula can be applied to find the K and L level energies in tungsten ($Z = 74$). Hence estimate roughly the wavelength of tungsten K X-rays.
3. The characteristic K X-rays from calcium have a wavelength of about 3.4 Å (0.34 nm). It has been shown that mesons with a single negative

charge and a mass of 300 electron masses may occupy K and L levels in atoms, and that their transitions from one level to another give rise to 'mesonic' X-rays. The measurement of these has been suggested as a means of determining whole-body calcium. Estimate their wavelength.

4. From the semi-empirical mass formula (Section 2.8) find the masses of $^{51}_{22}Ti$, $^{51}_{23}V$ and $^{51}_{24}Cr$. The principal decay mode of ^{51}V is by beta emission with $E_{max} = 2.13$ MeV, each beta particle being followed by a gamma ray of 0.32 MeV. Are these data in agreement with the mass difference? Calculate the energy loss associated with the electron capture process in ^{51}Cr. Could this nuclide emit positrons?
5. Refer to the example given in Section 2.19. The neutron capture cross-section of ^{198}Au is 26 000 barns. Find the rate of production of ^{199}Au at the end of the bombardment time (10^4 s).

   If a 1 g sample of ^{197}Au is bombarded in a flux density of 10^{16} neutrons/m^2/s, is the flux density significantly reduced by the sample itself? Would there be a significantly self-shielding effect if 1 kg of gold were being bombarded with a flux density of 10^{18} neutrons/m^2/s?
6. From the formula given in Section 2.20 find the range of (a) 40 MeV, (b) 30 MeV helium ions in (1) copper and (2) lead. (Note that the difference between the ranges at 30 and at 40 MeV gives the effective target thickness for those reactions which can take place in that energy interval.) Calculate the barrier height appropriate for helium ions incident upon lead. In what thickness of a lead target can nuclear reactions occur when it is bombarded with 40 MeV helium ions?
7. A small volume of solution contains two solutes which have the same concentration. They are to be separated by ion exchange in a column. The D_v values are 1C and 20 respectively with the exchanger, which has a void space i of 0.7. The solution is placed on the top of a column which has 50 theoretical plates and elution is carried out with pure solvent. Estimate the volume of eluate which would be used when the solution emerging from the bottom of the column has equal concentrations of the two solutes. Estimate the fraction of the less absorbed solute which has passed through the column at that point and the fraction of the other solute which contaminates it.

   Would it be better to (a) use a column for which $P = 100$, or (b) use an exchanger for which the D_v values were 10 and 40, instead of the one actually used?

2.34 Bibliography

BURCHAM, W.E. (1963) *An Introduction to Nuclear Physics* (Longmans, London)

FRIEDLANDER, G., Kennedy, J. and MILLER, J.M. (1964) *Nuclear and Radiochemistry* (John Wiley, New York)

WANG, Y. (1969) *Handbook of Radioactive Nuclides* (Chemical Rubber Company, Cleveland, Ohio)

LEDERER, C.M., HOLLANDER, J.M. and PERLMAN, I., *Table of Isotopes* (John Wiley, New York)
TUBIS, M. and WOLF, W. (1976) *Radiopharmacy* (John Wiley, New York)
Radionuclide Topic Group of Hospital Physicists' Association *The Hospital Preparation of Radiopharmaceuticals* (F.J. MILNER, Brentford, Middlesex)
LAWRENCE, J.H. (ed.) *Recent advances in Nuclear Medicine* Vol. 3. (1971) Chapter 3 by LAUGHLIN J.S., TILBURY R.S. and DAHL, J.R.

3
Radiation Measurement

INTRODUCTION

3.1 General Principles

In this chapter we consider the interaction processes by which alpha particles (helium ions), beta particles and electromagnetic radiations (X-rays and gamma-rays) affect matter. By far the most significant process, from our point of view, is the one called *ionisation*, in which the energy of the particle or photon, as it passes through matter, is used up in dislodging electrons from their atomic, or molecular, 'orbits'. Each such ionising event produces a pair of ions, viz. the dislodged electron and the residual positively charged atom (or molecule). Chemical changes often occur as a result of ionisation, and also through *excitation* processes in which the disturbed electron is not completely ejected from its parent atom or molecule, the latter remaining for a measurable time in an 'excited state'. The study of these chemical effects is the domain of radiation chemistry, which is clearly one of the basic sciences underlying radiobiology. It is convenient to begin with a discussion of ionisation processes since most forms of detector depend on them, although in some kinds of scintillation detector, excitation processes are paramount (see Section 3.18).

Detection techniques have been developed quantitatively to provide us with the means for measuring (a) the strength of a radioactive source, i.e. the numbers of events in which particles and/or photons are emitted per unit time, or the disintegration rate, and (b) the total energy being carried by the emitted particles and/or photons. The first of these is the activity of the source in becquerels; it is not always the same as the number of particles or photons emitted per second as this depends also on the details of the decay scheme. For a given source, the product of the number of disintegrations and the energy radiated (in all forms) per disintegration is clearly the total energy radiated, and the surrounding medium is said to be *exposed* to this radiant energy. When the medium is air, experiment shows that an average of about 34 eV of energy needs to be absorbed in order to create one pair of ions. Since it is easier in practice to measure the amount of ionisation in a gas than the energy absorbed by it, exposure is traditionally measured in terms of the ionisation it produces. The unit of exposure is the roentgen (symbol R) whose somewhat long-winded official definition (see Appendix 1) means that exposure which results in the production of ions, of each sign, to the extent of 2.58×10^{-4} coulombs(C) of charge released per kilogram of dry air being irradiated. Necessary conditions are that the sample of irradiated air is at standard temperature and pressure and is surrounded only by air. The latter condition means that ions formed in any wall material in which the sample is contained must be prevented from entering the sample, unless the wall material is equivalent to air itself in some appropriate way. The rather curious figure of 2.58×10^{-4} C arises from the fact that the original definition specified a charge of 1 electrostatic unit produced in 1 cubic centimetre, or 0.001 293 g, of air.

Another closely related radiation unit is the *gray*. This is a measure of the energy actually absorbed from the beam of radiation as it passes through any given medium. The gray is that energy absorption which amounts to 1 joule absorbed per kilogram of absorbing material; its symbol is Gy. Before the SI units were introduced the energy absorption unit was the rad, which was 0.01 Gy, so chosen because the passage of 1 R of radiation into air leads to the absorption of approximately 1 rad. More exactly, it follows from the definition of the roentgen and the fact that the electrons and positive ions carry charges of $\pm 1.6 \times 10^{-19}$ C that an exposure if 1 R leads to the production of $2.58 \times 10^{-4}/(1.6 \times 10^{-19})$ ion pairs per kilogram of air. Using the experimental figure of 33.9 eV needed on average to create an ion pair, we find the energy absorbed to be

$$\frac{2.58 \times 10^{-4} \times 33.9}{1.6 \times 10^{-19}} \quad \text{eV kg}^{-1}$$

$$= 2.58 \times 10^{-4} \times 33.9 \text{ J kg}^{-1}$$

$$= 0.008\,76 \text{ Gy}$$

The approximate numerical equality between the absorption of energy in rads from a radiation exposure in roentgens (in air) seems in the past to have led to confusion as to the meaning of ‘absorption’ and ‘exposure’. The two quantities are in fact quite distinct. Moreover, in media containing elements of higher atomic number than oxygen and nitrogen, an exposure to 1 R of radiation may give an energy absorption much greater than 0.008 76 Gy. This topic is taken up again in Section 3.11.

The biological effects of radiation constitute a topic too wide to be afforded more than a passing reference here. Whenever such effects are measured quantitatively, the question may arise as to whether their extent is proportional to the dose of absorbed radiation in grays. It may in fact depend on other factors such as the fractionation of the dose in time, or the density of ionisation along the path of the radiation beam. For radiation protection purposes it has been found convenient therefore to introduce the concept of the *dose equivalent*, whose unit is the sievert (symbol Sv). For ‘hard’ X-rays and electrons as the radiation considered, the dose equivalent of 1 Gy is just 1 Sv, but in general a dose of D Gy has a dose equivalent of H Sv where

$$H = DQN,$$

and Q is a numerical factor which allows for different biological effectiveness of different radiations, or 'quality', and N is a numerical factor which relates to other conditions such as fractionation. At the moment, when accurate data are unavailable, Q is taken to be 10 for neutrons and 20 for alpha particles, and N is taken to be 1. Since Q and N are pure numbers it follows that sieverts are, like grays, measured in joules per kilogram (J kg^{-1}).

We turn now to the ionisation phenomena shown by alpha particles, beta particles and X and gamma photons in passing through matter. We begin with a brief discussion of the dynamics of collision processes in general and then apply the general equations in turn to the particle collisions and to the photon interactions.

COLLISION PROCESSES

3.2 General Dynamics of Collisions

We will assume that a particle with kinetic energy E and momentum p makes a collision with a second particle, viz. an electron, which is initially at rest. The collision is *elastic* if the total energy of the particles after collision is equal to the kinetic energy of the incoming particle. This is a good enough approximation if the energies involved are all rather high. Also the principle of the conservation of momentum tells us that (i) the sum of the momenta of the particles after collision in the direction of motion of the incoming particle is equal to the momentum of that particle, and (ii) the sum of the momenta of the particles after collision at right angles to that direction is zero. We denote by θ the angle through which the incoming particle is deflected, and by ϕ the angle which defines the path of the struck particle, as shown in Fig. 3.1. (We may note as an aside that if the particles are identical, i.e. both are electrons, we have no experimental or theoretical means of distinguishing which is which after collision.)

The implicit assumption in the above that the struck particle (electron) is free to move introduces no significant error if the electron is in one of the outer orbits in an atom. But, particularly when K electrons are involved, we should remember that some, or even all, of the kinetic energy available from the incoming particle may be used up in overcoming the binding force on the electron, before it becomes free to move. Strictly speaking it is the total of kinetic energy plus binding energy which is conserved in the collision. Similarly the momentum conservation equations should take into account any momentum carried by the rest of the atom. Except in one important case (photoelectric absorption of X- and gamma-rays) the momentum imparted to

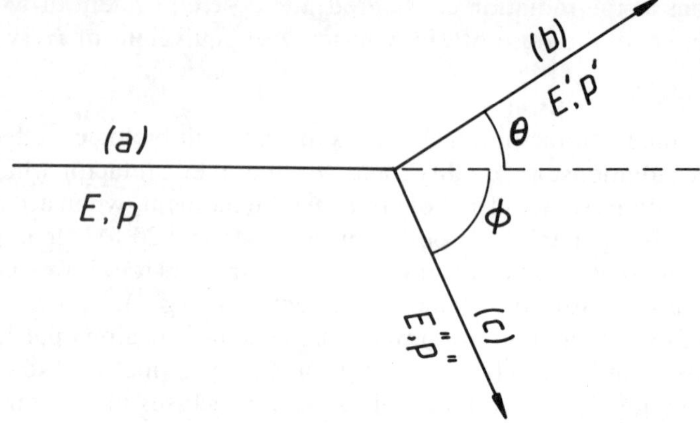

Fig. 3.1 Two-particle collision: (a) is the path of the incident particle, (b) that of the scattered particle, and (c) that of the struck particle.

the atom itself is negligible. For simplicity we will, therefore, regard the struck electron as being 'free', and allow for the effect of atomic binding only when it is sufficiently strong to affect the kinetic energy of the electron.

We can then write, with reference to Fig. 3.1, an equation representing energy conservation and two representing momentum conservation, as follows:

$$E = E' + E''$$
$$p = p' \cos(\theta) + p'' \cos(\phi)$$
$$0 = p' \sin(\theta) - p'' \sin(\phi)$$

Before we apply these equations to actual cases we can note two extreme kinds of collision:

(i) In the extreme when p'' is nearly zero, i.e. there is very little transfer of momentum, and therefore energy, to the struck particle, we must have that $p' \sin(\theta)$ is small so that θ is nearly zero; this means that the incident particle is scarcely deflected at all and we can speak of a 'glancing collision'.

(ii) In the other extreme case we find the condition for there to be a maximum transfer of energy to the struck particle. Solving the second and third equations for $p'' \cos(\phi)$ and $p'' \sin(\phi)$ respectively, squaring, and adding, we find, since $\cos^2(\phi) + \sin^2(\phi) = 1$, that

$$(p'')^2 = p^2 - 2pp' \cos(\theta) + (p')^2$$

It can be shown from this that p'' can have a maximum value when $\theta = 0$ or π (depending on whether p' is positive or negative). If this is so, it is clear from the third equation above that $\phi = 0$. We speak of this as a 'head-on' collision, the struck particle moving 'in the forward direction'. The incident particle must also move in this direction, or in the opposite direction. The maximum momentum transfer, $p - p'$ or p'', cannot be determined solely from the momentum equations, but can be found when the energy equation is taken into account.

In between the two extremes, when θ and ϕ are not zero, the struck particle may have any momentum value (and therefore any energy value) between zero and some upper limit. Its direction always has a component in the forward direction, i.e. $\phi \leq 90°$. If a large number of collisions are examined it will be found that there is a continuous distribution of events in which certain scattering angles θ and certain energies of the scattered particle occur. The relative frequency with which scattering at different angles occurs is difficult to calculate and depends on the nature of the particles, but in general one can say that when high energies are involved the distribution of scattered particles tends to be more in the forward direction than elsewhere. It tends to the isotropic (i.e. equally distributed in all directions) for low-energy particles, and when the struck particle is very massive.

3.3 Collisions of Alpha Particles with Electrons

We now write the energy and momentum equations using expressions for energy and momentum appropriate for alpha particles colliding with electrons. Since alpha particles (at least those from radioactive sources) do not travel very fast, we can use the non-relativistic expressions (Appendix 2)

$$E_\alpha = \tfrac{1}{2}m_\alpha v_\alpha^2$$
$$E_\alpha = \tfrac{1}{2}m_\alpha (v_\alpha')^2$$

for the energy before and after collision. The momenta will be

$$p_\alpha = m_\alpha v_\alpha$$
$$p_\alpha' = m_\alpha v_\alpha'$$

For the electron we will assume that ordinary classical mechanics is also applicable, and write

$$E'' = \tfrac{1}{2}m_e v_e^2$$
$$p'' = m_e v_e$$

Substituting in the three basic equations we get

$$\tfrac{1}{2}m_\alpha v_e^2 = \tfrac{1}{2}m_\alpha (v_\alpha')^2 + \tfrac{1}{2}m_e v_e^2$$
$$m_\alpha v_\alpha = m_\alpha v_\alpha' \cos(\theta) + m_e v_e \cos(\phi)$$
$$0 = m_\alpha v_\alpha' \sin(\theta) - m_e v_e \sin(\phi)$$

From these we can now calculate, for example, the energy and momentum of electrons produced at any specified angle. As shown in the previous section, the electrons will have a range of momenta, and therefore energies, from zero to some limiting values, depending of course on the initial alpha particle energy. The maximum energy transfer occurs in a head-on collision, for which

$\theta = \phi = 0$; calling this ΔE_α we easily find

$$\frac{\Delta E_\alpha}{E_\alpha} = \frac{\frac{1}{2}m_\alpha\{v_\alpha^2 - (v_\alpha')^2\}}{\frac{1}{2}m_\alpha v_\alpha^2}$$

$$= \frac{(v_\alpha - v_\alpha')(v_\alpha + v_\alpha')}{v_\alpha^2}$$

Now we know that the alpha particle is very much more massive than the electron, so its velocity will change very little in a collision. We may therefore replace $v_\alpha + v_\alpha'$ in the above by $2v_\alpha$ and get approximately

$$\frac{\Delta E_\alpha}{E_\alpha} = \frac{2(v_\alpha - v_\alpha')}{v_\alpha}$$

$$= \frac{2(m_\alpha v_\alpha - m_\alpha v_\alpha')}{m_\alpha v_\alpha}$$

$$= \frac{2m_e v_e}{m_\alpha v_\alpha}$$

the last step depending on the use of the second basic equation with cos (θ) and cos (ϕ) both equal to 1. We can put the last equation into a more useful form as follows. Squaring both sides, we have

$$\left(\frac{\Delta E_\alpha}{E_\alpha}\right)^2 = \frac{4m_e^2 v_e^2}{m_\alpha^2 v_\alpha^2}$$

$$= \frac{4m_e}{m_\alpha} \cdot \frac{\frac{1}{2}m_e v_e^2}{\frac{1}{2}m_\alpha v_\alpha^2}$$

$$= \frac{4m_e}{m_\alpha} \cdot \frac{E_e}{E_\alpha}$$

$$= \frac{4m_e}{m_\alpha} \cdot \frac{\Delta E_\alpha}{E_\alpha}$$

Therefore

$$\frac{\Delta E_\alpha}{E_\alpha} = \frac{4m_e}{m_\alpha}$$

Now the mass of the electron is only about 1.3×10^{-4} times that of the alpha particle. Thus for, say, a 5 MeV alpha particle the maximum energy a projected electron can have is $4 \times 1.3 \times 10^{-4} \times 5$ MeV or about 2.7 keV. This figure justifies our initial assumption that classical mechanics is adequate for the electron, its velocity being only 5×10^5 m s^{-1} or about 0.1% of the velocity of light. The maximum electron energy is thus only about 0.5% of that of the alpha particle; the *mean* energy is much less still. This, perhaps, surprising result stems from the necessity for conservation of both energy and momentum in the collision. If the alpha particle could be conceived of as colliding with an atom as a whole, much higher transfer of momentum, and, therefore, energy, would be theoretically possible.

3.4 Alpha Particle Tracks

As will be discussed in more detail in Section 3.5, electrons set free by collisions with alpha particles will themselves produce further ionising events, setting free more electrons—of progressively lower energy—until all the energy is dissipated. The alpha particle is scarcely deflected by the many collisions it makes, except near the very end of its range when its own energy is small. The electrons it releases have relatively short ranges, because of their low energy. The passage of an alpha particle through air therefore results in the formation of a narrow straight column of positive and negative ions. Alpha particle *tracks* can in fact be made visible to the human eye by placing alpha-emitting isotopes in special apparatus called a *cloud chamber*, in which under suitable conditions a tiny drop of water can be made to condense on each individual ion. Photographs of such tracks are reproduced in Fig. 3.2.

Fig. 3.2 Tracks of alpha particles from a mixed source of ^{212}Bi and ^{212}Po in a Wilson cloud chamber, showing two distinct ranges. (Reproduced from *Radiations from Radioactive Substances* (1930) by Rutherford, Chadwick and Ellis, by permission of Cambridge University Press.)

Under high magnification it is possible to count the individual droplets. From a knowledge of the energy dissipated by the alpha particle along the whole track, the average energy per pair of ions is then easily calculated. Quantitative mesurements on alpha particle tracks enable the following statements to be made:

(a) With a few rare exceptions all the alpha particle tracks generated from a given pure radionuclide are of very nearly the same length. This is taken to mean that all the alpha particles leave their parent atoms with exactly the same energy, although there are two reasons why they show slight variability in their ranges. One reason is that the source material itself can never be prepared in a quite weightless state; alpha particles which have to traverse specks of dust or other impurities before reaching the air in the cloud chamber will be slightly reduced in energy and have a shorter range. However nearly weightless the source, it is still found that some 'straggling' in the ranges occurs, and this is put down to the fact that by accident a few alpha particles will make more collisions than the average, and a few will make fewer collisions. The second reason for straggling is thus essentially statistical. The effect is illustrated in Fig. 3.3*a* in which N, the number of particles which travel a distance of at least R is plotted against R. If a tangent is drawn to the steepest part of the curve, it intersects the R axis at a point R_e, referred to as the 'extrapolated range'. The differential coefficient dN/dR for all points on this curve is plotted against R in Fig. 3.3*b*. This gives the differential range curve; points on it show the relative numbers of particles which travel a distance between R and $R + \Delta R$. The differential range curve is very nearly gaussian in shape and reaches a peak at a value of R which is practically indistinguishable from the *mean range* $\bar{R}$. The difference between R_e and $\bar{R}$ is a simple measure of the effect of straggling and is called the straggling

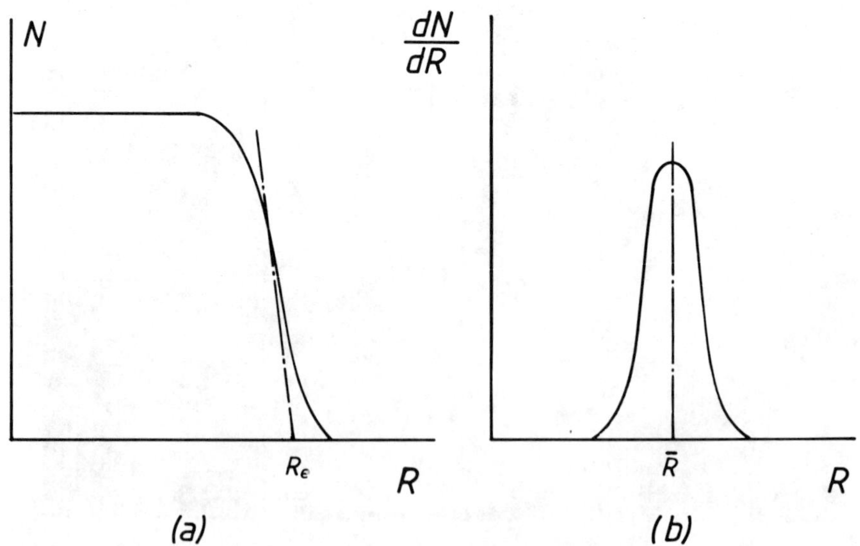

Fig. 3.3 Straggling of alpha particles: (a) integral range, (b) differential range.

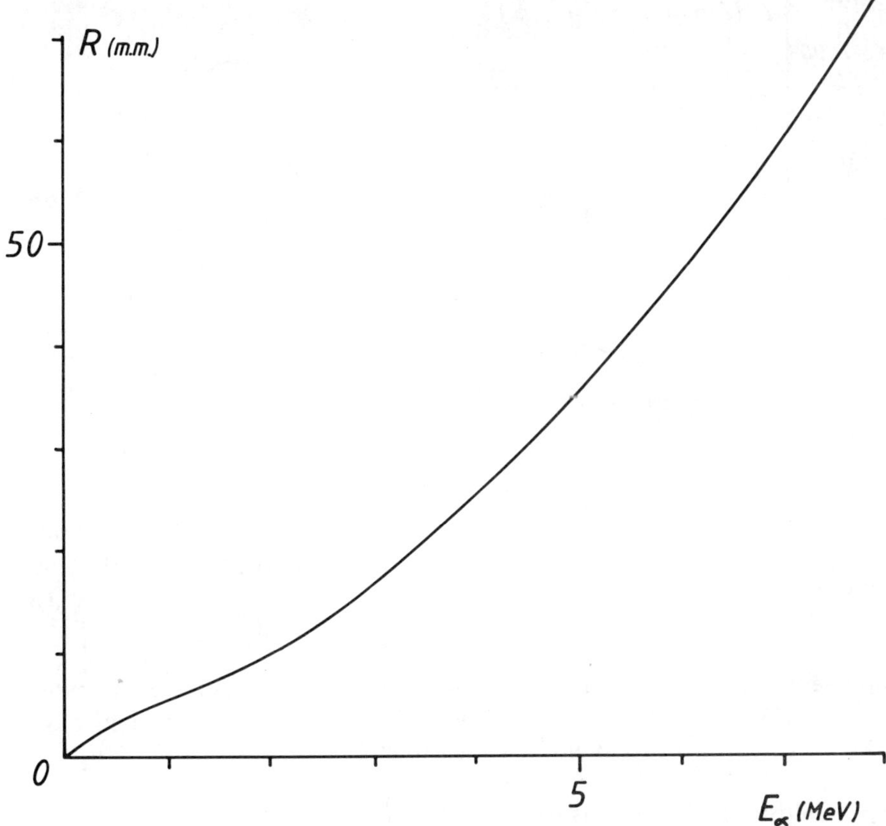

Fig. 3.4 Mean range $\bar{R}$ (mm) of alpha particles of energy E_α (MeV) in dry air.

parameter S. It is numerically about 1% of $\bar{R}$, and about 0.53 times the FWHM of the differential range curve.

Values of $\bar{R}$ for alpha particles in air are plotted as a function of energy E_α in Fig. 3.4. (For helium ions over a much greater energy range, see Section 2.20.)

(b) The ion pairs formed along the length of the tracks are not uniformly spaced; the ionisation density I, i.e. the number of ion pairs per millimetre of track, increases steadily towards the end, where it falls off again abruptly. This is illustrated in the *Bragg curve*, named after the discoverer of the effect, in Fig. 3.5.

(c) Finally we may note that the range of alpha particles in tissue is much the same as that in air if measured in terms of weight of material traversed, rather than in millimetres. Since tissue is of the order of 1000 times as dense as air, the range of alpha particles in tissue is very small—roughly $5E_\alpha$ μm. But the total ionisation along the track is still nearly the same. Such very dense ionisation may result in severe biological damage confined to a very small volume of tissue, e.g. a single cell. Quite clearly also the presence of

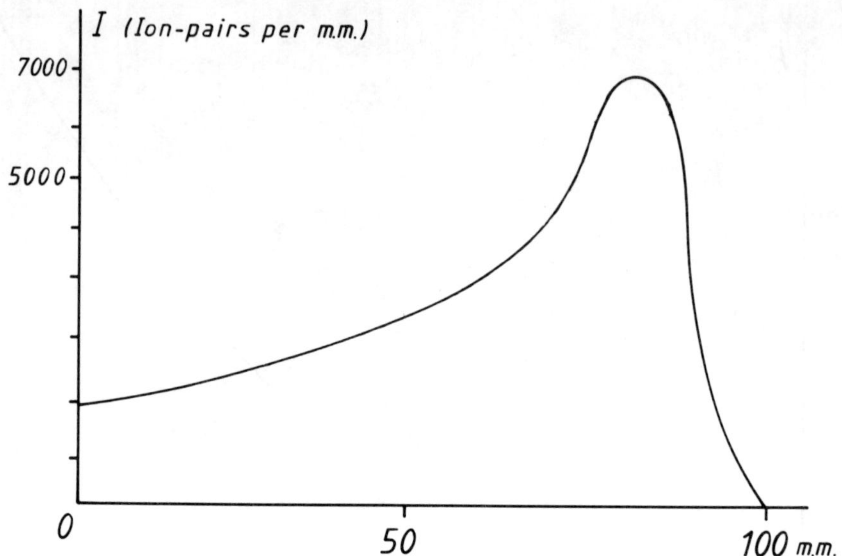

Fig. 3.5 Bragg curve; the ionisation density I along the length of an alpha particle track.

alpha-emitting radionuclides within the body cannot be detected by external devices. For these reasons workers who handle these radionuclides have to take very great care to avoid ingesting them.

3.5 Collisions of Electrons with Electrons

The dynamics of electron–electron collisions is quite different from that of alpha particle–electron collisions for the following reasons.

(a) Relativistic relationships must be used because we need to consider beta particles whose velocities may approach that of light. The energy equation, which equates the kinetic energy of the incoming electron to the sum of the energies of both electrons after collision reads

$$(m_e c^2 - m_0 c^2) = (m_e' c^2 - m_e c^2) + (m_e'' c^2 - m_e c^2)$$

or

$$m_e = m_e' + m_e'' - m_0$$

The relativistic masses m_e, m_e' and m_e'' must also be used in the momentum equations; m_0 is the electron rest mass (see Appendix 2).

(b) We cannot make the simplifying assumption that the incident particle is much more massive than the struck particle. Their rest masses are the same, of course; the incoming particle will have a much larger relativistic mass only if its velocity approaches that of light.

The fact that the two particles are identical means that in a 'head-on' collision there is complete transfer of momentum from the incoming to the struck particle. For the momentum equations read

$$m_e v = m_e' v' \cos(\theta) + m_e'' v'' \cos(\phi)$$
$$0 = m_e' v' \sin(\theta) - m_e'' v'' \sin(\phi)$$

and putting in $\theta = \phi = 0$ in the first of these to make it apply to a head-on collision we get

$$m_e v = m_e' v' + m_e'' v''$$

Now it is possible for the velocity of the scattered electron, v_1, to be zero, which would mean that the momentum of the incoming electron would just equal the momentum of the *struck* electron. For this equality implies both equal velocity and equal mass, which means $m_e = m_e''$; and clearly the energy equation is satisfied if both $m_e = m_e''$ and also m_e' is just the rest mass m_0. In electron-electron collisions it is, therefore, possible for the incoming electron to communicate any fraction of its energy, including the whole of it, to the struck electron. It will be found that all values of the angles θ and ϕ are possible in the range 0° to 90°—that is to say both electrons move after the collision with a velocity component in the forward direction. For any given energy of the incoming electron and a chosen angle of scatter θ, the three basic equations enable us to calculate the energy of the scattered electron and the energy and direction of the struck electron.

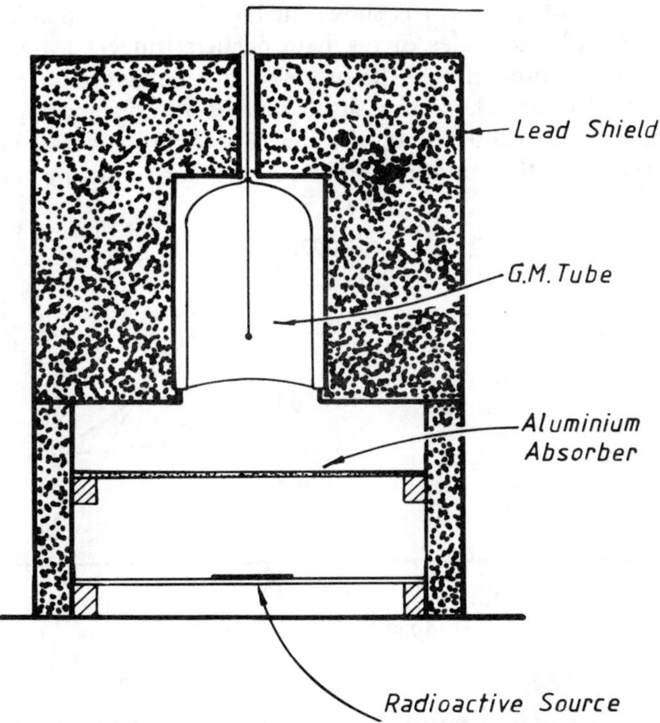

Fig. 3.6 Arrangement for range measurement of beta particles.

Experiments show, as might be expected because the electron has a much smaller radius than the alpha particle, that the ionisation density along the path of a high-energy electron is much less than that in an alpha particle track. The primary electron suffers many large-angle scattering encounters and many of the electrons it collides with may be energetic enough to make several ionising collisions before all their energy is dissipated. Electron tracks in a cloud chamber are therefore very diffuse. Nevertheless, for an electron starting with a given total energy, it is a practical possibility to define a range in matter beyond which no further ionisation occurs. Electron ranges are most conveniently measured, not in air, but in a light element such as aluminium. A very simple experiment which demonstrates the principle uses a radioactive source emitting beta particles placed in a fixed position with respect to the window of an end-windowed Geiger–Müller counting tube. Interposed between them is a sheet of aluminium foil (Fig. 3.6). The experiment consists of recording the numbers of electrons detected by the Geiger–Müller tube per unit time, for a set of aluminium foils of different thicknesses. (Corrections have to be made for 'background' radiation by repeating the measurements in the absence of the source. Corrections may also have to be made if the source decays appreciably during the experiment.) The aluminium thicknesses are usually expressed as the mass of aluminium per unit area, rather than directly in millimetres, as the former quantity is easier to measure experimentally. A plot of the corrected count-rate C, on a logarithmic scale, against aluminium thickness t is shown in Fig. 3.7a. This simple technique gives fairly reliable estimates of the beta particle range because the curve turns downwards quite strongly. Its precise shape depends somewhat on the width and thickness of the source itself and on the distances between the source, absorbing foil, and counter window. For accurate work, the thickness of the air space and the counter window, in the same units of mass per unit

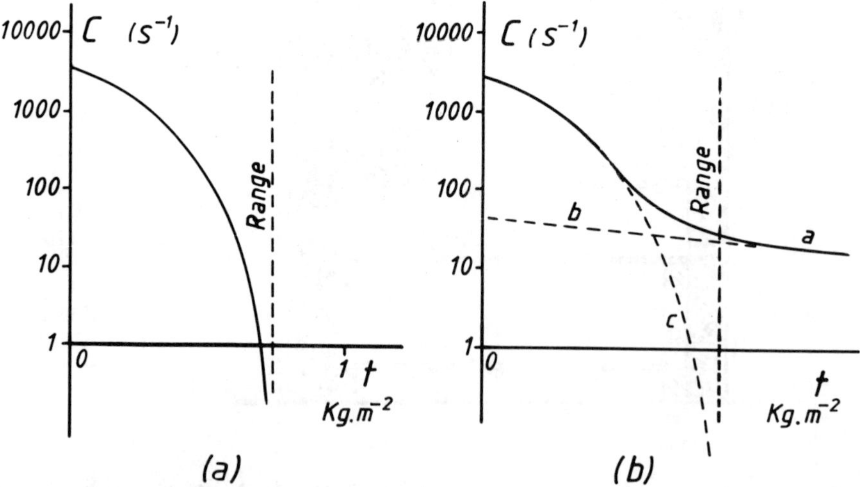

Fig. 3.7 Estimation of beta particle range: (a) pure beta particle source, (b) source emitting gamma-rays as well as beta particles.

area, should be added on to the estimated range in the aluminium alone. A complication arises if the source emits gamma rays as well as beta particles, because of the appreciable sensitivity of the counter to gamma rays. The curve then appears as in Fig. 3.7b, in which the portion of the curve labelled 'a' represents the contribution to the count-rate from the gamma rays. Extrapolating this (nearly) straight portion backwards along 'b' we can easily subtract this contribution to the original curve and arrive at the corrected curve 'c' from which the range may be estimated as before. If a very strong source of a pure beta emitter is examined, what appears to be a low-intensity gamma-ray 'tail' will be found; its effect can be allowed for in the same way. It is not in fact due to gamma rays, but to rather rare events in which beta particles pass close to atomic nuclei and are slowed down in doing so with the production of X-rays (the so-called bremsstrahlung, or 'slow-down radiation'). If a source is very weak, it may be impracticable to take measurements with the thicker absorbing foils because the count-rate becomes too low. A rough estimate of the range can be made by finding the half-thickness (a term familiar to radiographers as the half-value layer or HVL) and multiplying this by 13. This is in fact the thickness which will reduce the counting-rate of the bare source by a factor 2^{13} or approximately 10^4. The result will be accurate to about 10%.

Beta particle sources emit particles with a wide spectrum of energy, from zero upwards to an upper limit characteristic of the particular radionuclide. The range, as just measured, corresponds to electrons of this maximum

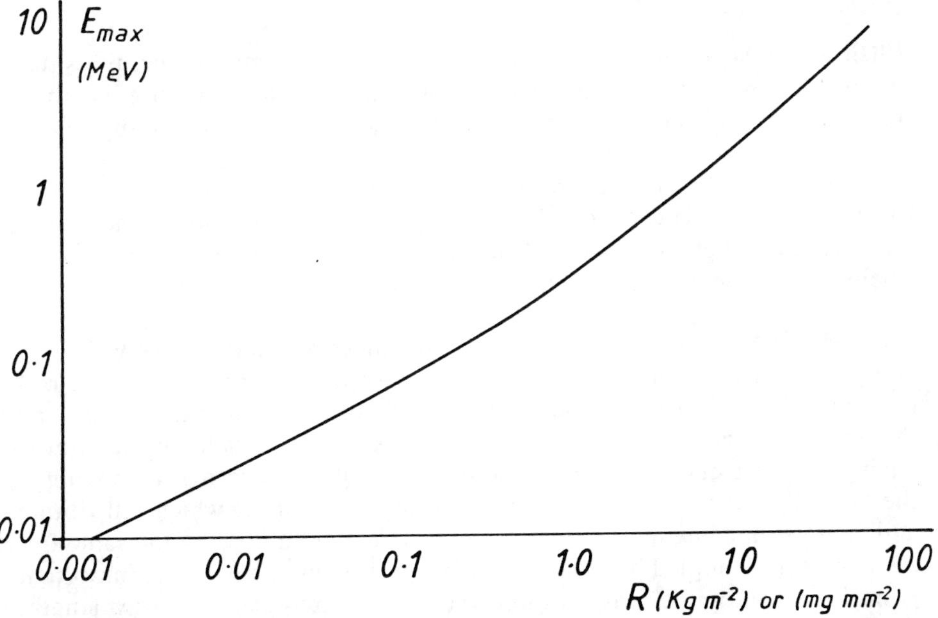

Fig. 3.8 Range/energy relationship for electrons (β particles) in aluminium. (Reproduced, converted to SI units, from G. Friedlander, J.W. Kennedy and J.M. Miller, *Nuclear and Radiochemistry*, 2nd ed., by permission of John Wiley & Sons, Inc.)

energy, which by chance suffer relatively fewest collisions leading to large energy loss or large angular deflections. The exact relationship between maximum beta-particle energy and range is not easy to calculate, depending as it does on chance encounters. However, the relationship has been examined very thoroughly in experiments with radionuclides for which the E_{max} has been accurately determined by magnetic deflection methods. A range/energy curve is shown in Fig. 3.8, relating electron range in kilograms per square metre with energy in MeV. An empirical equation which is accurate to a few per cent in energy is

$$E_{max} = \text{antilog}\,[-0.461 + 0.753 \log R + 0.1008 (\log R)^2 + 0.0064 (\log R)^3 - 0.00835 (\log R)^4]$$

The range of positrons is probably a few per cent larger than that of negative electrons. Beta particle ranges in tissues are roughly the same as in aluminium, in units of kg m^{-2}. This means that for a fairly energetic beta emitter such as ^{32}P (E_{max} = 1.7 MeV), the range in tissue is about 8 mm. It must be remembered however that the *mean* electron energy is only about 0.6 MeV, and that most of the beta particles from this source have ranges of less than 3 mm. As with alpha particle emitters, radiation damage from ingested beta emitters is virtually confined to the organ or organs which take up the radionuclide, and it is not generally possible to find where the radionuclide is in the body by using external detectors.

3.6 Photon–Electron Interactions

When we consider gamma photons, or X-rays, traversing matter, the situation is different from the alpha-particle and electron cases because we know that the photons possess wave-like characteristics and it hardly seems appropriate to use the term 'collision'. Nevertheless, although we now use the more general term 'interaction', there are some features which are common to a particle–particle collision. The plain fact is that Nature sometimes works one way and sometimes another, and we have to alter our descriptions to suit! There are four main types of interaction we may observe, as follows;

(*a*) *Classical Scattering*. This is a process, important particularly with low-energy photons, in which they show essential wave-like properties. We picture an atomic electron in the path of a photon, absorbing its energy and thereby being raised to an excited state, and then re-radiating a photon having the same energy as the initial one, although not necessarily moving in the same direction. A good analogy is that of a violin string which will vibrate and emit its fundamental tone when it resonates with a tone, of the same frequency, being sounded by another nearby musical instrument. The important point is that in neither case is there a change in frequency, or wavelength. Since for the photon the energy carried is proportional to frequency, this means that there is no energy lost in the scattering process.

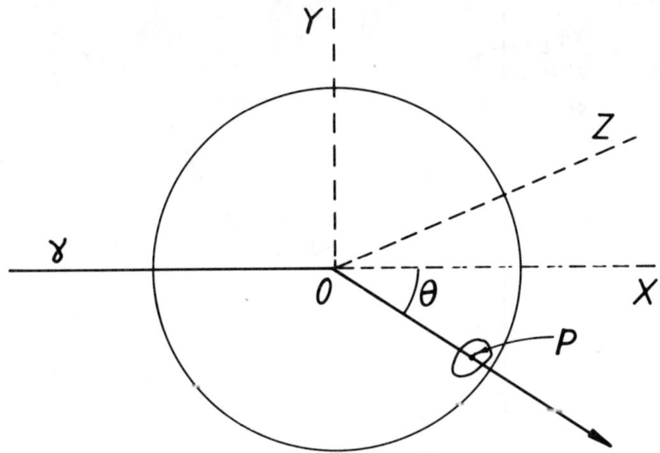

Fig. 3.9 Classical scattering of photons.

The theoretical treatment of classical scattering was first given by J. J. Thomson who derived an expression for the relative number of photons scattered (per unit solid angle) at an angle θ to the direction of the incident photons. Thus in Fig. 3.9 we suppose that at the origin of coordinates there is a 'scattering centre', which could be any suitable particle. We describe a sphere of radius r about this point. Photons approach the origin along the X axis and are scattered from 0 in all directions. Let us choose a point P on the sphere and draw round it an element of area; the angle P0X is θ. If, for a given number of incident photons, the number of scattered photons which pass through the element is $N(\theta)$, Thomson found that

$$N(\theta) = k\{1 + \cos^2(\theta)\}$$

k being a constant. The value of k is easily appreciated because if P is on the X axis, $\theta = 0$ and $N(\theta) = N(0)$, i.e. the number of electrons making a 'glancing' collision. Then we have

$$N(0) = k\{1 + \cos^2(0)\} = 2k$$

so

$$\frac{N(\theta)}{N(0)} = 0.5\{1 + \cos^2(\theta)\},$$

The ratio $N(\theta)/N(0)$ is plotted against θ in Fig. 3.10. Since the photons are all of the same energy, the curve also represents the variation of scattered intensity (numbers times energy) as a function of angle. Thomson's theory also gives the fraction of the incident intensity I_i which is scattered through an angle θ; this fraction is, per unit solid angle,

$$f = \frac{I(\theta)}{I_i} = \frac{e^4}{2r^2m^2c^4}\{1 + \cos^2(\theta)\}$$

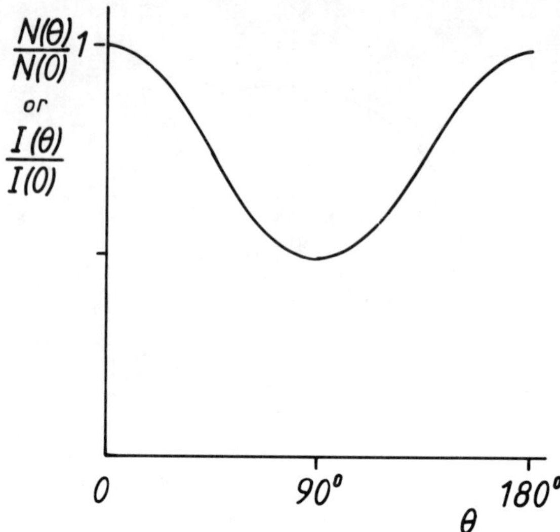

Fig. 3.10 Relative numbers of scattered photons, $N(\theta)/N(0)$, or relative scattered intensity, $I(\theta)/I(0)$, as a function of angle of scatter, θ.

It is a basic assumption of the Thomson classical scattering theory that the scattering centre is very massive. Experiments show that the above equations are borne out if the scattering occurs at the atomic nucleus (now called Thomson scattering). However, when the scattering centre is an orbital electron, the Thomson equations have to be modified; the new equations are consistent with the experimental observation that the scattered intensity in the forward direction is very much enhanced. This kind of scattering is called Rayleigh scattering. Both kinds occur mainly when the photon energy is below about 100 keV; but at all energies at least one of the other effects described in the following paragraphs occurs more frequently.

(*b*) *The Photoelectric Effect.* In this process the photon interacts with a bound atomic electron, and the energy transferred is used to eject the electron altogether from its parent atom. The atomic system is not set into oscillation so there is no loss of energy in the form of a scattered photon. The energy, and momentum, transferred are shared only by the ejected electron and the residual positive ion, but the latter is so massive compared with the electron that its motion can usually be ignored. The binding energy of the electron is not however negligible and we may write

$$E_e = E_\gamma - E_b$$

to mean that the electron energy E_e is less than the incoming photon energy E_γ by an amount E_b, the binding energy. When the incoming quanta are those of visible light it may in fact be impossible for the photoelectric effect to occur. Millikan showed that red light (wavelength 700 nm, photon energy 1.5 eV) will not release electrons from a clean surface of sodium or potas-

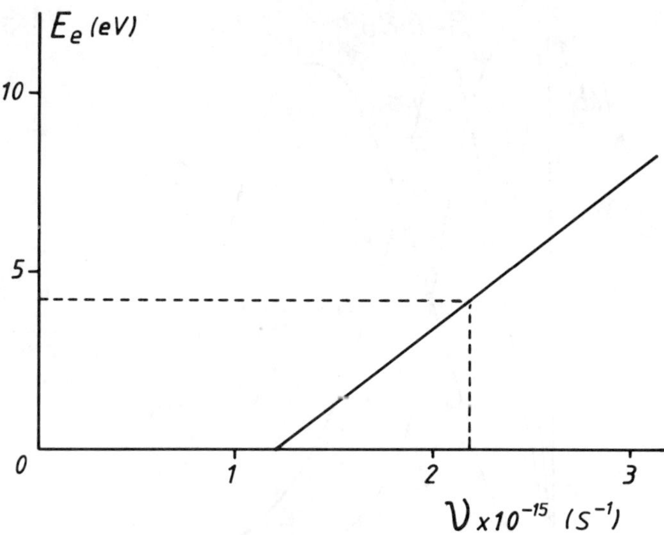

Fig. 3.11 Energy of photoelectrons.

sium, whereas blue light (wavelength 400 nm, photon energy 3 eV) will do so. The energy of the ejected electrons can be measured for different wavelengths of photons, and a plot of E_e against the photon frequency ν, or c/λ, is shown in Fig. 3.11. No electrons are ejected for values of ν less than a certain limit; above this limit the energy is found to be linearly related to the frequency. It is evident that the photon energy is just $h\nu$, h being a constant of proportionality. The experimental results therefore confirm Einstein's expectation that $E_\gamma = h\nu = hc/\lambda$. In the figure the slope of the line is $4.14/10^{15}$ eV s, from which the value of Planck's constant is found to be $h = 4.14 \times 10^{-15} \times 1.6 \times 10^{-19} = 6.63 \times 10^{-34}$ J s.

As with scattered photons, photoelectrons are by no means emitted in one particular direction. In the theoretical limit of zero photoelectron energy, they would be emitted equally in all directions, i.e. isotropically. Thus the number $N(\phi)$ emitted, per unit solid angle, through an elemental area on a sphere, as in Fig. 3.9, would be proportional to $\sin^2(\phi)$ where ϕ is the angle of emission. However for photoelectrons whose velocity is not negligible a relativistic treatment gives a very much more complicated result. A reasonable approximation, for not too high energy, is

$$N(\phi) = N_p \frac{\sin^2(\phi)}{\{1 - \beta \cos(\phi)\}^4}$$

where β as usual means v/c and N_p is a constant. In Fig. 3.12 the ratio $N(\phi)/N_p$ is plotted as a function of ϕ for the three cases of $\beta = 0$, 0.548 and 0.863 corresponding to electron energies of 0, 100 and 500 keV respectively. It will be seen that for zero-energy photoelectrons the most common angle of emission is 90°. However at 100 keV the most frequent angle is only 40° and at 500 keV only about 18° to the direction of the incident photon.

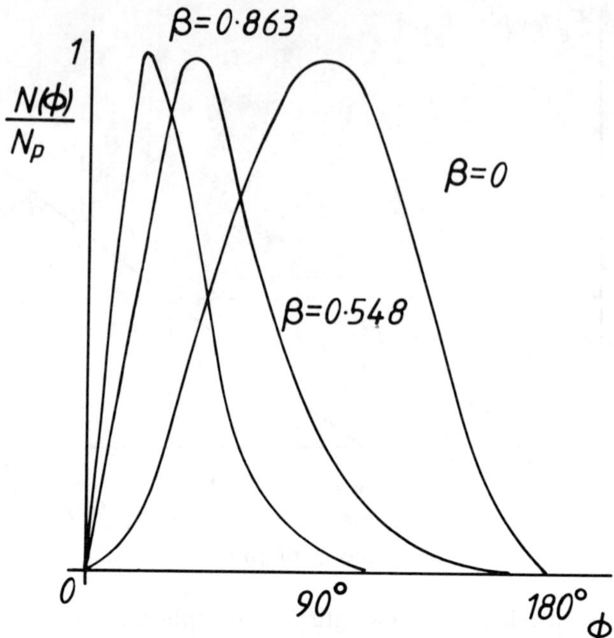

Fig. 3.12 Relative numbers of photoelectrons, $N(\phi)/N_p$, at angle ϕ.

(*c*) *The Compton Effect.* Here the photon shows the kind of interaction which most clearly resembles a collision, as the experimental observations are most easily interpreted in terms of the energy and momentum equations used already. We have to use not only the expression $E_\gamma = h\nu$ for the photon energy but also $P = h\nu/c$ to represent its momentum. (The idea of a weightless photon possessing momentum which is defined for massive particles as the product of mass and velocity, can only be explained in wave-mechanical terms. The fact that experimental data about the Compton effect agree so well with theoretical expectations is indeed the most convincing evidence that photons do have the momentum $h\nu/c$.) In the Compton effect the incident photon interacts with a single electron, whose binding energy we can usually ignore in comparison with the photon energy. However, only a portion of the available energy is given to the electron, the remainder being used to create a *secondary* photon—sometimes called the Compton scattered photon. In a way, the Compton process is a hybrid of scattering (but with an increase in wavelength) and electron ejection. It is depicted in Fig. 3.13, the scattered photon being emitted at an angle θ and the electron at an angle ϕ to the direction of the primary photon.

Using symbols with single primes to represent the secondary photon, the energy and momentum equations read

$$E_\gamma = E_\gamma' + E_e \qquad \text{or} \qquad h\nu = h\nu' + E_e$$

$$h\nu/c = (h\nu'/c)\cos(\theta) + mv_e\cos(\phi)$$

$$0 = (h\nu'/c)\sin(\theta) - mv_e\sin(\phi)$$

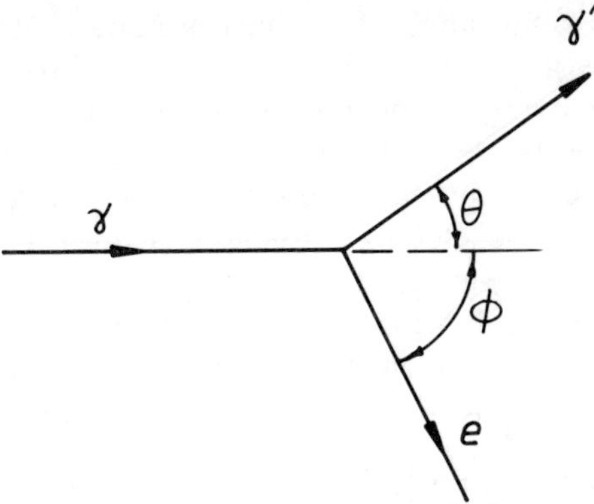

Fig. 3.13 The Compton process.

Now if we eliminate ϕ from the second and third equations we get (compare Section 3.2)

$$m^2 v_e^2 = h^2\nu^2/c^2 - 2(h\nu/c).(h\nu'/c)\cos(\theta) + h^2\nu'^2/c^2$$

or

$$m^2\beta^2c^2 = h^2/c^2\{\nu^2 - 2\nu\nu'\cos(\theta) + \nu'^2\}$$

Also by writing for the electron energy $E_e = mc^2(1/\sqrt{(1-\beta^2)} - 1)$ we can rewrite the energy equation as

$$h\nu = h\nu' + mc^2(1/\sqrt{(1-\beta^2)} - 1)$$

It is now possible to eliminate β from these two last equations, and when this is done we get

$$1/\nu' - 1/\nu = (h/mc^2).\{1 - \cos(\theta)\}$$

Then writing λ for c/ν and λ' for c/ν'

$$\lambda' - \lambda = (h/mc)\{1 - \cos(\theta)\}$$

—a surprisingly simple result for the difference in wavelength between the scattered and incident photon, in terms of the angle of scatter. The value of h/mc, in convenient units, is 0.002 42 and we have that $\Delta\lambda$, the increase in wavelength, is given by

$$\Delta\lambda = \lambda' - \lambda = 0.002\,42\{1 - \cos(\theta)\}\ \text{nm}$$

We note that the maximum possible wavelength change (an increase of 0.004 84 nm) occurs when $\theta = \pi$—that is to say, when the scattered photon moves exactly in the backward direction. This is also the condition for maximum transfer of momentum, and therefore energy, to the electron. As with the alpha-particle case, there is, therefore, a whole range, or spectrum, of electron energies, ranging from zero up to a certain limit, dependent on the incident photon energy. The limit is called the Compton 'edge'. Its value is easily calculated. For instance, if we have a photon of energy 124 keV, its wavelength will be given as $hc/E \approx 1.24/124 = 0.010$ nm. The Compton scattered photon moving in the backward direction will therefore have a wavelength $0.010 + 0.0048 = 0.0148$ nm. Consequently its energy will be $1.24/0.0148 = 84$ keV. The scattered electron will therefore have an energy $124 - 84 = 40$ keV. The electrons released at all other angles must have energies lower than this figure.

While the angle θ can have any value, ϕ cannot exceed 90°. The relative numbers of photons scattered (per unit solid angle) at an angle θ was investigated by Klein and Nishina. Their results can be expressed by the formula

$$N(\theta) \propto \frac{1 + \cos^2(\theta) + \alpha^2\{1 - \cos^2(\theta)\}/[1 + \alpha\{1 - \cos(\theta)\}]}{[1 + \alpha\{1 - \cos(\theta)\}]^2}$$

where the parameter α is $h\nu/m_0c^2$, or in energy units $E_\alpha(\text{MeV})/0.51$. It will be seen at once that for zero-energy photons $\alpha = 0$ and the expression reduces to just the classical scattering case. For photons with energy above zero, the number scattered at very small angles remains at almost the value expected for classical scattering, but becomes markedly less for greater angles. Fig. 3.14 expresses the variation in numbers of scattered photons with angle (continuous lines). Since the radiation intensity in a beam of photons is given by the product of numbers of photons and energy carried by each, one can also find the intensity of scattered-photon radiation at different angles, as a fraction of the 'forward' intensity. Scattered intensities are also shown in Fig. 3.14 by dashed lines. The graphs make it clear that for photons in the range 100 keV to 1 MeV—and this includes the gamma rays from most of the useful radionuclides—the scattered radiation intensity of photons is far from

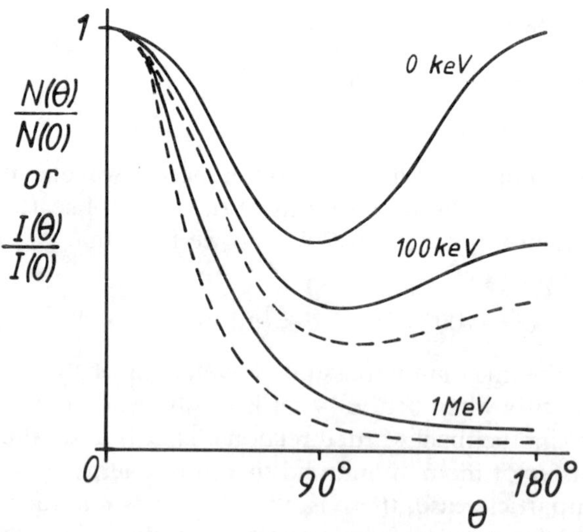

Fig. 3.14 Relative numbers of scattered photons, $N(\theta)/N(0)$ (full lines), and relative intensity of radiation, $I(\theta)/I(0)$ (broken lines), as a function of the angle of scatter, θ.

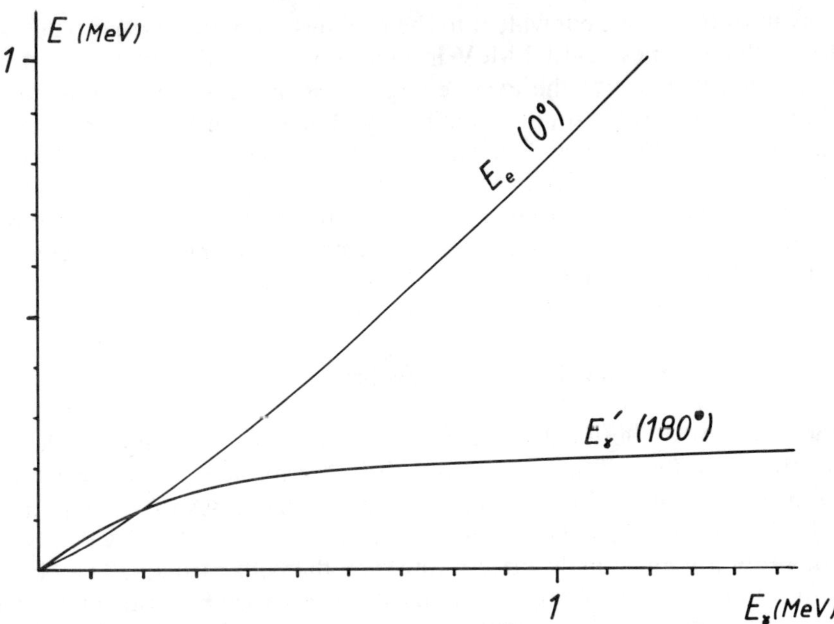

Fig. 3.15 Energy of the Compton edge, $E_e(0)$, and back-scattered photons, $E_\gamma'(180°)$, as a function of the primary photon energy, E_γ.

isotropic but tends to be greatest in the forward direction. Nevertheless, even at the highest energy considered here the intensity in the backward direction is not negligible, and this fact must not be forgotten in the provision of shielding materials for personal protection.

To conclude this section we draw attention to a useful graph (Fig. 3.15) which shows the energy of the Compton edge (most energetic electrons) and of the back-scattered secondary photons, as a function of the energy of the incident photons. The graph makes it clear that for incident gamma photons of about 300 keV and above, the back-scattered photons at 180° have energies within a quite narrow range (about 200 ± 20 keV).

The energy is not very much less even at 150°. Back-scattered photons of about 200 keV are therefore of quite frequent occurrence and many experimenters who have examined the spectra of gamma-rays from weak sources placed within lead shielding walls in an effort to reduce background radiation have been convinced of the presence of 200 keV photons in the sources. But in fact radiation at this energy has been caused by gamma-rays of much higher energy entering the lead shielding and producing back-scattered photons.

3.7 Pair Production

The fourth process we will consider is that in which a photon vanishes entirely and an electron-positron pair is created. For this process to occur the photon energy must be at least equivalent to the rest mass of both these particles, viz. $2m_0c^2$, which comes to 1.02 MeV in energy units. If the photon energy is greater than this figure, the extra energy is carried away as kinetic energy shared, not necessarily quite equally, by the two electrons. The process occurs only in the presence of matter—i.e. not in free space—and most often very close to an atomic nucleus, which appears to take up part of the available momentum and therefore a small part of the energy. More rarely an electron–positron pair appears as a result of a photon interacting with an electron, which is also set in motion with high kinetic energy. This rare 'triplet' production occurs only with photon energy above about 4 MeV.

3.8 The Attenuation of X- and Gamma-Rays

If the experiment outlined in Section 3.5 (absorption of beta particles in aluminium, see Fig.3.7) is attempted with a source which emits only gamma-rays, it will be found that very much greater thicknesses of aluminium are required to give significant changes in the count-rate recorded. Evidently, the gamma-rays are much more penetrating than the beta particles. The diminution in count-rate when thick aluminium sheets are used is caused by the facts that some of the initial photons are entirely removed by the photoelectric process, and some are reduced in energy and changed in direction (by the Compton and possibly pair-production process). While most of the electrons released by these processes are absorbed by the aluminium, we prefer to use the more general term *attenuation* for the whole phenomenon, as the count-rate will be affected to some extent by scattering effects. Some gamma photons from the source which would have entered the detector in the absence of the attenuator may in fact be scattered through such a large angle that they do not arrive there; and the converse effect can happen too. Moreover the count-rate obtained is hard to interpret because a Geiger–Müller tube is not equally sensitive to gamma-rays over the whole spectrum of energy (it is more sensitive to lower-energy photons). Also it will respond with high efficiency to electrons set free in the attenuator close to the surface facing the counter window.

In order to investigate the attenuation effects we have therefore to visualise a somewhat idealised experiment in which we have a detector which is sensitive only to those photons which penetrate the attenuator unchanged. Any interaction such as the photoelectric, Compton, pair production and even classical scatter is regarded as catastrophic in the sense of removing the affected photon altogether from the emergent beam. Then if N photons are incident, per unit area, on an attenuator of thickness Δx we would expect the number of photons which are lost, $-\Delta N$, to be proportional to N and also to the number of atoms in the path of the radiation. The latter is just proportional to the thickness Δx (see Fig. 3.16). Using μ to represent the constant of proportionality we have

$$-\Delta N = \mu N \Delta x$$

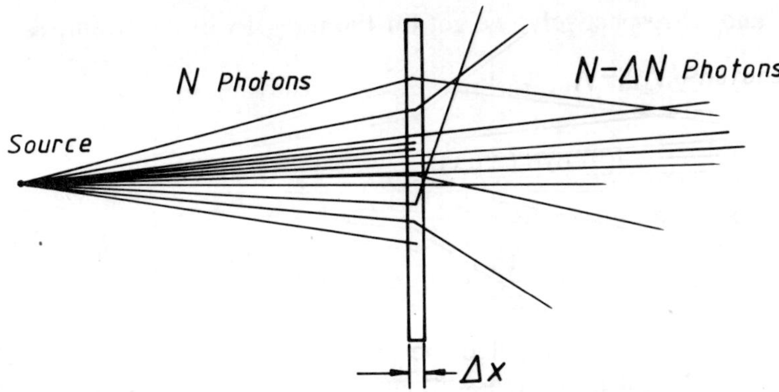

Fig. 3.16 Attenuation of photons: some of the photons are scattered and some are absorbed.

The value of μ will of course be different for different attenuating materials. The equation will be accurate in the limit as Δx is taken to be indefinitely small. Then we can write

$$\frac{dN}{dx} = -\mu N$$

from which

$$N = N(0).\exp(-\mu x)$$

—an exponential law. We may now easily derive the following conclusions:

(a) There is no such thing as a *range* of X or gamma photons, as there is for beta and alpha particles.

(b) By putting in $N = N(0)/2$ and solving for x we find the thickness $x_{1/2}$ of material for which exactly half of the photons are removed from the original beam. This half-thickness, or half-value layer (HVL) is numerically

$$x_{1/2} = \text{HVL} = 0.6932/\mu$$

In practice the HVL is measured experimentally and μ is calculated from the equivalent relationship

$$\mu = 0.6932/\text{HVL} = 0.6932/x_{1/2}$$

(c) Solving the equation for μ we find

$$\mu = -\frac{1}{N}\frac{dN}{dx} \approx \left(\frac{-\Delta N}{N}\right) \cdot \frac{1}{\Delta N}$$

This gives us an insight into the physical meaning of μ; it is the fractional decrease in N per unit thickness traversed.

(d) If we multiply the relevant equations throughout by E_γ, the energy carried by each photon, and identify the products NE_γ and ΔNE_γ as I

and ΔI respectively, we get for the intensity of the beam, I,

$$\frac{dI}{dx} = -\mu I$$

$$I = I(0) \exp(-\mu x)$$

and for μ,

$$\mu \approx \left(\frac{-\Delta I}{I}\right) \cdot \frac{1}{\Delta x}$$

The quantity μ is called the linear attenuation coefficient. Closely related to it is the mass attenuation coefficient μ/ρ where ρ is the density of the material. Clearly the product μx is equal to $(\mu/\rho \times (\rho x)$, and we can identify ρx as the thickness of material measured as mass per unit area. In the older literature, the units used were milligrams per square centimetre; here we use kilograms per square metre. Notice that the equation $\Delta N/N = -\mu\Delta x$ can be applied even to a single photon if we interpret $\Delta N/N$ as the probability that the photon will interact in some way; this probability is just proportional to the thickness of material traversed (if this is small).

3.9 Coefficients for the Attenuation Processes

The processes we have been considering do not occur with equal probability. Also they are completely independent of one another, and any one X- or gamma-ray might interact in matter in any one way. It is convenient to define τ, σ and κ as the probability (per unit thickness traversed) that a photon undergoes photoelectric absorption, Compton scattering, or the pair-production process respectively. Properly we ought to define a probability for classical scattering too, but for gamma-rays and 'hard' X-rays this is one or two orders of magnitude smaller than τ and for our present purpose can be neglected. Then we have

$$\mu = \tau + \sigma + \kappa$$

and also

$$\mu/\rho = \tau/\rho + \sigma/\rho + \kappa/\rho$$

These quantities are clearly the linear, or mass, attenuation coefficients associated with the three main processes of interaction. Results of measurements of the relative frequency with which the processes occur, over a range of photon energy, are conveniently summarised graphically, as in Fig. 3.17. This applies for air as the attenuating medium. There are many points of interest in this graph and we summarise the most important ones as follows:

(a) Photoelectric absorption is by far the most common event at energies up to about 30 keV but its coefficient falls off almost linearly on this (log–log) scale. Approximately $\tau \propto E_\gamma^{-3.3}$.

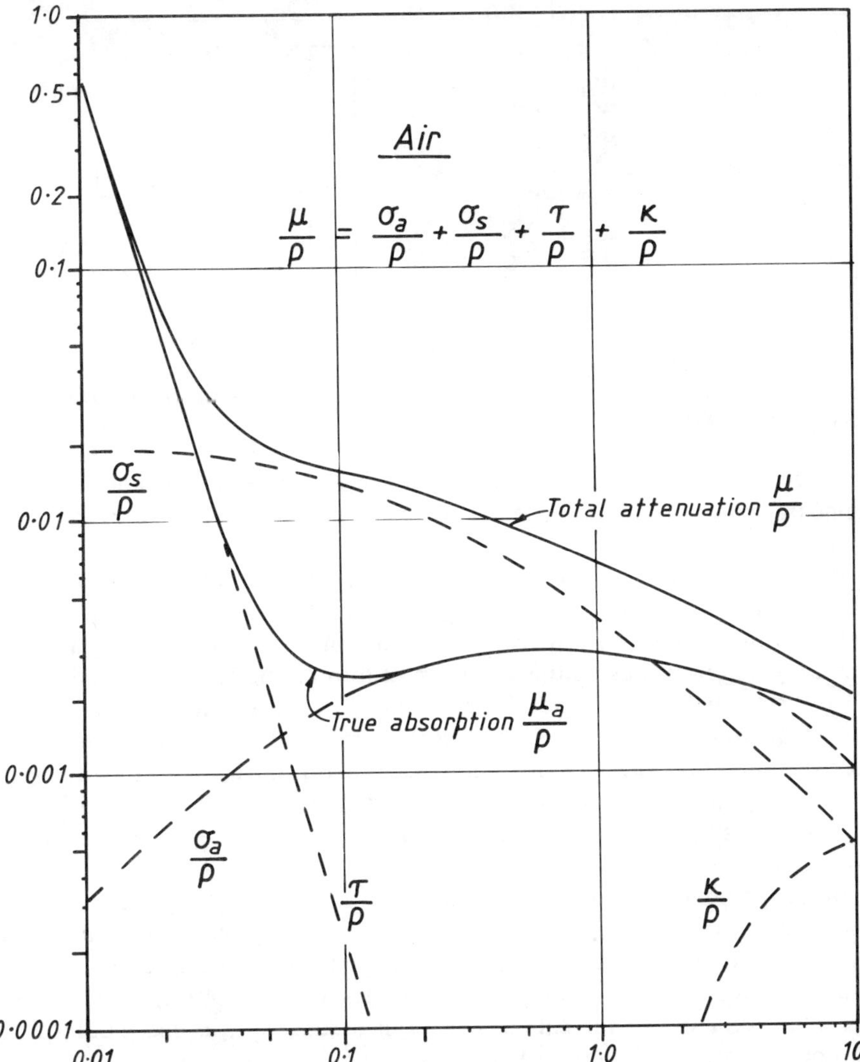

Fig. 3.17 Mass-attenuation coefficients of γ-photons in air. (σ_a, σ_s.τ and k are attenuation coefficients for true absorption, Compton scattering, photoelectric process and pair production respectively.) (Reproduced from G. Hine and G. Brownell, *Radiation Dosimetry*, by permission of Academic Press.)

(b) Pair production begins to be appreciable at energies above about 2 MeV and predominates over the others above about 10 MeV.

(c) At intermediate energies the Compton process is the most frequently occurring.

It will be seen that three curves, labelled σ_T/ρ, σ_a/ρ, and σ_s/ρ are displayed and it is important to realise the significance of these. Experimentally they come about because of the two extremes in geometrical conditions under

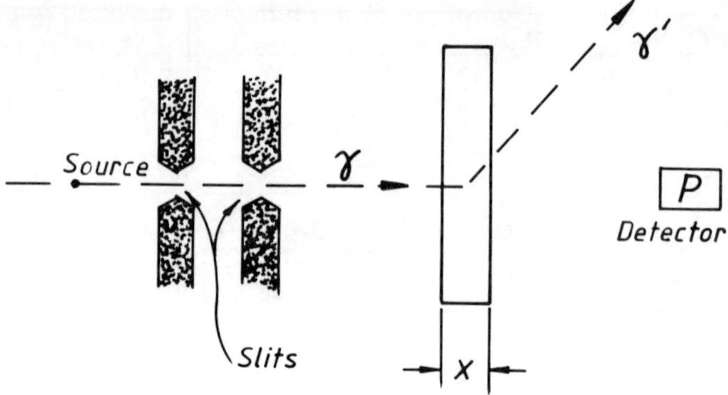

Fig. 3.18 Attentuation in narrow-beam geometry; experimental arrangement (Compton scattering only).

which measurements may be taken. Thus in Fig. 3.18 we consider a narrow pencil of photons (defined by lead slits) incident on an attenuating material of thickness x.

If we now take measurements of the intensity of radiation at a point p which is in the beam direction, using any suitable form of radiation detector (see Section 3.13 *et seq.*) we shall find that any Compton event in the attenuating material causing the secondary photon to be emitted outside a very narrow cone will reduce the observed intensity. In the limit of indefinitely narrow beam width, the measurements give the decrease in beam intensity in the exact direction of the beam. This condition is referred to as 'narrow beam' or 'good' geometry. Intensity readings for different values of thickness x, or ρx, enable us to make a plot of log $(I/I(0))$ against x which will appear as curve (a) in Fig. 3.19.

In this geometry the reduction in intensity at the detector is due to the fact that some of the original energy is absorbed (as Compton electrons) in the attenuator and some is carried away at an angle. We can therefore deduce a *total* attenuation coefficient σ_T or σ_T/ρ by calculating 0.6932/HVL. σ_T is in fact what we have so far called σ without a subscript, as the whole of the Compton process is taken into account.

On the other hand we could carry out an exactly similar experiment but without collimating the gamma-rays from the source, as is shown in Fig.3.20. This diagram shows that it is now possible for a photon which was not directed initially towards the detector to undergo scattering at such an angle that the scattered photon can contribute to the observed intensity. For measuring the intensity in practice it is now essential to have a detector equally sensitive to direct and scattered photons. The decrease in intensity due to the presence of the attenuator is a measure of energy truly absorbed (as Compton electrons). The fact that photon scattering occurs in the attenuator has no effect on the intensity at P because events like those illustrated in Fig. 3.18 will be compensated for by those shown in Fig. 3.20. Intensity measurements for different attenuator thicknesses will still show an

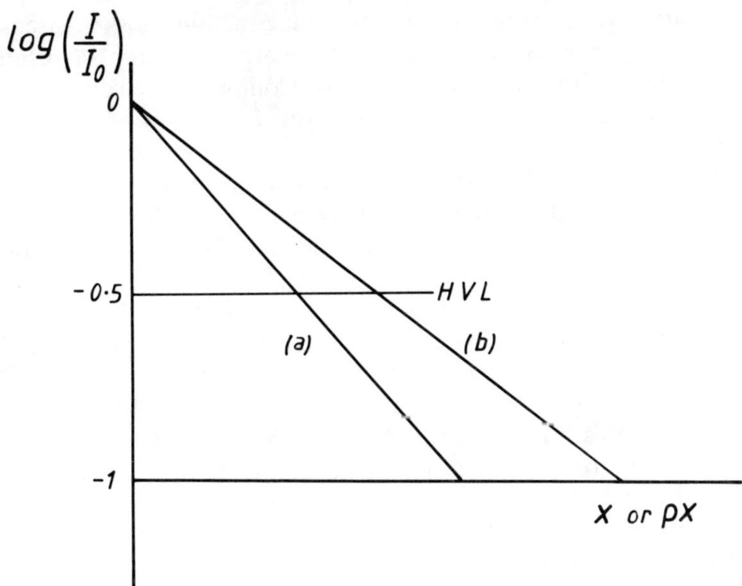

Fig. 3.19 Attenuation curves for photons in (a) narrow-beam, (b) wide-beam geometry.

exponential decrease with thickness but the slope of the plot of log $(I/I(0))$ against x will be less steep. The HVL is thus larger and the coefficient $\sigma_a = 0.6932/\text{HVL}$ will be smaller. This coefficient (or σ_a/ρ, depending on how the thickness is measured) represents the Compton coefficient for absorption alone. The difference between measured values of σ_T and σ_a gives $\sigma_s = \sigma_\tau - \sigma_a$ and this (or σ_s/ρ) is called the Compton scattering coefficient. (Sometimes, and rather confusingly, σ_a has been called the *true* absorption coefficient because it relates to the energy truly absorbed within the attenuator; one must be careful not to interpret the subscript τ as meaning 'true', thus writing σ_τ when σ_a is meant.)

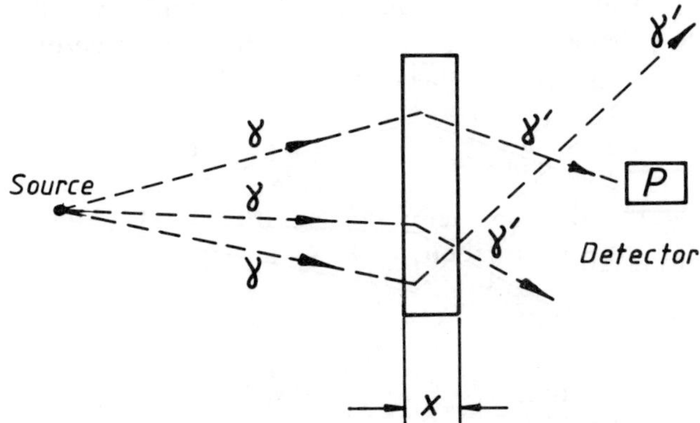

Fig. 3.20 Attenuation in wide-beam geometry (Compton scattering only).

In summary, σ_T and σ_T/ρ represent the total fractional loss of intensity of a beam of photons, per unit thickness of attenuating medium traversed. They take into account absorption of energy (of Compton electrons) and energy scattered out of the beam. They are derived from narrow-beam measurements.

σ_a and σ_a/ρ represent the fractional loss of intensity of a beam of photons, per unit thickness of attenuating medium traversed, which is accounted for by creation of Compton electrons. They are derived from wide-beam measurements.

σ_s and σ_s/ρ represent the fractional loss of intensity of a beam of photons, per unit thickness of attenuating medium traversed, which is accounted for by Compton scattering out of the beam direction. They are derived from the other two coefficients by subtraction.

A good deal of care is needed in selecting the proper Compton coefficient to use for calculation purposes. For instance in comparing the relative frequencies with which the main processes of photoelectric absorption, Compton scattering, and pair production occur at some defined energy, it is correct to compare the values of τ, σ_T and κ. But in estimating the thickness of shielding to use to give a specified degree of reduction of radiation intensity when a gamma-emitting radionuclide is to be stored behind a lead wall, it is essential to use the value of σ_a appropriate to lead. The geometry is clearly 'wide-beam'. The use of σ_T instead of σ_a in the calculation might lead to a serious underestimation of the thickness required for personal safety. Problems of shielding of detection equipment are taken up in Section 3.21.

3.10 Exposure Rate in Air from a Gamma-Ray Emitting Source

Suppose we have a radioactive source of unit activity (1 Bq), each disintegration giving a single gamma-ray of energy E_γ. For simplicity we will regard the source as a geometrical point and neglect self-absorption in it. Then the flux density of radiated energy on the surface of a sphere of radius r centred on the point is $E_\gamma/4\pi r^2$ MeV m^{-2} s^{-1} or $E_\gamma \times 1.6 \times 10^{-13}/4\pi r^2$ J m^{-2} s^{-1}. To find the rate of energy absorption per kilogram of air we multiply by the appropriate attenuation coefficient (μ/ρ). Since the roentgen is the exposure which gives an energy absorption of 0.008 76 Gy or 0.008 76 J kg^{-1} we find

$$\text{exposure rate} = \frac{E_\gamma \times 1.6 \times 10^{-13}}{0.008\,76 \times 4\pi r^2}\left(\frac{\mu}{\rho}\right) \qquad \text{R s}^{-1}$$

Fig. 3.17 shows that over an energy range of 0.1 to 2 MeV the value of (μ/ρ) is 0.0028 m^2 kg^{-1} to within about 10%. As a rough estimate we then have

$$\text{exposure rate} = 4.07 \times 10^{-15}\,\frac{E_\gamma}{r^2} \qquad \text{R s}^{-1}$$

Of course most radionuclides have decay schemes which are much more complicated than that in our initial assumption. In these cases, it it necessary to sum the contributions to the exposure rate from all the gamma-rays emitted per disintegration, due allowance being made for their relative abundances. Some care is needed in interpreting published decay scheme data to take proper account of conversion-electron emission. This process and that of electron capture produce vacancies in the K and more rarely the L and M levels and therefore give rise to X-ray emission. If the X-rays are of more than 10 keV they are conventionally regarded as contributing to the gamma-ray exposure but if of lower energy they are ignored. This is because their attenuation in matter is so strong that they deposit all their energy in a distance comparable with beta-ray ranges; it is therefore convenient when calculating exposure, e.g. in tissue, to add their contribution to that from beta particles. The calculation of exposure from X-rays must also take account of fluorescence yield (Section 2.14). A table of exposure rates for point sources of common radionuclides at a distance γ of 1 m is given here (Table 3.1).

Table 3.1 Exposure Rates (R s^{-1} from 1 Bq at 1 m)

Nuclide	Half-life	Rate $\times 10^{14}$	Nuclide	Half-life	Rate $\times 10^{14}$
^{3}H	12.6 yr β	0.0	^{81m}Kr	13 s IT	0.06
^{11}C	20.4 min β^{+}	0.43	^{85}Kr	10 yr β^{-}	0.001
^{14}C	5700 yr β^{-}	0.0	^{86}Rb	18.6 d β^{-},γ	0.04
^{15}F	110 min β^{+}	0.43	^{85}Sr	65 d EC,r	0.22
^{22}Na	2.7 yr β^{+},γ	0.89	^{87m}Sr	2.8 h IT	0.16
^{24}Na	15.0 h β^{-},γ	1.34	^{95}Nb	35 d β^{-},γ	0.32
^{32}P	14.3 d β^{-}	0.0	^{99}Mo	67 h β^{-},γ	0.12
^{35}S	86 d β^{-}	0.0	^{99m}Tc	6.05 h IT	0.056
^{40}K	1.3×10^{10} yr β^{-},γ	0.02	^{113}Sn+^{113m}In	120 d EC,γ	0.19
^{42}K	12.4 h β^{-},γ	0.10	^{113m}In	100 min IT	0.09
^{45}Ca	160 d β^{-},γ	0.42	^{123}I	13 h EC,γ	0.13
^{47}Ca+^{47}Sc	4.5 d+3.4 d β^{-},γ	0.46	^{124}I	40 d EC,β^{+},γ	0.49
^{47}Sc	3.4 d β^{-},γ	0.04	^{125}I	60 d EC,γ	0.09
^{51}Cr	27.8 d EC,γ	0.01	^{131}I	8.05 d β^{-},γ	0.19
^{55}Fe	2.6 yr β	0.0	^{132}I	2.3 h β^{-},γ	0.96
^{54}Fe	45 d β^{-},γ	0.46	^{133}I	20 h β^{-},γ	0.25
^{56}Co	77 d EC,β^{+}	1.36	^{127}Xe	36.4 d EC,γ	0.16
^{57}Co	270 d EC	0.06	^{131m}Xe	11.8 d IT	0.09
^{58}Co	71 d EC,β^{+}	0.41	^{133}Xe	5.27 d β^{-},γ	0.04
^{60}Co	5.3 yr β^{-},γ	0.97	^{133m}Xe	2.26 d IT	0.05
^{64}Cu	12.8 h β^{-},β^{+}	0.09	^{130}Cs	30 min β^{+},EC,β^{-}	0.23
^{65}Zn	245 d EC,β^{+}	0.22	^{131}Cs	9.7 d EC,γ	0.05
^{67}Ga	78 h EC	0.07	^{137}Cs+^{137}Ba	30 y+2.6 min β^{-},γ	0.25
^{68}Ga	68 min β^{+}	0.42	^{198}Au	2.7 d β^{-},γ	0.18
^{74}As	17.5 d β^{+},β^{-}	0.37	^{197}Hg	65 h EC,γ	0.05
^{75}Se	120 d EC	0.52	^{203}Hg	47 d β^{-},γ	0.11
^{82}Br	36 h β^{-},γ	1.09	^{203}Pb	52 h EC,γ	0.13

They have been calculated using the following empirical equation for (μ/ρ) which fits the curve in Fig. 3.17 to better than 2% over the range 0.1 to 2 MeV:

$$\mu/\rho = \text{antilog}\,(-6.903 - 3.276 \log E_\gamma) + \text{antilog}\,\{-2.551 - 0.1715 \log E_\gamma - 0.3062\,(\log E_\gamma)^2\}$$

For most radionuclides the contribution to exposure rate from X-rays is comparatively small; K X-ray energies have been calculated from the empirical relationship

$$E_K = -1.2 + 0.103\,33Z + 0.006\,125Z^2 + 0.000\,049\,17Z^3$$

which gives values correct to better than 1% (for the $K\alpha_1$ X-ray) for values of Z for the emitting atom between 20 and 90. The K-shell fluorescence yield ω_K has been calculated from the empirical equation

$$\omega_K = -2.6587 + 0.121\,818Z - 0.001\,583Z^2 + 0.000\,007\,125Z^3 + 29.057/(Z + 10)$$

It has been assumed throughout that only K shell events occur, L and M X-rays being ignored. As some electron capture does actually involve L and M electrons the results should err on the side of overestimating exposure rates. Decay scheme data have been taken almost entirely from the compilation *Table of Isotopes* (Lederer, Hollander and Perlman, 6th edition). Most of the values given have been confirmed by experimental measurement to within a few per cent. There is something to be said for using a table of self-consistent values rather than individual experimental determinations.

Earlier compilations of exposure rates were based on the older units, in particular the 'specific gamma-ray dose constant Γ' which was the exposure-rate in roentgens per hour with a source of one millicurie at a distance of 1 cm. Values of Γ in the literature may be converted to the present SI-based units by multiplying them by 7.5×10^{-16}. Still another widely used unit was the '*k*-factor', which was the exposure-rate in milliroentgens per hour with a source of one millicurie at a distance of one metre. For these units the conversion factor is 7.5×10^{-17}.

3.11 Attenuation Coefficients in Materials Other Than Air

The attenuation of X and gamma photons in water, aluminium, sodium iodide and lead has been extensively studied. The main results are that the mass-attenuation coefficients for the photoelectric and pair-production processes are progressively larger the higher the Z (or average Z) of the material. τ/ρ varies approximately as Z^3 and κ/ρ approximately as Z. However the Compton coefficients σ_s/ρ and σ_a/ρ decrease slightly with Z. Figs. 3.21 to 3.24 show some essential data. It will be noticed that there are discontinuities in the curves for the photoelectric attenuation coefficients for sodium iodide and

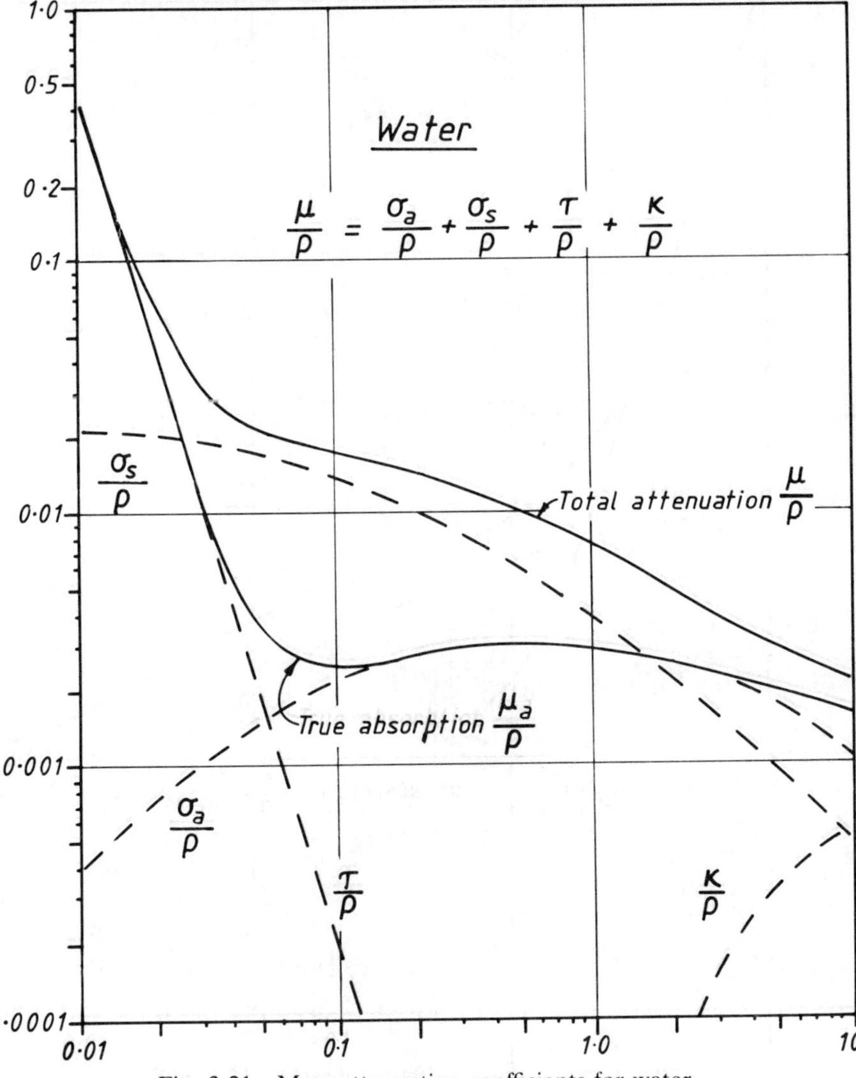

Fig. 3.21 Mass-attenuation coefficients for water.

for lead. Similar discontinuities would also be evident on the curves for air, water and aluminium if they had been extended to lower energies. To explain how these arise will take the abrupt change in (τ/ρ) value for sodium iodide, occurring at a photon energy of about 30 keV, as an example. Photons of energy less than this value are unable to eject K electrons from iodine atoms because these have a slightly higher binding energy, and the photoelectric coefficient depends entirely on the interactions with less strongly bound electrons in the medium. However, with photons whose energy exceeds 30 keV, the K electrons in iodine can also be dislodged and the value of the photoelectric coefficient rises accordingly. The sharp rise in the curve is

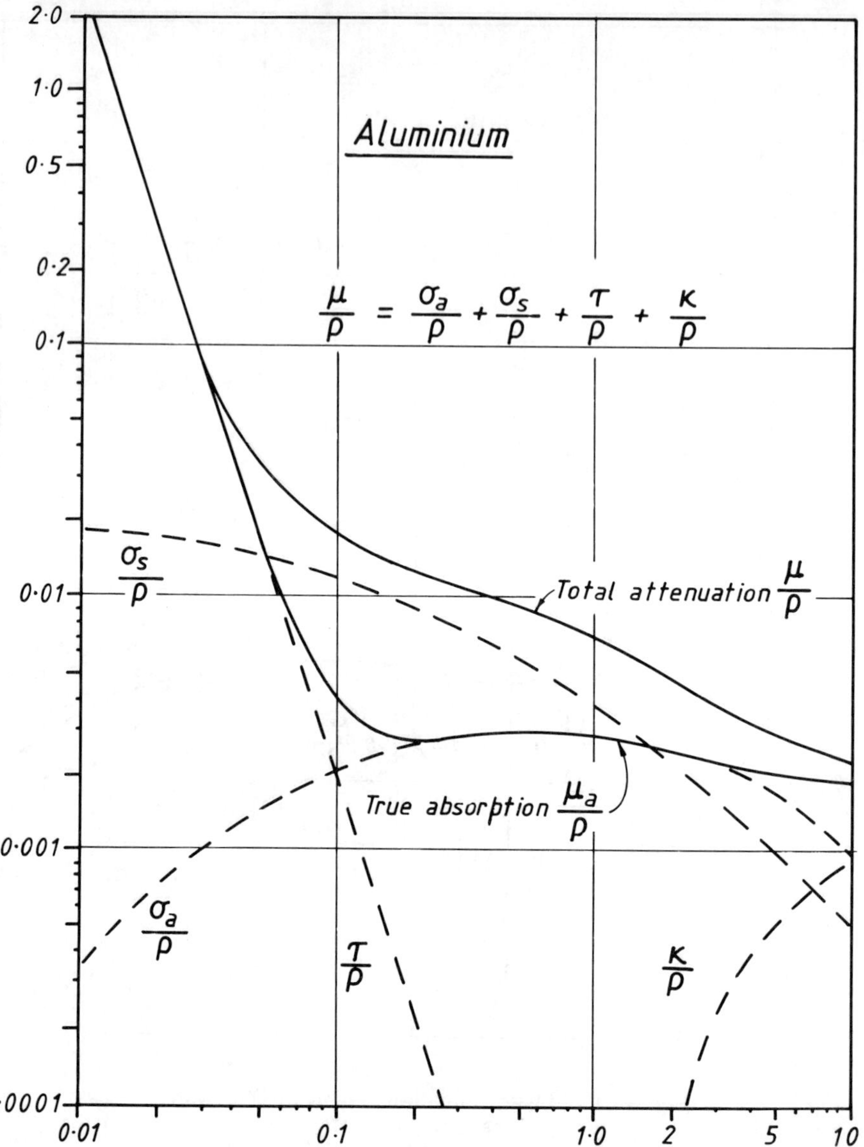

Fig. 3.22 Mass-attenuation coefficients for aluminium. (Reproduced from G. Hine and G. Brownell, *Radiation Dosimetry*, by permission of Academic Press.)

labelled 'K-edge' to denote this effect. The K-edge for lead occurs at about 87 keV and the curve also shows L-edges for photon energies of about 15 keV as L electrons are involved in a similar way.

The attenuation properties of many other materials have also been examined and it is possible to get estimates of the coefficients for all the elements, either from direct measurements or from interpolations between

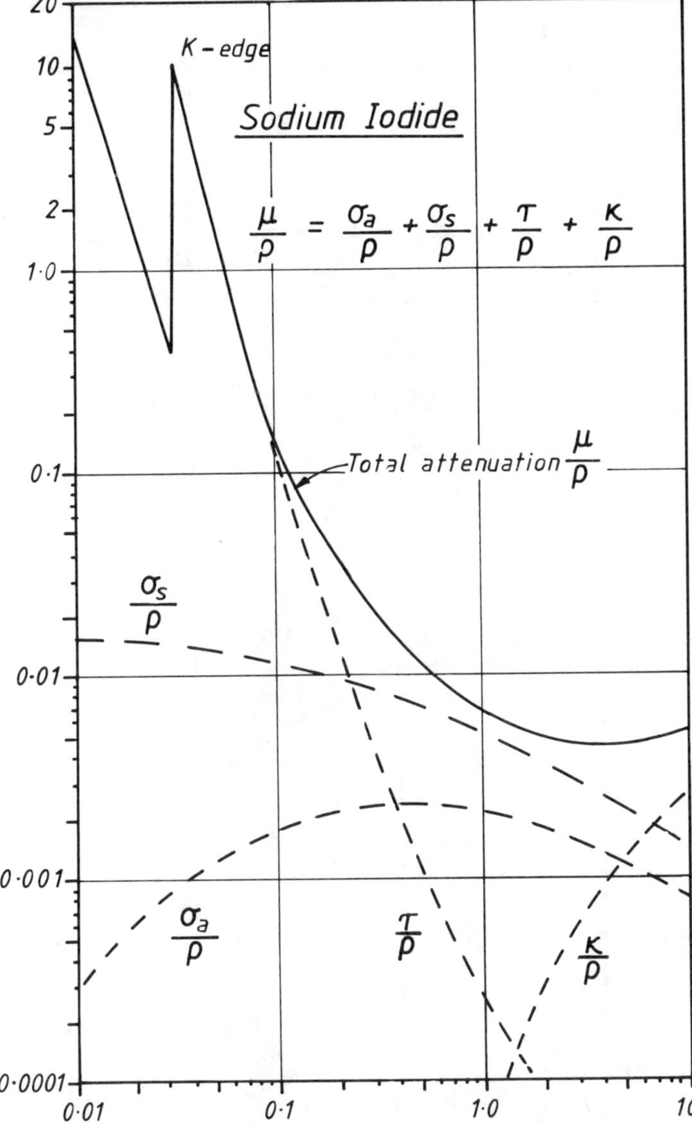

Fig. 3.23 Mass-attenuation coefficients for sodium iodide.

neighbouring elements. If the atomic composition of a complex substance such as muscle is known, it is then a relatively easy matter to work out the coefficients for that, too, by summing the contributions of the constituents. These estimates make possible a very important step forward which we illustrate by means of an example. We find that for photons of 20 keV the sum of the mass-attenuation coefficients for energy absorption (very largely τ/ρ) for bone is very close to 4.8 times that for air. Now 1 R of exposure produces

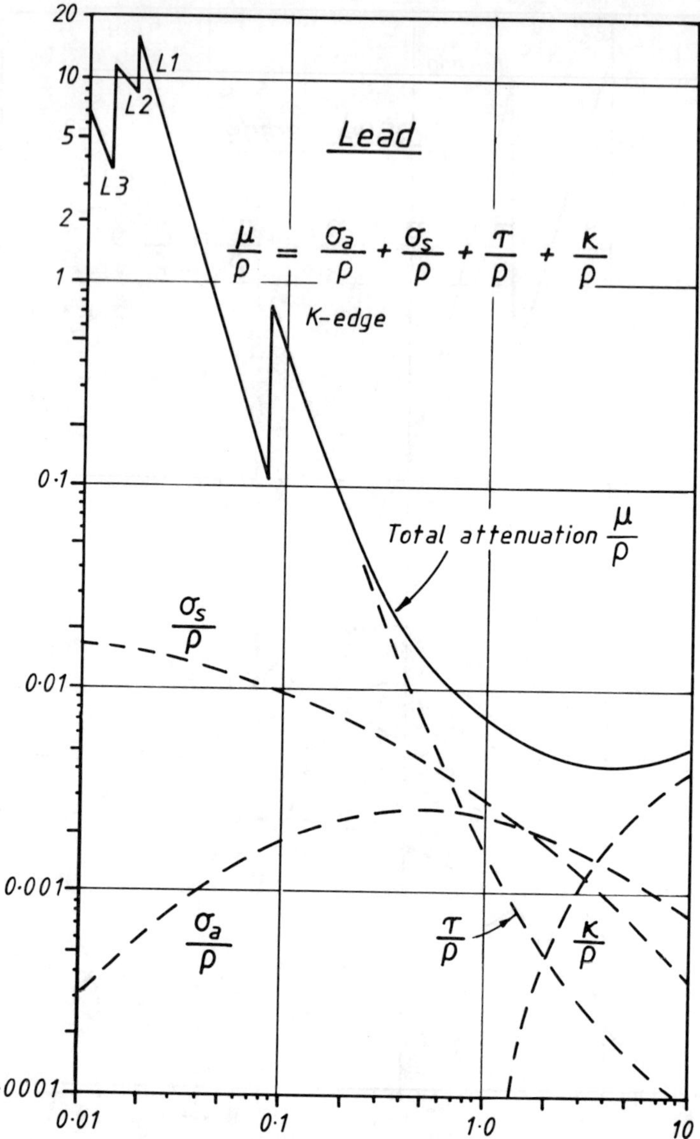

Fig. 3.24 Mass-attenuation coefficients for lead. (Reproduced from G. Hine and G. Brownell, *Radiation Dosimetry*, by permission of Academic Press.)

in air an energy absorption of 0.008 76 Gy; therefore, in bone 1 R of exposure at 20 keV will give an absorbed dose of 0.008 76 × 4.8 = 0.043 Gy. The number of grays per roentgen for a particular material and photon energy is called the *f* value. (In older literature it appears as the number of rads per roentgen.) Some values for biologically important materials are given in Table 3.2.

Table 3.2 *f* **Values; Absorbed Dose in Grays per Roentgen**

(Reproduced from W.J. Meredith and J.B. Massey, *Fundamental Physics of Radiology, by permission of John Wright & Sons.*)

Photon energy (keV)	Air	Water	Muscle	Bone
20	0.008 76	0.008 79	0.009 17	0.0423
40	0.008 76	0.008 79	0.009 20	0.0413
60	0.008 76	0.009 05	0.009 29	0.0219
80	0.008 76	0.009 32	0.009 40	0.0191
100	0.008 76	0.009 49	0.009 49	0.0146
150	0.008 76	0.009 62	0.009 56	0.0105
200	0.008 76	0.009 73	0.009 63	0.00977
500	0.008 76	0.009 65	0.009 57	0.00925
1000	0.008 76	0.009 65	0.009 57	0.00919

RADIATION MEASUREMENTS BASED ON IONISATION

3.12 Ionisation Chambers

An enclosed space containing air, or some other suitable gas, and fitted with electrodes for the collection of ions created in the space, is termed an ionisation chamber. When such a chamber is irradiated with X- or gamma rays, or a beta or alpha particle source placed within it, primary ionisation will occur in the gas and also within the chamber walls and electrodes if these are within the paths of the incident photons or particles. Some of the primary electrons set free near the inner surfaces of the walls and electrodes may find their way into the gas space; conversely some electrons set free in the gas may be absorbed in the walls. The primary electrons cause secondary ionisation and the final set of ions remaining in the gas phase following a single ionising event depends in a very complicated way upon the geometry of the system and the nature of the gas and the wall and electrode materials. In the absence of an electric field, the ions in the gas phase will recombine, but if a sufficiently high potential is maintained, the negative ions will drift towards the positive electrode, and vice versa. As the negative ions approach the positive electrode they induce a positive charge upon it, or, what comes to the same thing, they set free electrons within the electrode which may flow to the supply of potential and constitute a current; it is important to realise that the current flows *during* the motion of the ions being collected, and stops when they have all arrived (on both electrodes). The total duration of the pulse of current due to electron collection is typically a microsecond or less, and depends on the magnitude of the applied electric field. The collection time for the positive ions is much longer, of the order of milliseconds. An important feature of ion chambers and other gas-filled detectors is that the electrons remain as such only in the absence of electronegative gases such as oxygen and the halogens; if these are present in the gas, negative ions such as O_2^- are readily formed and these are also collected but only in millisecond times. If it is required to detect individual ionising events, i.e. to use the chamber in 'pulse mode', it is essential to exclude these gases and count the 'sharp' pulses produced from electron collection. The pulses are of small magnitude and some form of pulse amplification with very high gain is necessary (Section 3.14). However, an ionisation chamber can be used to measure radiation from rather strong radioactive sources, and from X-ray machines, by using it in the 'continuous mode'. The gas filling the chamber can be air, or indeed almost any gas; it is assumed that the exposure is great enough, and continued over sufficient time, to produce a steady *ionisation current*. For most chambers the current is roughly proportional to the radiation exposure, and of course in principle a chamber of any design can be calibrated in known radiation fields. There are, however, two particular kinds of chamber for which the relationship between radiation exposure and response can be found theoretically and which can therefore be used to give absolute measurements. These are the 'free-air' chamber and the 'air-equivalent wall' or 'thimble' chamber.

In the free-air chamber the ionisation is created in a certain volume of air, called the *active volume*, which is entirely surrounded by air. The peculiar geometry which allows this to be done is illustrated in Figs. 3.25 and 3.26. Ionising radiation from the source (e.g. a radioactive source or an X-ray tube

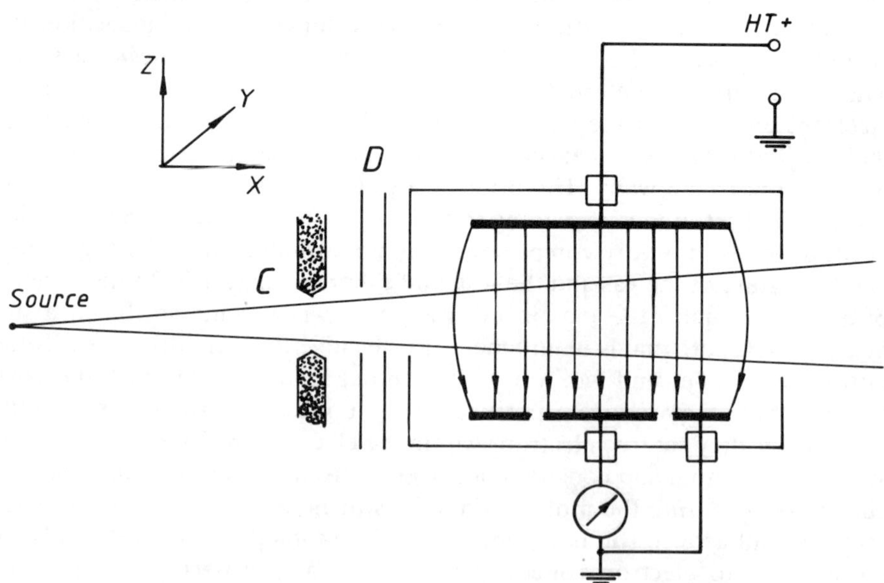

Fig. 3.25 The free-air ionisation chamber (section). (Reproduced from G. Hine and G. Brownell, *Radiation Dosimetry*, by permission of Academic Press.)

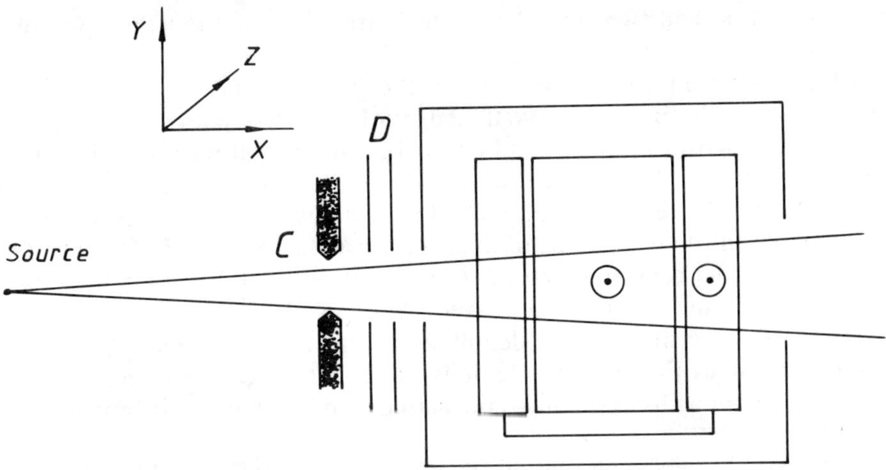

Fig. 3.26 The free-air ionisation chamber (plan).

target) enters the chamber through a defining hole C in a lead wall. This means that the volume of air in which primary ionisation occurs is defined in the Y and Z directions (Fig. 3.27) entirely by the geometry of the source and the aperture. In order to prevent radiation scattered from the edges of the aperture C from entering the chamber, there are further diaphragms D between them. The beam is thus confined to the cone with apex at C. It does not strike chamber walls or electrodes, and photoelectrons ejected from the edges of C have too short a range to enter the main part of the chamber. All the primary ionisation in the middle part of the chamber therefore truly originates in the air itself. As shown in the diagrams the collecting electrodes are parallel plates well outside the beam path. One of them is split into a

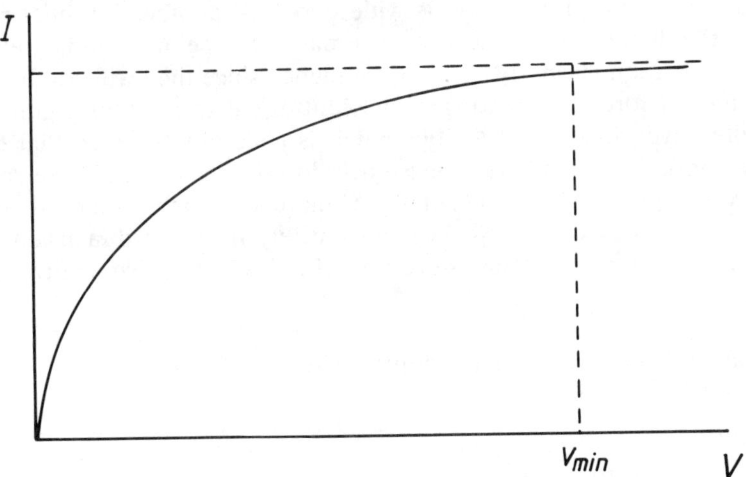

Fig. 3.27 Characteristic curve of ionisation chamber.

central portion and two side-plates which are connected together (see plan view).

Electrostatic lines of force between the plats are shown in Fig. 3.25 by arrowed lines; those which leave the edges of the plates show some curvature but they are sensibly straight and parallel in the central region of the chamber. The current which is measured is that carried between the single-plate electrode and the central region only of the split electrode. This arrangement ensures that the active volume of the chamber in the X direction is defined by the length of the central electrode in that direction. All three dimensions of the active volume are therefore fixed. The chamber, therefore, fulfils the requirement, specified in the definition of the roentgen unit, of having a volume of air as the ionising medium which is surrounded entirely by air, while still enabling us to calculate the actual amount of air contributing to the current.

It has still to be shown that all the secondary ionisation will be collected. One requirement for this to be the case is that the chamber dimensions should be so large that no secondary electron should be able to reach the walls or electrodes. If the active volume is large, a high proportion of them will come to a halt within it, but inevitably some will escape into the surrounding spaces. Let us consider the X, Y and Z directions in turn. There will be a flux of secondary electrons leaving the active volume in the X direction, and these will fail to contribute to the measured current. But unless there is appreciable attenuation in the active volume itself, this flux will be balanced almost exactly by electrons entering the active volume from the left (in Figs. 3.25 and 3.26). (The same argument applies to primary electrons moving on the X direction.) There will also be a balancing flux of secondaries in the *negative X* direction. There is thus no resultant loss of current through electron escape in the X direction. As for electrons escaping from the active volume in the Y direction, Fig. 3.26 makes it clear that there need be no loss of ionisation current if the central electrode is wide enough—another essential requirement of the dimensions of the system. Finally, escape of secondary electrons in the Z direction is clearly of no consequence since they will still be carried along lines of force terminating of the central electrode. Our arguments do not quite cover all the possibilities but is is generally believed that the few free-air ionisation chambers in existence do give reliable absolute radiation measures. In principle the mass of air, M kg (correcting to standard temperature and pressure) can be worked out entirely from the dimensions of the instrument; then a steady exposure rate of r R s^{-1} will give a current I of

$$I = Mr \times 2.58 \times 10^{-4} \quad \text{A}$$

Alternatively we may say that when a current of I amperes is observed, the exposure rate r is

$$r = \frac{I}{2.58 \times 10^{-4} M} \quad \text{R s}^{-1}$$

An assumption that we have implicity made is that the recombination of positive and negative ions has been prevented. An experimental test that this is so can be carried out by measuring the current, under constant irradiation conditions, as a function of applied potential. A typical current–voltage characteristic is illustrated in Fig. 3.27. The current rises steadily as the voltage is increased from zero and approaches a limiting value asymptotically. Clearly, it is advisable to use a working voltage such that the current is within say 99% of the saturation value. This working voltage is, however, different for different levels of radiation exposure rate, a greater voltage being needed the greater the rate. For a chamber filled with air at normal pressure, one may calculate the ratio f of the ionisation current actually observed to the theoretical limit from the relation

$$f = 1/(1 + \eta)$$

where η is a function given approximately by

$$\eta = 2 \times 10^{12} \frac{ID^3}{AV^2}$$

and I = ionisation current, D = distance between plates, A = plate area, V = applied voltage. From these formulae it can be shown that the working voltage of a chamber with plates 0.1 m by 0.1 m and plate spacing of 0.1 m with an ionisation current of 1 μA will be about 14 kV. Typical values for the dimensions of the active volume of the chamber would be 0.03 m by 0.1 m. Then since 1 m^3 of standard air weighs 1.293 kg, the exposure rate r would be

$$r = \frac{10^{-6}}{2.58 \times 10^{-4} \times 3 \times 10^{-4} \times 1.293}$$

$$\approx 10 \text{ R s}^{-1}$$

This is quite a high exposure rate; a typical X-ray therapy unit will give an exposure rate an order of magnitude smaller.

In the second kind of ionisation chamber we are discussing there is no attempt to ensure that all the ions collected arise from ionising events in air. The chamber is quite small and is often referred to as a thimble chamber, and although some designs are larger than ordinary sewing thimbles others are much smaller. The body of the chamber is of some conducting material and forms one electrode (usually earthed). The second electrode is a short aluminium rod entering the chamber through a high-quality insulator (see Fig. 3.28). When a chamber of this type is placed within a radiation field it is clear that a substantial fraction of the ionisation formed within the gas will originate from primary ions created initially in the walls, not in the air itself. It follows that the ionisation current, although it may be proportional to the radiation intensity, is not simply related to the mass of air within the chamber. There is, however, one important circumstance in which the chamber does act as an ideal air–wall chamber. This is when the wall material is chosen to have

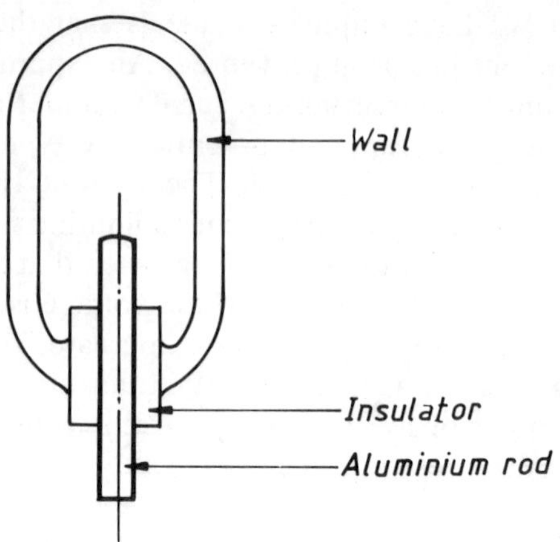

Fig. 3.28 Thimble ionisation chamber.

the same effective atomic number as air itself—a condition which is met approximately if the material is graphite. Special graphite-impregnated plastics incorporating a small addition of a heavy element are produced commercially for the manufacture of 'air-wall-equivalent' chambers. If the wall thickness is greater than the range of the primary electrons formed in it, it can be shown that the ionisation in the cavity is the same as it would be if the cavity were surrounded only by air. (This is the so-called Bragg-Gray principle.) That this must be so can be appreciated by considering the paths traced out by the ion-pairs in the two cases, viz. an air cavity surrounded by air-equivalent walls, and an identical air cavity surrounded by air only. If we imagine the surrounding air in the second case to be compressed uniformly until it had the same density as the real walls but without changing the cavity itself, it will be seen that the tracks within the compressed air will be shortened but not changed in numbers or direction between collisions. The tracks within the chamber will be identical with those formed when the walls were in the uncompressed state. This principle is illustrated in Fig. 3.29. For work with X-rays and gamma-rays of up to a few hundred kilo-electron volts, thimble-chamber wall-thickness needs to be of the order of 1 mm. They should not of course be too thick or they will cause appreciable attenuation of the radiation being measured.

For some purposes it is convenient to use a thimble chamber in a discontinuous mode, by placing an electrostatic charge on the central electrode, exposing the chamber for a fixed time in a radiation field, and then measuring the remaining charge. In this mode the chamber is acting essentially as a capacitor which is being partially discharged by the collection of ions on its plates. From the familiar expression for the charge Q carried by a capacitor having a capacitance C,

$$Q = CV$$

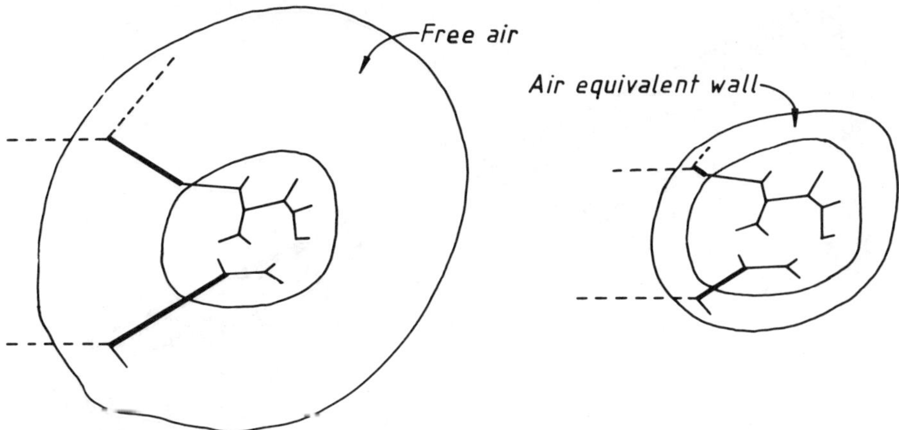

Fig. 3.29 The Bragg–Gray principle: broken lines, incident X- or gamma radiation; thick full lines, track of primary electrons; thin full lines, track of secondary electrons.

we easily find that the collection of a charge ΔQ will cause a change of voltage ΔV where

$$\Delta Q = C\Delta V$$

Then if m is the mass of air within the chamber, the exposure in roentgens will be

$$\text{exposure} = \frac{\Delta Q}{m} \times \frac{1}{2.58 \times 10^{-4}}$$

$$= \frac{C\Delta V}{m \times 2.58 \times 10^{-4}} \quad \text{R}$$

A precaution that needs to be taken here is that the electrical potential across the electrodes remains high enough during the whole irradiation to ensure efficient ion collection. The voltage remaining after irradiation can conveniently be measured using a quartz-fibre electroscope, whose own capacitance has to be allowed for. With this system an exposure of 100 R can easily be measured to an acuracy of 1–2%.

A popular but much less precise form of 'condenser' chamber is the so-called 'pen-monitor' which can be clipped into the pocket on the clothing and used for personnel monitoring. The construction of one of these is shown diagrammatically in Fig. 3.30. The essential feature is an electrode system which can be charged at the connection E and which is supported on a transparent insulator. The electrode bears a quartz-fibre electrometer F. An image of the fibre is focused onto a translucent graduated screen S by the lens L_1. The image superimposed on the graduations may be viewed through the eyepiece lens L_2. The image takes up different positions on the scale depending on the charge on the electrode; the scale is commonly graduated to read directly in milliroentgens from 0 to 200 mR.

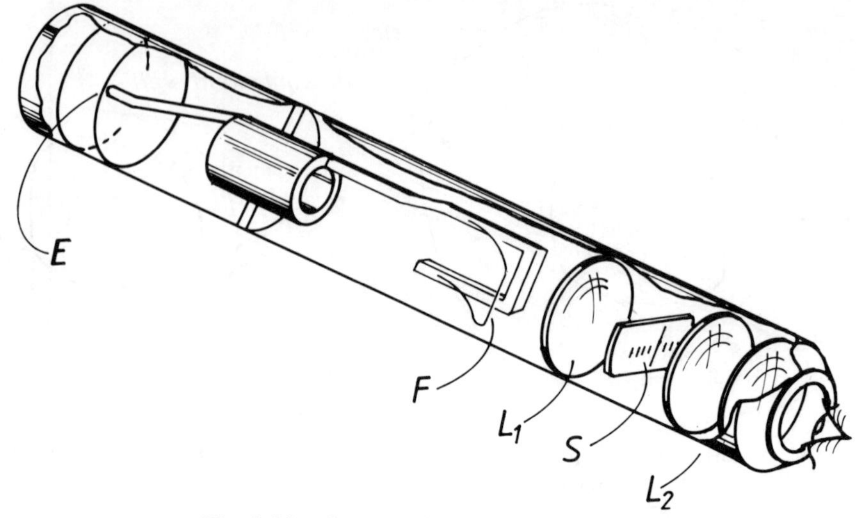

Fig. 3.30 Construction of a 'pen-monitor'.

Aside from the above kinds of ionisation chamber there are many other designs, there being very few restrictions on the nature of the filling gas and the materials of the walls of a chamber whose output one is prepared to calibrate by comparison with a free-air or air-equivalent-wall chamber. Thus many portable 'survey meters' have brass or light-alloy walls and may be filled with argon at high pressure to increase their sensitivity. These are hardly very precise instruments but give, or should give, high reliability in service as environmental monitors.

3.13 Pulse-Mode Detectors for Single Ionising Events

In order to make useful measurements of the radiation from weak radioactive sources, it is necessary to have detectors working in the pulse mode, so that the ionisation produced by individual particles or photons can be detected and the number of these detection events occurring in a given time can be recorded. Now a beta particle of say 1 MeV will generate about 3×10^4 ion-pairs along its track in a suitable medium such as argon at atmospheric pressure. If these are collected on electrodes which have an effective electrical capacity of say 20 pF they will change the potential across the electrodes by $3 \times 10^4 \times 1.6 \times 10^{-19}/(20 \times 10^{-12}) = 2.4 \times 10^{-4}$ V. This is very small compared with the potential which has to be maintained in order to ensure rapid and complete ion collection and is impossible to measure directly. It is common practice to connect the detector in a circuit as shown in Fig. 3.31, where R has very high resistance and C is a coupling capacitor. The circuit is also applicable to many forms of solid-state detectors. The operation of the circuit may be described as follows. As electrons are collected by the anode of the chamber they induce positive charges on it, or in other words release electrons which flow to the coupling capacitor and partially discharge it. If the

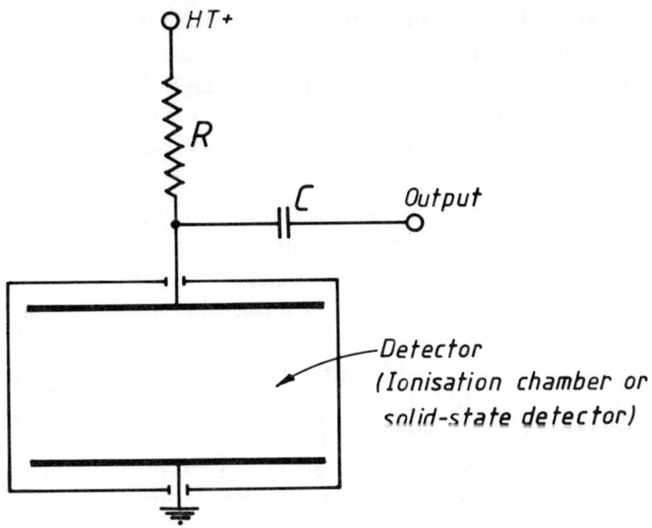

Fig. 3.31 Basic deflector circuit.

resistance of R is high enough (10^6–10^9 Ω), relatively few electrons will be able to flow through it to the positive potential supply before electron collection is complete. Thereafter the capacitor (and the chamber) will slowly become fully charged again as current flows through R. The whole process therefore gives rise to a voltage pulse at the output side of the coupling capacitor; this is negative-going and attains its maximum excursion in approximately the time taken for electron collection, which may be much less than 1 μs. The pulse dies away much more slowly, with a time constant determined by the value of R and of the effective elective electrical capacitance in the circuit. The positive ions formed in the chamber will also generate a positive-going pulse but its rise-time will be at least a thousand times longer than the electron-collection pulse because of the very much smaller mobility of the positive ions, and its amplitude will be low because current can flow through R quickly enough to neutralise the induced charges. The device thus acts as a voltage generator giving negative pulses with relatively rapid rise-time and slow fall-time, with an amplitude of the order 10^{-4} V, and of course with a very high output impedance. The first stage of an amplifying system often takes the form of a single-stage preamplifier constructed integrally with the chamber itself. It may have quite a low voltage gain—indeed this is often less than unity, its main function being to provide an output signal with a low output impedance.

3.14 Auxiliary Electronic Equipment

Following the preamplifier, the signals are sent to a main amplifier whose design frequently incorporates a circuit with fixed voltage gain and variable attenuators so that the overall gain may be adjusted at will. The original signal amplitude may thus be increased by a factor of up to 10^6. The final pulses will have the same waveform as the original pulses if the whole system is truly *linear* and has a good frequency response up to at least 10 MHz, so that it can follow the fast rise-time (of the order 100 ns or less) of the detector signal.

Output pulses from the main amplifier may be dealt with in various ways. The simplest technique is to feed them into a scaling unit which counts them electronically, displaying the total received in a given time as determined by an electronic clock. This number may usefully be typed out on an electric typewriter, or punched out on a card or paper tape. For fast counting, recording on a magnetic tape is advisable. For some purposes the output train of pulses needs to be sorted according to the pulse amplitudes. A single-channel analyser will take the pulse train as input and give output pulses, of a standard waveform suitable for sending to a scaler, only for those inputs whose magnitude lies between predetermined limits E and $E + \Delta E$. The limits are in actual fact voltages set by adjusting potentiometers, but if we can assume linearity in the amplifying system we may as well use the letters E and ΔE because the pulse voltages are just proportional to the energy deposited in the initiating event in the detector. In other words, the potentiometers can be calibrated to read directly in MeV or KeV instead of voltage. We can refer to E and $E + \Delta E$ as the lower and upper discriminator levels respectively, and to ΔE as the window width. Multichannel analysers perform the same function as a set of many–e.g. 1000–single-channel analysers, giving separate output trains within the limits E and $E + \delta E$, $E + \delta E$ and $E + 2\delta E$, $E + 2\delta E$ and $E + 3\delta E$, and so on, where δE (a small increment in E), is the channel width; in addition the numbers of pulses sorted into the channels are stored in separate memories for subsequent print-out. Oscilloscope displays on these instruments enable the contents of all the memories to be displayed simultaneously at the end of, or even during, a recording interval, giving a picture of the whole of the energy spectrum. Other facilitics may include that of printing out the sum of the contents of a chosen group of channel memories. Another important use of multichannel analysers is to store counts from a detector obtained serially in time, successive memories being used at fixed time intervals. The intervals being electronically controlled, it is easy to record data over very short times, for instance to determine half-lives as short as a few milliseconds, or even microseconds if special techniques are used to handle very short pulses.

The need sometimes arises to have a continuous analogue signal as a record of count-rates as a function of time. The output train of pulses from an amplifier or, preferably, a single-channel analyser should then be passed to some form of rate-meter. In a common design of this instrument, a small charge is generated as each pulse is received, and fed through a diode into a capacitor which is by-passed by a high resistance. The voltage developed on the capacitor attains a steady value, for a constant input rate, when the rate of charge accumulation is just balanced by the leakage rate. This voltage is read on a voltmeter or used to drive a pen-recorder. If the count-rate is fluctuating, the indicated voltage is proportional to the mean rate over the previous period of about 2τ, τ being the time-constant RC of the resistance–capacitance system. Adjustment of R allows a fast count-rate to be displayed with a short time-constant and a slow count-rate to be displayed with a long time-constant, thereby 'smoothing' an otherwise irregular record.

Scintillation detectors operate on quite different principles from ionisation chambers and require light-sensitive amplifiers. By far the most commonly used of these are photomultiplier tubes, a discussion of which is deferred until Section 3.19. A complete description of the design and operation of amplifying and recording systems is outside our present scope.

3.15 Proportional Counters

The size of the voltage pulse developed from an ionisation chamber even when a rather large number of ions have resulted from the initiating particle (as in Section 3.14) is inconveniently small; the power available from such a pulse is only about an order of magnitude greater than that originating in thermal noise in the resistor. Alpha particles, whose energy is commonly about 5 MeV, will of course generate larger pulses, but a way of increasing the ionisation from low-energy beta particles or photoelectrons or Compton electrons would obviously be useful. Such an increase, which is known as 'internal amplification', is achieved in the proportional counter by a modification of the detector geometry. Instead of a planar configuration as is used in typical ionisation chambers, the proportional counter possesses a hollow cylindrical cathode and a very fine coaxial wire as anode. The gas in the chamber must not contain electronegative elements because it is essential for the operation of the chamber that the electrons created in the ionisation should retain their high mobility. As is well known from electrostatics the electric field strength at a distance r from a line carrying charge is proportional to $1/r$ and therefore tends to infinity as r tends to 0. In practice, with a wire anode of say 0.05 mm in diameter, fields of several thousand volts per millimetre are easily achieved very near the wire although the potential across the electrodes is only a few hundred volts. When ionisation occurs within the gas, electrons attracted to the wire are accelerated in this field to the point where they can initiate further ionisation. In this way an 'avalanche' of electrons can be built up and the final number reaching the wire can be 10^3 or 10^4 times as many as there were in the original ionisation. The multiplication factor is almost entirely dependent on the voltage applied and not at all on the initial number of ions. Therefore, the amplitude of the pulse output is just proportional to the initial amount of ionisation, and the constant of proportionality can be changed at will by changing the applied voltage. A chamber operating under these conditions is called a proportional counter (although strictly speaking it is a detector, not a counter).

3.16 Geiger–Müller Counters

If the applied voltage on a proportional counter is steadily increased there comes a stage when a single ionising event initiates not just a single output pulse but the passage of continuous current. The chamber is said to be 'breaking down'. It appears that the bombardment of the anode from the initial avalanche can, if the voltage is high enough, produce ultraviolet photons which in turn produce photoelectrons at points remote from the original event and so start further avalanches. The discharge becomes self-perpetuating and spreads over the whole of the anode wire. It has been found however that under certain conditions it is possible to allow the discharge to grow to this extent but then to be terminated, or 'quenched'. The chamber is then performing as a Geiger–Müller counter. At least two factors are involved in the quenching process:

(a) The composition of the gas filling can be adjusted to exert some control. The main gas constituent is generally an inert gas such as helium or argon, but a small proportion of a 'quenching gas' such as alcohol vapour or bromine is also present. The mechanism of quenching is essentially that the quenching gas can absorb the ultraviolet photons without the production of photoelectrons. An excessive amount of quenching agent would prevent the propagation of the avalanche altogether, but the small amount actually used seems to allow limited growth and give time for other processes to come into effect.

(b) The effect of the discharge spreading all over the anode is to create a space-charge which reduces the effective field on electrons still some distance from it. The tendency to form new avalanches is thus reduced.

The effectiveness of these quenching mechanisms can be enhanced by deliberate reduction in the anode voltage. This can be achieved by the use of an external quenching circuit triggered by the early part of the discharge. After a time interval of some 200–400 microseconds during which the original ions—including the positive ions—are collected, the full voltage is again applied. The length of the time interval is entirely dependent on the characteristics of the quenching circuit, and is referred to as the 'dead-time'. It is important to note that the passage of a particle, which normally would initiate an avalanche but which traverses the chamber during one of these intervals, will not be recorded as a separate 'count', nor will it affect the duration of the interval. The number of events recorded by a Geiger–Müller counter is therefore always somewhat less than the number actually occurring. A correction for the counts 'lost' during the dead-time should be applied (see Section 3.30). This is trivial for count-rates of the order 10 s^{-1} but is a major effect if the count-rate exceeds say 1000 s^{-1}.

For any given radiation source the count-rate observed with a Geiger–Müller counter as a function of the voltage applied to is of the form shown in Fig. 3.32. The working voltage of the counter is usually chosen to be about

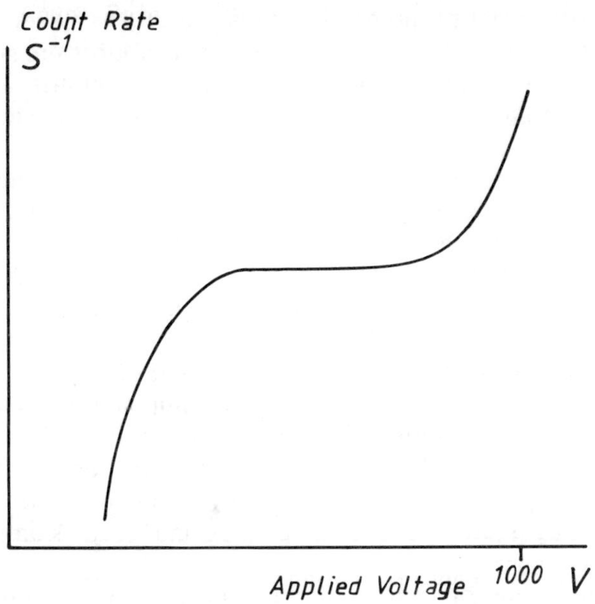

Fig. 3.32 Count-rate as a function of applied voltage for a Geiger–Müller counter.

half-way along the nearly horizontal part of the curve—the so-called 'plateau'. For a good counter the plateau extends over some 50–100 V. To the left of the plateau, some events are not recorded because of failure of the discharge to develop pulses large enough to operate the recording device; when the applied voltage on the other hand is too high the quenching processes become ineffective and the counter tends to go into continuous discharge to develop pulses large enough to operate the recording device; is of the order 0.1% per volt; it is therefore unnecessary to stabilise the voltage supply to better than a few volts unless extremely high reproducibility in performance is required. All Geiger–Müller counters gradually deteriorate owing to changes in the composition of the gas filling; the effect is to shorten and steepen the plateau region. In this respect the modern halogen-quenched tubes are much to be preferred to the original designs in which the quenching agent was alcohol vapour. Halogen-quenched counters will record at least 10^{11} counts before becoming unserviceable. Their working voltage is usually in the range 300 to 600 V.

It should be noted that a Geiger–Müller counter differs from a proportional counter in that there is no longer a proportionality between the energy deposited in the gas filling and the amplitude of the output pulse. Proportional counters need to be quite large, or to be operated at high gas pressure, in order to contain all the ionisation generated by an energetic primary particle, if they are to be used as spectrometers. But as Geiger–Müller counters do not have the proportionality property they cannot be used as spectrometers and there is no point in a large size. Many commercially available tubes have volumes of about 50 ml and they can be made much smaller still for special purposes. The pressure of gas within them is often considerably less than atmospheric. The number of ion pairs generated by a fast electron traversing the gas may therefore by only about 100; nevertheless the very high internal amplification factor, which is commonly in the range 10^7 to 10^8, results in output pulses of 10 V or more. The power in these pulses is sufficient to operate a scaling circuit without further amplification. At any one applied voltage the pulses are all nearly the same amplitude, showing that each primary ionising event in the chamber produces nearly the same number of secondary ion-pairs. The total charge passing through the chamber is typically that carried by 10^{10} electrons per initial event, or 1.6×10^{-9} C. When the count-rate is 10^3 s^{-1} the counter is passing a quasi-steady current of 1.6×10^{-6} A which is easily measurable on a microammeter. Its robustness and the simplicity of the auxiliary electronic equipment required makes the Geiger–Müller counter almost the ideal basis for portable instruments useful in radiation survey work. It should be remembered however that although these counters respond to at least 99% of any beta particles traversing them, the corresponding figure for medium energy gamma rays is only of the order 1%.

One common type of Geiger–Müller tube has been mentioned already (Section 3.5) and another is described in Section 3.26 (Fig.3.44). In spite of their advantages some research workers who are concerned with high-precision measurements have in the past few years discarded them and instead use tubes for counting beta particle sources which superficially resemble Geiger–Müller tubes but which are in fact operated in the proportional region. They are constantly flushed through with a gas mixture of argon and methane at atmospheric pressure. They have the disadvantage of requiring external amplifiers but this is offset by dead-times of only one or two microseconds, which enable them to handle much higher count-rates than would be possible with Geiger–Müller tubes.

3.17 Semiconductor Detectors

In contrast to gas-filled detectors, semiconductor detectors are fabricated from solid materials. Usually they take the form of thin slabs or wafers with metallic contacts on opposite sides to act as collecting electrodes. If such a device could be made with absolutely pure silicon or germanium it could possibly act in the same way as a gas-filled detector, in that ionising events within the material would give detectable pulses as the electrons set free were collected on one electrode and electrons from the other diffused into the material to neutralise the positive ions. But the ideally pure materials are not available and instead we use materials which, if not pure, at least have impurity contents whose nature and relative amounts are under control. During manufacture, samples of silicon and germanium of the very highest purity achievable are prepared. To these are added, in various ways, very small amounts of such elements as phosphorus, gallium, arsenic and antimony. The mixture is formed into large single crystals by melting and slow cooling; the crystals must be as uniform as possible in every sense. The added 'impurities' constitute a few atoms per million of the 'host' material, which means that every cube of host atoms, 100 atoms on a side, has a good chance of containing one or two, but not too many, impurity atoms. Exceptionally, lithium impurity atoms can be introduced into pure germanium crystals after these have been formed from the melt, by a diffusion process at high temperature. The added impurities are of two kinds; 'p-type' impurities consist of 'acceptor' atoms such as gallium which, having only three outer (valency) electrons each instead of the four of silicon or germanium atoms, can in some circumstances accommodate an extra electron, or in other words act as a 'positive hole'. The 'n-type' impurities, such as antimony and arsenic, are 'donors' because, having five valency electrons, they can easily make one available to the rest of the crystal lattice. It is the migration of these electrons, and the apparent migration of the positive holes as electrons move into them, under the influence of an electric field which is responsible for electrical conduction in the material. The conductivity is now some orders of magnitude greater than in the purest available host material.

A semiconductor material which contains a region with a p-type impurity in contact with a region with n-type impurity shows rectifying properties at the junction of the regions, electrons flowing easily from n-type to p-type but not in the reverse direction. If a p–n junction is reverse-biased, that is to say a potential is applied in the sense that the *n*-type material is *positive*, electrons are drained away from the junction on that side, leaving a region called the 'depletion layer'. This is positively charged as electrons have been removed from it, but as it contains no further mobile charge-carriers it is an insulator. Hence the rectifying property of the junction. It is an essential feature of the use of the device as a radiation detector that if an ionising event occurs within the depletion layer, the electrons set free in it can migrate to the anode. Simultaneously electrons from the cathode can drift through the p-type region and neutralise the positive ions. While this picture of the process is very much over-simplified it brings out the essential feature that it is the depletion layer, not the whole slab of material, which is playing the part of an ionisation chamber. It is important too to realise that the width of the depletion layer increases with the applied voltage, so that we have the analogue of an ionisation chamber with variable volume.

There are two main kinds of solid-state detector in common use. The first we consider takes the form of a thin slice of 'doped' silicon which during manufacture is ground perfectly flat and etched to give a mirror-like surface. One side is coated with gold by vacuum evaporation; some workers coat the surface with a thin layer of evaporated germanium beneath the gold, but it seems that a layer of silicon oxide (from exposure to air) on the silicon surface is the prime factor in the formation of a rectifying junction at that surface. The other side of the slice is aluminium coated. The gold and aluminium faces form the anode and cathode respectively of the device, connections being made through a cold-setting silver paste. On applying a potential of 10–100 V, depending on the thickness and other properties of the material, a depletion layer is formed which is a small fraction of 1 mm deep but still sufficient to contain all the track of an alpha particle or a proton accelerated to a few MeV. The mean energy needed to produce an ion-pair in silicon is only about 4 eV (as against 34 eV in air) so that the charge collected from the ionisation is eight times what it would be in an air-filled ionisation chamber. This means that the energy carried by the alpha particle, or proton, can be much more accurately measured and for this reason solid-state detectors have entirely displaced ion chambers for precise energy spectrometry of densely ionising particles. Indeed the energy deposited in only a small portion of such a track can be measured by using a shallow depletion layer. Moreover the direction of a particle in space can be found by using two or more thin detectors one behind another in a straight line so that the particle passes through all of them. For this purpose, detectors whose thicknesses are only 25 μm have been manufactured; these are of course extremely fragile. On the other hand detectors of 2–3 mm thickness are produced for beta-particle detection; they may need a polarising voltage of as much as 2 kV to utilise a high fraction of their geometrical thickness. Unfortunately, still thicker silicon detectors which would be needed for gamma-ray detection with a reasonably high intrinsic efficiency seem beyond technical capability and even if this were not so they would need inconveniently high operating voltages.

Solid-state detectors with sufficient absorption of gamma rays to ensure fair efficiency are however available based on germanium as the detecting medium. Crystals of germanium with lithium diffused into them and weighing up to 100 g or so can be obtained. Their effective volume is somewhat less than their actual size, and they have the disadvantage, unlike the silicon detectors which work at room temperature, of having to be maintained at −190°C continuously during their working life. Usually they are cooled by liquid nitrogen in a cryostat containing 100–200 litres of the refrigerant. Maintenance of a very high vacuum (10^{-4} Pa) in the space surrounding the detector is advisable to avoid having to replenish the liquid nitrogen almost daily. These Ge–Li detectors are characterised by very high ionisation yield, an ion-pair being produced for an average of only 2.9 eV of energy absorbed. The energies of the photoelectrons created by gamma-ray absorption can therefore be measured with very high precision. (For a discussion of the energy resolution of these and of scintillation detectors, see Section 4.18.)

3.18 Scintillation Detectors

In these, the secondary electrons arising from an initial ionising event are detected through the photons of visible light (scintillations) which they generate in suitable materials. The light emission from a single ionising event is generally far too faint to be seen by the unaided eye, and the scintillator has to be optically coupled to a photomultiplier tube. (Exceptionally, the scintillations produced by alpha particles in crystals of zinc sulphide are just visible under the microscope to the dark-adapted eye, as was discovered by Crookes, Elster and Geitel in 1903.) Very many other materials scintillate to some degree, but rather few happen to emit radiation in the limited range of wavelength to which the eye is sensitive. Commercial photomultiplier tubes have a maximum sensitivity in the region of 400 nm, so that the choice of scintillators is in practice limited to those which emit light near this wavelength. They must have properties such as chemical stability, transparency and high yield of light photons. There are five main kinds which are easily available, and these are now outlined.

(*a*) Organic solid compounds such as anthracene, which is distinguished by its relatively high light output. In this kind of scintillator, individual molecules may become excited by an electron colliding with them. The energy imparted to the molecule is insufficient to cause ionisation but it becomes polarised with a temporary rearrangement of its conjugated single and double bonds. Excited molecules return to their normal ground state by releasing their excitation energy as photons; it so happens in anthracene and similar molecules that these fall into the visible region of the electromagnetic spectrum.

(*b*) Liquid organic scintillators, which are solutions (in solvents such as toluene) of certain materials which are also scintillators in the solid form. The advantage of the solution form, as compared with the solid, is that radioactive materials can also be dissolved, or suspended as a fine powder, in the medium. Aqueous solutions of radioactive compounds can be incorporated into the medium by forming an emulsion with the aid of suitable detergents. These techniques result in very high intrinsic detection efficiency and are invaluable in detecting electrons which have very short ranges in matter, such as the beta particles from ^{3}H and ^{14}C, and the conversion electrons from ^{125}I. The amount of 'foreign' material so added to the scintillator must be strictly controlled if it is not to interfere too seriously with the light emission process. A typical scintillator will contain a *primary solute*, such a *p*-terphenyl (Fig. 3.33a) at a concentration of about 20 mmol l^{-1}, which is actually the source of the light photons. Also there may be a *secondary solute*, such as 2,2′-*p*-phenylene-*bis*-(5-phenyloxazole), conveniently known by the acronym POPOP (Fig. 3.33b). The secondary solute will be at a much lower concentration. It does not scintillate itself but has the function of absorbing light

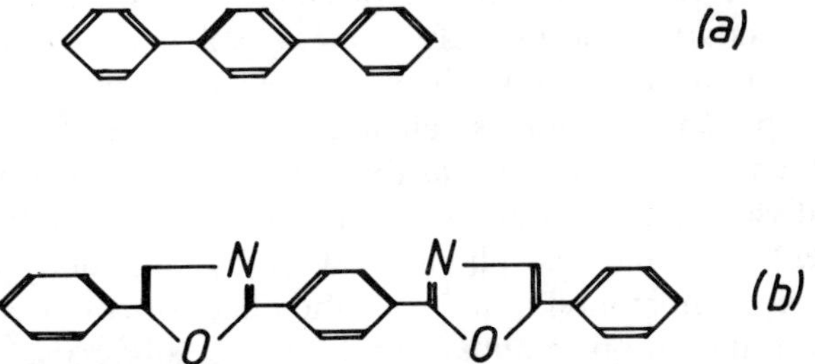

Fig. 3.33 Structural formulae of typical liquid scintillator solutes: (a) *para*-terphenyl, (b) POPOP.

photons from the primary solute and re-radiating light at a somewhat longer wavelength which is more nearly matched to the sensitivity of the photomultiplier tube. Recent developments in the manufacture of photomultiplier tubes are obviating the need for these secondary solutes.

In use, the scintillating solution is placed in a transparent vial in optical contact with the photomultiplier tube. The vial may have inlet and outlet tubes so that it can be filled and emptied *in situ*. In some systems, two photomultiplier tubes are used, an event being recorded only when both receive some light photons within a very short time interval. The output pulse from the system is obtained by summing the pulses from both photomultiplier tubes. This device gives larger signals than would be given by either tube alone. It also has the effect of reducing the 'background' of pulses caused by thermal noise, because such pulses rarely occur in both tubes almost simultaneously. Background due to thermal noise is also reduced in most modern equipment by maintaining the tubes and scintillators at 4 °C.

In view of the fact that only about one in 300 of the molecules in a liquid scintillator is actually capable of emitting visible light, it is a little surprising that the quantum emission per initial event is as high as that observed; it is typically some 10% of that given by anthracene. The high light yield can be explained only on the basis of good utilisation of energy imparted to the solvent molecules. Along the track of the initial particle in the medium many ionising encounters will occur. It is possible then for ionised solvent molecules in colliding with their neighbours to exchange charges so that eventually solute molecules become ionised in turn, and in recapturing an electron are able to produce their characteristic scintillation. Alternatively ionised solute molecules might diffuse through the liquid and exchange their charge with solute molecules. Both mechanisms would however be much too slow to account for the observation that light production occurs typically within 10^{-9} s of the primary event. The actual timescale is in much closer accord with the postulate of the handing-on of excitation energy only, in a chain of collisions among the solvent and solute molecules. The vibrational energy transferred in this process remains as a constant quantity throughout and is referred to as a 'phonon'. Transfer of energy from excited solvent molecules directly to the solute as electromagnetic radiation would imply a much shorter timescale and is also unlikely because of the small probablity of the absorption of radiation at the required wavelength by solute molecules. The concept of intramolecular transfer of excitation energy is also used to explain one of the mechanisms of quenching in liquid scintillators, in which the light output per event is much reduced by the presence of quite small quantities of impurities such as oxygen and water. Quenching agents such as these interfere by absorbing the excitation energy, dissipating it as random thermal motion before handing it on to another solvent molecule; the phonon is thereby degraded. This is not the only quenching phenomenon; another is the effect of coloured impurities in absorbing the scintillation quanta. Curiously the solute itself exhibits self-quenching if present beyond an optimum concentration, by a mechanism which is still obscure. From a practical point of view the quenching effects in a given liquid scintillator in which a radioactive substance is dissolved, thereby introducing impurities, can be assessed by careful measurement of the energy spectrum of the emitted light. Calibrations with standards with known concentrations of quenching impurities enable one to estimate these concentrations in an unknown sample and hence apply a correction to the count-rate over the whole spectrum. Another useful technique is to take measurements of the count-rate obtained with a given scintillating sample both before and after mixing it with an additional activity of known strength. Still another check on the performance of a liquid scintillation system is afforded by using an external gamma-ray source under fixed geometrical conditions. Modern commercially available equipment designed for routine measurements on large numbers of samples may have this check as a standard facility and can correct the count-rate of the samples automatically.

(*c*) *Plastic scintillators*, which resemble the liquid scintillators except that the solvent is styrene or one of its derivatives which has been polymerised. The polymer is a transparent plastic resembling 'Perspex' but usually has a faint blue colour. It is available in large blocks which may be machined to form cylinders, tubes, and indeed quite complicated shapes for particular applications. In this respect it is much more convenient than anthracene, but has the disadvantage of giving only about 10% of the light output of that material.

(*d*) *Inorganic scintillators*, some of which happen to occur naturally in small pieces. They include such diverse materials as zinc sulphide and barium platinocyanide, but by far the most commonly used examples are sodium iodide and to a lesser extent caesium iodide. These latter are manufactured under very carefully controlled conditions and may be obtained as quite large blocks, many centimetres in each dimension. They have controlled amounts of deliberately added minor constituents; for example a typical sodium iodide crystal will contain 0.1% by weight of thallium. The thallium atoms within the sodium iodide lattice provide localised energy levels some 3 eV above the band of valency electrons. Electrons set free by an ionising event in the crystal lose energy by collisions in the lattice. Their successive energy states are depicted schematically in Fig. 3.34. At (a) an electron set free in an ionising event leaves a vacancy in the filled valency band (f). Successive collisions follow, the electron energy falling as shown at (b). The losses are irregular; the final loss (c) may leave the electron in a trap (t) caused by the presence of an activator (thallium) atom in the crystal, and (d) represents the final transition to a vacancy in the valency band. In this final transition the energy lost is radiated as a photon in the visible spectrum. Since the traps are all at the same energy level all the scintillations will be at the same wavelength, which happens to be quite close to the optimum as required by the photomultiplier. Some of the collisions may themselves be ionising, and although some electrons may reach the ground state without passing through a trapping centre, a single energetic electron can generate several photons. On average one photon is produced for roughly 30 eV of energy dissipated in the crystal.

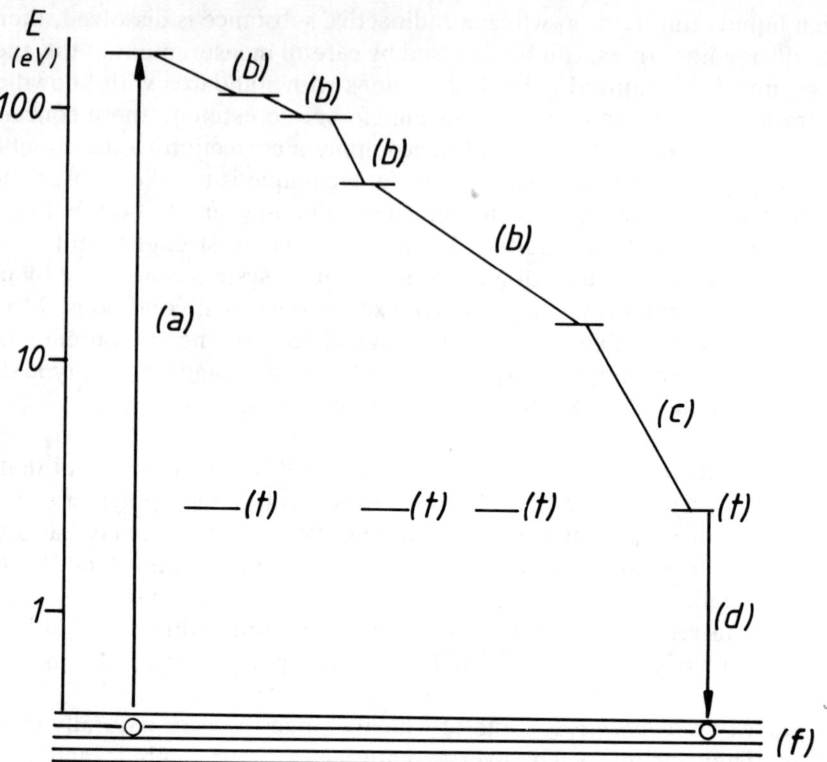

Fig. 3.34 Schematic energy diagram for an electron set free in an inorganic scintillator.

Sodium iodide would be an almost ideal scintillation material except for the fact that it is hygroscopic and has to be encapsulated to prevent access of atmospheric air. A typical crystal is pictured in Fig. 3.35, which shows the conventional aluminium can, lined on the inside with magnesium oxide to reflect any light incident on it from the scintillations, and a glass window on one face to allow light to escape into the photomultiplier tube. Caesium iodide crystals are not hygroscopic and therefore do not need encapsulation, but unfortunately they are very expensive.

(*e*) *Thermoluminescent materials.* It is characteristic of the scintillation materials mentioned so far that the light photons are produced in a quite short time after the initiating event (microseconds for inorganic crystals, nanoseconds for anthracene and other organic scintillators). In contrast, crystals of lithium fluoride (suitably activated) are able to 'store' excited states for many days or even weeks. This property makes them invaluable as integrating radiation detectors. In one popular form, the thermoluminescent dosimeter or 'TLD badge', a few milligrams of the material is selated in a thin plastic capsule or envelope which can be worn on one's finger or strapped to

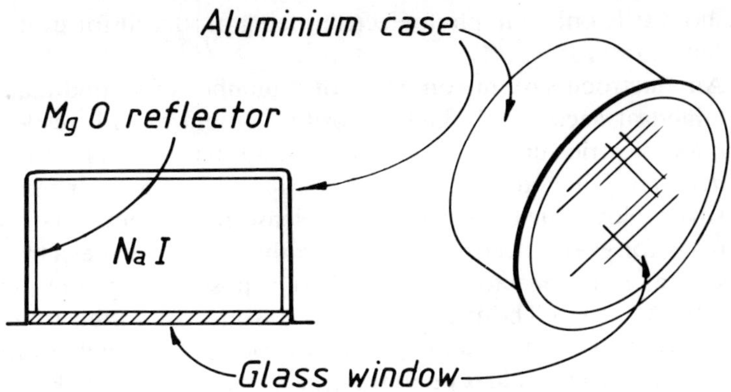

Fig. 3.35 Section and external appearances of sodium iodide scintillator.

surfaces such as laboratory walls and benches. In order to measure the absorbed radiation dose integrated over say a fortnight's exposure the lithium fluoride is inserted into a special furnace where it is rapidly heated to a predetermined temperature (250 °C.) This causes lattice vibrations which greatly facilitate the return of electrons to the ground state. Photons in the visible spectrum are produced within a few seconds. The light intensity, as measured with a photomultiplier tube, is proportional over a range of at least 10^5 to the total radiation dose absorbed. Calibrations are done with lithium fluoride exposed under known conditions, e.g. by being placed at a fixed distance from a radium source for a measured time.

3.19 Photomultiplier Tubes

The process of radiation detection using a scintillator is not really complete until a usable electrical signal has been made available, and for this purpose a photomultiplier tube in optical contact with the scintillator is almost invariably used.

The essential features of a typical tube are as follows:

(a) An evacuated glass envelope having at one end an optically flat surface, or window, through which light from the scintillator passes. Good optical contact between the two must be maintained. Solid scintillators and vials containing liquid scintillator may be placed in contact with the photomultiplier face with an intervening film of silicone oil or grease, which has a refractive index close to that of the mating surfaces. For some particular purposes, which need not concern us here, an optically clear slab of Perspex may be interposed between the scintillator and photomultiplier tube. This is referred to as a 'light guide'.

(b) A thin film of photoelectric material such as caesium–antimony alloy on the inside surface of the window. Light photons falling on this from outside the tube generate photoelectrons which travel to electrodes further down the tube. The photoefficiency of this surface is about 0.1, only one photoelectron being produced for every 10 incident photons.

(c) An electrode system consisting of a number of slotted plates maintained at successively higher positive potential with respect to the photoelectric surface. Commonly used tubes have 11 or 14 of these *dynodes*. They are coated with caesium–antimony, or magnesium–silver alloy which is designed to release rather more than two electrons for every electron incident on them. The final electrode in the system is a collector connected to a positive high-tension supply (800–1500 V.) The individual dynodes are connected to a voltage-dividing chain of resistors. Most commercially available tubes work well with equal values of all the resistors except for the lowest one in the chain which may have twice the resistance of the others. This gives a relatively higher voltage between the photocathode and the first dynode and a higher multiplication factor for the rather few electrons arriving at the first dynode. It is also advantageous to load the two uppermost dynodes with small capacitors to help maintain their potential when passing a large current pulse. Fig. 3.36 gives a diagram of the tube and associated circuits; (a) is the photocathode, (b) the dynodes, (c) the collector, R and R' the dynode resistors and C the loading capacitors.

In the actual tube all the electrical connections, the photocathode, dynodes and collector, are brought out of the tube at the end remote from the photocathode. The dynode resistor chain can easily be mounted directly above the tube and the whole assembly, complete with scintillator held at the other end with a spring-loaded collar, fitted into a snugly fitting lead radiation shield. Care must be taken to

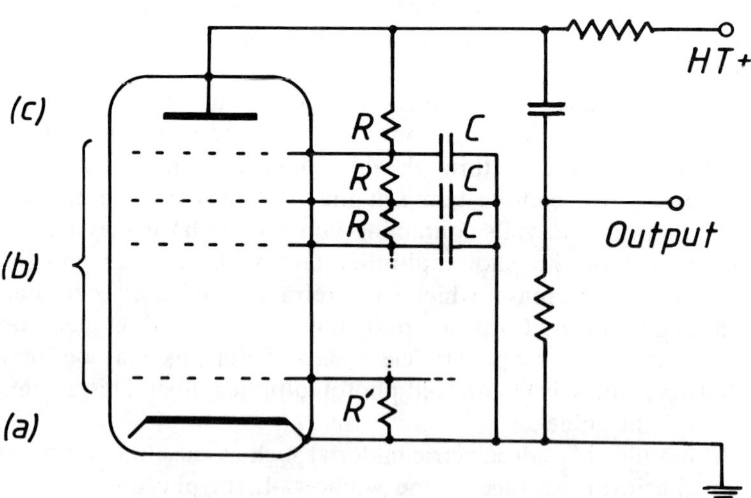

Fig. 3.36 Photomultiplier tube and associated circuit.

ensure that no stray light can reach the photomultiplier tube through fixing-screw holes, ill-fitting joints, etc., in the outer casing.

It will be realised that if there are n dynodes, the first releasing $2p$ electrons and the others p electrons per incident electron, the overall multiplication will be $2p^n$. This factor will in practice be in the range 10^6 and 10^8 and will be constant to within about 1% if the overall high voltage is constant to 0.1%.

3.20 Pulse-Height Distributions from the Detectors

If we are interested merely in the *number* of ionising events occurring in a certain time in a detector, we may pass the pulses after suitable amplification and possibly pulse shaping to a scaler or rate-meter. For all the detectors except Geiger–Müller tubes, however, it is much more informative to sort them according to pulse height, using a single- or preferably multichannel analyser (Section 3.13). We may then choose to count those pulses whose amplitude lies between certain limits, corresponding to particular kinds of event in the detector. Ideally, if a number of electrons all of the same energy were injected into a detector the output pulses would all be at the same height (i.e. voltage amplitude), and this height would be proportional to the electron energy. But in practice pulse spectra are very much more complicated. Thanks to thermal noise in the early stages of amplification there are always low-voltage pulses independent of the presence of a source of radiation; these clearly must be discriminated against. Then there is the fact that the initial electrons do not all produce the same number of secondary ions; there is an *average* number but some events give more secondary ionisation and some

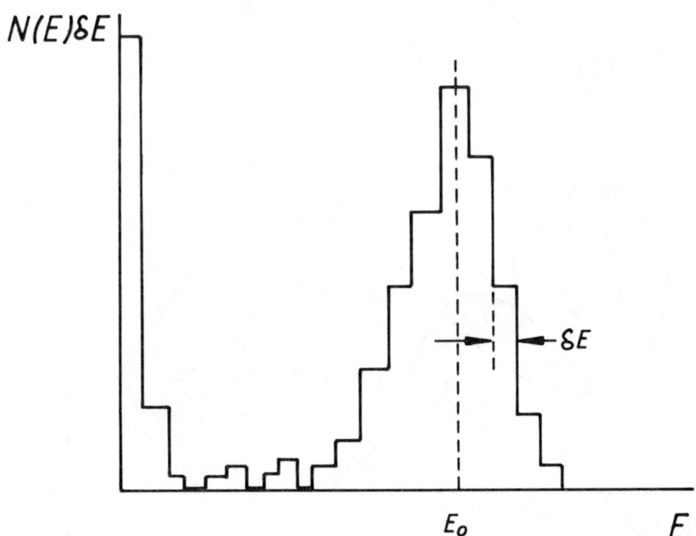

Fig. 3.37 Spectrum of pulses from a proportional counter with a source of monoenergetic electrons.

give less than the average. Also in scintillation detectors there are fluctuations from event to event in the numbers of photons received at the photocathode. Finally the spectrum will be degraded to some extent by wall effects in the detector; that is to say by events in which some loss of energy occurs in the walls, thereby reducing the amount available to generate a signal. With these effects in mind we can expect the pulse spectrum to have a form like that shown in Fig. 3.37. Here one can clearly distinguish noise pulses, genuine pulses due to the electron source having a well marked peak but with a noticeable spread, and pulses of intermediate amplitude showing the wall effect.

When a beta-particle emitting source is examined with an anthracene crystal or plastic scintillator, or dissolved in a liquid scintillator, the pulse spectrum roughly corresponds with the shape of the beta spectrum as determined with much greater precision with a magnetic spectrometer. The effect of quenching on the spectrum from a liquid scintillator is shown schematically in Fig. 3.38. (In this and later spectra, a smooth curve is drawn for simplicity, although strictly speaking they should all be histograms as in Fig. 3.37.) It must be remembered that both excitation-transfer and colour quenching can occur. When detectors are used for X-rays and gamma photons, the spectral distribution of the pulses is still more complicated because the secondary ionisation does not necessarily correspond to the energy carried by the photons. Thus, for example, if a xenon-filled proportional counter is used to detect monoenergetic X-rays of less than about 35 keV, a simple spectrum like that of Fig. 3.37 will result. However, if photons of say 50 keV are used, there will be events in which K electrons in xenon (whose binding energy is 35 keV) will be ejected with energies of about 15 keV. Immediately following

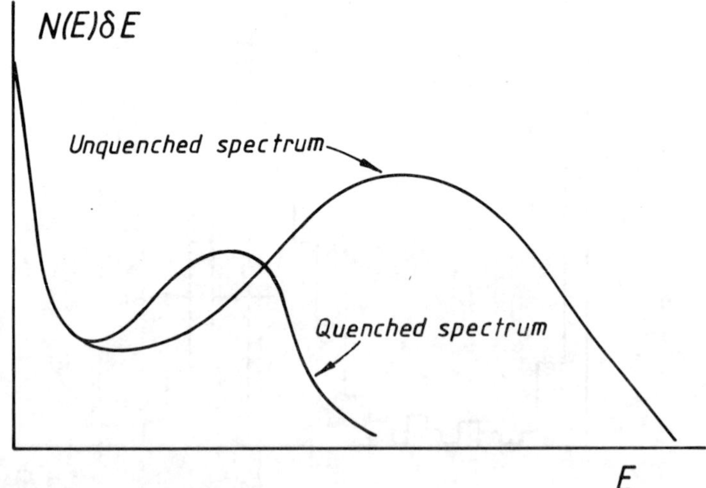

Fig. 3.38 Effect of quenching on the pulse spectrum obtained with a liquid scintillator.

these, the ionised xenon atoms will emit their characteristic K X-rays of about 30 keV and L X-rays of about 5 keV. If the X-rays are absorbed within the gas, the total energy deposited will be 15 + 30 + 5, i.e. the original 50 keV. Now xenon, in common with all elements, has a rather low absorption coefficient for its own K X-rays. Therefore it can happen quite frequently that the 30 keV X-ray can escape from the counter without being detected, so that only 20 keV of energy is deposited. The spectrum will then show an 'escape peak' at 20 keV as well as a 'full-energy peak' at 50 keV. The escape peak will be more noticeable the smaller the linear dimensions of the counter, and the lower the gas pressure.

Similar effects are possible with scintillation and solid-state detectors, although in actual practice these are generally 'thick' in this respect and escape of K X-rays is negligible. However, when these detectors are used, as they commonly are, for gamma-rays of more than about 100 keV it must be remembered that the Compton process becomes the predominant interaction mechanism. The following possibilities then arise, for gamma-rays of energy E_γ:

(a) Photoelectric events, giving rise to a 'full energy' peak corresponding to E_γ.

(b) Compton events, releasing electrons with energy anywhere between zero and the Compton edge. The spectrum will show a continuous distribution of pulses extending from height zero up to a value corresponding to the maximum energy possible for a Compton scattered electron.

(c) Compton events as in (b), but where in addition the scattered Compton gamma-rays are also totally absorbed in the detector. The total energy deposited in these events will again be E_γ. For gamma-rays betweeen 300 and 2000 keV and for detectors with linear dimensions of a few centimetres, this last composite process occurs more frequently than the others, but since the energy deposited is E_γ it is easy to get the mistaken impression that the photoelectric process is the main mechanism of absorption of the primary gamma-rays. A typical spectrum of pulses from a source emitting monoenergetic gamma-rays which are detected in a sodium iodide crystal is shown in Fig. 3.39.

Again there are noise pusles at (a); (c) is the full-energy peak, often loosely but incorrectly called the photoelectric peak. At (b) there is the continuous spectrum of Compton electrons with a fairly well marked upper edge. There may be a small 'hump' in this part of the curve due to back-scattered gamma-rays of about 200 keV, as mentioned in Section 3.6. Note the 'valley' between the end of the Compton spectrum and the rising part of the full-energy peak. Compare this figure with that of Fig. 2.7, which is a spectrum obtained with a Ge–Li detector. The latter shows much sharper energy resolution (Section 4.18).

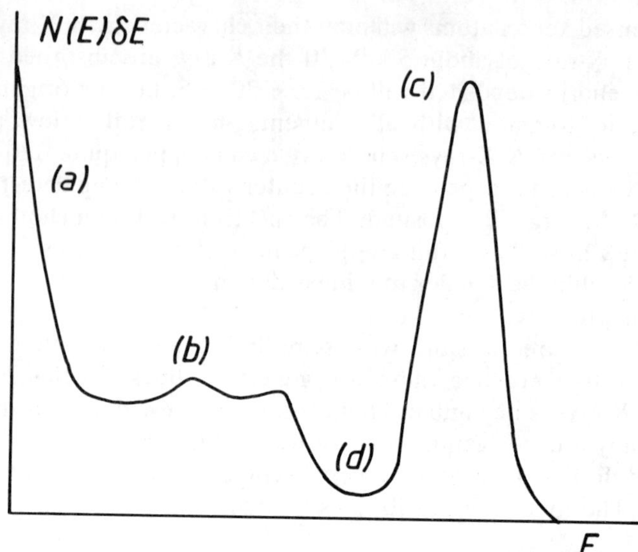

Fig. 3.39 Pulse spectrum for a scintillation detector, with a monoenergetic gamma source.

Many nuclides emit more than one set of gamma-rays, and their gamma-ray spectra are therefore compounded of two or more curves like the above but with different horizontal and vertical scales. Resolution of a complex spectrum into its components is not difficult, given sufficient data about the shapes of spectra from monoenergetic sources, and is an essential first step in the investigation of decay schemes. The same numerical techniques are useful in analysing a mixture of two or more nuclides; for a simple example of this see Section 3.28. However complicated the spectral curve for a particular pure nuclide, it is obvious that a good detection system should be able to reproduce the same curve on different occasions. A 'standard source' of a long-lived radionuclide, e.g. ^{137}Cs or ^{60}Co, is therefore an invaluable aid in checking not only the detection efficiency but energy measurement (full-energy peak height) and energy resolution (peak width). A well engineered scintillation system will work consistently well for several years. Gradual deterioration of the photomultiplier tube will necessitate raising the applied voltage; poor optical contact between crystal and photomultiplier tube caused by disappearance of the coupling fluid will result in worsening resolution. A crack in the crystal or a hole developing in the encapsulation allowing the ingress of moist air will also cause poor resolution.

3.21 Radiation Shielding for Detectors

All instruments used for radiation detection are affected by radionuclides present as impurities in structural materials and the environment, and by radiation originating in outer space, i.e. cosmic radiation. Also under laboratory conditions many radioactive samples may be present in the same room as the detection equipment. Particularly when the level of activity from the source being examined is low, it is essential to reduce these 'background' effects to a minimum. We cannot overemphasise the importance of establishing sound laboratory practice in order to prevent accidental exposure of sources to the detectors. But in view of the penetrating nature of the background radiation, the provision of adequate shielding must be carefully considered too. The traditional shielding material is lead, combining relative cheapness with ease of fabrication, chemical stability, and low penetrability for X- and gamma-rays. The latter characteristic is evident from Fig. 3.24, but a rather more useful presentation of the same data is shown in Fig. 3.40, which is a plot of half-value layers (HVL) in lead for photons of various energies. For calculation purposes the following empirical equation is useful, for gamma-ray energies between 0.1 and 1.5 MeV:

$$\text{HVL(mm)} = 2.39E_\gamma + 36.90E_\gamma^2 - 39.20E_\gamma^3 + 12.86E_\gamma^4 \qquad E_\gamma \text{ in MeV}$$

Thus the HVL for gamma rays from ^{99m}Tc (140 keV) is about 3.3 kg m^{-2}, i.e. a thickness of about 0.29 mm. A shield of 10 mm thickness is thus approximately 35 HVLs and will reduce the radiation intensity by a factor 2^{35} or 3×10^{10}. However for ^{99}Mo (E = 740 keV) the HVL is 6.4 mm, so that a shield of 10 mm thickness will reduce the intensity of this radiation by less than a factor 4. Clearly it would be unwise to work with strong sources of

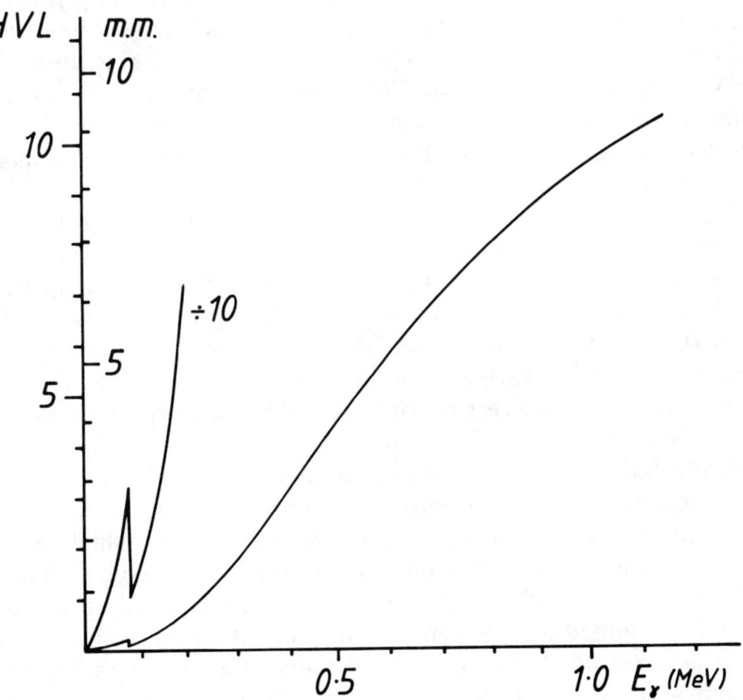

Fig. 3.40 HVL values for gamma-rays in lead.

^{99}Mo in close proximity to a detector unless the shielding thickness were many centimetres of lead.

As a general guide one would recommend a shielding thickness for detectors of 25 mm, where these are used in a room from which extraneous sources of more than 100 MBq can be excluded. In a laboratory where gigabecquerel gamma-ray sources of high energy, such as ^{99}Mo, are used, it will be advisable for the source itself to be shielded by some 150 mm of lead as radiation protection for the staff. This will reduce background level to the point at which accurate measurement of a small fraction of a megabecquerel is possible without significant interference. Even then the detector should be as far as possible from the source and will need its own 25 mm shield. An important factor in the design of the laboratory itself is the radiation shielding afforded by the walls. Ordinary concrete has a HVL for 140 keV radiation of about 20 mm, and for 740 keV radiation about 60 mm. A conventional 15 inch brick wall will reduce the radiation intensity from a ^{99}Mo source by a useful factor of about 20. The use of plaster containing barium on the walls of 'isotope' laboratories is scarcely worthwhile, although it is so valuable in diagnostic X-ray departments; this is because by far the greatest shielding problem in an isotope department is occasioned by the high-energy radiation against which a few millimetres of barium plaster is hardly effective.

An important source of background radiation within buildings is the ^{40}K (E_γ = 1.46 MeV) in the brick and plaster. In some recent buildings there is a detectable contamination by ^{60}Co (E_γ = 1.1 and 1.3 MeV) in the steel frame, which arises from this radionuclide being added to the steel during manufacture as a means of monitoring the behaviour of blast furnace walls. The HVL for gamma-rays from ^{40}K is about 10 mm in lead, so that lead shielding for an installation intended for the measurement of the ^{40}K content of the whole (human) body would be prohibitively expensive. This point is taken up again in Section 5.2.

Background from cosmic ray effects can be much reduced by disposing other shielded detectors in the spacc surrounding the detector actually being used for the samples. These extra detectors should in particular occupy the space above and below the main one. The detectors are connected to an anticoincidence circuit which rejects events in which the main detector and any one of the others is triggered simultaneously. In this way an ionisation occuring in the main detector which is caused by a highly penetrating cosmic ray has only a small probability of being recorded.

With adequate lead shielding and good 'housekeeping' one should achieve background count-rates in conventional end-window Geiger–Müller counters of about 10 events per minute, and with anticoincidence shielding about 2 events per minute. The background count-rate for a large sodium iodide crystal is of the order 1000 events per minute over the whole spectrum. When a source is presented to a detector we may call the observed count-rate the 'gross' count-rate; when the background count-rate is subtracted we have the 'nett' count-rate, which is due to the source alone. It is, of course, the latter quantity which is usually of most interest and in this book we often omit the word 'nett' when it is obvious from the context that background has been allowed for.

MEASUREMENT OF RADIOACTIVE SOURCES

3.22 The Assay of Radionuclides

The measurement of the number of radioactive nuclei in a given source is regarded as an *absolute* assay. For most nuclides it can be achieved only by measuring the absolute disintegration rate, which implies detecting the radiations emitted in all directions in space, or at any rate within a known solid angle. The intrinsic efficiency of the detector must be known, either on the basis of reliable calculation or of independent measurement. Absorption of radiation in the source itself must be allowed for, or the source must be so thin that self-absorption losses can be shown to be negligible. In general, therefore, absolute measurements are possible only when we have good control over the physical form of the source and of its geometrical relationship to the detector, and can evaluate the detector efficiency. Further, the decay scheme must be understood, so that we can relate the number of particles and/or gamma-rays detected to the number of disintegrations producing them; only for the simplest decays does one disintegration give rise to one beta particle or one gamma-ray.

For most investigations it is fortunate that only a relative assay is required; we may need to compare the strength of one source only relative to that of a second, or 'standard', source, or to compare a source at one instant of time with the same source at some other time. Clearly relative assays are much the easier to carry out, since only a fraction of the emitted radiation needs to be detected and it is not necessary to evaluate that fraction so long as it remains constant throughout the experimental measurements. For experiments of a purely physical nature this can be ensured to a high accuracy by careful replication of geometrical conditions as between 'unknown' and 'standard'. When the 'unknown' is a biological sample such as a body organ in a human subject, whose geometry is not at all well defined, there can be large errors. This problem is taken up again in Chapter 5.

3.23 Absolute Assays: Weighing

There is one radionuclide, namely ^{226}Ra, which however has been traditionally assayed by weight. One gram of the element has an activity (within 0.5%) of 3.7×10^{10}Bq and this activity was used for many years as the basic unit (the curie). Radium is weighed not as the element, which is highly reactive chemically, but as the much more stable compounds $RaSO_4$ or $RaBr_2$. Other naturally occurring elements of longer half-life (uranium and thorium, which are mixtures of isotopes) and the synthetic transuranium elements (e.g. plutonium) are also available in weighable amounts. As a general rule, however, weighing is a quite impractical means of assaying the nuclides of importance as tracers. The mass of a nuclide having an activity of 1 MBq, mass number A, and half-life $t_{1/2}$, is easily shown to be

$$m = At_{1/2} \times 2.07 \times 10^{-13} \text{ g} \qquad (t_{1/2} \text{ in seconds})$$

or

$$m = At_{1/2} \times 1.79 \times 10^{-8} \text{ g} \qquad (t_{1/2} \text{ in days})$$

Thus a gigabecquerel (10^9 Bq) of ^{131}I(half-life 8 d), even if it could be prepared as the insoluble compound AgI quite carrier-free would weigh less than half a microgram, and the heating effect on the sample due to its radiation would make weighing on an ordinary balance prohibitively difficult.

3.24 Absolute Assays: High-Efficiency Counting Methods

These methods depend on the use of a detector which responds to all the radiation emitted from the source. The most accurate teechnique is that of so-called 4π beta counting, in which the source is mounted on a thin foil placed within a pair of proportional counters (Fig. 3.41). The foil is carried on a slide so that it can be inserted in the midplane of a spherical counting chamber, each hemisphere having an anode in the form of a loop of fine wire. The chamber is continuously flooded with an argon-methane gas mixture.

With sources weighing a few micrograms as a thin layer on gold-plated plastic mounting films of about the same weight, it has been shown that 99.9% of all beta particles emitted will produce a recorded count, even when the E_{max} of the beta-particle spectrum is quite low. Nuclides decaying by

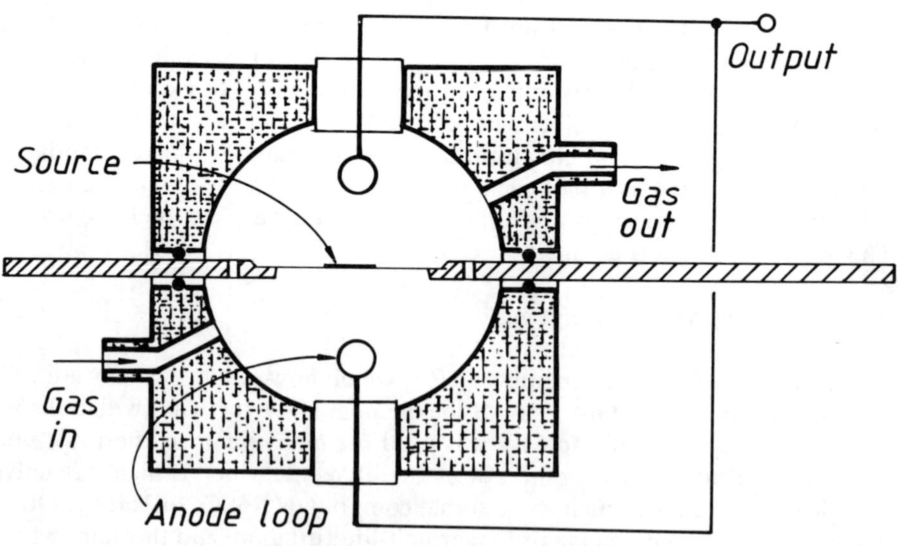

Fig. 3.41 A 4π beta counter (schematic).

isomeric transition can also be assayed with this technique through counting the conversion electrons, although there are difficulties if these have energy less than a few kilo-electron volts. It should be realised that gamma-rays, X-rays, and electrons emitted in place of X-rays (Auger electrons), which follow the beta decay do not give an additional count unless they are delayed by a time longer than the resolving time of the detector (about a microsecond). The method is reliable and very accurate but demands high technical skill in specialised techniques of electrodeposition, electrospraying, evaporation under high vacuum, and the handling of minute amounts of material.

3.25 Absolute Assays: Low-Efficiency Counting Methods

Many attempts have been made to derive accurate values for source-strengths using detectors which respond to radiation emitted into only a small solid angle. In the first of these we consider that the solid angle is defined simply by the system geometry, i.e. it is the angle subtended at the source by the window of a Geiger–Müller or proportional counter, or by the surface of a scintillation crystal. For best results in beta counting, the area of the window is circumscribed by a diaphragm (Fig. 3.42a) which defines the solid angle more precisely than does the window itself. The diaphragm, however, brings in some uncertainty due to scatter of electrons at its edges; to some extent the difficulty is overcome by taking separate measurements with diaphragms of different size and relating the difference in count-rates observed to the difference in solid angles subtended. If the source is close to being a geometrical

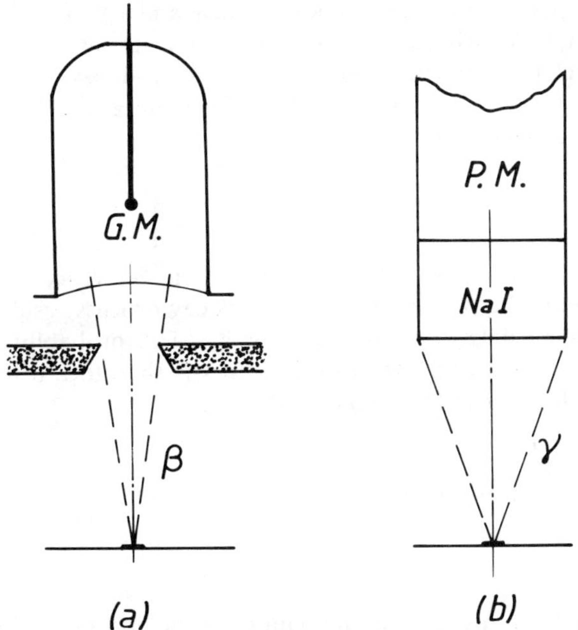

Fig. 3.42 Solid angle subtended at source by β and γ detectors.

point, the estimation of the solid angle is straightforward, but in practice it is more likely to be a thinly spread disc, and the effective solid angle has to be estimated by numerical integration. Corrections to observed count-rates have to be applied if (as is often the case) the source emits gamma-rays as well as beta particles. It is hardly worthwhile using diaphragms to define the geometry when measuring the gamma radiation from radioactive sources, because these would have to be so thick that scattering at their edges would be a large effect. It is clear too from Fig. 3.42b that gamma-rays entering the crystal near the edge of its front face are less likely to be detected that those entering near the centre. Intrinsic detection efficiencies for bulk sodium iodide can however be estimated from a knowledge of crystal thickness and the relevant attenuation coefficient. Tables have been prepared which take into account edge effects and give the overall detection efficiencies for point sources at fixed distances along the axes of some standard-size sodium iodide crystals (see *Applied Gamma Ray Spectrometry*, referred to in the Bibliography). These data refer to the whole spectrum of observed pulses, not just those in the full-energy peak. The results of these small-solid-angle measurements can be accurate to 1–2%

The second low-efficiency method we consider involves a coincidence technique. It can be applied to sources in which each decay gives rise to a beta particle and a gamma-ray almost simultaneously. Sources which emit pairs of gamma-rays can also be assayed by a modification of this method. For beta–gamma sources the principle of the method is to present the source to a pair of detectors, of which one is sensitive to beta particles only and the other to gamma-rays only. The detectors are linked by a coincidence circuit which will record those events in which a beta particle and a gamma-ray are detected 'simultaneously'. We will assume that a source of strength A Bq—i.e. source in which A nuclei are transformed per second—produces A beta particles per second, each of which is accompanied by one gamma-ray. The count-rates in the two detector channels will be

$$C_\beta = A\varepsilon_\beta$$
$$C_\gamma = A\varepsilon_\gamma$$

where ε_β and ε_γ are the overall efficiencies of the detectors. Now $\varepsilon_\beta = C_\beta/A$ simply represents the probability that any decay process results in the detection of a beta particle, and similarly for ε_γ. The probability that a decay process simultaneously affects both detectors is therefore the product $\varepsilon_\beta\varepsilon_\gamma$. We may therefore write for the coincidence rate

$$C_c = A\varepsilon_\beta\varepsilon_\gamma$$

Solving these equations we easily have

$$A = \frac{C_\beta C_\gamma}{C_c}$$

Similar equations apply for gamma-gamma coincidences, and slightly different ones can be deduced for cases in which the decay scheme is more complicated. The method does not work for the pair of gamma-rays (annihilation radiation) emitted when the positrons from a positron-emitting source combine with electrons, because the annihilation radiations are correlated in direction whereas it is implicit in the above derivation that all radiations are emitted in random directions. If positron emission is accompanied by a gamma-ray of different energy, one can proceed by measuring coincidences between this and one of the annihilation quanta.

Although an elegant method in principle, the coincidence counting technique presents some practical difficulties if it is to give accurate results, mainly on three grounds, as follows:

(a) No beta-particle detector is absolutely insensitive to gamma-rays; conversely beta-particle sources produce bremsstrahlung radiation which will be detected in the gamma detector. Therefore the count-rates C_β and C_γ have to be corrected for these effects; the corrections can be worked out from measurements taken on sources known to emit only beta particles or only gamma-rays.

(b) Although a source may emit beta particles and gamma-rays very nearly simultaneously, the time duration of signals in either channel is not negligibly small. For any experimental equipment, there is an associated time interval τ such that the detection of a beta particle and a gamma-ray within it will be regarded as coincident. Any such pair may indeed by truly coincident in the sense that they originated in the same nucleus, but a 'coincidence count' will also be recorded if a beta particle is detected within a time τ of the detection of a gamma-ray from a different nucleus. The 'true' coincidences are therefore accompanied by 'accidental' coincidences. Their number is obviously larger, the longer the resolving time, τ. Coincidence circuits therefore usually incorporate a facility for varying τ, by stretching the signals in either channel. One can then measure the coincidence count-rate over a range of τ values and extrapolate to $\tau = 0$. Alternatively the accidental coincidence count-rate can be calculated if τ is known accurately.

(c) Of much practical significance is the fact that rather high count-rates have to be used in the beta and gamma channels in order to get a coincidence count-rate high enough for reasonable statistical accuracy. This is because ε_β and ε_γ are often rather small because of the physical difficulty of arranging two detectors, shielded from each other, close to the source. The coincidence count-rate depends on the product $\varepsilon_\beta\varepsilon_\gamma$ and is necessarily smaller still. This consideration makes it rather impractical to use a Geiger–Müller counter as the beta-particle detector because of the long dead-time (which of course implies a long resolution time too).

An interesting application of coincidence counting techniques has been to check the performance of a 4π beta counter by placing a scintillation counter close by and measuring the coincidences. In the simple case when each beta particle is accompanied by just one gamma-ray, the gamma count-rate and the coincidence count-rate should be identical. The main use of coincidence techniques has been in the elucidation of decay schemes. Also by introducing electronic delays in one or other channel it is possible to measure the half-lives of excited states in nuclei which are of the order of only nanoseconds.

3.26 Relative Assays: Counting Methods

The intercomparison of two sources (of the same radionuclide) is, of course, considerably simpler than an absolute measurement. Intrinsic efficiencies are the same for both sources and it is generally easy to make geometrical factors, i.e. self-absorption effects and disposition of the sources relative to the counter, the same also. Gamma-ray emitting sources of only a few becquerel can be compared in well–crystal sodium iodide detectors; the sources may comprise an aqueous solution occupying typically 4 or 5 ml in volume (see Fig. 3.43). Errors in the volume of the samples of a few per cent will generally have a negligible effect on the count-rate.

Accurate intercomparisons among beta-particle sources in the form of dry powders can be made by spreading them uniformly on planchets and using end-window GM or proportional counters in a fixed geometry. For best results all sources should be of nearly the same weight and very uniformly spread so that self-absorption effects are the same in each. They should be mounted on planchets of uniform weight so that the contribution to the count-rate from beta particles entering them and being back-scattered into the detector is always the same fraction of the source strength (see Fig. 3.44).

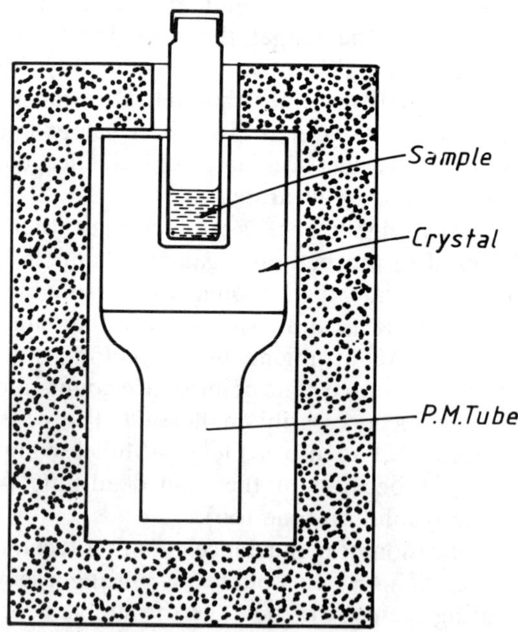

Fig. 3.43 Typical well–crystal detector.

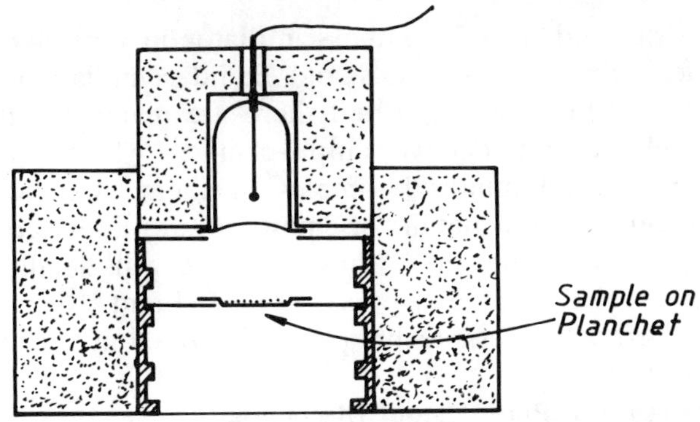

Fig. 3.44 End-window counting of beta particle source.

Energetic beta-particle emitters (e.g. ^{32}P) can be counted in solution in annular-liquid GM tubes. In these, the tube anode takes the form of a spiral of flat tape. There is considerable self-absorption of the beta particles within the liquid, and some beta particles are absorbed in the thin glass envelope of the tube and in the anode, but the overall efficiency is still some 10% and of course is reasonably reproducible (see Fig. 3.45).

Low-energy beta-emitting sources (^{3}H and ^{14}C) are not conveniently counted on end-windowed counters because self-absorption would be very high, as well as beta-particle absorption in the counter window. They are usually assayed by a scintillation technique in which the active material is

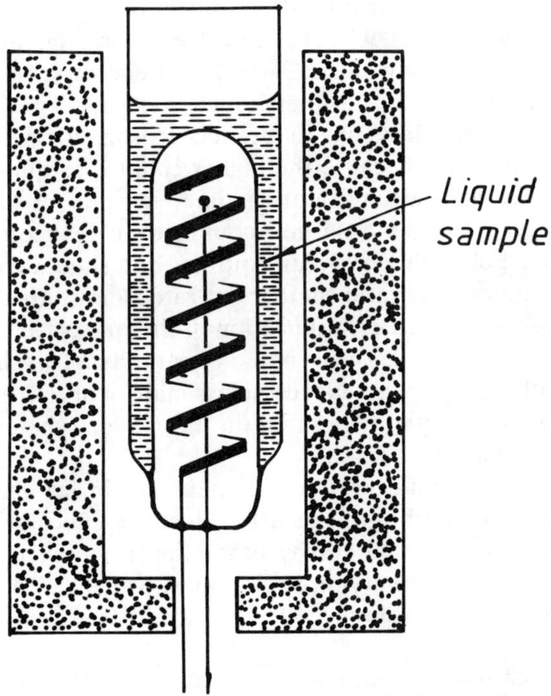

Fig. 3.45 Annular GM tube for liquid samples.

dissolved or dispersed in an organic scintillator in optical contact with a photomultiplier tube. Well–crystals and liquid scintillation systems lend themselves well to mechanisation and there are many manufacturers who supply highly reliable and effective sample-changers which will automatically take counts on several hundred samples, including standards and backgrounds, and will print out details of counting rates after subtracting background. The more sophisticated of these machines will handle samples with two radionuclides simultaneously present (e.g. ^{3}H and ^{14}C), deriving assays for each by a spectral analysis technique as described in Section 3.28.

3.27 Relative Assays: Pulse-Height Discrimination in Scintillation Counting

As shown in Section 3.20 the spectrum of pulses obtained when a scintillation crystal is used to examine a gamma-ray source is quite complicated, even when the source is monoenergetic. For estimations of the relative strength of sources of the same nuclide, one needs to decide what part of the spectrum to select in order to get maximum reliability in the result, i.e. the ratio of nett count-rates. Two extreme regimes suggest themselves. One may set a discriminator level E_0 as low as possible while still eliminating low-energy noise, and count all pulses of greater magnitude. This technique is referred to as bias-level counting. It has the advantage of obtaining the highest possible count-rate, for fixed geometry, but at a cost of including background which extends over the entire spectrum. Alternatively one may set a discriminator level somewhere in the 'valley' between a Compton edge and a full-energy peak, and to use a window width ΔE which just encompasses the full-energy peak. The background effect is then much reduced. The alternatives are illustrated in Fig. 3.46. There are theoretical reasons for preferring the first alternative (Section 4.16) because the higher count-rate will make for better statistical accuracy in a given counting time, in spite of the higher background, *except* for sources whose nett count-rate is low compared with the background rate. For such very weak sources, one should choose a window width which maximises the ratio (nett count-rate from source)^2/(background rate)—the so-called S^2/B rule. Application of the rule calls for the narrowest possible window, that is to say one should count only a few channels δE near the top of the full-energy peak. Nevertheless many workers prefer to use the 'narrow-window' technique illustrated in Fig. 3.46, whatever the source strength, for the following reasons:

(a) Recording of counts in the full-energy peak, as against the whole spectrum, gives a built-in guard against the effect of accidental contamination of the sources by other lower-energy nuclides, and reduces the effect of varying background which can occur when large sources of other radionuclides are moved from one site to another nearby.

(b) The use of too narrow a window centred on the full-energy peak is liable to cause large errors if there is 'drift' in the high tension applied

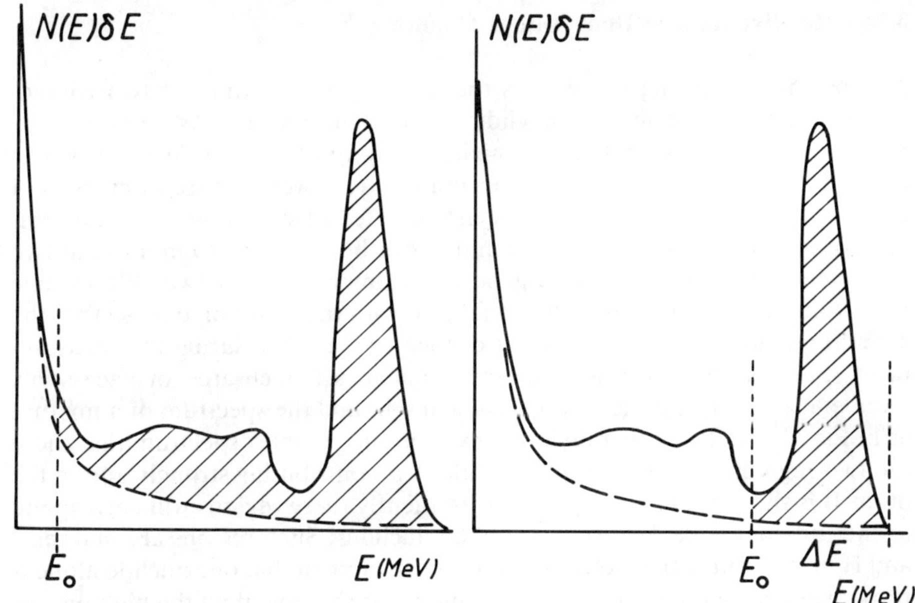

Fig. 3.46 Energy discrimination in gamma counting (background shown as broken line).

to the photomultiplier or in the bias level or window width. As is clear from Fig. 3.46, bodily movement of the peak across the window, or small changes in ΔE, will have very little effect on the total count-rate over the window (shaded area).

(c) If the detector is collimated so as to examine only part of a widely distributed source (as happens with *in vivo* organ counting), the narrow-window count-rate will be less likely to be affected by scattered radiation originating from outside the field of view, and therefore by changes in the amount of scattering material. (This point is taken up in more detail in Chapter 5.)

(d) There is always the possibility of a sample having an unexpectedly low count-rate, in which case a narrow-window regime is theoretically justifiable as giving better statistical accuracy. If a sample is unexpectedly strongly active, the background rate is negligible anyway and it does not matter which regime is used.

If the sources to be compared emit two or more gamma-rays per disintegration, one may have a choice between setting a window over one, or another, full-energy peak or even of having a window wide enough to encompass two peaks if these are reasonably close together. The relative peak heights in a spectrum will depend not only on the relative abundance of the gamma-rays but also on the size of the crystal, so that the best conditions are often to be found only by experiment with standard sources. In general, the larger the crystal the better, both because of higher sensitivity and because a larger proportion of counts appear in full-energy peaks.

3.28 Relative Assays: Dual Isotope Counting

A constantly recurring problem is the analysis of a mixture of two components, which we shall call nuclide A and nuclide B. As explained in Section 4.9 such mixtures can be analysed by resolution of decay curves, a method which is inherently time-consuming. If, however, pure samples of A and B are known to give gamma spectra of different shapes we can utilise this fact to analyse a mixture. The technique can be extended to more than two components but we give here a general treatment for only two. We assume that both half-lives are very long compared with the counting time so that the actual amounts of A and B do not change appreciably during the measurement. Suppose the spectra of the pure nuclides (which are not necessarily monoenergetic) appear as in Fig. 3.47a and b, and the spectrum of a mixture as in Fig. 3.47c. It is generally obvious from the complex spectrum that there are two regions, in one of which nuclide A is contributing strongly and in the other B is the more important. Most frequently these regions will correspond to the full-energy peak, or peaks, of the nuclides. Such regions are marked I and II in the figure. In favourable cases it can happen that one nuclide alone is responsible for all the counts in one region; for example if all the high-energy part of spectrum (a) had been absent, nuclide B would have contributed all the activity in region II.

Suppose now we set up windows on single-channel analysers to count pulses from the two regions in turn. We will refer to the counts as being in channel I or channel II. We count a pure sample of A in both channels; this sample we refer to as 'standard A'. It becomes our unit of activity for this nuclide. We do not need to know it in absolute measure such as megabecquerels; it could be specified as a fixed volume of a stock solution, or any such convenient measure. We will call the observed count rates in the two channels $C_{\mathrm{A}}^{\mathrm{I}}$ and $C_{\mathrm{A}}^{\mathrm{II}}$. Now we count a pure standard of nuclide B under the same geometrical conditions, obtaining rates $C_{\mathrm{B}}^{\mathrm{I}}$ and $C_{\mathrm{B}}^{\mathrm{II}}$.

Finally we count a mixture of A and B, which will contain say x units of A and y units of B. This sample must again have exactly the same geometrical form as the standards. It will give say M^{I} counts in channel I and M^{II} counts in channel II in the same counting intervals. Then we must have

$$M^{\mathrm{I}} = xC_{\mathrm{A}}^{\mathrm{I}} + yC_{\mathrm{B}}^{\mathrm{I}}$$
$$M^{\mathrm{II}} = xC_{\mathrm{A}}^{\mathrm{II}} + yC_{\mathrm{B}}^{\mathrm{II}}$$

Solving for the unknowns x and y we easily have

$$x = \frac{M^{\mathrm{I}}C_{\mathrm{B}}^{\mathrm{II}} - M^{\mathrm{II}}C_{\mathrm{B}}^{\mathrm{I}}}{C_{\mathrm{A}}^{\mathrm{I}}C_{\mathrm{B}}^{\mathrm{II}} - C_{\mathrm{A}}^{\mathrm{II}}C_{\mathrm{B}}^{\mathrm{I}}}$$

$$y = -\frac{M^{\mathrm{I}}C_{\mathrm{A}}^{\mathrm{II}} - M^{\mathrm{II}}C_{\mathrm{A}}^{\mathrm{I}}}{C_{\mathrm{A}}^{\mathrm{I}}C_{\mathrm{B}}^{\mathrm{II}} - C_{\mathrm{A}}^{\mathrm{II}}C_{\mathrm{B}}^{\mathrm{I}}}$$

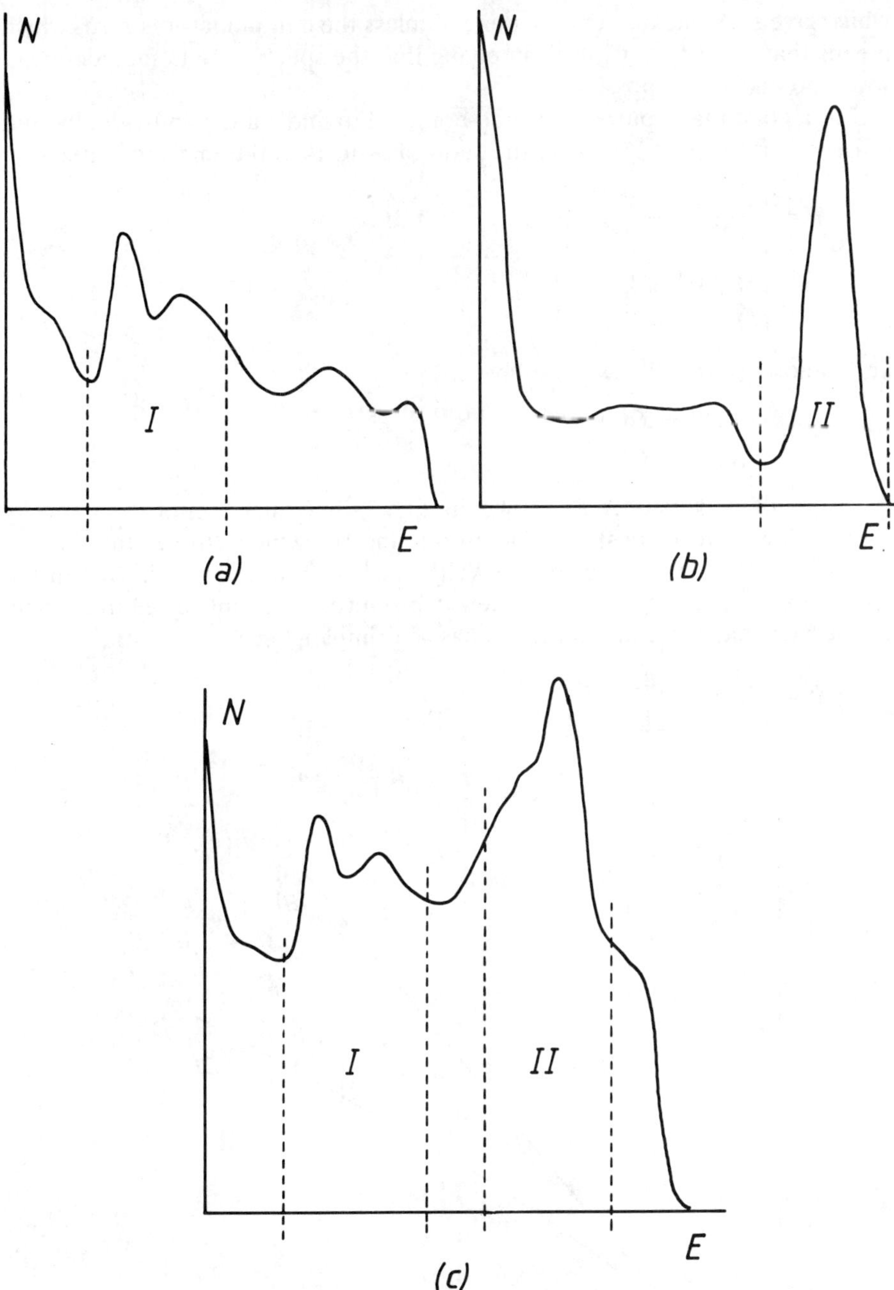

Fig. 3.47 Gamma-ray spectra: (a) and (b) of pure nuclides A and B, and (c) of mixture.

which give sensible answers for x and y unless the denominator is zero, which means that $C_A^I/C_A^I = C_B^I/C_B^{II}$, implying that the spectra are in fact identical, when no analysis is possible.

For a good many purposes we do not need to find x and y individually, but only their ratio. Let $Y = x/y$, the ratio of A to B in the mixture. Then

$$Y = -\frac{M^I C_B^{II} - M^{II} C_B^I}{M^I C_A^{II} - M^{II} C_A^I}$$

$$= \frac{(M^I/M^{II})C_B^{II} - C_B^I}{(M^I/M^{II})C_A^{II} - C_A^I}$$

and denoting M^I/M^{II} by X we have

$$Y = -\frac{XC_B^{II} - C_B^I}{XC_A^{II} - C_A^I}$$

A graph of Y versus X is shown in Fig. 3.48 (which omits theoretically possible negative values). It is a curved line (b) which crosses the X axis, where $Y = 0$, at a point given by $XC_B^{II} - C_B^I = 0$, i.e. $X = C_B^I/C_B^{II}$. In the special case when $C_A^{II} = 0$, i.e. when A produces no counts at all in channel II, we find that the equation for Y has the simpler form

$$Y = \frac{C_B^{II}}{C_A^I} X - \frac{C_B^I}{C_A^I}$$

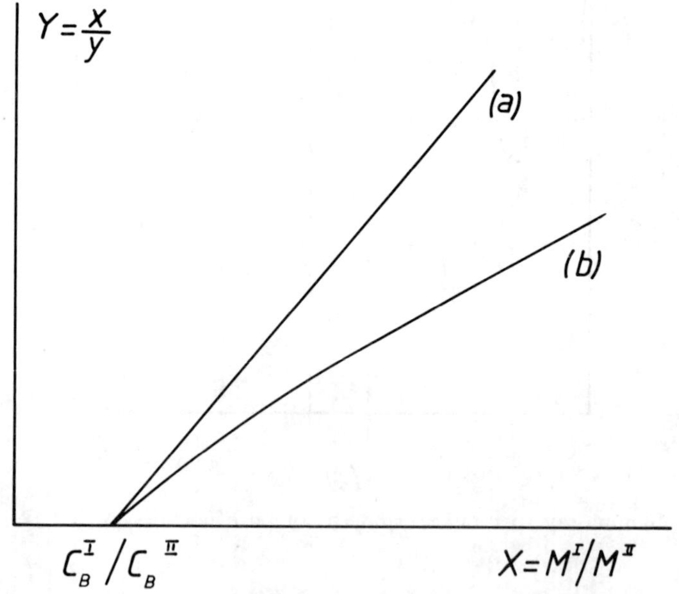

Fig. 3.48 Relationship between the ratio $Y = x/y$ of two nuclides A and B, and the ratio of the counts in the two channels, $X = M^I/M^{II}$.

This is the equation to a straight line with slope C_B^{II}/C_A^I and intercept $-C_B^I/C_A^I$ on the Y axis, and is shown as line (a) in the figure. The intercept on the X axis is still C_B^I/C_B^{II}. Calculations of Y, when many samples have to be dealt with, is of course very conveniently done on a programmable desk calculator.

3.29 Dead-Time Corrections

The disintegrations in a radioactive source occur randomly in time and for source strengths above a certain value a noticeable effect is that the detector fails to respond to all of them. Thus if say N ionising events take place in a detector in a time of 1 s, the number of counts recorded will be say C, where $C < N$, and the *loss* in counts $N - C$ is greater, the greater the value of N. The simplest case of counting loss is that in Geiger–Müller counting. An ionising event in a GM tube will trigger the quenching circuit which imposes a dead-time τ during which the occurrence of a second event cannot produce an avalanche and will therefore fail to be detected. In scintillation counting a dead-time is effectively set by the analyser circuit, whose output is typically a stream of flat-topped pulses all about 4 μs in duration; during the existence of any of these an extra event in the scintillator would not give rise to an extra output pulse. Also if two events occur in the scintillator within a fraction of a microsecond of each other, the photomultiplier tube may produce a single pulse which the analyser will interpret as if it were due to a single event. Then a single count will appear in a high-energy channel instead of two in lower-energy channels.

Most theoretical treaments of counting losses due to dead-time assume that there are two extreme classes of effects within a counting system, according to whether the system is regarded as non-paralysable, or as paralysable. In the first class it is assumed that the occurrence of a 'countable' event imposes a dead-time τ on the whole system, during which the occurrence of a second (or indeed any subsequent) event produces no additional recorded count, nor does it affect the length of the dead-time already existing. In the second, paralysable, class, an additional event occurring during an existing dead-time does not produce a count, but it adds its own dead-time τ. This means that if two events occur within a time interval of τ', say (where $\tau' < \tau$), only one count will be recorded and the system will be inoperative for a time $\tau + \tau'$. Geiger–Müller detectors and mechanical registers are often quoted as examples of non-paralysable systems, and the combination of amplifier and analyser for scintillation counters as an example of paralysable systems. Detailed examination of many scintillation counter systems, however, seems to show that they may behave in some intermediate way, which is determined by details of the electronic circuitry. We will examine the theoretical relationship between N and C, as used above, for each of the cases mentioned before discussing how accurate values of N may be obtained from tthe observed values of C on the basis of experiment.

(*a*) *The Non-paralysable Case.* An elementary treatment of a non-paralysable system is to say that since C counts are observed within a unit time (1 s), each creating a dead-time τ, the system is effectively capable of recording events only for a time $(1 - \tau C)$ second. Since C counts are recorded, the 'true' count-rate is $C/(1 - \tau C)$, which must be equal to N. Therefore

$$
\begin{aligned}
N &= C/(1 - \tau C) \\
&= C(1 - \tau C)^{-1} \\
&= C(1 + \tau C + \tau^2 C^2 + \tau^3 C^3 + \cdots)
\end{aligned}
$$

by expanding the binomial term in series. We may also arrive at the relation between N and C on the assumption that the events which are *not* recorded, i.e. $(N - C)$, have the same rate of occurrence as those which *are* recorded. Since the $(N - C)$ unrecorded events occupy a time τC, we have

$$(N - C)/\tau C = C/(1 - \tau C)$$

which reduces to the same expression as before.

It is clear from the above expressions that the count-rate C will steadily increase if N is increased, but cannot exceed a maximum value of $1/\tau$ which should theoretically be attained when N becomes infinite.

(*b*) *The Paralysable Case.* In this extreme case, each *event* produces a dead-time τ, independently of whether another event occurs in that time. Thus a pair of events which occur within a time τ' of each other will produce just one recorded count if τ' is less than τ, but the 'composite' count will occupy a dead-time of $\tau' + \tau$. To find the new relationship between N and C, proceed as follows. Let us consider any instant of time within a unit interval during which N events occur randomly. The probability that any one event does *not* create a dead-time interval overlapping this instant is $(1 - \tau)$. As there are N events the probability that *none* of them overlap the instant is $(1 - \tau)^N$, since the individual probabilities are independent. (For a fuller statement of the principles involved here see Section 4.12.) This expression gives the probability that the detector is 'live' at the particular instant, and is therefore equal to the time, during the interval, for which the detector can respond to events. It therefore replaces $(1 - \tau C)$ in the equations derived for the non-paralysable case. Therefore

$$N = C/(1 - \tau)^N$$

or

$$C = N(1 - \tau)^N$$

It can be shown from these that as N is increased, C goes through a maximum and then decreases again to zero as N tends to infinity and the detector becomes 'dead' for the whole of the time. However it can usually be assumed that τN is rather less than 1. Since τ is usually a very small fraction of the unit counting interval, we may replace $(1 - \tau)^N$ with trivial loss of accuracy by $\exp(-\tau N)$. Then

$$N = C \exp(\tau N)$$

and

$$C = N \exp(-\tau N)$$

On expanding the exponential in the expression for N we have

$$N = C\left(1 + \tau N + \frac{\tau^2 N^2}{2} + \frac{\tau^3 N^3}{6} + \cdots\right)$$

To make this comparable with the forms we have found for the non-paralysable case we replace N wherever it occurs on the right-hand side by the approximation

$$N = C\left(1 + \tau N + \frac{\tau^2 N^2}{2}\right)$$

Then

$$N = C\left\{1 + \tau C\left(1 + \tau N + \frac{\tau^2 N^2}{2}\right) + \frac{\tau^2 C^2}{2}\left(1 + \tau N + \frac{\tau^2 N^2}{2}\right)^2 + \frac{\tau^3 C^3}{2}\left(1 + \tau N + \frac{\tau^2 N^2}{2}\right)^3 + \cdots\right\}$$

$$= C\left\{1 + \tau C + \tau^2\left(CN + \frac{C^2}{2}\right) + \tau^3\left(\frac{CN^2}{2} + C^2 N + C^3\right) + \cdots\right\}$$

and proceeding in this way, we get

$$N = C\left\{1 + \tau C + \frac{3}{2}\tau^2 C^2 + \frac{7}{2}C^3\tau^3 + \cdots\right\}$$

This expansion gives N in terms of C correct to less than 1% as long as τC is less than 0.2. As with non-paralysable systems, much equipment in common use becomes unreliable if the attempt is made to use it at much higher count-rates than this implies.

The above treatment shows that for both the ideal paralysable and non-paralysable cases the estimation of the true count-rate N from the observed rate C depends in a rather complicated way on the dead-time τ. It is of course always possible to display on an oscilloscope the actual pulses occurring within a counting system and measure their duration in time. Unless one is familiar with the detailed electronics of the system there is always doubt as to whether any such measurement gives the effective dead-time, and it always seems preferable when one has to make corrections to an experimentally observed count-rate, to do so on the basis of counting experiments with some kind of standard source. It is common practice to use two sources A and B, together with a 'dummy" of zero activity, all sources being as nearly as possible of the same geometrical form. Counts are taken with source A and the dummy, and then source B and the dummy, in exactly reproducible positions near to a detector. If the sources are such as to give counts N_A and N_B on an ideal counting system with no dead-time, then the count-rates on the real system will be C_A and C_B where

$$N_A = C_A[1 + \alpha C_A\tau + \beta C_A^2\tau^2 + \gamma C_A^3\tau^3 + \cdots]$$

$$N_B = C_B[1 + \alpha C_B\tau + \beta C_B^2\tau^2 + \gamma C_B^3\tau^3 + \cdots]$$

and the coefficients α, β, γ, . . . are chosen in accordance with how it is considered the system operates. Finally the detector is presented with sources A and B together in the positions previously occupied by one of them and the dummy. The use of the dummy ensures that scattering effects caused by one source in the presence of the other are the same in all three counting experiments. For the composite source we shall have a count-rate C_C given by

$$N_A + N_B = C_C[1 + \alpha C_C\tau + \beta C_C^2\tau^2 + \gamma C_C^3\tau^3 + \cdots]$$

These three equations contain the three unknowns N_A, N_B and τ. A large number of more or less accurate methods have been recommended for solving them; the approximations made in some of them, together with the uncertainties in what the best values of α, β, γ really are for any given counting system, can lead to estimates of τ differing by as much as a factor of two, as has been shown in some recently published work.

Another point which needs emphasis is that when using the two-source method as outlined above in scintillation counting, correct values of the dead-time can be expected only if one counts over the whole extent of the energy spectrum or makes a suitable correction if this is not done. In practice, of course, one often wishes to use a scintillation counter with a window setting to include only a photoelectric peak, so that only a fraction, say f, of the counts selected by the analyser is recorded. The true count-rate over the peak is then

$$\begin{aligned} fN &= fC[1 + \alpha C\tau + \beta C^2\tau^2 + \gamma C^3\tau^3 + \cdots] \\ &= fC[1 + \alpha(fC)(\tau/f) + \beta(fC)^2(\tau/f)^2 + \gamma(fC)^3(\tau/f)^3 \cdots] \end{aligned}$$

If follows that any two-source method for determining the dead-time, in which the observed counts are quantities (fC_A), (fC_B), etc., will in fact determine τ/f which will appear to be the dead-time. The situation is even more complicated when a multichannel analyser is used, because these instruments invariably operate by digitising each pulse to be analysed. Digitisation requires a longer time, the greater the incoming pulse amplitude. Measurements using paired-source counting can, therefore, yield only a weighted mean value of the dead-time.

From a pragmatic point of view, of course, it does not matter, for any given counting system, what the actual value of the dead-time is; all that we are interested in is having a means of calculating the 'true' count-rate N for a source, having measured its apparent count-rate C. Although the two-source method of obtaining a correction is simple, it is clearly not wholly satisfactory, and it may be preferable to calibrate a counting system by using a larger number of sources, particularly if their relative strengths are already known *a priori*. Such sources can be prepared for example by dispensing accurately known, but different, volumes of a homogeneous stock solution into a set of identical vials. Each sample is then made up to the same volume by addition of pure solvent. The samples are then presented, under identical geometrical condition, to the detecting equipment. In a variant of this technique, a single sample having a relatively short half-life is used, this one sample being counted at different times during its decay. Then regarding the quantities $p = \alpha\tau$, $q = \beta\tau^2$, $r = \gamma\tau^3$, . . . as unknowns, we would have, for four sources or for a single source at different times,

$$
\begin{aligned}
N_A &= C_A[1 + pC_A + qC_A^2 + rC_A^5 \cdots] \\
N_B = k_1 N_A &= C_B[1 + pC_B + qC_B^2 + rC_B^3 \cdots] \\
N_E = k_2 N_A &= C_C[1 + pC_C + qC_C^2 + rC_C^3 \cdots] \\
N_D = k_3 N_A &= C_D[1 + pC_D + qC_D^2 + rC_D^3 \cdots]
\end{aligned}
$$

In these equations, where k_1, k_2, k_3 are known and C_A, C_B, C_C, C_D are measured, there are only the four unknowns N_A, p, q, r. (A fifth source would be needed to determine the coefficient of a term in C^4, and so on.) Having determined p, q and r, it is then a simple matter to compute the correction to be applied to any observed count C, from the relation

$$N - C = pC^2 + qC^3 + rC^4$$

It will be seen that this technique obviates the need to assume any particular mode of counting loss due to dead-time.

3.30 Relative Assays: Ionisation Current Method

Sources of more than a few megabecquerels of a gamma-ray emitting nuclide are not conveniently measured by the methods so far discussed unless one is prepared to use a system with very low geometrical efficiency, e.g. a small scintillation crystal at a distance of a metre or so from the source. Such a system is quite practical if space is available, and has the advantage that small displacements of the source have little effect on the observed count-rate. This means in turn that the source need not approximate to a geometrical point, but may be a few centimetres in extent, although, depending on the gamma-ray energy, small self-absorption effects might then occur and require corrections to be made.

An alternative technique is to use an ionisation chamber in the continuous mode. Arguments similar to those used in Section 3.10 show that the ionisation current which can be obtained when a mass of 1 kg of standard air all at 1 m distance from a source of 1 MBq emitting a gamma-ray of 1 MeV per disintegration is theoretically 1.01×10^{-12} A. It is of course not a practical

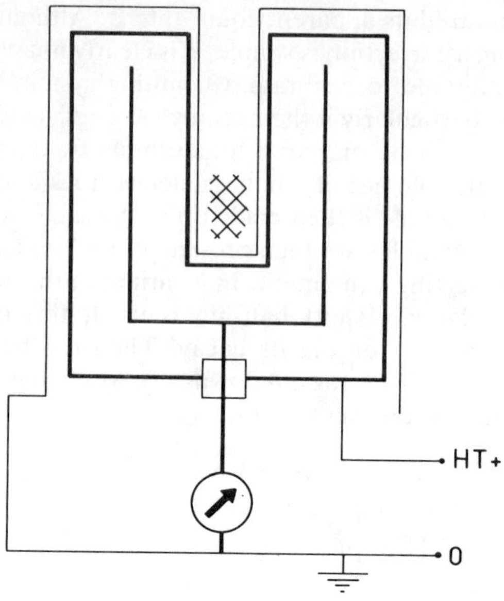

Fig. 3.49 Re-entrant ionisation chamber.

proposition to have 1 kg of air all at a distance of 1 m, but an ionisation chamber with a volume of a few litres is quite easy to construct. A simple practical design is shown in section in Fig. 3.49.

The chamber has a hollow cylindrical geometry, the outer wall being at high potential. The inner electrode is a cylinder closed at the lower end and supported on a high-quality insulator. The whole is enclosed in an earthed outer case to permit safe handling. Current collected by the inner electrode goes to earth via a sensitive electrometer. The shape of the chamber gives rise to the name of re-entrant ionisation chamber. Careful attention to the relative dimensions ensures that there is quite an appreciable space (shown cross-hatched in the figure) within the central re-entrant wherein a point source can be moved with very little effect on the ionisation current it produces. The chamber can therefore be used for accurate assays of samples with volumes of several millilitres provided the material does not absorb the radiation appreciably, after calibration with a point source of known strength. Better still is to calibrate with a known source having the same physical form as the samples to be used (e.g. aqueous solutions all of the same volume and contained in identical vials). The ionisation current produced in a typical chamber is of the order 10^{-11} A per megabecquerel of medium-energy gamma emitter, so that an amplifier with a current gain of some 10^6–10^8 will enable measurements to be made on a microammeter. An improvement in sensitivity can be achieved by operating the chamber at elevated pressure and using a gas with greater atomic number than air. A commercially available chamber may use argon at a pressure of 10–12 times atmospheric pressure, giving a current 20 times what it would be with air filling at ordinary pressure.

So far we have implied that the ionisation current produced by a source, like the exposure in roentgens, is proportional to the product AE (A being the source strength in becquerels and E the gamma-ray energy in MeV), or to the sum of terms like these if the source emits several gamma-rays of different energy per disintegration. This rule is indeed obeyed fairly well if E is greater than about 0.2 MeV. However when the gamma-ray energy is between 0.2 and 0.1 MeV a chamber is usually found to be much more sensitive than would be expected from the simple linear relationship. A typical sensitivity curve is shown in Fig. 3.50. The raised sensitivity in the range 0.1–0.2 MeV comes about through the large fraction of the ionisations (originating in the gas filling but more particularly in the chamber walls) which give rise to electron tracks *ending* in the gas space. These contribute a large current due to the relatively dense secondary ionisation. For higher energy gamma-rays, most of the photoelectrons and Compton electrons will have ranges large compared with the chamber dimensions, so that relatively few of these will come to the end of their range in the gas. The eventual fall-off in sensitivity, for the values of E less than 0.05 MeV, is caused by absorption of gamma-rays and their ionisation products in the chamber walls.

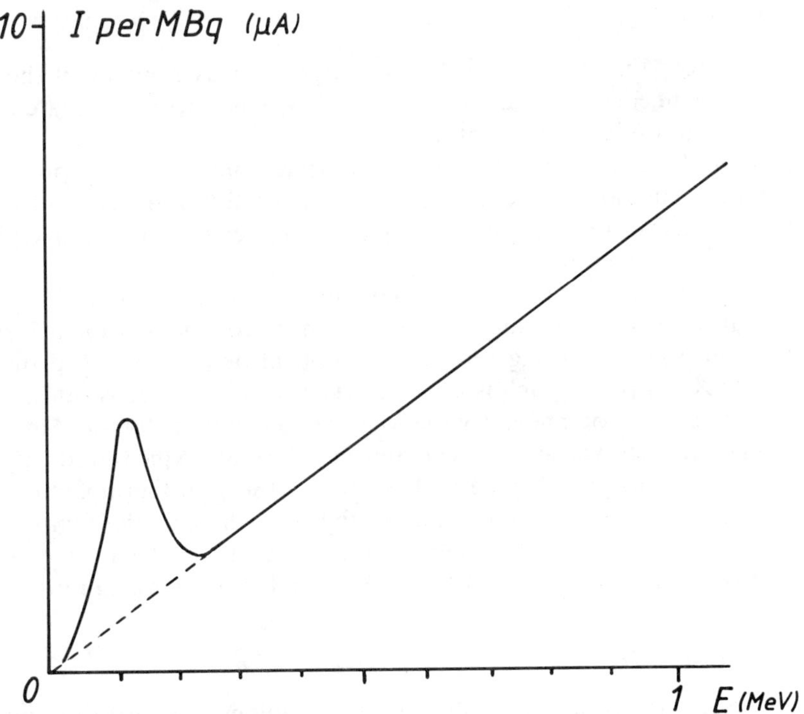

Fig. 3.50 Sensitivity of a re-entrant ionisation chamber.

3.31 Photographic Method

Photographic films and plates contain grains of silver compounds. Each grain can be regarded as a three-dimensional lattice of silver ions Ag^+ and negative ion $Cl-$, Br^- and S^{2-}. The effect of an ionising event (photoelectric or Compton effect) in a grain results in one or more silver ions capturing a freed electron and becoming uncharged silver atoms. When the film is 'developed' those grains which have not been affected by radiation remain unchanged but any grain containing a few free silver atoms undergoes further chemical reduction and many more silver atoms are produced. The final number of silver atoms in a grain is determined by the development time, temperature and strength and composition of the developing solution, all of which have to be under strict control. On 'fixation', the development process is stopped and unchanged silver salts are dissolved out of the film, the remaining silver atoms constituting the final image. The image is blackest where the radiation intensity was highest.

The degree of blackening can be measured with a densitometer; if visible light of intensity F_0 is incident on the image and it is found that the intensity transmitted is F_1, we define the optical density D of the film as

$$D = \log_{10} (F_0/F_1)$$

It is clear from this definition that only relative measurements of the light intensity are needed to evaluate D, and that a perfectly transparent film would have an optical density of zero.

For a set of exposures all of the same exposure time, and with reproducible development conditions, the density of an exposed film depends of course upon the intensity of the radiation being used, the relationship being typically of the form expressed in Fig. 3.51.

In this figure the intensity I of the radiation is in arbitrary units. The curve is reasonably linear over a range of optical density commonly from 0.3 to 1.5, which is the useful working range. (The optical density averaged over an acceptable X-ray photograph is about 1.) The working range is sufficient for accurate assessment of intensity variations over a factor of 10 or so. However, radiation intensities whose ratio is greater than 100 can be measured only with some loss in accuracy (compare with Section 3.18(e) on thermoluminescent materials). The slope of the linear part of the curve is called the *gamma value* of the film. The *contrast* between two areas of film which have optical densities of D_1 and D_2 is defined as $D_1 - D_2$ and in the working range this is given by

$$\text{Contrast} = D_1 - D_2 = \gamma \log_{10} (I_1/I_2)$$

Some idea of the sensitivity of films used for radiographic exposure may be gained from the following rough calculation. It is found that a 'diagnostic' exposure of 0.1 s duration with a beam current of 100 mA and a voltage of

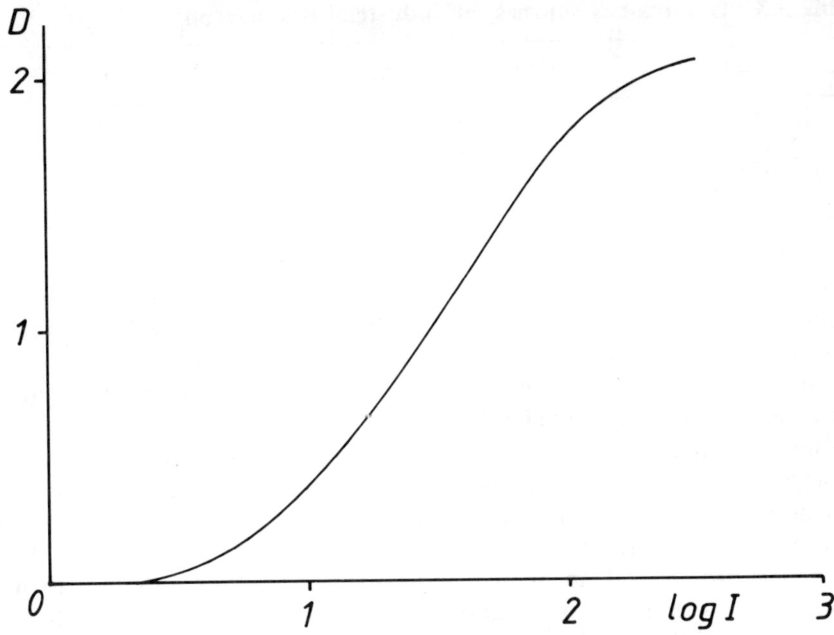

Fig. 3.51 Optical density of exposed film.

100 kV will give an optical density of about 1 with a film at 2 m distance. The energy carried by the electron beam in the X-ray tube is the product of exposure time, current and voltage, which comes to 10^3 J. The efficiency of conversion of this energy to X-rays is of the order 1%, so that the useful radiated energy is about 10 J. Assuming isotropic X-ray emission, the radiation energy traversing 1 mm^2 of film will then be $10/(4\pi \times 2000^2) \approx 2 \times 10^{-7}$ J. Assuming a mean photon energy in the X-ray beam of 50 keV, or 8×10^{-15} J, we find that the photon fluence is about 2.5×10^7 mm^{-2}. (This is of course in the absence of a subject being X-rayed; during a chest X-ray the number of photons transmitted is about 10% of those incident, but for a lateral pelvis view the figure is only about 0.1%.) Multiplying the energy fluence of 2×10^{-7} J mm^{-2} by the mass attenuation coefficient for air, which is 0.0028 m^2 kg^{-1} or 2.8×10^3mm^2 kg^{-1} we find an energy absorption in air of 5.6×10^{-4}J kg^{-1}, i.e. 5.6×10^{-4}Gy. This calculation is intended to indicate numerical values which are fairly representative of common practice, and in particular to show that a reasonable exposure of X-ray film requires a photon fluence of the order 10^7mm^{-2}. There is a wide range of sensitivity in commercial films, of course, and sensitivity can be enhanced by the use of intensifying screens.

It is of interest in this connection to consider the use of geometrcially small ('point') sources of radionuclides for radiographic purposes. Thus a point source of 1 GBq at a distance of 100 mm from a film would give a photon flux density of $10^9/(4\pi \times 100^2) \approx 8 \times 10^3$ mm^{-2} s^{-1}. Again if a fluence of photons of 2×10^7 mm^{-2} is needed for reasonable optical density on the film,

Table 3.3 Gamma-ray sources for Industrial Radiography

Nuclide	Gamma-ray energies (MeV)	Half-life
^{60}Co	1.1, 1.3	5.3 yr
$^{137}Cs + ^{137}Ba$	0.66	30 yr
^{182}Ir	0.316, 0.47, 0.30	74 d

the exposure time must be 2500 s. Such exposure times would be too long for human subjects but are quite convenient for industrial radiography. Suitable sources are listed in Table 3.3; the choice of source depends of course on the thickness and density of the material to be radiographed. Source strengths of many gigabecquerels are available.

Radiation measurements by photographic means has an important practical application in the well known 'film badge' as used to monitor personal exposure. This is a plastic holder, some $50 \times 35 \times 9\ mm^3$, shaped to contain a piece of dental X-ray film $45 \times 30\ mm^2$ in a paper envelope. The sides of the badge provide (a) an open 'window' which enables radiation to have unobstructed access to the film and also displays a printed serial number and the identity of the person wearing the badge, (b) a plastic window about 0.5 mm thick, (c) a plastic window about 3 mm thick, (d) a composite window of plastic and light alloy (Dural), about 1 mm thick, and (e) a composite window of plastic, tin 0.5 mm thick and lead 0.2 mm thick. The windows are not all of the same size and shape, as an aid in identifying the pattern of blackening on the film after exposure. 'Soft' radiation such as beta particles and low-energy photons will affect the film under the open window to a much greater extent than elsewhere, whereas 'hard' radiation, i.e. gamma photons of 500 keV and above will affect all regions of the film almost equally. Under the tin-lead window the blackening is very nearly the same, for a fixed dose of gamma radiation, over an energy range of 100 keV to several mega-electronvolts.

When a batch of films are issued to a group of radiation workers, samples are reserved for exposure under controlled conditions to radiation from a variety of sources. These control films then enable dosages received by the personnel during a two-week, or four-week, working period to be estimated fairly accurately in the range 20 mR to about 2 R. Higher doses give an optical density over 2. Study of the relative densities over the window areas gives a good indication of the nature of the incident radiation, and this is useful for instance in determining whether a film has been exposed to a direct intense beam of X-rays or gamma-rays for a short time, or to a weak field of scattered photons for a long time. Occasionally bizarre effects are observed which are attributable to contamination of the film or its holder with radioactive material. This might indicate careless handling of radionuclides or the unexpected—and potentially highly dangerous—presence of radioactive dust in the laboratory.

A further utilisation of the photographic effect is exemplified by the technique of microautoradiography. Thin slices of tissue containing compounds labelled with ^{3}H, ^{14}C or ^{35}S, when laid down in contact with X-ray film, produce remarkably clear autoradiographs. The high resolution is the result of the small range of the beta particles in the emulsion. Other nuclides which can be used are ^{32}P, which does not give such good resolution because of the greater range of its beta particles, and ^{51}Cr, which emits many very-low-energy Auger electrons and gives particularly sharp pictures. Special emulsions are available which are relatively insensitive to electrons but show individual alpha particle tracks, from, for example, ^{226}Ra, very clearly. In all these examples the autoradiograph is conveniently examined under a magnification of ×100 or more, and under good conditions it is possible to locate the site of a particular concentration of radionuclide to a few tenths of a micrometre. A disadvantage of the technique is its low sensitivity, typical exposure times for long-lived kilo-becquerel β-emitting sources extending to several weeks. As discussed in Section 2.14, ^{99m}Tc gives a rather low yield of conversion electrons (except for the very-low-energy conversion electrons in the 0.002 MeV transition). As the half-life is so short, it is unlikely that good microautoradiographs can be obtained from this nuclide unless source strengths of several megabecquerels are used.

3.32 Radiation-Chemical Method

The energy absorbed in matter from a beam of radiation can give rise to chemical changes in the material, either as a result of ionisation or, perhaps as importantly, as a result of molecules being put into excited states but without the physical loss of an electron. Ionisation occurring in the K or L shells of atoms within a molecule is liable to lead to chemical reaction because vacancies will be filled by electrons descending from outer electronic levels until valency electrons are displaced. A valency bond may then be broken, leading to chemical change. Chemical change is not, however, inevitable because the affected molecule may be able to recapture an electron set free elsewhere in the material before such a change can occur. Chemical effects of this nature constitute what is known as radiation chemistry. We shall consider only the quantitative aspect of some simple systems, but the subject is clearly of basic importance in the study of the effects of radiation on living systems (radiobiology).

The radiation chemistry of such an apparently simple substance as water has been the subject of hundreds of research papers over many years. The primary products are believed to be free hydrogen atoms and free hydroxyl radicals, chemically symbolised as H and OH respectively. (It is possible that hydrogen atoms in water are bound in some way to water molecules and could be more accurately symbolised as H_3O, but for our purposes the simpler form is adequate to explain the phenomena under discussion.) The reaction products may be thought of as being produced either by a direct decomposition of excited molecules which we symbolise in the form

$$H_2O \rightsquigarrow H + OH$$

or by a mechanism in which a water molecule is first ionised,

$$H_2O \rightsquigarrow H_2O^+ + e$$

In each of these changes the arrow with the zigzag indicates that it is caused by the absorption of radiation, the distinguish it from ordinary chemical reactions. The second of these changes is followed by the decomposition of the ionised water molecule

$$H_2O^+ \rightarrow H^+ + OH$$

and by the migration of the electron through the medium (perhaps causing further ionisation and excitation on its journey) until it meets and combines with one of the H^+ ions normally present in the medium.*

$$H^+ + e \rightarrow H$$

It seems likely that the first mechanism predominates; the evidence is that it will give rise to H and OH species in close physical proximity and so account for the fact that, in very pure water, very little overall chemical change occurs, implying a highly effective recombination mechanism

$$H + OH \rightarrow H_2O$$

However, in even the purest water, a series of other reactions does occur, which lead to the decomposition of water into hydrogen and oxygen—as might be expected—but also to the production of hydrogen peroxide, H_2O_2. Moreover, under conditions of steady irradiation these reactions proceed at a steadily accelerating rate for some time, finally decelerating again and becoming steady as the water becomes saturated with dissolved oxygen and the concentration of hydrogen peroxide becomes constant. To explain these phenomena, it is postulated that initially small amounts of oxygen are formed by reactions such as

$$OH + OH \rightarrow H_2O + O$$
$$O + O \rightarrow O_2$$

The initial rate of production of molecular oxygen is slow because of the difficulty of transferring a hydrogen atom from one OH radical to the other in the first of these reactions. Once formed, however, oxygen molecules react very readily with hydrogen atoms set free elsewhere in the liquid, forming

*In pure water there is a dynamic equilibrium

$$H_2O \rightleftharpoons H^+ + OH^-$$

which gives rise to a standing concentration of H^+ of 10^{-7} mol 1^{-1}, i.e. a pH of 7.

first the hydroperoxyl radical and then hydrogen peroxide.

$$H + O_2 \rightarrow HO_2$$
$$H + HO_2 \rightarrow H_2O_2$$

It is, therefore, of fundamental importance that the irradiation of water leads to the production of hydrogen peroxide. Clearly the conversion of water to hydrogen peroxide can take place only with the simultaneous liberation of hydrogen atoms in accordance with the overall reaction

$$2H_2O \rightarrow H_2O_2 + 2H$$

Therefore in the early stages of irradiation, while the hydrogen peroxide concentration is building up, the ratio of hydrogen to oxygen evolved is greater than 2:1.

If water which has not been especially purified from dissolved oxygen is irradiated, the build-up of hydrogen peroxide starts rapidly until a steady state is reached identical with that attained only slowly with pure water as starting material. The explanation of the eventual constancy of hydrogen peroxide concentration (under steady irradiation) is to be sought in terms of a balance between the above production reaction and others in which it is destroyed. Destruction reactions such as the following are believed to occur:

$$H_2O_2 + H \rightarrow H_2O + OH$$
$$H_2O_2 + OH \rightarrow H_2O + HO_2$$
$$H_2O_2 + HO_2 \rightarrow H_2O + OH + O_2$$

The last reaction in particular provides a direct mechanism for the regeneration of oxygen. Whatever the mechanisms for regulating the hydrogen peroxide concentration, it is clear that once dynamic equilibrium has been attained the overall effect of continuing radiation is to generate hydrogen and oxygen in the exact 2:1 ratio. If the radiation intensity changes from one steady level to another the hydrogen peroxide concentration will slowly readjust itself, and during this time the generation of hydrogen and oxygen will not be in the ratio 2:1. From the point of view of the measurement of a dose of radiation, the collection of the evolved hydrogen and oxygen is an experimental possibility, but even if the obvious technical difficulties of such a method were overcome it would be almost impossible to apply it to irradiations on a short timescale or if the radiation intensity were fluctuating.

Fortunately, there are many other systems which are more easily exploited for radiation measurement. The most widely used of these employs a solution of ferrous ammonium sulphate in dilute sulphuric acid and is known as the 'ferrous sulphate dosimeter', or 'Fricke dosimeter', after its originator. Ferrous ammonium sulphate has the advantage over other ferrous compounds of being readily available in a pure crystalline non-hygroscopic state and as such is easily weighed out accurately. The ammonium ion plays no part in the radiation chemistry, nor does the sulphuric acid, but the latter prevents hydrolysis of the ferrous ions and eventual precipitation of ferric hydroxide. In a ferrous sulphate solution initially free of dissolved oxygen, the effect of irradiation is oxidation of ferrous ions to ferric ions, through interaction with the OH radicals formed in the radiolysis of the water

$$H_2O \rightsquigarrow H + OH$$
$$Fe^{2+} + OH \rightarrow Fe^{3+} + OH^-$$

If (as is generally the case in practice) the solution is oxygenated, ferrous ions can also be oxidised by reaction with HO_2:

$$Fe^{2+} + HO_2 \rightarrow Fe^{3+} + HO_2^-$$

The last reaction produces hydrogen peroxide which can in turn oxidise still more ferrous ion:

$$HO_2^- + H^+ \rightleftharpoons H_2O_2$$
$$2Fe^{2-} + H_2O_2 \rightarrow 2Fe^{3+} + 2OH^-$$

The presence of oxygen in the solution thus theoretically enables four times as much ferrous ion to be oxidised as would be the case in oxygen-free conditions. It is not clear that this ratio has been exactly confirmed by direct experiment. Nevertheless, the Fricke dosimeter does give highly reproducible results in that the amount of ferrous ion oxidised is strictly proportional to the total absorbed does, independently of dose-rate or fractionation in time. For good reliability it is essential to ensure very high purity of the irradiated solution; organic impurities in particular must be avoided. The ferrous ion concentration should be between 10^{-3} and 10^{-4} mol l^{-1} and the sulphuric acid concentration between 0.2 and 1.5 mol l^{-1} (preferably 0.8 mol l^{-1}). Some investigators add sodium chloride to the solution at the same molar concentration as the ferrous ammonium sulphate. The yield of ferric ion can be measured spectrophotometrically at a wavelength of 304 μm, corrections being made for variations of the extinction coefficient with temperature. The yield of ferric ion is independent of the temperature during irradiation, at least between 0 and 50 °C, and the proportionality between yield and absorbed dose extends up to at least 2000 Gy if the solution is kept saturated with oxygen. A convenient measure of the yield is the so-called *G* value, which is the number of ferric ions produced per 100 eV of deposited energy. This has been measured by many workers and the commonly accepted value is 15.5 and appears to be virtually identical whatever the nature of the radiation (beta particles or X- or gamma-rays). Alpha particle irradiation may give a lower value. A service from the National Physical Laboratory is now available in which quartz ampoules containing ferrous ammonium sulphate are sent out to customers who may irradiate them and return them for spectroscopic measurement.

The difficulties of measuring very small amounts of ferric ion mean that the dosimeter is not very sensitive and can hardly be used for absorbed doses of much below 10 Gy. Its sensitivity can be enhanced by complexing the ferric ion with a reagent such as α-phenanthroline, the complex giving much stronger absorption of light. There is some slight loss in reproducibility. Some investigators add the complexing agent post-irradiation, just before the measurement; others add a reagent (e.g. xylenol orange) before the irradiation, which is perhaps a little more convenient in practice since a large 'stock' solution can be subdivided into several identical portions for replicate experiments. The presence of complexing agent in the solution during irradiation however alters the *G* value slightly.

Among other well known systems should be mentioned the oxidation of Ce^{3+} to Ce^{4+}, which has the advantage that the yield is independent of the oxygen concentration. It is unfortunately not very sensitive, the *G* value being only 3.2, i.e. 3.2 Ce^{3+} ions are oxidised per 100 eV absorbed. Solutions of chloroform and other chlorinated hydrocarbons in water show the production of H^+ with yields proportional to dose. The reaction mechanism seems to involve three kinds of step:

(a) Initial radiation effect, giving the radical $CHCl_2$:

$$CHCl_3 \rightsquigarrow CHCl_2\text{—} + Cl$$

(b) Chain propagation, in which long-chain molecules are formed:

$$CHCl_3 + CHCl_2\text{—} \rightarrow CHCl_2\text{—}CCl_2\text{—} + HCl$$
$$CHCl_3 + CHCl_2\text{—}CCl_2\text{—} \rightarrow CHCl_2\text{—}CCl_2\text{—}CCl_2\text{—} + HCl$$
$$\vdots \qquad\qquad\qquad \vdots$$

One molecule of hydrochloric acid is produced in each step.

(c) Chain termination, which can occur when one chain 'meets' another, or reacts with a free chlorine atom arising in the initial radiation step.

Clearly the yield of acid product, per initial radiation step, depends on the chain length. This can be influenced by the presence of other substances in the solution, and indeed the deliberate addition of certain alcohols has been recommended as exerting control over chain length. However other substances present only adventitiously, i.e. very small amounts of unknown impurities can be responsible for erratic results. The liberation of free acid as a result of irradiation is conveniently measured by colour changes in dye indicators. *G* values, i.e. numbers of acid molecules per 100 eV absorbed, are 100 or more, depending on conditions, so that the system is inherently an order of magnitude more sensitive than the Fricke dosimeter but much less reproducible.

As discussed so far, radiation chemical effects have been related to the energy deposited in a system from a beam of radiation, rather than to the strength of a radioactive source. The data of Section 3.10 make it clear that even for terabecquerel sources (1 TBq = 10^{12} Bq), many hours of exposure with quite large amounts of irradiated material are needed to produce easily measurable chemical change and indeed the interest in radiation chemistry revolves much more about the nature of the changes and their quantification than on the possibility of measuring source strengths. On the other hand, the administration of radioactive material to human or animal subjects at only the gigabecquerel level can have damaging, or therapeutic, effects because quite small chemical changes in sensitive regions of living cells can have profound biological consequences.

3.33 Calorimetric Methods

At first sight the possibility of measuring the strength of radioactive sources by the heat they produce is attractive and the technique has been developed extensively on an experimental basis. Unfortunately thermal effects are not large; for example 1 TBq of a source emitting a total radiation energy of 1 MeV per disintegration which is ultimately degraded to heat generates $10^{12} \times 10^{6} \times 1.6 \times 10^{-19}$ J s^{-1}, i.e. 0.16 W. Moreover, if the radiation is penetrating (hard gamma-rays), the mass of absorbing material would have to be of the order of a kilogram in order to develop a reasonably large fraction of this power, so that the rate of rise in temperature of the absorber would be of the order 10^{-3} °C s^{-1}. The heat generated by a source of ^{60}Co of about 100 TBq has indeed been measured essentially by observing the rate at which ice in contact with water at 0°C was melted in a modern version of Bunsen's ice calorimeter. But for sources of only a few gigabecquerels, calorimetry is used routinely only for alpha and beta particle sources which can be placed in calorimeters weighing only 1 g or so. These have walls thick enough to contain the alpha or beta particle tracks, while the contribution to the heat dissipated by any gamma-rays is negligibly small. In a successful technique, a pair of identical calorimeters is used, one containing the source itself and the other an electrical heating coil. A thermistor is attached to both calorimeters and connected in a bridge circuit which controls the power in the heating coil so as to maintain the calorimeters at an exactly equal temperature. In this way thermal losses from the calorimeters are compensated and the power in the heater is exactly equal to that generated by the source. The method is satisfactory for sources of the nuclides ^{226}Ra and ^{32}P.

An interesting recent development has been the measurement at the National Physical Laboratory of the intensity of electron beams at 2 MeV by exposure of a calorimeter filled with a solution of ferrous ammonium sulphate, thus combining the calorimetry and radiation chemistry techniques. Corrections have to be made to the observed heat generation from the heat of the reaction $Fe^{2+} \rightarrow Fe^{3+} + e$ and when this is done there is good agreement between the values of dose calculated from the temperature rise and from the chemical yield.

3.34 Exercises

1. Obtain the relation between the momentum $p = mv$ of a relativistic particle and its total and rest energy (see Appendix 2).
2. Calculate the energy of a helium ion ($^{4}He^{2+}$) which has a relativistic mass just twice that of an alpha particle. What is the maximum energy

this ion could transfer to an electron at rest, assuming an elastic collision?

3. A 5 MeV electron collides with an unbound electron at rest. After the collision one electron is found to have an energy of 2 MeV. Calculate (a) the energy of the other, (b) the momenta of the electrons, (c) the angles that their trajectories make with the forward direction.
4. Estimate the range of beta particles in aluminium from ^{131}I (E_{max} = 0.6 MeV), ^{35}S (E_{max} = 0.167 MeV), ^{3}H (E_{max} = 0.0186 MeV).
5. Calculate the energy of Compton scattered gamma rays at angles 30°, 60°, 90°, 120°, 150°, 180° for incident gamma rays of (a) 140 keV, (b) 350 keV.
6. Estimate the thickness of a slab of sodium iodide needed to absorb 95% of the energy in an incident beam of gamma rays of (a) 140 keV (^{99m}Tc), (b) 320 keV (^{51}Cr), and 1.29 MeV (^{59}Fe). In each case make the extreme assumptions (a) that none of the Compton scattered photons escape, (b) that all of the Compton scattered photons escape.
7. A point source of ^{22}Na, which emits an average of 0.9 positrons and 1.0 gamma rays of 1.28 MeV per disintegration, is placed in a spherical lead shield 100 mm thick. Estimate the ratio of the 1.28 MeV photons to the annihilation photons which penetrate the shield unchanged. (Assume that each positron gives rise to two annihilation photons each of 0.51 MeV.)
8. Verify the exposure dose rates given for ^{99}Tc and ^{99m}Mo in Table 3.1, using decay scheme data.
9. Confirm the result quoted in Section 3.30 for the ionisation current generated in air from a gamma-ray source.
10. Calculate the weight of ferrous iron oxidised in a Fricke dosimeter if 200 Gy are absorbed in a mass of 0.5 kg of solution.
11. A nearly weightless source of a radionuclide emitting 5 MeV is prepared in the form of an aqueous solution weighing 10 g, initially at 20 °C. The strength of the source is 1 TBq. Assuming no heat losses, and that the specific heat of the solution is 4200 J K^{-1} kg^{-1}, calculate how long it takes the solution to boil.

4
Laws and Statistics of Radioactive Decay

RADIOACTIVE DECAY LAW

4.1 Definitions

In this chapter we study the decay of a collection of radioactive nuclei which we refer to as the 'source'. This might be a chemical preparation of a radionuclide or a mixture of radionuclides expressly made for an investigation of the decay, or it might be a 'dose' of a radiopharmaceutical either before or after administration to a subject. We shall assume we have a 'detector', i.e. a device which will give a recordable signal, or 'count', whenever it is triggered by a particle or quantum of radiation emitted from the source. The detector will have a natural background (Section 3.12) and in what follows we will assume that background counts have been allowed for.

We use the symbol $N(t)$ to mean the number of nuclei present in the source at an instant of time t following some arbitrary instant $t = 0$; sometimes the single letter N will be used for $N(t)$ when there is no risk of confusion. The number of nuclei present at time zero will be denoted $N(0)$. Similarly we write A or $A(t)$ to mean a *decay-rate.* (In this chapter we have consistently used the term 'decay-rate' instead of the exactly equivalent 'activity' because the latter is often used colloquially in several different connections. For our present purposes it is essential to distinguish between 'decay-rate' and 'count-rate'. The count-rate given by a source is of course an *indication* of its activity but is not the same thing; our terminology removes all possibility of confusing the two.) In general the detector will not give a recorded count for each decay event, for two reasons: first, the radiations from the source are emitted in all directions in space, whereas the detector, unless of very special design, does not completely surround the source; and secondly the detector may not be triggered by every particle or photon incident upon it. We shall assume that the extrinsic (geometrical) and intrinsic efficiencies implied here are constant.

Then if the recorded (nett) count-rate is $C(t)$ we can write

$$C(t) = \varepsilon A(t)$$

where ε is the overall counting efficiency (assumed constant). We now wish to examine the functional form of $N(t)$, $A(t)$ and $C(t)$.

4.2 Fundamental Decay Law

If $N(t)$ is the number of nuclei at time t, the number at a slightly later time $t + \Delta t$ is $N(t + \Delta t)$ and the change in this number is $N(t + \Delta t) - N(t) = \Delta N$ say. ΔN is in fact negative as N is decreasing with time. The rate of decay is then $-\Delta N/\Delta t$ approximately, at time t; proceeding to the limit as $\Delta t \to 0$ we get

$$A(t) = -\mathrm{d}N(t)/\mathrm{d}t$$

All experimental evidence goes to show that this rate, for a single species of isotope, is just proportional to $N(t)$ itself. Writing λ as the constant of proportionality we get

$$A(t) = -\mathrm{d}N(t)/\mathrm{d}t = \lambda N(t)$$

giving

$$\mathrm{d}N(t)/\mathrm{d}t + \lambda N(t) = 0$$

whose solution is

$$N(t) = N(0) \exp(-\lambda t)$$

The decay is said to be exponential with a decay constant λ. The dimension of λ is $(\text{time})^{-1}$; convenient units are s^{-1}, h^{-1}, d^{-1}, etc. The value of λ for any radionuclide can be found in principle from the relation

$$\lambda = A(t)/N(t)$$

For example, a sample of 1 mg of ^{239}Pu is found to emit 2.26×10^6 alpha particles per second. Since 1 mole = 239 g of Pu, comprising 6.02×10^{23} atoms, $N(t) = 10^{-3} \times 6.02 \times 10^{23}/239$. Assuming the radionuclide emits one alpha particle per decay,

$$A(t) = 2.26 \times 10^6 \text{ s}^{-1}$$

Then

$$\lambda = \frac{2.26 \times 10^6 \times 239}{10^{-3} \times 6.02 \times 10^{23}} = 9.0 \times 10^{-13} \text{ s}^{-1} = 2.83 \times 10^{-3} \text{ yr}^{-1}$$

In practice it is difficult to apply this method to rdionuclides whose half-lives are less than about a hundred years, because of the technical problems of obtaining and weighing the samples which would be needed. These problems are avoided by measuring the half-life and calculating λ from this (Section 4.4).

4.3 Application to a Single Nucleus

The decay law $N(t) = N(0) \exp(-\lambda t)$ applies to a *collection* of nuclei, initially numbering $N(0)$; but since there is no discernible means of communication between one nucleus and another, we are driven to conclude that there must be some fundamental property, inherent in each nucleus, which determines the timescale of its own decay. We can get some idea of the nature of this property by considering the fate of a single nucleus. The probability that it will survive for a time of at least t from some arbitrary instant will be some function of t; let us call this $P(t)$. Clearly $P(0)$ will be indefinitely close to unity and $P(t)$ will be a steadily decreasing function, ultimately becoming zero as t tends to infinity. Also $P(t + \Delta t)$ will represent the probability of survival for at least a time $t + \Delta t$. Therefore, the expression $P(t + \Delta t) - P(t)$ will represent the probability that the nucleus will survice for a time t but fail to survive for a time $t + \Delta t$, i.e. the probability that decay will occur within the interval Δt immediately following t (see Fig. 4.1). Now let us make the *assumption* that a nucleus is just as likely to decay at any one instant as at any other, that is to say that decay occurs purely randomly in time. Then the probability of decay in an interval Δt is just proportional to the length of the interval; let us call this $k\Delta t$. This probability can be evaluated in terms of the behaviour of a large number of nuclei, in the same way that we could determine the probability of throwing a 6 with dice by making a large number of throws and recording the number of times a 6 appears. The probability we want is just the fraction, out of a large number of nuclei, which *do* decay in a time Δt, viz.

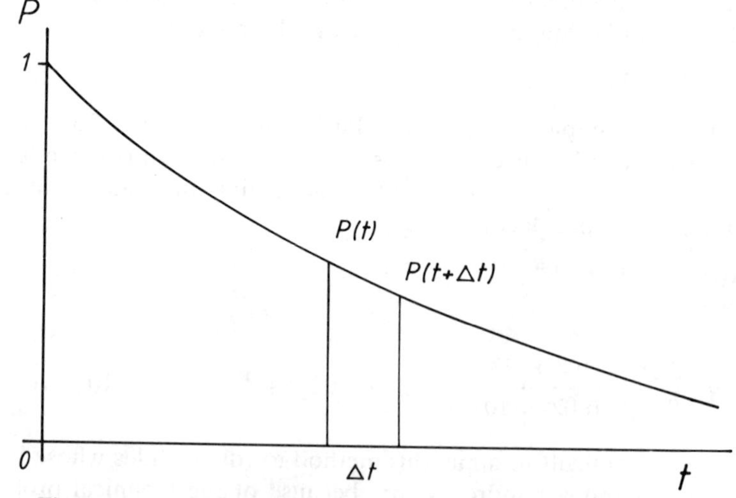

Fig. 4.1 Probability of decay within an interval Δt at a time t.

$\Delta N/N$. We find therefore that

$$\Delta P(t) = P(t) - P(t + \Delta t)$$
$$= k\Delta t$$
$$= \Delta N/N$$

and the last two equalities lead at once, as $\Delta t \to 0$, to

$$N(t) = N(0) \exp(-kt)$$

We are therefore bound to conclude that k and the decay constant λ are in fact identical and that the assumption of decays occurring randomly in time is justified. The significance of λ, expressed, for example, in units of s^{-1}, with respect to a single nucleus is that it is the probability per second that decay will occur. It seems curious at first sight that the behaviour of a single nucleus is purely random, yet when we consider a very large number we find an apparently reliable way of calculating the number remaining at fixed times. But indeed there is no real contradiction; when we consider large numbers of nuclei we can make quantitative statements about these numbers which are true enough for practical purposes, in the same way that we can say that of the population of a large city nearly 50% will be male and 50% will be female; whereas for small numbers of nuclei we can expect somewhat erratic and unpredictable behaviour, in the same way that among families of four children we often find for example three boys and one girl—by no means always two of each.

4.4 Quantities Derived from the Decay Constant

A graph of $N(t)$ plotted against t is shown in Fig. 4.2. The graph shows how we may define the *half-life*, $t_{1/2}$, as that value of t for which $N(t)$ falls from $N(0)$ to $N(0)/2$. To evaluate this in terms of λ we have

$$N(0)/2 = N(0) \exp(-\lambda t_{1/2})$$

whence

$$\exp(-\lambda t_{1/2}) = 0.5$$

and

$$t_{1/2} = \frac{\ln 2}{\lambda} = \frac{0.6932}{\lambda}$$

or

$$\lambda = \frac{0.6932}{t_{1/2}}$$

The value of the decay constant is thus easily obtained from a measurement of the half-life, and this is the method adopted except when the half-life is very long. The physical significance of the latter is perhaps a little easier to grasp than that of the decay constant and for this reason many lists of the properties

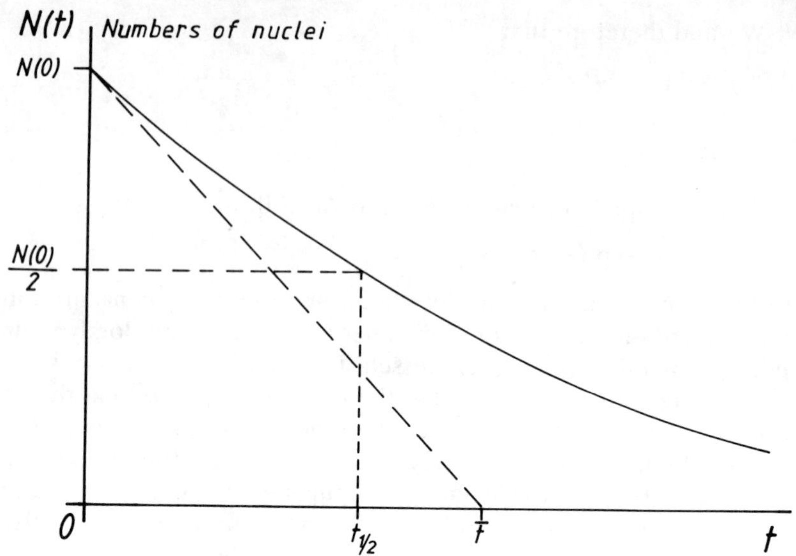

Fig. 4.2 Radioactive decay $N(t) = N(0) \exp(-\lambda t)$.

of radionuclides quote values of $t_{1/2}$ rather than λ. The range of observed half-lives extends from less than 10^{-12} s to more than 10^{12} yr.

Another useful result is derived as follows. The *initial* rate of decay (at time zero) is

$$\begin{aligned}[\mathrm{d}N(t)/\mathrm{d}t]_{N(t)=N(0)} &= [\mathrm{d}\{N(0)\exp(-\lambda t)\}/\mathrm{d}t]_{t=0} \\ &= -\lambda N(0)\end{aligned}$$

Since there were $N(0)$ nuclei to begin with, this means that if the decay-rate had remained constant at its initial value, all $N(0)$ nuclei would have decayed in a time $1/\lambda$. Also if $N(t)$ nuclei survive to a time t, the *mean lifetime* must be

$$\bar{t} = \int_0^\infty N(t).t\,\mathrm{d}t \bigg/ \int_0^\infty N(t)\mathrm{d}t$$

and this is also found to be equal to $1/\lambda$. Clearly then

$$\bar{t} = \frac{t_{1/2}}{0.6932} = 1.442 t_{1/2}$$

The meanlife $\bar{t}$ is shown also in Fig. 4.2.

Finally we have from the expression

$$N(t) = N(0) \exp(-\lambda t)$$

that

$$\log_{10} N(t) = \log_{10} N(0) - 2.303\lambda t$$

Therefore a plot of $\log_{10} N(t)$ against t gives a straight line with intercept $\log_{10} N(0)$ on the ordinate and a slope of -2.303λ.

4.5 Decay-Rates and Count-Rates

From the equation

$$N(t) = N(0) \exp(-\lambda t)$$

we get the decay-rate by differentiation, i.e.

$$A(t) = dN(t)/dt = \lambda N_0 \exp(-\lambda t)$$

and since $\lambda N(0)$ is the initial decay-rate,

$$A(t) = A(0) \exp(-\lambda t)$$

Further, by multiplying both sides of this equation by the counting efficiency ε we get

$$C(t) = C(0) \exp(-At)$$

These equations are of exactly the same form as the decay law itself and show that the half-life and mean life could have been defined just as well in terms of decay-rates or count-rates instead of in terms of the numbers of radioactive nuclei.

4.6 A Useful Practical Rule

Since

$$dN(t)/dt = -\lambda N(t) = -\frac{0.6932N(t)}{t_{1/2}}$$

we have

$$dC(t)/dt = -\lambda C(t) = -\frac{0.6932C(t)}{t_{1/2}}$$

or approximately

$$\frac{\Delta C(t)}{C(t)} = \frac{\Delta N(t)}{N(t)} = -\frac{70\Delta t}{t_{1/2}} \quad \%$$

This means that the percentage rate of decrease of strength of a radioactive source, or of the count-rate given by it, per unit of time is approximately $70/t_{1/2}$. For example, the activity of a sample of ^{99m}Tc (half-life 6.0 h) diminishes by 70/6, or nearly 12% per hour. Similarly ^{131}I, with a half-life of 8 d, decays away at a rate of 70/8 or approximately 9% per day.

4.7 Successive Decays

Many cases are known in which radioactive nuclei of one species decay into a second species which is also radioactive—indeed chains of 10 or more members are exemplified by the very heavy elements. It is customary to refer to pairs of such genetically related nuclei as 'parent' and 'daughter' sources. Using subscripts 1 and 2 to denote such a pair, we have for the parent

$$dN_1(t)/dt = -\lambda_1 N_1(t)$$

which leads to

$$N_1(t) = N_1(0) \exp(-\lambda_1 t)$$

as before. For the daughter the fundamental decay law will have to be modified because new daughter nuclei are being formed, as they are the decay products of the parent. Hence for the daughter, we write for the rate of change of $N_2(t)$ with time,

$$dN_2(t)/dt = \lambda_1 N_1(t) - \lambda_2 N_2(t)$$

where the first term on the right-hand side expresses the rate of growth of the daughter (i.e. the rate of decay of the parent). Substituting $N_1(0) \exp(-\lambda_1 t)$ for $N_1(t)$ in this equation and using standard methods we easily find

$$N_2(t) = \frac{\lambda_1}{\lambda_2 - \lambda_1} N_1(0)\{\exp(-\lambda_1 t) - \exp(-\lambda_2 t)\} + N_2(0) \exp(-\lambda_2 t)$$

where the two terms give respectively the growth-plus-decay of the daughter originating from the parent nuclei, and the decay of $N_2(0)$ nuclei of the daughter present initially at $t = 0$. If the source were an initially pure sample of the parent alone, $N_2(0)$ would be zero.

From the equations for $N_1(t)$ and $N_2(\mathrm{t})$ we easily get the corresponding decay rates:

$$A_1(t) = \lambda_1 N_1(0) \exp(-\lambda_1 t) = A_1(0) \exp(-\lambda_1 t)$$

$$A_2(t) = \lambda_2\left[\frac{\lambda_1}{\lambda_2 - \lambda_1} N_1(0)\{\exp(-\lambda_1 t) - \exp(-\lambda_2 t)\}t + N_2(0) \exp(-\lambda_2 t)\right]$$

$$= \frac{\lambda_2}{\lambda_2 - \lambda_1} A_1(0)\{\exp(-\lambda_1 t) - \exp(-\lambda_2 t)\} + A_2(0) \exp(-\lambda_2 t)$$

and the count-rates:

$$C_1(t) = \varepsilon_1 A_1(0) \exp(-\lambda_1 t) = C_1(0) \exp(-\lambda_1 t)$$

$$C_2(t) = \frac{\varepsilon_2 \lambda_2}{\lambda_2 - \lambda_1} A_1(0)\{\exp(-\lambda_1 t) - \exp(-\lambda_2 t)\} + \varepsilon_2 A_2(0) \exp(-\lambda_2 t)$$

Notice that the first term in the expression for $C_2(t)$ involves a multiplication of ε_2 with $A_1(0)$, giving a product which is *not* equal to the initial count-rate of either radionuclide. It is, however, true that $C_2(t)$ can always be written in the form

$$C_2(t) = P \exp(-\lambda_1 t) + Q \exp(-\lambda_2 t)$$

where P and Q are constants, although not always identifiable with initial count-rates.

For simplicity in what follows, we will assume that $N_2(0)$ is zero, i.e. that we have an initially pure parent. Then there are three cases of parent-daughter relationship which are of general interest, depending on the relative values of λ_1 and λ_2 in the simplified equations for, e.g. $A_1(t)$ and $A_2(t)$;

$$A_1(t) = A_1(0) \exp(-\lambda_1 t)$$

$$A_2(t) = \frac{\lambda_2}{\lambda_2 - \lambda_1} A_1(0)\{\exp(-\lambda_1 t) - \exp(-\lambda_2 t)\}$$

These cases are as follows:

(*a*) *Secular Equilibrium.* This is the situation in which the parent's half-life is extremely long compared with that of the daughter. λ_1 is then taken to be very small so that for all practical values of t, $\exp(-\lambda_1 t)$ remains very nearly 1 and $A_1(t)$ is virtually a constant, $A_1(0)$. If we neglect λ_1 in comparison with λ_2 and replace $\exp(-\lambda_1 t)$ by 1 in the equation for $A_2(t)$ we get

$$A_2/(t) = A_1(0)\{1 - \exp(-\lambda_2 t)\}$$

This decay-rate is depicted in Fig. 4.3. The decay-rate shows an exponential rise, tending towards the value $A_1(0)$; the rate reaches half its final value in a time $0.6932/\lambda_2$, three-quarters of its final value in twice this time, and so on. The limiting decay-rate is exactly that of the parent; hence the expression 'secular' equilibrium. If the radiations from the parent and daughter are detected with the same efficiency, ε_1 is equal to ε_2, and the count-rates also become equal after a very long time.

Again for the condition of λ_1 being negligibly small compared with λ_2, we find for the number of daughter nuclei, $N_2(t)$ at time t,

$$N_2(t) = \frac{\lambda_1}{\lambda_2} N_1(0)\{1 - \exp(-\lambda_2 t)\}$$

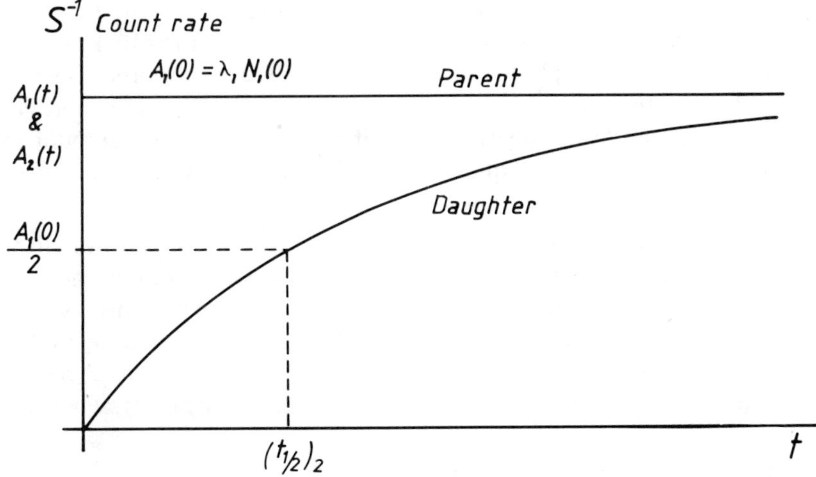

Fig. 4.3 Secular equilibrium.

so that after a very long time exp $(-\lambda_2 t)$ tends to zero and

$$N_2(\infty) = \frac{\lambda_1}{\lambda_2} N_1(0)$$

and so

$$N_1(0)/N_2(\infty) = \lambda_2/\lambda_1 = (t_{1/2})_1/(t_{1/2})_2$$

This last equation means that the ratio of the numbers of parent nuclei to those of daughter tends to equal the ratio of the half-lives. This rule applies over a whole chain of radioactive decays in which the half-life of the parent is very much longer than that of any daughter in the chain. For example the half-life of uranium (^{238}U) is 4.5×10^9 yr which is some orders of magnitude longer than that of any of its radioactive descendants. One descendant is radium (^{226}Ra) whose half-life is 1.6×10^3 yr. Therefore in a geologically old uranium ore, 1 tonne (10^3 kg) of uranium will contain

$$\frac{10^3}{238} \times 6.02 \times 10^{23}$$

atoms of ^{238}U and these will be in secular equilibrium with

$$\frac{10^3}{238} \times 6.02 \times 10^{23} \times \frac{1.6 \times 10^3}{4.5 \times 10^9}$$

radium atoms. These will weigh

$$\frac{10^3}{238} \times 226 \times \frac{1.6 \times 10^3}{4.5 \times 10^9} \text{ kg} \equiv 300 \text{ mg}$$

It says much for the genius and hard work of P. and M. Curie that they succeeded in extracting the radium present in two tonnes of pitchblende ore with a chemical efficiency of some 25%. In a similar way radium itself can be regarded as the long-lived parent of its daughter radon (^{222}Rn), whose own half-life is 3.85 d. A sample of radium will scarcely decay appreciably during a human lifespan but will continuously generate radon. If the radon is separated from the radium at any instant, half of the equilibrium amount will be regenerated in 3.85 d, three-quarters in 7.7 d, and so on (see Section 2.23).

(*b*) *Transient Equilibrium.* In this case, the decay of the parent is slow, but is not inappreciable during experimental time. The decay-rate of the daughter is depicted in Fig. 4.4; at first it rises approximately exponentially as in the previous case, but soon falls away and eventually assumes a decay half-life which approaches that of the parent. The equation for the final part of the curve is obtained as before by neglecting the exp $(-\lambda_2 t)$ term, when we find

$$A_2(t) = \frac{\lambda_2}{\lambda_2 - \lambda_1} A_1(0) \exp(-\lambda_1 t)$$

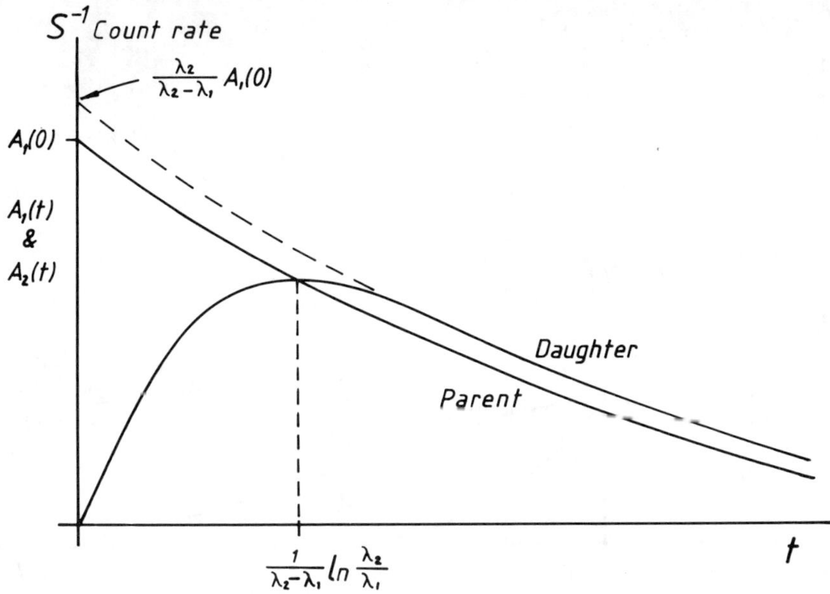

Fig. 4.4 Transient equilibrium.

Comparing this with the equation for the decay of the parent

$$A_1(t) = A_1(0) \exp(-\lambda_1 t)$$

we see that the daughter tends eventually to decay at a rate which is $\lambda_2/(\lambda_2 - \lambda_1)$ times that of the parent. It is, therefore, only in a rather curious sense that the two radionuclides can be said to be in equilibrium; the expression *transient* equilibrium implies that even this state is impermanent, as the parent itself decays to a negligible amount during experimental time. Again, if the radiations from the two radionuclides are counted with the same efficiency, we shall find that the daughter eventually has a faster count-rate than the parent; only at one instant in time are the decay-rates and count-rates of the radionuclides equal.

It is of interest to find the time at which $A_2(t)$ reaches its maximum. Differentiating the expression for $A_2(t)$ and setting the result to zero, we get

$$\frac{\lambda_2}{\lambda_2-\lambda_1} A_1(0)\{-\lambda_1 \exp(-\lambda_1 t) + \lambda_2 \exp(-\lambda_2 t)\} = 0$$

and solving for t it is easy to show that

$$t = \frac{1}{\lambda_2 - \lambda_1} \ln \left(\frac{\lambda_2}{\lambda_1}\right)$$

This relation is more conveniently expressed in ordinary logarithms and half-lives instead of decay constants, and then reads

$$t = \frac{3.322(t_{1/2})_1.(t_{1/2})_2}{(t_{1/2})_1 - (t_{1/2})_2} \log \left(\frac{(t_{1/2})_1}{(t_{1/2})_2}\right)$$

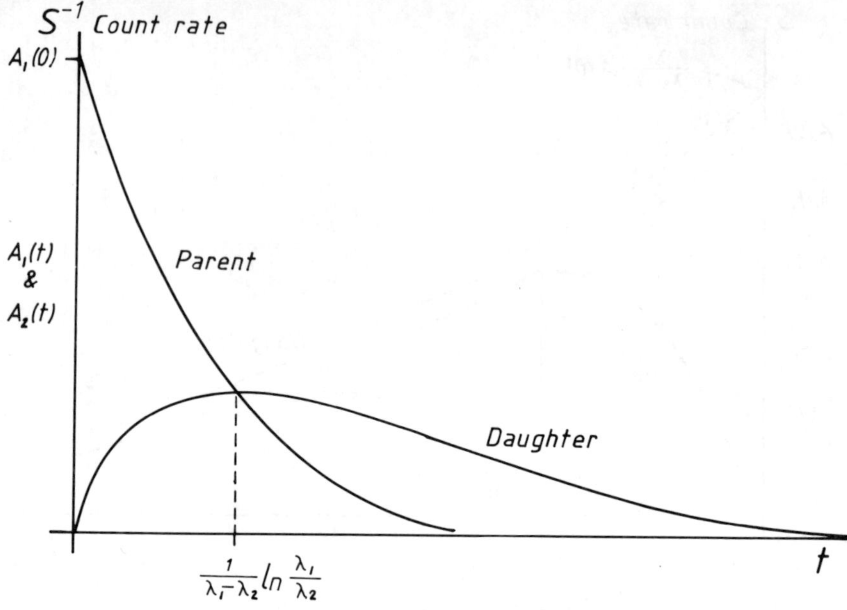

Fig. 4.5 Case of no equilibrium.

A good example of transient equilibrium occurs for ^{99}Mo which decays with a 67 h half-life to ^{99m}Tc which itself has a 6 h half-life. Starting with pure ^{99}Mo the time taken for the daughter ^{99m}Tc to attain its maximum activity is, according to the above formula, about 23 h. Beyond this time the activity falls, although the decay-rate is now slightly greater than that of the parent ^{99}Mo; the ratio of the decay rates tends to

$$\lambda_2/(\lambda_2 - \lambda_1) = (t_{1/2})_1/\{(t_{1/2})_1 - (t_{1/2})_2\} = 1.11$$

(*c*) *Case of No Equilibrium.* This case occurs when the parent has a half-life shorter than that of the daughter. Again the parent will generate daughter whose activity will rise to a maximum at a time given by the formula above. But now the parent will decay almost completely while daughter nuclei still survive, so that the daughter will decay away eventually with its own natural half-life (Fig. 4.5). This case is by far the most common among the known radionuclides. For instance, we may quote the chain of fission products ^{131}Sn (half-life 3.4 min), ^{131}Sb (25 min), ^{131m}Te and ^{131g}Te (25 min and 30 h), ^{131}I (8.05 d).

4.8 Decay Curves: Single Component

We have obtained an expression for the decay-rate of a single radioactive source at a time t by finding the differential coefficient $dN(t)dt$. In theory the count-rate is then $\varepsilon\ dN(t)/dt$. But in practice we measure a count-rate by recording the number of events observed by a radiation detector in a finite time interval Δt; in the language of elementary calculus, the measurement gives the slope of a secant, not of a tangent, to the curve of $C(t)$ plotted against t. However, it can be shown that the counting time interval Δt can be made quite long without introducing large errors. For instance if Δt is as long as one-third of one half-life, the quotient of the number of observed counts and the length of the interval is only about 1% greater than the theoretical true count-rate at the *middle* of the counting interval. Quite often in practice the counting interval is much shorter than this and no appreciable error is made by ascribing the observed count-rate (which is really a *mean* count-rate over the interval) to the instant at the *start* of the interval.

The counting of a radioactive source has usually one of two objectives: (a) to establish the count-rate which the source must have had at some previous time, i.e. the initial count-rate (ICR), and/or (b) to measure the half-life of the source. It is a common practice to plot the observed rates at different times on semi-log paper on which the points should give a straight line (Section 4.4), as illustrated in Fig. 4.6. Note that the count-rates appear to be subject to errors which are most noticeable for the lowest rates.

This graphical method is adequate for many purposes but can rarely be trusted to give an ICR value correct to better than a few per cent unless there are many experimental points including some very close to t_0. (A typical case in which there may be a considerable time interval between t_0 and the first counting interval is that of radionuclide production, when it is convenient to regard the end of bombardment times as t_0. Then, depending on the product sought and the method of chemically separating it from the target material, a half-life or more might elapse before it is possible to present a radiochemically pure sample to the detecting equipment.) In finding the ICR it is, of

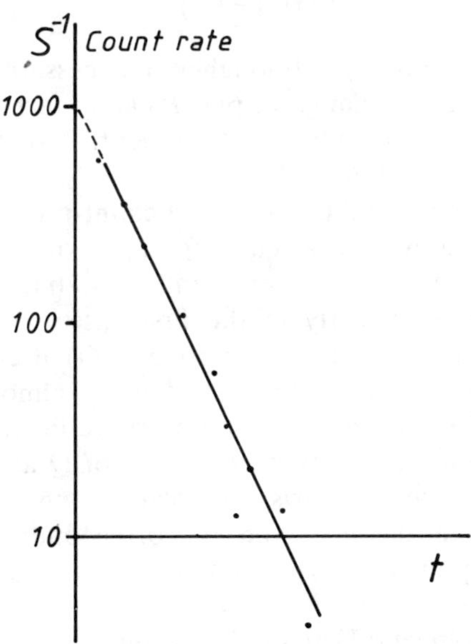

Fig. 4.6 Graphical representation of decay of a single radionuclide (semi-log plot).

course, very helpful if the half-life is already accurately known as this determines the slope of the line. If the half-life is not known it must be estimated from the observed slope. The decay must be followed for at least one half-life in order to obtain even a rough value. It may be noted in passing that if the purpose of the plot is just to measure the half-life, no errors at all are introduced by making the counting time intervals Δt of any arbitrary length, provided they are all the same. Really accurate half-life values can be obtained from observation of decay-rates only when these are extended over at least 10 half-lives, unless one can measure the mass of the radionuclide also. Many half-life values quoted in the older literature have later been found to err on the side of being too long because newly discovered radionuclides were frequently contaminated with longer-lived impurities. Even today with high-speed computers available to process vast amounts of data, the half-lives of many radionuclides even in common use are not known to an accuracy better than a few tenths of one per cent.

4.9 Decay Curves: Complex (Two or More Components)

If a source consists of two radionuclides with half-lives $(t_{1/2})_1$ and $(t_{1/2})_2$ the count-rate observed will be a function of t of the form

$$C(t) = C_1(0) \exp(-\lambda_1 t) + C_2(0) \exp(-\lambda_2 t)$$

with $C_1(0)$ and $C_2(0)$ the initial count-rates of the pair. If the radionuclides have a parent–daughter relationship the observed count-rate has the same form, so we need concern ourselves in general only with the equation

$$C(t) = P \exp(-\lambda_1 t) + Q \exp(-\lambda_2 t)$$

where P and Q are constants although not necessarily initial count-rates (Section 4.7). For a parent–daughter pair P can in fact be negative, but this makes no difference to the following treatment. Two typical plots of $C(t)$ versus t are shown in Fig. 4.7a and b.

Here again we have plotted the observed count-rate $C(t)$ on a logarithmic scale. In Fig. 4.7a we show a decay curve for a pair of radionuclides with very different half-lives ($\lambda_1/\lambda_2 = 0.1$) whereas in Fig. 4.7b the half-lives are nearly the same ($\lambda_1/\lambda_2 = 0.8$). Clearly in the first case one can see the decay approaching that of the longer-lived component (Q) after a time sufficient for the shorter-lived component to have died away almost completely. This suggests a method of *resolution* of the decay curve in which a straight line is fitted to the later points to represent the decay of Q alone. The line is then extrapolated back to the $C(t)$ axis. 'Resolved counts' are then found by subtracting ordinates on this line from the corresponding observed points at the earlier times. A plot of the resolved counts then gives the decay of the shorter-lived radionuclide (see Fig. 4.8).

In carrying out this 'peel-off' method of resolution, one often finds that the resolved counts do not appear to lie on a straight line, as they should do,

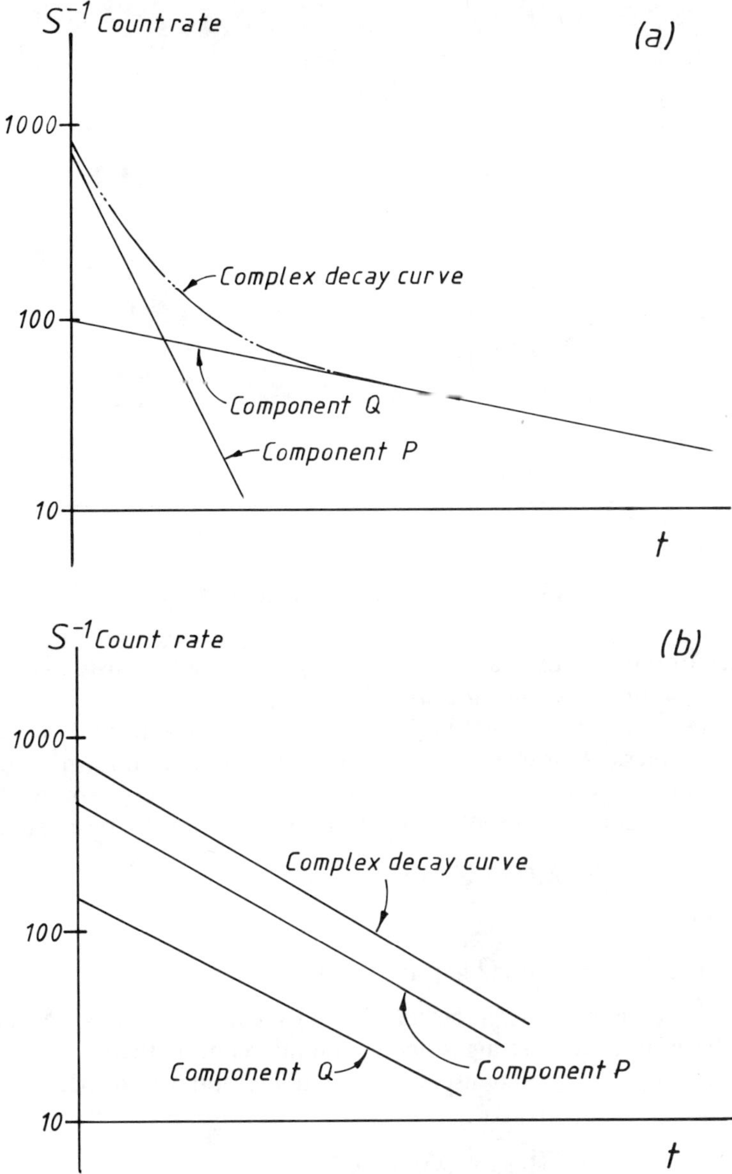

Fig. 4.7 (a) Two-component decay curves (ratio of decay constants $\lambda_1/\lambda_2 = 0.1$). (b) Two-component decay curves (ratio of decay constants $\lambda_1/\lambda_2 = 0.8$).

because of a small error in the positioning of the line for the longer component. For instance if this is set a little too high, the resolved counts will lie on a curve which becomes steadily steeper with time. A certain amount of trial-and-error is then necessary to get a satisfactory fit for both components. Within the inevitable limits of graphical methods, the peel-off technique is useful if the half-lives are in a ratio of at least 2:1, given enough data, and if

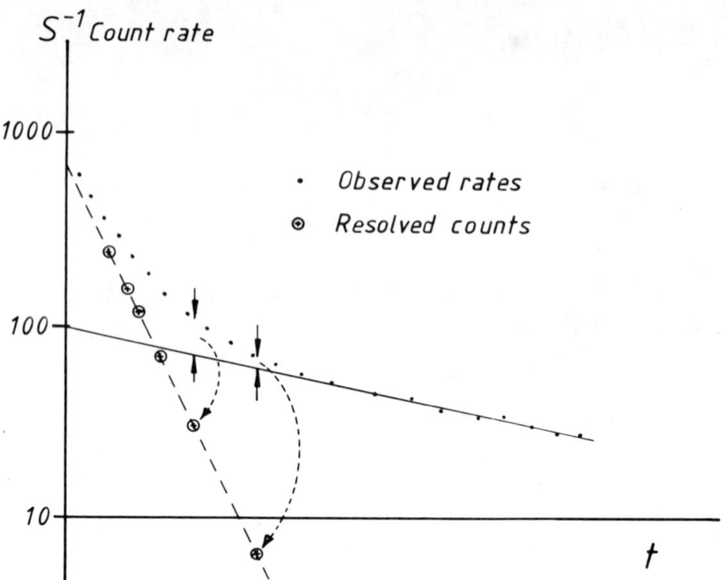

Fig. 4.8 Resolution of the curve of Fig. 4.7a.

the values of the half-lives are already known. It is notoriously inaccurate if the half-lives are not known *a priori.*

If the half-lives are too similar (Fig. 4.7b), the peel-off method becomes practically useless. However if the half-lives are already known so that the problem is just to estimate the values of P and Q, there are two quite effective alternative procedures. In the first of these, we rewrite the basic equation

$$C(t) = P \exp(-\lambda_1 t) + Q \exp(-\lambda_2 t)$$

in the form

$$C(t) \exp(\lambda_1 t) = P + Q \exp\{-(\lambda_2 - \lambda_1)t\}$$

A plot of $C(t) \exp(\lambda_1 t)$ as ordinate versus $\exp\{-(\lambda_2-\lambda_1)t\}$ as abscissa should, therefore, give a straight line whose intercept on the vertical axis is P and whose slope is Q. To use the second method we rewrite the basic equation as

$$1 = \frac{\exp(-\lambda_1 t)/C(t)}{1/P} + \frac{\exp(-\lambda_2 t)/C(t)}{1/Q}$$

A plot of $\exp(-\lambda_1 t)/C(t)$ as ordinate versus $\exp(-\lambda_2 t)/C(t)$ as abscissa should therefore give a straight line whose intercepts on the axes are $1/P$ and $1/Q$. The slope of this line is $-Q/P$, so the method is particularly convenient when, as is often the case, the ratio between the constants is all that is required.

In the general case in which P and Q and the two half-lives are all regarded as unknowns, there is no satisfactory graphical method of evaluating them if the half-lives are nearly the same. Several highly sophisticated computer programs have been written for this task; they are essentially iterative methods in which the value of some function defined as a measure of the overall 'error' is minimised as slight changes are made in assumed values for P, Q, λ_1 and λ_2. These programs can take into account the fact that there are statistical uncertainties in the observed $C(t)$ values, a point which is very difficult to allow for in a simple graphical method, and they have been extended to handle cases of up to six components instead of only two. The results obtained by the half-dozen or so computer programs that have been tried on test data do not agree very well, because different authors have chosen different criteria of what constitutes a 'best fit'. Graphical peel-off methods (but not the others mentioned above) may be used for mixtures of more than two components, but the results can be highly unreliable and even quite misleading, even if the half-lives are very different from one another.

4.10 Decay Curves: Compound Decays

We use the expression 'compound decays' to embrace cases in which radioactive nuclei are progressively removed from the source by mechanisms other than that of the radioactive process we are considering. Such a mechanism might be an alternative mode of decay; for example a few radionuclides decay both by emission of alpha particles and by emission of beta particles. We associate decay constants λ_α and λ_β with the competing processes (each constant representing the probability per unit time of decay by the alternative pathway). Then the *total* probability of decay will be $\lambda_\alpha + \lambda_\beta = \lambda$, say, and the half-life of the source will be $t_{1/2} = 0.6932/\lambda$. The decay rate of $N(t)$ nuclei will be $\lambda_\alpha N(t)$ as measured by a detector sensitive only to the alpha particles and $\lambda_\beta N(t)$ as measured by a detector sensitive only to the beta particles. The total decay rate will be $(\lambda_\alpha + \lambda_\beta)\ N(t)$. It must be emphasised that the only observable half-life is $t_{1/2}$; both the alpha decay rate and the beta decay rate fall off with time with this single half-life. The nuclide ^{212}Bi decays in this way; one-third of its nuclei decay by emitting alpha particles and the remaining two-thirds emit beta particles.

In living systems there are many cases in which radionuclides are removed progressively from a particular organ by the intervention of a biological process. The classical case is that of hyperthyroid patients in whom radioactive iodine initially concentrated in the thyroid gland is steadily discharged (in a form in which it is bound to protein) into the bloodstream. Within about 24 hours after the administration of radioiodine it is usually the case that uptake into the gland is complete. Thereafter measurements of the radiations emitted from the thyroid, regarded as a source, will in hyperthyroid patients show an abnormally rapid decrease. If we make the assumption that the discharge rate obeys an exponential law with time constant λ_{biol} we can say that in conjunction with the physical radioactive decay constant λ_{phys}, we have an effective time constant $\lambda_{biol} + \lambda_{phys}$. The count-rate observed over the organ will therefore be governed by

$$[C(t)]_p = [C(0)]_p \exp\{-(\lambda_{biol} + \lambda_{phys})t\}$$

where the outer subscript p means that the counts are taken on an organ of the patient. The effective (apparent) half-life is then $0.6932/(\lambda_{biol} + \lambda_{phys})$. Analogous to the physical half-life $0.6932/\lambda_{phys}$ we can define a biological half-life $0.6932/\lambda_{biol}$.

It is general practice when such measurements are made to prepare a 'standard' sample of the same radionuclide as that administered to the patient. A suitable physical form for this is a plastic or glass bottle, or ampoule, of roughly the same size and shape as the actual organ and containing an aqueous solution of the same activity of the radionuclide. The standard is counted at nearly the same time as the patient—ideally both before and afterwards, and the mean value calculated. Then for the standard, since it is affected only by physical decay, we have

$$[C(t)]_s = [C(0)]_s \exp(-\lambda_{phys}t)$$

using the subscript s to denote counts on the standard. We can now define the ratio $R(t)$ by

$$R(t) = \frac{[C(t)]_p}{[C(t)]_s} = \frac{[C(0)]_p}{[C(0)]_s} \exp(-\lambda_{biol}t)$$

$$= \text{constant.} \exp(-\lambda_{biol}t)$$

This means that the *ratio* of counts from the patient to counts from the standard, as a function of time, must follow a decay curve governed only by the biological half-life. It does not matter whether the physical half-life is known or not, but, of course, it is essential that each source (patient and standard) is counted with constant efficiency. This is easy to achieve with the physical standard but great care is needed to ensure that the patient is placed in an exactly reproducible position each time he is counted. Some further remarks on this topic are given in Section 5.1.

It is easy to show that if we define

$$\text{physical half-life} = (t_{1/2})_{phys} = 0.6932/\lambda_{phys}$$
$$\text{biological half-life} = (t_{1/2})_{biol} = 0.6932/\lambda_{biol}$$
$$\text{effective half-life} = (t_{1/2})_{eff} = 0.6932/(\lambda_{phys} + \lambda_{biol})$$

we have

$$(t_{1/2})_{eff} = (t_{1/2})_{phys} \cdot (t_{1/2})_{biol}/[(t_{1/2})_{phys} + (t_{1/2})_{biol}]$$

Thus, for instance, for a physical half-life of 8 d and a biological half-life of 5 d, the effective half-life will be 40/13 or very nearly 3 d. These figures would be typical of hyperthyroidism. In normal subjects the biological half-life for the discharge of tracer iodine from the thyroid is at least an order of magnitude longer than this.

4.11 Radioactive Growth

As shown in Section 2.16, when a target containing N_T nuclei each with a cross-section σ is bombarded for a time t in a flux density of particles f, the number of collisions which occur is $N_T\sigma ft \times 10^{-28}$. However, this number is not the same as the number of new nuclei available at the end of the time t, for two reasons. First, we should replace σ by a partial cross-section, appropriate to the particular nuclear reaction required (see Fig. 2.12). Secondly, depending on the half-life of the nuclide being produced, a certain fraction of the product nuclei will decay during the bombardment itself. Let us write R for the actual production rate, equal to $N_T\rho_p f \times 10^{-28}$, ρ_p being the partial cross-section. Then if N is the number of product nuclei which exist at any instant t, when their decay-rate is λN, their overall growth-rate is

$$\frac{dN}{dt} = R - \lambda N$$

If we assume a constant bombardment flux density, R is also constant with time. Integration of the above equation easily leads to

$$N(t) = \frac{R}{\lambda}[1 - \exp(-\lambda t)]$$

where $N(t)$ represents the number of nuclei present at the end of a bombardment time t. For a bombardment time short compared with the half-life, we may approximate to $\exp(-\lambda t)$ by expanding it to $1 - \lambda t + \cdots$. Then

$$N(t) \approx Rt$$

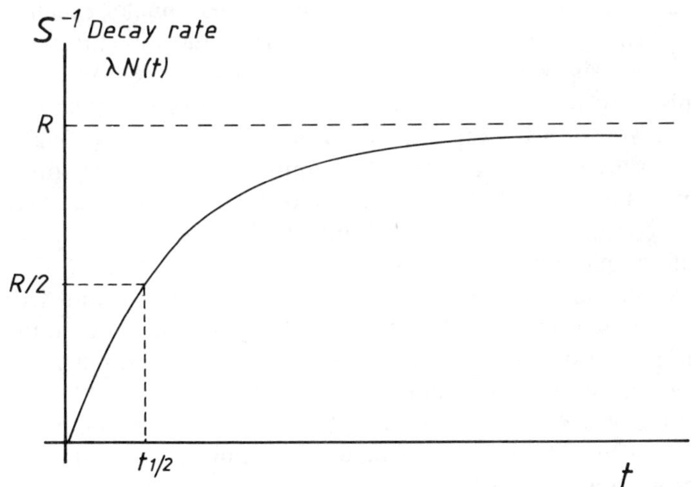

Fig. 4.9 Radioactive growth at a constant rate R.

On the other hand, for a *long* bombardment exp $(-\lambda t)$ will tend to zero and approximately

$$N(t) \approx R/\lambda$$

The activity, or decay-rate, at the end of a long bombardment (that is to say long compared with the half-life) is then $\lambda N(t)$, or just R. Clearly we have a saturation effect, the activity rising exponentially during a long bombardment until the decay-rate just equals the production rate. If the bombardment time t happens to equal the half-life, the final decay-rate will be just half the maximum possible, or saturation, rate. These relationships are illustrated in Fig. 4.9.

STATISTICS OF RADIOACTIVE DECAY

4.12 Decay Statistics

Let us now focus attention on a source of long half-life, such that the decay rate is not appreciably affected by the passage of time during which measurements are taken. Since we are no longer considering this rate to be a function of time, we will drop the time dimension and denote it simply by A. A detector viewing the source will give a nett count-rate C given by $C = \varepsilon A$, ε being the detector efficiency. Experiment shows that neither C nor A are exactly constant from one measurement to another. Nor could we expect them to be; for as we have seen in Section 4.3, the exact instant of decay of any nucleus is unpredictable; the constant λ only tells us the *probability* that out of a large number, N, of nuclei, a number ΔN will decay in a time Δt. This fraction only becomes precisely known when N and ΔN are very large numbers. We can no more be certain that out of 6000 nuclei *exactly* 1000 will decay in a certain period, than we can say that if dice are thrown, a total of 6000 throws will show exactly 1000 aces or threes or sixes. We would expect these numbers to be *nearly* 1000 in each case, unless the dice were loaded. Also the *fraction* of sixes thrown would be expected to approach 1/6 more and more closely as we increased the number of throws. It should also be remembered that the radiations from radioactive sources are (except under very special conditions) emitted randomly in all directions. Since most detectors do not completely surround the source, only a fraction of the radiations can be detected, and this fraction will not be exactly constant from trial to trial. Thus even if a source did emit exactly 1000 particles or quanta in a set of equal intervals of time and we had a detector with 10% geometrical efficiency, the number of 'detectable' radiations would not be exactly 100 in each interval. We express all this by saying that the number of recorded counts is subject to statistical fluctuations, which arise from the basic statistical nature of radioactive decay, both in time and angular distribution.

If we replicate a counting experiment say n times over, we may denote the numbers of counts recorded in equal time intervals by C_1, C_2, . . . ,

$C_r, \ldots, C_n$, and following Section 1.20, we can assume the most reliable estimate of the 'true' number of counts to be

$$\bar{C} = \frac{\sum_{r=1}^{n} C_r}{n}$$

and associate with this mean value a standard deviation

$$\sigma_C = \sqrt{\left(\frac{\sum (C_r - \bar{C})^2}{n}\right)}$$

$$= \sqrt{(\overline{C_r^2} - \bar{C}^2)}$$

Now, for the vast majority of physical measurements, the fluctuations that one finds among individual estimates are compounded from various instrumental imperfections and personal errors; they are inherent in the particular equipment used and the technique of using it. But with radioactive measurements, what is often the major source of the fluctuations is the nature of the decay process itself, and this means that the expected standard deviation of a counting experiment can be deduced from the nature of the underlying statistics, and does not have to be determined by taking replicate measurements. To show how this is done we need two elementary notions from probability theory which we summarise thus:

(i) We ascribe a probability p which is always positive and less than or equal to 1, to the occurrence of any specified event in a give time. If we know that the event never happens in that interval we write $p = 0$; if it is certain to happen we write $p = 1$. Generally p is some fraction; we also regard $1 - p = q$ as a fraction indicating the probability that the event does *not* occur.

(ii) If two independent events A and B have probabilities of occurring p_{A} and p_{B}, the probability of *both* occurring is the product $p_{\mathrm{A}}.p_{\mathrm{B}}$. The probability that neither will occur is $(1 - p_{\mathrm{A}}).(1 - p_{\mathrm{B}}) = q_{\mathrm{A}}.q_{\mathrm{B}}$. (The probability that either A or B will occur is $1 - p_{\mathrm{A}}.p_{\mathrm{B}} - q_{\mathrm{A}}.q_{\mathrm{B}}$.) Further, for any number N of independent events A, B, C, D, . . . , we would have, for example, that the probability of all of them happening is the product $p_{\mathrm{A}}.p_{\mathrm{B}}.p_{\mathrm{C}}.p_{\mathrm{D}} \cdots$ (N terms).

We now apply these basic ideas to the problem of throwing dice because this is a familiar example of independent and random events; independent in the sense that the number shown by any one die does not affect any of the others, and random in the sense that we have no *a priori* knowledge of what any die will show.

4.13 The Random Probability Distribution

Consider now a perfectly symmetrical die being thrown, and ask what the probability is that a six will be shown. The answer is clearly to be expressed as

$p = 1/6$; also the probability that any other number will be shown is $q = 5/6$, from the first principle given above.

Now suppose the die is thrown a large number N of times and we want to know the probability that in none of these trials does a six occur. Let us denote this probability by $W(0)$. Since there are N trials in each of which the required probability (no six) is 5/6, we have by the second principle that

$$W(0) = \frac{5}{6} \times \frac{5}{6} \times \frac{5}{6} \times \cdots (N \text{ items}) = \left(\frac{5}{6}\right)^N$$

Now let $W(1)$ denote the probability that in the N trials exactly one six is shown. At first sight this appears to be the same as the value for $W(0)$, except that one of the factors should be replaced by 1/6. However there are N trials and it is immaterial *which* of these shows the six, so that the expression we need should also include a factor N. Therefore

$$W(1) = N\left(\frac{1}{6}\right)\cdot\left(\frac{5}{6}\right)^{N-1}$$

We can pursue the argument to give $W(2)$; this will be given by the product of (a) a 'statistical' factor

$$\frac{N(N-1)}{1 \cdot 2} = \frac{N!}{2!(N-2)!}$$

which expresses the fact that it does not matter which two throws show sixes, and (b) a 'chance' factor

$$\left(\frac{1}{6}\right)^2\cdot\left(\frac{5}{6}\right)^{N-2}$$

which expresses the chance that two throws show sixes and all the others show some other number.

We can clearly generalise the argument. Thus if p is the probability of a certain event happening, and $q = 1 - p$ that of it not happening, then in N trials the probability of the event occurring just M times is

$$W(M) = \frac{N!}{M!(N-M)!}\, p^M q^{N-M}$$

The expression is just the binomial distribution which was discussed in Sections 1.21 and 1.22. One result obtained therein was the average value of M, called $\bar{m}$, which could be expected for a large number N of trials is

$$\bar{m} = \sum W(M).\ M = Np$$

so that in the dice problem with $p = 1/6$ we find that the number of times a six should appear is just $N/6$. More importantly we know the variance to be given by

$$V = \sum W(M) M^2 = Npq$$

So far $N = 6000$ trials, with $p = 1/6$, we find $\bar{m} = 1000$ and $V = 6000.1/6.5/6$, giving a standard deviation of $\sqrt{833} = 28$. We conclude that while we can hardly expect exactly 1000 sixes to appear in 6000 throws, the chances are 70.4% that the number of sixes will lie between $1000 - 28 = 972$, and $1000 + 28 = 1028$. Similarly the chances are 95.5% that the number of sixes will lie between 944 and 1056. A number of sixes outside these limits should occur on average only once for every 20 times that a trial of 6000 throws is made.

4.14 Standard Deviations in Counting Measurements

Instead now of enquiring into the probability of observing sixes when N dice are thrown, let us ask how many events will be recorded by a detector with efficiency ε when a source of $N(0)$ nuclei is allowed to decay for a time T. We know from the decay law that the number of surviving nuclei $N(T)$ is given by

$$N(T) = N(0) \exp(-\lambda T)$$

so that the probability of *decay* must be

$$\frac{N(0) - N(T)}{N(0)} = 1 - \exp(-\lambda T)$$

We can then write for the probability of detection

$$p = \varepsilon\{1 - \exp(-\lambda T)\}$$

and the probability that a nucleus does not give rise to a recorded event

$$q = 1 - p = 1 - \varepsilon + \varepsilon \exp(-\lambda T)$$

Using the results of the previous section we find:

$$\text{expected number of counts} = N(0)p = N(0)\varepsilon\{1 - \exp(-\lambda T)\}$$

$$\begin{aligned}\text{standard deviation} &= \sqrt{[N(0)p.q]}\\ &= \sqrt{[N(0)\varepsilon\{1 - \exp(-\lambda T)\}\{1 - \varepsilon + \varepsilon \exp(-\lambda T)\}]}\end{aligned}$$

These equations are exact; if we now assume, as in Section 4.12 that the decay period T is short compared with the half-life, i.e. that the source strength does not diminish appreciably during the measurements, we may approximate by replacing $\exp(-\lambda T)$ by $1 - \lambda T$ and find:

$$\text{expected number of counts} = N(0)\varepsilon\lambda T = \varepsilon\Delta N$$

where ΔN is the number of nuclei decaying in the interval—a result wholly to be expected. Also we have

$$\begin{aligned}\text{standard deviation} &= \sqrt{[N(0).\varepsilon.\lambda T(1 - \varepsilon\lambda T)]}\\ &\approx \sqrt{(\varepsilon\Delta N)}\end{aligned}$$

—the standard deviation is just the square root of the number of recorded counts *to be expected*. In practice, it is often assumed that the standard deviation is estimated accurately enough by taking the square root of the counts *obtained* in a single interval T, and indeed this is justifiable if the number is not too small. For example, suppose a source were to give a 'true' count of 100 events; this is to say that the average count determined from measurements over a large number of intervals was 100. The standard deviation is then $\sqrt{100} = 10$, so that we would expect 95% of all the trials to give a count somewhere between 80 and 120 (two standard deviations). Therefore, if the source had in fact been counted only once, the chance would be only one in 20 that we would have estimated the standard deviation to be as low as 9 ($\sqrt{80}$) or as high as 11 ($\sqrt{120}$).

An interesting special case occurs when we have a detector with $\varepsilon = 1$ (i.e. a '4π' counter with 100% intrinsic efficiency) and we count for a time interval very long compared with the half-life. This means that we will detect virtually all the decays in the source, and application of the formula above in the exact form shows that we have a standard deviation of zero, instead of the square root of the number of counts. The result of the counting experiment is in fact the value of $N(0)$, the number of active nuclei originally present in the source, which of course is fixed and not subject to statistical fluctuations.

4.15 Standard Deviation of Count-Rates

It must be emphasised that we have just found the standard deviation of the number of counts—which may denote by c—obtained from a source in an interval T. T is usually measured in seconds or minutes, and what we have hitherto called a count-*rate*, in so many counts per second or counts per minute is C or c/T. Since as a general rule the standard deviation in c is just $\sqrt{c}$, it follows that the standard deviation in C is $\sqrt{c}/T$ and we may write

$$\text{count-rate} = C \pm \sigma_c = \frac{c}{T} \pm \frac{\sqrt{c}}{T}$$

Alternatively the standard deviation in C, σ_c, is $\sqrt{(C/T)}$; notice that it is *not* just $\sqrt{C}$. As an example of this important point, if one is told that a certain source gives a count-rate of 100 counts per second, one cannot infer that this figure implies $(100 \pm 10)\ \text{s}^{-1}$. For if the source had been counted for 1000 s the total count would have been about 10^5, whose square root is about 320. One would then be justified in requiring a place of decimals in the count-rate and expressing it as, for example, $(100.4 \pm 0.3)\ \text{s}^{-1}$. (And if this were true it might be worth enquiring whether other sources of uncertainty in the measurement were more significant than the statistical fluctuations.)

4.16 Combinations of Count-Rates

We very often have to assess the standard deviations of sums, differences, quotients, etc., of counts or of count-rates. The standard deviations applicable to the counts or count-rates can be worked out as above, and then we apply a rule which applies to many, though not all, common cases. It says that if $X, Y, Z, \ldots$ are quantities which have independent standard deviations $x, y, z, \ldots$, then the standard deviation f in any function $F \equiv F(X, Y, Z, \ldots)$ is given by

$$f = \sqrt{[(\partial F/\partial X)^2 x^2 + (\partial F/\partial Y)^2 y^2 + (\partial F/\partial Z)^2 z^2 + \cdots]}$$

(The standard deviations $x, y, \ldots$ should be small compared with $X, Y, \ldots$. Also F should not become infinite, as it could do if the function chosen were $F = x/y$ and the range of Y included zero.) For example, if we choose the function $F = X + Y$ and let X and Y be counts, or count-rates, we can calculate f which will be the standard deviation of the sum of these counts, or count-rates. Performing the partial differentiations we get $(\partial F/\partial X) = 1$ and $(\partial F/\partial Y) = 1$, so that $(\partial F/\partial X)^2 = (\partial F/\partial Y)^2 = 1$. Then clearly $f = \sqrt{(x^2 + y^2)}$. Notice that we get the same result for the difference $F = X - Y$, because then $(\partial F/\partial Y) = -1$ but $(\partial F/\partial Y)^2 = 1$.

As an example of this result, let us consider a detector which in the absence of a radioactive source gives a 'background' count of c_b events in a time T_b seconds; the background count-rate is then

$$\frac{c_b}{T_b} \pm \frac{\sqrt{c_b}}{T_b}$$

When a source is present we may denote the count obtained in a time T_s by c_s, so that the count-rate with the source present is

$$\frac{c_s}{T_s} \pm \frac{\sqrt{c_s}}{T_s}$$

Application of the result for the difference of the two rates gives easily

$$\text{nett rate} = \frac{c_s}{T_s} - \frac{c_b}{T_b} \pm \sqrt{\left(\frac{c_s}{T_s^2} + \frac{c_b}{T_b^2}\right)}$$

or, calling the rates C_s and C_b respectively,

$$\text{nett rate} = C_s - C_b \pm \sqrt{\left(\frac{C_s}{T_s} + \frac{C_b}{T_b}\right)}$$

Sometimes it is of value to work out the *ratio* of two count-rates. For example a source X may give a nett count-rate C_x, equal to $C_{sx} - C_{bx}$, where the extra subscript denotes measurements on source X; for brevity we will write x for the corresponding standard deviation

$$\sqrt{\left(\frac{C_{sx}}{T_{sx}} + \frac{C_{sx}}{T_{bx}}\right)}$$

Similar notation will be used for a second source Y. Then (unless C_y is near zero) we can define the ratio

$$F = \frac{C_x}{C_y}$$

from which

$$\partial F/\partial C_x = \frac{1}{C_y} \quad \text{and} \quad \partial F/\partial C_y = -\frac{C_x}{C_y^2}$$

The standard deviation in F is then

$$\sqrt{\left(\frac{x^2}{C_y^2} + \frac{y^2C_x^2}{C_y^4}\right)}$$

so that we can write for the complete expression for the ratio

$$F \pm f = \frac{C_x}{C_y} \pm \frac{\sqrt{(x^2C_y^2 + y^2C_x^2)}}{C_y^2}$$

with of course the appropriate values for x and y. If it happens that the rates C_x and C_y are so high that the background count-rates can be neglected, we can put

$$x = \sqrt{\frac{C_x}{T_x}} \quad \text{and} \quad y = \sqrt{\frac{C_y}{T_y}}$$

and the ratio becomes

$$F \pm f = \frac{C_x}{C_y} \pm \frac{\sqrt{(C_y^2/T_x + C_yC_x^2/T_y)}}{C_y^2}$$

It is often important to be able to estimate how one should divide the total time available between the several counting measurements that need to be made. For example one might have a total time τ available during which a 'source' and a 'background' must be counted. Suppose that a time t is spent on counting the source so that a time $\tau - t$ is left for the background count. For source and background rates of C_s and C_b we then find a nett rate

$$C_s - C_b \pm \sqrt{\left(\frac{C_s}{t} + \frac{C_b}{\tau - t}\right)}$$

Obviously the standard deviation in the nett count-rate will theoretically become infinite for $t = 0$ and $t = \tau$, and we can expect a minimum value at some intermediate time. To seek a minimum in the standard deviation, or more simply its square, the variance, we put

$$\frac{\mathrm{d}}{\mathrm{d}t}\left(\frac{C_s}{t} + \frac{C_b}{\tau - t}\right) = 0$$

i.e.

$$-\frac{C_s}{t^2} + \frac{C_b}{(\tau - t)^2} = 0$$

i.e.

$$\frac{t}{\tau - t} = \sqrt{\frac{C_s}{C_b}}$$

and solving for t, we get

$$t = \frac{\tau}{1 + \sqrt{\frac{C_b}{C_s}}}$$

These equations mean that the total time should be divided in the ratio of the square root of the count-rates of the source and the background. Since the source rate is always higher than the background (since the source rate actually includes the background), it is always necessary to count the source for longer than the background; the times become more nearly equal as we consider weaker and weaker sources.

It must be said that while the above embodies a good general principle, there are several practical considerations which may modify it:

(a) The assumption of a fixed available time τ may be unrealistic because the counting equipment is not used for *sources* for much of the 24-hour day. It is the practice of some laboratories to keep all the counting equipment running 'on background' whenever it is not in actual use for counting sources. This may mean that background rates are somewhat biased towards values appropriate to night hours, which are likely to be a little low perhaps because the switching of heavy electrical loads during the day produces mains-borne interference which is responsible for 'spurious' counts. The continuous recording of background has much to commend it, as sudden and inexplicable changes in it give warning of imminent malfunctioning or unsuspected presence of extraneous sources, etc.

(b) In clinical practice the time spent on counting *in vivo* is limited by patient toleration—and sometimes by patient load—more than by any other factor.

(c) It is often better, and especially when counting *in vivo*, to count for, say, four separate 100 s intervals than for a single interval even if this is a long as 500 s. This is because if the four separate counts are self-consistent one has some assurance that the equipment is operating correctly and that the source—e.g. a patient—has not moved with respect to the detector. If the counts disagree by more than two standard deviations they must obviously be repeated more carefully. A single long count gives neither of these safeguards and its apparent better statistical accuracy might be misleading.

A similar problem is the allocation of time spent on each of two sources in order to optimise the accuracy of the ratio between their count-rates. The problem is quite difficult if the sources are weak, but if the count-rates are high compared with background the following treatment is adequate. Again suppose we have an available time τ during which source X is counted for a time t so that a time $= \tau - t$ is available for source Y. Then the standard deviation in the ratio C_x/C_y becomes

$$f = \frac{\sqrt{\left[\frac{C_x C_y^2}{t} + \frac{C_y C_x^2}{(\tau - t)}\right]}}{C_y^2}$$

so that we clearly have to find the minimum value of the expression under the square-root sign. Putting

$$\frac{d}{dt}\left(\frac{C_x C_y^2}{t} + \frac{C_y C_x^2}{(\tau - t)}\right) = 0$$

we get

$$-\frac{C_x C_y^2}{t^2} + \frac{C_y C_x^2}{(\tau - t)^2} = 0$$

i.e.

$$\frac{t}{\tau - t} = \sqrt{\frac{C_y}{C_x}}$$

a result which has a superficial resemblance to that for the difference of two count-rates. But now we have that if $C_y > C_x$, then $t > \tau - t$; in other words we should spend the longer time counting the *weaker* source, a conclusion directly opposite to the previous one. Evidently there is no universal 'blanket' rule to cover all cases, and if one is attempting to derive a quantity whose value depends on a more complicated function of two count-rates the choice of optimum counting intervals can be quite difficult. This applies *a fortiori* when the function involves more than two count-rates.

4.17 Choice of Counting Systems

Another practical problem in which statistical considerations offer some guidance is that of choosing between two counting systems, one of which has a high counting efficiency ε_1 but gives a high background rate C_{b1}, while the other has a low efficiency ε_2 but a low background rate C_{b2}. Let us suppose we have a source of strength X becquerel and a time τ available on either system. We will take as the criterion for the better system, that which gives the smaller relative standard deviation in the nett count-rate (i.e. ratio between standard deviation and the count-rate itself).

For the first system, we shall clearly have a nett count-rate of $\varepsilon_1 X$ above background, so that when the source is counted the observed count-rate will be $C_{s1} = \varepsilon_1 X + C_{b1}$. For optimum accuracy, the counting interval will be divided in the ratio $\sqrt{C_{s1}/C_{b1}}$, a time t_1 being spent on counting the source and a time $\tau - t_1\sqrt{(C_{b1}/C_{s1})}$ being spent on counting the background. The square of the standard deviation, or variance, in $C_{s1} - C_{b1}$ is then obtained, following the first part of Section 4.5, as

$$\text{variance in } C_{s1} - C_{b1} = \frac{C_{s1}}{t_1} + \frac{C_{b1}}{t_1}\sqrt{\frac{C_{s1}}{C_{b1}}}$$
$$= \frac{1}{t_1}[C_{s1} + \sqrt{(C_{b1}C_{s1})}]$$
$$= \frac{1 + \sqrt{(C_{b1}/C_{s1})}}{\tau}[C_{s1} + \sqrt{(C_{b1}C_{s1})}]$$
$$= \frac{(\sqrt{C_{s1}} + \sqrt{C_{b1}})^2}{\tau}$$

The relative standard deviation in the nett count-rate is then

$$\frac{\sqrt{C_{s1}} + \sqrt{C_{b1}}}{\sqrt{\tau}(C_{s1} - C_{b1})}$$

The expression for the second system is the same, with 2 as subscript in place of 1 throughout.

The first system is to be preferred if

$$\frac{\sqrt{C_{s1}} + \sqrt{C_{b1}}}{C_{s1} - C_{b1}} < \frac{\sqrt{C_{s2}} + \sqrt{C_{b2}}}{C_{s2} - C_{b2}}$$

or

$$\frac{\sqrt{C_{s1}} + \sqrt{C_{b1}}}{\varepsilon_1 X} < \frac{\sqrt{C_{s2}} + \sqrt{C_{b2}}}{\varepsilon_2 X}$$

or

$$\frac{\varepsilon_2}{\varepsilon_1} < \frac{\sqrt{C_{s2}} + \sqrt{C_{b2}}}{\sqrt{C_{s1}} + \sqrt{C_{b1}}}$$

or

$$\frac{\varepsilon_2}{\varepsilon_1} < \frac{\sqrt{(\varepsilon_2 X + C_{b2})} + \sqrt{C_{b2}}}{\sqrt{(\varepsilon_1 X + C_{b1})} + \sqrt{C_{b1}}}$$

The criterion thus depends on the value of X. For very low activity, for which both $\varepsilon_2 X$ and $\varepsilon_1 X$ are small compared with the background rate, system 1 is the better if

$$\frac{\varepsilon_2}{\varepsilon_1} < \frac{\sqrt{C_{b2}}}{\sqrt{C_{b1}}}$$

or

$$\frac{\varepsilon_2}{\sqrt{C_{b2}}} < \frac{\varepsilon_1}{\sqrt{C_{b1}}}$$

This means that of two systems we should choose that for which the ration $\varepsilon/\sqrt{C_b}$, or ε^2/C_b, has the higher value. Since, for any given *source*, the nett count-rate on different systems is proportional to the counting efficiency, this rule is often quoted as the requirement to maximise the ratio (source count)^2/background, or S^2/B.

However, when the nett rates $\varepsilon_1 X$ and $\varepsilon_2 X$ are both large compared to the background rates, we find the criterion reduces to

$$\frac{\varepsilon_2}{\varepsilon_1} < \sqrt{\frac{\varepsilon_2}{\varepsilon_1}}$$

which means

$$\varepsilon_2 < \varepsilon_1$$

and we are to prefer the system which has the higher efficiency. Thus our choice may not be the same as it would be for weak sources.

As a numerical example, consider counting systems with the following characteristics:

1. 'high efficiency'; $\varepsilon_1 = 0.5$, background $C_{b1} = 50.0\ s^{-1}$
2. 'low efficiency'; $\varepsilon_2 = 0.1$, background $C_{b2} = 1.0\ s^{-1}$

System 1 has $\varepsilon_1^2/C_{b1} = 0.005$; system 2 has $\varepsilon_2^2/C_{b2} = 0.01$. System 2 is therefore preferable for low count-rates. We find, indeed, by straightforward application of the above formulae that for a source strength of 10 Bq and $\tau = 100$ s, system 1 gives a relative standard deviation of 0.29 whereas system 2 gives 0.24. However, for source strengths of 30 Bq and larger, system 1 gives the lower relative standard deviation. These figures are only illustrative, but it does seem to be sound advice to most workers to obtain the most sensitive equipment available and to regard a high background as a secondary consideration. This is because it is reasonable to suppose that most of the samples one needs to count are not so very weak; high-efficiency equipment can deal with all of these in a short time, leaving something extra for the weak samples. If one's experiments do seem to involve counting many weak samples, one ought to be thinking very seriously about how to improve the technique of the rest of the procedure in order to provide stronger samples.

4.18 Application to Dual Isotope Counting

Referring back to Section 3.28 on dual isotope counting it will be remembered that we derived a formula for Y, the ratio of the strengths of a pair of radionuclides A and B in a given mixture, which reads

$$Y = \frac{C_B^{II}}{C_A^{I}} \cdot X - \frac{C_B^{I}}{C_A^{I}}$$

where the C's represent counts in channels I and II from standards of pure A and B, and X is the ratio of counts in the two channels from the mixture. Difficulties arise when the unknown mixture happens to be almost pure B because then X becomes nearly equal to C_B^I/C_B^{II} and Y becomes the difference between two nearly equal quantities. A numerical example will make this clear.

Table 4.1 Standard Deviations for a Dual Isotope Counting Measurement

Sample	Counts in channel I	Counts in channel II	$X = M_1/M_2Y$	Y	σ_y	$\sigma_y/Y(\%)$
1 MBq pure A	$C_A^I = 10\,000$					
1 MBq pure B	$C_B^I = 1000$	$C_B^{II} = 10\,000$				
Mixture P	$M_1 = 10\,000$	$M_2 = 10\,000$	1	0.9	0.023	2.5
Mixture Q	$M_1 = 8000$	$M_2 = 40\,000$	0.2	0.1	0.025	25

We give in Table 4.1 a set of results which show the counts for the pure standards and also for two mixtures P and Q, all obtained in equal counting intervals. For the mixtures we tabulate the ratio X of observed counts in the two channels and the calculated value of Y. Also the formulae of Section 4.15 have been used to obtain the standard deviation σ_y in Y; the last column shows the relative standard deviation σ_y/Y for both mixtures.

It will be seen that the standard deviations in Y are nearly the same for the two mixtures, but that the relative standard deviation for mixture Q is 10 times larger than that for mixture P. This is in spite of the fact that more total counts have been recorded for mixture Q. The large difference in relative standard deviation is due to the low value of Y for mixture Q (0.1). Clearly a better result could only be obtained by recording many more counts, and in a case like this it is essential to find out which samples need a longer counting interval. In the present example it can be shown that it is most profitable to determine the counts in channel II, for both pure standards and the mixture, with greater statistical accuracy by counting for a longer time.

4.19 The Energy Resolution of Detection System

For all detection devices working in the pulse mode (except for the Geiger–Müller tube) we have shown that output pulses are produced whose voltage amplitude is proportional to the initial amount of ionisation and therefore to the energy of the initiating electron (Compton or photoelectron in a device for detecting gamma-rays), but with some variability which is statistical in origin. The variability arises from fluctuations in (a) the number of ion-pairs generated in the primary ionisation and (b) in the numbers of electrons and/or photons taking part in the subsequent processes of amplification. We briefly discuss these effects for solid-state and scintillation detectors.

(*a*) *Solid-State Detector.* For a germanium crystal detector the mean energy needed to create an ion pair is only about 3 eV. A gamma ray from ^{137}Cs (^{137}Ba) of 660 keV can therefore produce some 2.2×10^5 ion-pairs, a number which we might expect to have a standard deviation of $\surd(2.2 \times 10^5)$, giving a relative standard deviation of $\surd(2.2 \times 10^5)/(2.2 \times 10^5)$, or

$1/\surd(2.2 \times 10^5)$. However this is too high an estimate, because no account has been taken of the fact that the absorption of energy from the initial electron also causes a good deal of mechanical vibration in the germanium lattice—the so-called 'phonons'. The total number of 'events' in which statistical fluctuations occur, i.e. ionisations plus phonon generation, is therefore many times larger than 2.2×10^5 and the *relative* standard deviation of this larger number is correspondingly smaller. Calculations by U. Fano introduce a factor F (the Fano factor), which is about 0.15, so that the corrected relative standard deviation becomes

$$\sqrt{\frac{F}{2.2 \times 10^5}} = 8 \times 10^{-4}$$

Assuming a gaussian distribution of pulse amplitudes, we therefore have a FWHM, for a gamma-ray energy of 660 keV, of $2.36 \times 660 \times 8 \times 10^{-4} = 1.3$ keV. This figure is referred to as the *energy resolution* of the detector. Experimental values are often quite close to this figure, there being no significant additional statistical effects in the rest of the detection system, e.g. amplifiers. The low value of the resolution enables energy measurements to be made with much higher precision than are possible with scintillation detectors, and this accounts for the fact that solid-state detectors have revolutionised gamma-ray spectroscopy in the last decade.

(*b*) *Scintillation Detector.* Here the primary event is essentially the creation of photons of visible light, and in sodium iodide crystals the average energy expenditure is about 33 eV per photon produced. On this account alone the scintillation detector must have a poorer resolution than the solid-state device. Other parameters which have to be taken into account are as follows:

(i) f, the mean efficiency for collection of light from the crystal—this is nearly 1 for a crystal with highly reflective encapsulation and good transparency;

(ii) $\bar{p}$, the average number of photoelectrons generated per incident light photon at the photocathode—this is commonly about 0.1;

(iii) $\bar{n}$, the average electron multiplying factor on the dynodes, say about 3.

With these factors taken into account it can be shown that the FWHM for an initiating electron of E keV is

$$\text{FWHM} = 2.36\sqrt{\left[\frac{E}{33f\bar{p}} \cdot \frac{\bar{n}}{(\bar{n}-1)}\right]} \approx 0.4\sqrt{\left[\frac{E\bar{n}}{f\bar{p}(\bar{n}-1)}\right]}$$

Thus gamma-rays of 660 keV being absorbed by photoelectric processes should give a photoelectric peak with FWHM of about 40 keV. A good-quality crystal will indeed show a resolution approaching this figure, and a 50 keV resolution is not exceptional.

It must not be forgotten that the spectrum of pulses actually observed includes a contribution from the detector's natural background. The back-

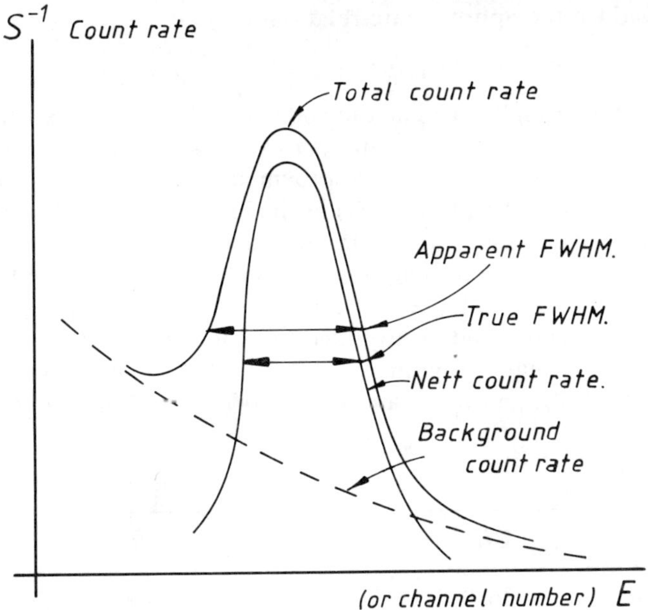

Fig. 4.10 Apparent broadening of the FWHM by the inclusion of background to a photoelectric peak.

ground effect in solid-state detectors may be very low but is a few counts per second per channel when a typical sodium iodide crystal is examined on a multichannel analyser. The background counts arise partly from extraneous radioactive material, e.g. ^{40}K as an impurity in the sodium iodide, and partly because of electrons being released thermally in the photomultiplier tube. If the source is rather weak, the apparent FWHM will be somewhat greater than the true value, as is demonstrated in an exaggerated way in Fig. 4.10.

4.20 Exercises

1. Photographic emulsions are available which are sensitive to alpha particles. One drop of a solution of uranium nitrate was placed on one of these, allowed to dry, and stored for 1 week. After developing and fixing, examination under a high-powered microscope showed the presence of 9000 tracks whose length corresponded to an energy of 4.2 MeV. Chemical analysis of the emulsion showed that it contained 1.2 micrograms of ^{238}U (E = 4.2 MeV). What is the half-life of this nuclide?

   In addition to these tracks, how many should be observed from the ^{234}U in secular equilibrium with the ^{238}U? Given that the half-life of ^{235}U is 7×10^8 yr and that its natural abundance is 0.72%, how many alpha tracks from this nuclide would be expected?
2. A radioactive source is counted for a period τ which is comparable with the half-life $t_{1/2}$, and gives C counts in that period. Find a formula for the time t, after the start of the interval, at which the decay rate is equal to

C/τ. Deduce the approximate relation

$$\tau/2 - t = 0.028(\tau/t_{1/2}).\tau$$

3. A sample of a single pure radionuclide gave the count-rate $C(t)$ shown in Table 4.2. Estimate the half-life and $C(0)$ graphically. Regard your estimates of $t_{1/2}$ and $C(0)$ as first approximations. From them *calculate* the values of the counting rates $C'(t)$ at the given times t and compare these with the original data. Then try the effect of small variations in your estimates, e.g. by increasing or decreasing them by 1% and calculating new values $C''(t)$, $C'''(t)$, etc. Can you get a 'better fit'? Decide what criterion you should adopt to judge the goodness of fit.

   Write a computer program to read in $C(t)$ and t and also first estimates of $t_{1/2}$ and $C(0)$, and to obtain improved values of these quantities.

Table 4.2

t(s)	$C(t)$
100	3630
150	2271
200	1380
250	890
300	558
350	355
400	220
450	125
500	91
550	55
600	25
650	23

Table 4.3

t	$C(t)$
100	47034
150	26178
200	15499
250	9770
300	6514
350	4545
400	3277
450	2418
500	1810

4. A mixture of two radionuclides P and Q gave the count-rates shown in Table 4.3 at times t(statistical errors are negligible).
   Resolve the complex decay curve given that P has half-life 45 s and Q has half-life 131 s. Use (1) the peel-off method and (2) the two alternative methods given in Section 4.9.

   Write a computer program or use a programmable calculator to deal with part (2).
5. The concentration of thyroxine in blood (T4 value) is a valuable diagnostic parameter. One technique for measuring it (THYOPAC 4, Code 1M.64, The Radiochemical Centre, Amersham, Bucks) involves the measurement of activity in a sample obtained from a human subject and comparing this with that from two standard samples, which are designated 'high' and 'low'. The standards contain typically 1 μg T4/100 ml and 16 μg T4/100 ml. Use the symbols: C_L = count-rate of low stand-

ard; C_H = count rate of high standard; C_U = count-rate of unknown sample; ν_L = T4 content of low standard; ν_H = T4 content of high standard; ν_V = T4 content of unknown sample.
We may define

$$S = \frac{(C_L/C_H) - 1}{\nu_H - \nu_L} \quad \left(\approx \frac{(C_L/C_H) - 1}{15}\right)$$

and

$$\nu_U = \nu_L + \frac{(C_L/C_U - 1)}{S}$$

Find an expression for the standard deviation in ν_U as a function of C_L, C_H, C_U and their own standard deviations.

Write a program to read in the values of C_L, C_H, C_U, ν_H, ν_L and to print out ν_U.

Write a program to read in, in addition, the counting intervals T_L, T_H, T_U and to print out the standard deviation in ν_U.

6. Construct a table of the activity of ^{99m}Tc which accumulates during hourly intervals in a source of ^{99}Mo of strength 100 MBq at time zero.

   Write a program for a desk calculator or computer to carry out this calculation for genetically related radionuclides with any given half-lives.

7. A nuclide decays by emission of one beta particle and one gamma ray in coincidence per disintegration. Measurements on a source gave 60 000 particles detected in a proportional counter and 30 000 gamma rays detected in a scintillation crystal during one minute. In the same time 10 coincident events were recorded. The resolution of the coincidence circuit was 5 μs.

   What was the accidental coincidence rate? What was the standard deviation of this figure? Estimate the strength of the source, and the standard deviation of your estimate. For how long would you need to record coincidences in order to be able to estimate the source strength with a relative standard deviation of 2%?

8. The half-life of ^{99m}Tc is 6.0 h. 1.4% of the disintegrations give a gamma ray of 0.142 MeV and also conversion electrons. What is the partial decay constant for this branch?

   For every gamma ray of this energy emitted from the nucleus there are 80 K-shell conversion electrons and 30 L-shell electrons. What is the partial decay constant for the gamma ray alone?

   Rough formulae for the gamma ray partial decay constants for E3, M3, E4 and M4 transitions are:

$$\lambda(\text{E3}) = 55E^7A^2\ \text{s}^{-1}$$
$$\lambda(\text{M3}) = 15E^7A^{4/3}\ \text{s}^{-1}$$
$$\lambda(\text{E4}) = 1 \times 10^{-5}E^9A^{8/3}\ \text{s}^{-1}$$
$$\lambda(\text{M4}) = 5 \times 10^{-6}E^9A^2\ \text{s}^{-1}$$

E being the transition energy in MeV and A the mass number of the nuclide. Which formula best fits the 0.142 MeV gamma decay? Given that ^{99m}Tc has a spin and parity of (9/2, +) deduce the probable spin and parity of ^{99g}Tc.

The remaining 98.6% of ^{99m}Tc decays proceed via an electric octupole (E3) transition with energy 0.002 MeV. Estimate the partial half-life for gamma emission. By comparing this with the partial half-life for the branch as a whole, estimate the total electron conversion coefficient.

9. The half-life of ^{252}Cf is 2.65 yr. It decays by alpha emission in 96.9% of cases and by spontaneous fission in the remaining 3.1%. The average number of neutrons per fission is 3.5. Calculate the rate of neutron production from one gram of the nuclide. Calculate also the maximum activity of ^{131}I generated in it, assuming a fission yield of 2.9%.

4.21 Bibliography

FRIEDLANDER G., KENNEDY J.W. and MILLER J.M. *Nuclear and Radiohemistry* (John Wiley)

HOEL P.G. *Introduction to Mathematical Statistics* (John Wiley)

MOORE, P.G., SHIRLEY E.A. and EDWARDS D.E. *Standard Statistical Calculations* (Pitman)

5
Radionuclide Uptake and Imaging

UPTAKE

5.1 General Principles of Uptake Measurements

Measurements of radionuclides *in vivo*, in animals and humans, are generally carried out with one of the following objectives:

(a) the determination of the total activity in the whole body,

(b) the determination of the activity in one or more organs of the body, (In this chapter we also review some measurements on body fluids *in vitro*.) and

(c) the determination of the spatial distribution of activity within particular organs.

In view of the complex geometrical form of the body and its organs, and the scattering and absorption of radiation from any radionuclide present within it, it is clear that none of these determinations can be truly absolute. Nevertheless, careful attention to geometrical factors enable us to make relative measurements of gamma radiation emitted from the body which are not too inaccurate to be of clinical value, and the judicious use of artificial models of body organs, known as 'phantoms', allows us to compare this radiation from that of known sources under roughly similar geometrical conditions. Various kinds of equipment for whole-body counting have been described and are briefly reviewed in Section 5.2. Uptake in individual organs has been traditionally measured using scintillation detectors with collimators attached so as to select radiation from the organ in question and exclude that from neighbouring organs. More and more in recent years this task has been taken over by gamma-camera systems with electronic devices to select radiation from only those parts of the body lying within a region of interest. It is however still important to understand the geometry of the 'single-hole collimator' as this gives an insight into the design problems of gamma-camera and scanner collimators. These two latter instruments can provide a visualisation, in two dimensions, of the spatial distribution of radionuclides within an organ. Here one is no longer very concerned with absolute measurements, the main interest being in the relative activity from point to point in the radiograph, or photoscan.

5.2 The Whole-Body Counter

A very rough measure of the activity in the whole body of a human subject can be made by placing a large unshielded scintillation detector about 30 cm in front of the chest while the subject is seated. This technique is certainly inexpensive but would give useful information about the changes with time of the body burden of a radionuclide only if its distribution in the body remained the same. However, changes in its spatial distribution would change the count-rate. In other words, such a simple system does not have the desirable property of *uniformity of response.* Attempts to improve the uniformity of response, e.g. by using a scintillator at much greater distances, say 3 m, bring the disadvantage of loss of sensitivity. Intermediate geometries, with the subject lying on a curved couch so that all parts of the body are about 1 m from a detector, are acceptable if the subject is a volunteer but are hardly to be recommended for hospital patients. Modern whole-body counters seek to achieve good uniformity of response and high sensitivity, the latter requirement also necessitating low background count-rates, but the installation is necessarily quite sophisticated.

There are two main types of whole-body counter which have stood the test of time. In a typical example of one of these, eight large (150 mm diameter) sodium iodide crystals are used. The subject lies supine on a flat couch, four crystals being suspended above him in a line along the body axis and four supported in corresponding positions below him. Phantoms are used to find the horizontal and vertical dispositions of the detectors which give the most uniform response (see Fig. 5.1). The HT supplies to the detector photomultiplier tubes are adjusted so that the pulse height spectra obtained for a standard gamma-ray source are exactly matched. When the whole-body

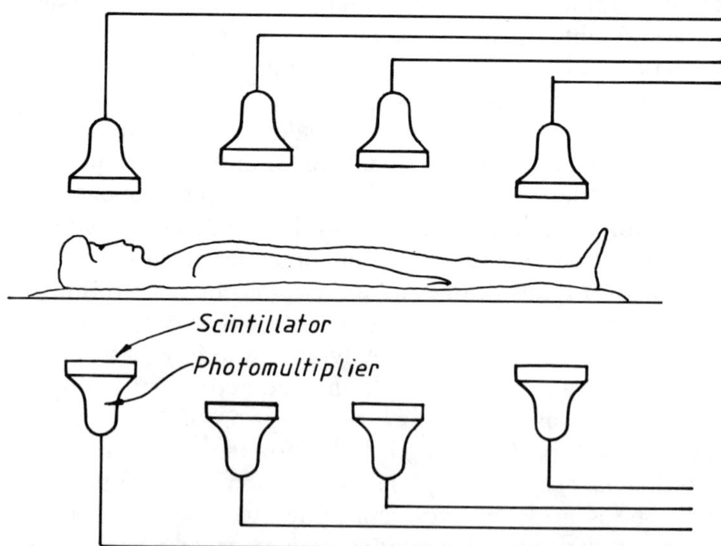

Fig. 5.1 Whole-body counter using sodium iodide detectors (side view).

counter is in use, the trains of pulses generated by the detectors can then be combined and the spectra displayed on a multichannel analyser just as if only a single crystal were used. The system has the advantage of good energy resolution, but clearly the geometrical efficiency is low—of the order of 2%.

The second kind of apparatus uses liquid organic scintillator contained in the annular space between two large tubes, with the couch for the subject within the inner space, as shown in Fig. 5.2. The scintillator is viewed by a large number of photomultiplier tubes. The geometric efficiency is about 90% but the intrinsic efficiency is only about 5% for medium-energy gamma rays and the energy resolution is poor.

In either system, with such a large mass of scintillating material, the sensitivity to radiation is quite high in spite of the low overall efficiency. Let us for instance consider the activity due to the naturally occurring radionuclide ^{40}K. In adult males the potassium content of the body is about 50 millimoles per kilogram body weight. Therefore, an adult male of 80 kg will contain $80 \times 50 \times 6 \times 10^{23} \times 10^{-3}$ atoms of potassium. Since the isotopic abundance of ^{40}K is 0.0012 and the half-life is 1.3×10^9 y or 4.0×10^{16} s, the

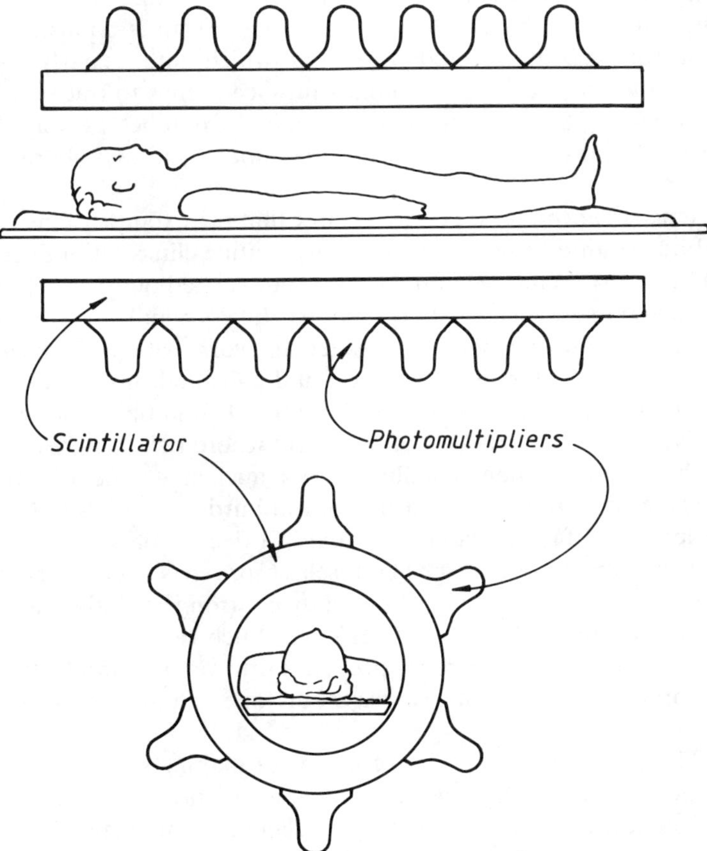

Fig. 5.2 Whole-body counter using liquid organic scintillator.

decay rate will be $(80 \times 50 \times 6 \times 10^{20} \times 0.0012 \times 0.6932)/(4.0 \times 10^{16})$ or $5.0 \times 10^4\ s^{-1}$. Finally, since 11% of the disintegrations give a gamma ray (of 1.46 MeV), a system with an overall detection efficiency of 2% will give a count-rate of $5 \times 10^4 \times 0.11 \times 0.02$ or about $100\ s^{-1}$. Counting times of some 100 to 1000 s will, therefore, give a total count of good statistical accuracy, if the background count-rate can be reduced to a comparable level.

Suitably low background levels can be achieved by careful siting of the installation, avoidance of extraneous activity in constructional materials, and the provision of shielding walls. All igneous rocks contain appreciable amounts of uranium and thorium and their decay products, but limestone rock is much less strongly radioactive and an underground room in such rock has much to commend it. Structural materials such as wall-plaster which normally contains potassium, and newly smelted steel which is contaminated with ^{60}Co should be avoided. Sodium iodide crystals themselves ordinarily contain small amounts of potassium as an impurity and a special grade of sodium iodide is essential. If the counter is to be installed in a basement room in an existing building, it will be found that heavy shielding walls, roof and possibly floor are needed. For this purpose, lead is uneconomical as compared with steel, of which a thickness of some 30 cm is typical. Again this steel must not be contaminated with ^{60}Co, which enters nearly all modern steel as this radionuclide is used in blast-furnace linings to trace the progress of wear in them. Such sources of old steel as derelict railway lines and salvaged battleships from World War I have been successfully pressed into service.

Experience over the years seems to show that the expense of installation of a whole-body counter is not justified by any routine clinical usefulness. This is not to say that invaluable scientific work is not carried out by a relatively few centres which possess such a facility. This work falls mainly under four heads:

(i) The monitoring of staff engaged in work with radionuclides as a check on contamination of skin and hair, and on ingestion.

(ii) The monitoring of people under normal conditions, as well as those living in certain regions of the world where risk is higher, as regards their body-burden of fall-out from nuclear explosions. In this respect, it is gratifying that the typical burden of ^{137}Cs in humans has decreased fairly steadily over the last decade or so.

(iii) The carrying out of specific research projects such as the measurement of ^{49}Ca following whole-body neutron irradiation, as a method of determining whole-body calcium. Such research may or may not lead to the procedure of proven worth as a routine test on large populations. Until it is carried out, one cannot assess its potential value.

(iv) The provision of alternative means of making some other measurements, such as those which involve estimation of activity in excreta. Sometimes it may be difficult to collect excreta completely and even when this is ensured one still has to relate counts on the samples to the initial activity administered.

Excreta and administered activity are inevitably counted in different geometries, so that a calibration factor has to be determined and this is a possible source of error. Serial whole-body counts are thus much more convenient and more reliable as a measure of activity lost in excreta, if this is not too small.

The calibration of whole-body counters can be carried out in several ways. The average whole-body potassium content of humans can be assessed from the body-weight and this gives a helpful calibration point. Another is got by administration to human subjects of known activities of ^{42}K (half-life 12.6 h) which happens to have very nearly the same gamma-ray energy as that of the naturally occurring ^{40}K and which seems to exchange with body potassium and reach sensibly equal bodily distribution in a few few hours. Other means of calibration involve the use of phantoms in the form of plastic bottles which mimic the size and volume of the human body and which can be filled with aqueous solutions containing known activities of a variety of other radionuclides. All such measurements are of limited precision because phantoms cannot exactly reproduce the distribution of radionuclide in the human body, which in any case is likely to vary somewhat from subject to subject. Measurements on any one human are, therefore, bound to carry some uncertainty in absolute terms, although serial counting on the subject is much more reliable. In summary, it seems fair to say that whole-body counting has not proved itself to be of general use but for some rather well defined special purposes it is quite invaluable.

5.3 Uptake Measurements with Collimated Single Sodium Iodide Scintillators

Uptake measurements in discrete organs of the body, e.g. thyroid gland, liver, spleen, are possible using sodium iodide scintillators attached to collimators having cylindrical geometry as shown in Fig. 5.3. The collimators are invariably of lead and serve to reduce extraneous radiations from the rest of the body and the environment. It is general practice to construct the collimator so as to provide a supporting sleeve for the crystal, photomultiplier and the chain of dynode resistors. The collimator also excludes daylight, which should never be allowed to reach the photomultiplier tube when the high voltage is on.

In order to assess the performance of a collimated detector, it is essential to consider some geometrical relationships. To begin with we will consider a single plane containing radioactive material at a distance F, and in air, below the end-plane of the collimator. Other parameters are the radius of the exposed face of the crystal, r, the radius of the open end of the collimator, r', and the 'depth' of the collimator, t (see Fig. 5.4). In this figure, AD represents the diameter, $2R$, of a circle called the *circle of view*, outside of which radiation from the plane source cannot enter the crystal except by penetrating the collimator. BC represents the diameter, $2R'$, of a smaller circle, the *visual field*; radiation emitted from within this circle may enter the crystal

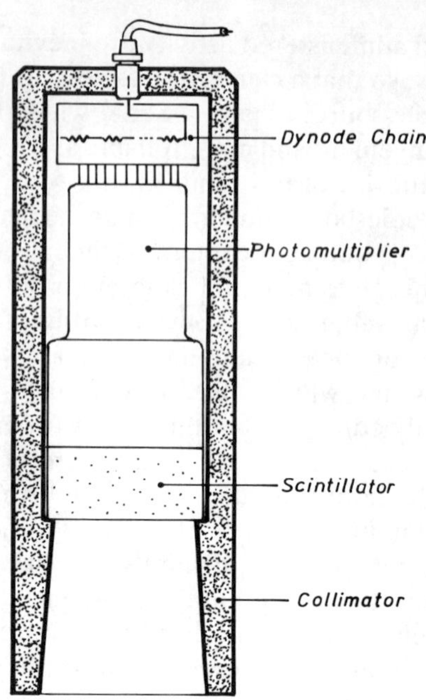

Fig. 5.3 Collimated detector for uptake measurements.

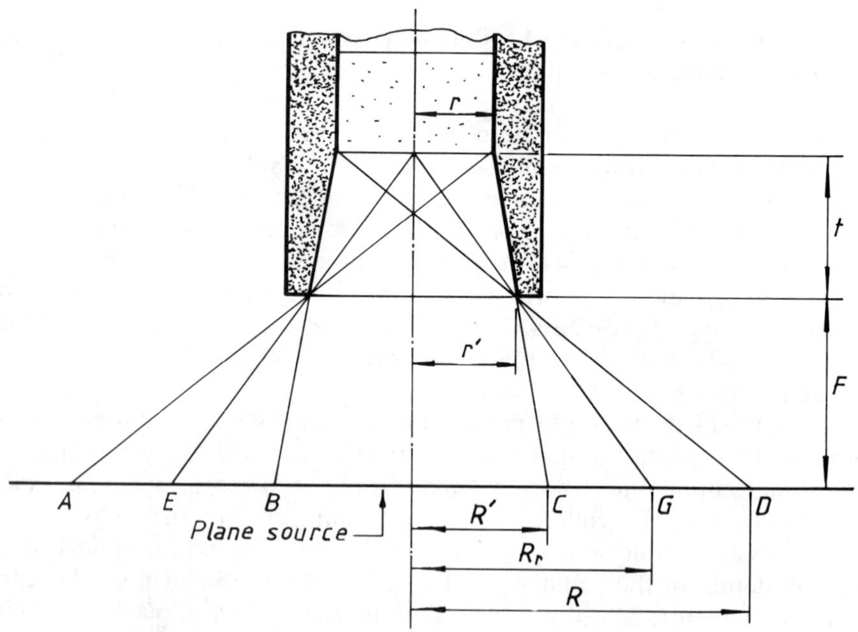

Fig. 5.4 Collimated detector and plane-source geometry.

unhindered. In the annual region between these circles and represented by AB and CD, we have what is commonly but quite incorrectly called a penumbra; from any point in this the crystal is partly obscured by the edge of the collimator. It is easy to prove, from the geometry of the similar triangles in the figure, that

$$R = \frac{r'(t + F) + rF}{t}$$

and

$$R' = \frac{r'(t + F) - rF}{t}$$

Also the *mean* of these radii, R_r, is

$$R_r = \frac{r'(t + F)}{t}$$

and is the radius of the circle with diameter EG, i.e. that circle in the source plane which forms the base of the cone just fitting within the collimator and whose apex is the centre point of the exposed face of the crystal. R_r is commonly called the *resolution radius* and the circle itself the *circle of resolution*.

Of course, the body organs being measured exist in three dimensions and are never exactly circular in cross-section. However, as a rough guide, one should try to choose collimator dimensions, and position with respect to the subject, such that the circle of resolution at the mid-plane of the organ has about the same area as the organ in that plane. If the organ is small and uptake in neighbouring tissue is low, as with radioiodine in the thyroid, the circle of resolution can be with advantage somewhat larger, because, then, small inaccuracies in the positioning of the detector in the horizontal plane are unimportant. If the organ is large (e.g. the liver) and uptake in a neighbouring organ (e.g. the spleen) is considerable, it is preferable to err on the side of too small a collimator, so that the resolution circle is within the organ. Subjects vary a great deal from one to another, and it is hardly surprising that results on, for example, the ratio of activity in the liver to that in the spleen as obtained in different laboratories show some differences even when the subjects are clinically fairly comparable.

When serial counts have to be taken on any one subject, it is obviously important to maintain constant geometry from one measurement to another. For thyroid counting it is a common practice to use a simple optical system. A focused beam of light is projected at an angle from each of two light sources on opposite sides of the collimator and the beams are arranged to cross at a point some 250 mm away. The detector and subject are then positioned so that the crossed beams coincide at a fixed point on the neck (usually 10–20 mm below the projection of the thyroid cartilage). This geometry is easily reproduced with a standard phantom. For large and deep-seated organs the edge of the collimator may touch the skin which is then marked round with a felt-tipped pen. Sometimes it is useful to have the axis of the detector system angled to the vertical and a simple form of protractor, or a device which ensures that only a limited choice of angles is possible, is convenient. It is highly desirable to arrange for all measurements on any one subject to be carried out by the same technician, thereby avoiding subtle personal errors.

5.4 Physical Assessment of Single-Collimator Performance

Although we have defined the circle of resolution with reference to a flat planar source, such sources are not very easy to prepare and in any case tests of collimator performance call for other kinds of source. One test consists of taking a source of as small linear dimensions as possible (a 'point source') and recording the (nett) counts from it when placed at a variety of points under the collimator. Since collimators invariably have cylindrical geometry these points can all be in a single plane passing through the collimator axis. For each point, the distance below the collimator opening and the distance from the central axis are noted. If the points are closely enough spaced, one can find by interpolation the coordinates of all points for which the counts are some definite fixed value. Lines joining such points are *isocount* lines. A line very close to the collimator opening is usually taken as a standard of 100% and counts along other lines expressed as a percentage. Typical isocount lines are shown in Fig. 5.5

With the open-ended collimator as discussed so far, the isocount lines show a little curvature within the visual field, but of course bend very strongly just outside it. Some commercially available collinators are fitted with a wide-

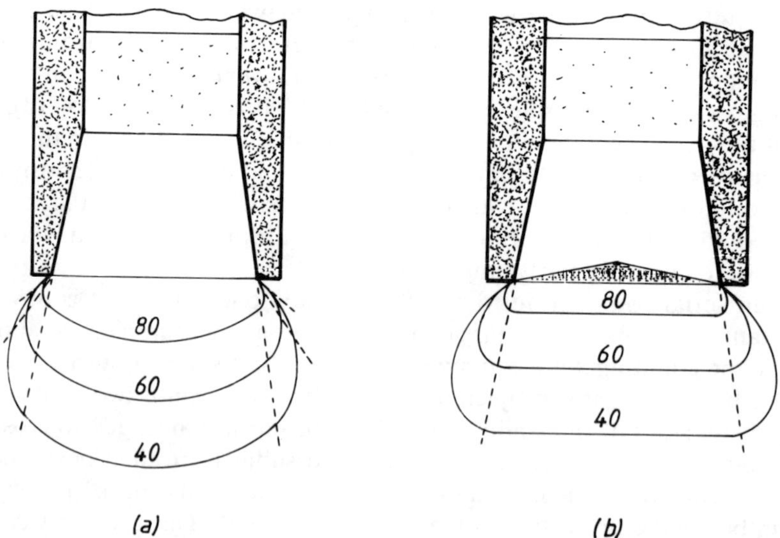

Fig. 5.5 Isocount lines for a collimator (a) without and (b) with absorbing cone.

angled cone of a suitable absorbing material, as shown at (b). This has the effect, for one particular gamma-ray energy, of reducing the radiation intensity from points near the axis and thus straightening the isocount lines. It is doubtful whether this refinement has any practical value.

The derivation of the isocount lines for a collimator is tedious experimentally and an alternative technique which gives almost as good a visual impression of the geometrical properties of the equipment uses a 'line source' of activity instead of a point. A line source may physically be a straight narrow thin-walled capillary tube filled with a solution of a suitable radionuclide. This is placed at various fixed positions in planes perpendicular to the collimator axis and counts are recorded. In a more sophisticated version of this experiment, the line source is moved at right angles to itself on a carriage driven by a constant-speed motor, as shown diagrammatically in Fig. 5.6. During any one run, within the line-source moving in a plane at distance F from the collimator, counts are stored and printed out at intervals of one or

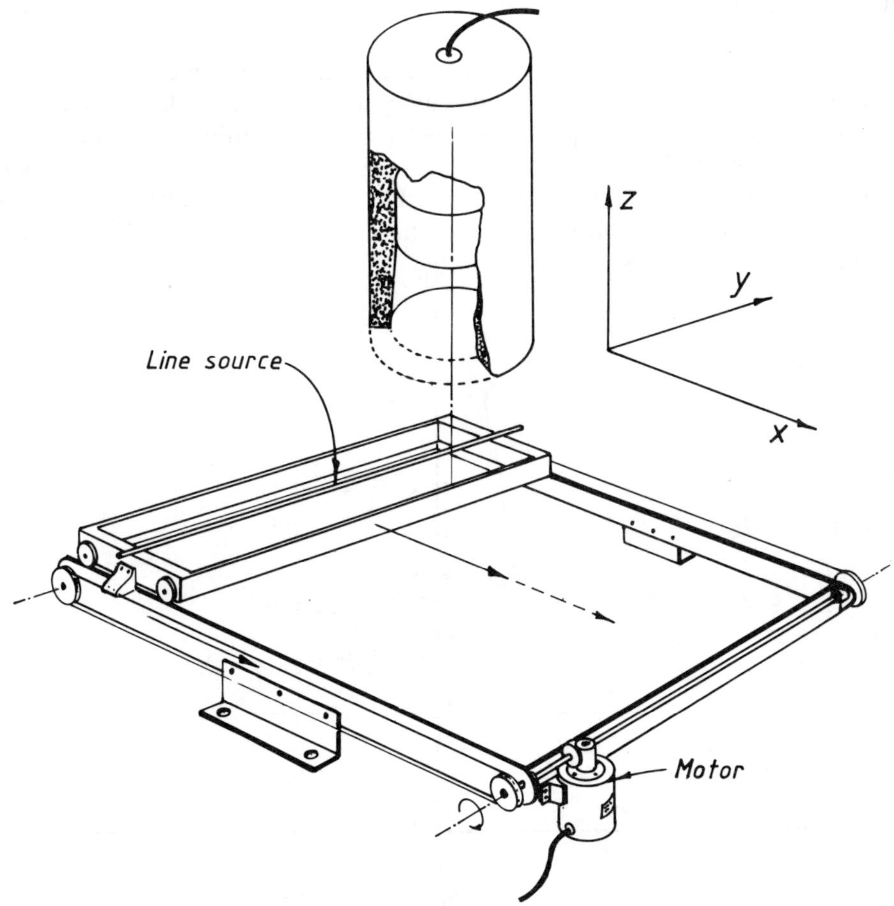

Fig. 5.6 Experimental arrangement to measure line-spread functions.

two seconds. The speed of the motor drive is measured and then it is a simple matter to derive the mean count-rate over each small distance Δx travelled during the counting intervals. Plots of the nett count-rate against lateral displacement of the source are called line-spread functions (LSF). (Strictly speaking the data represent a histogram of discrete values, but we usually arrange to have very closely spaced readings so that we can regard the LSF as a continuous function.) Typical LSF plots are given in Fig. 5.7; the origin of coordinates in the x direction corresponds to the collimator axis.

The curves approach zero count-rate for lateral displacements $x = \pm R$, the radius of the circle of view. A useful parameter is the width across each curve at half the peak count-rate—the FWHM; its value for $F = 0, 0.1, 0.2$ m is shown in the figure. The FWHM is often loosely referred to as the *resolution* of the collimator at the corresponding F distance, and half the FWHM as the resolution radius, although the latter is not identical with R_r as defined geometrically. It is perhaps better to avoid this latter usage, although there is good justification for using the FWHM as a measure of resolution since two line-sources whose spacing on the carriage is significantly greater than this value will give a double-humped LSF, and so can be recognised as separate.

It should be noted in passing that the same kind of experiment can be done with a point-source instead of a line-surce, although this is less convenient because of the difficulty of preparing the source itself. Point-source functions (PSF) defined in the same way as line-spread functions are of the same general shape but give somewhat higher FWHM values. We shall use LSFs throughout this book, although for certain calculations it would be more correct to use the PSFs.

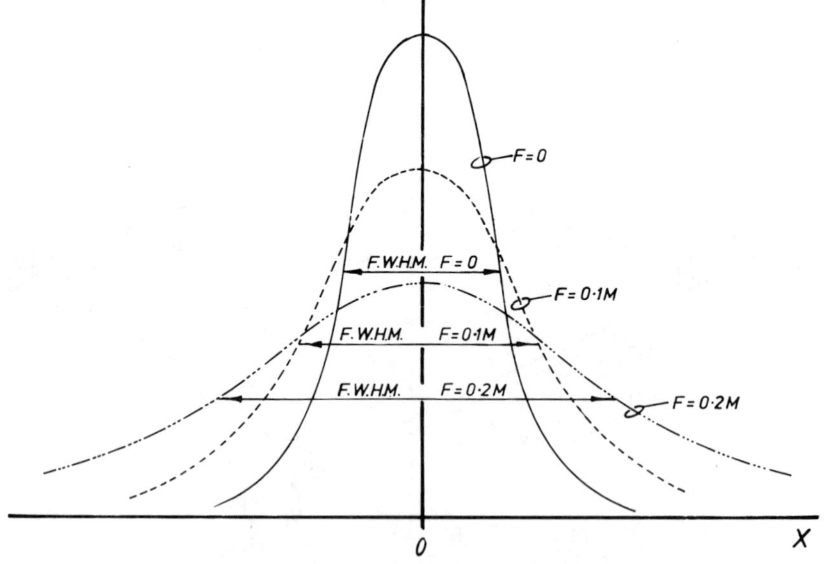

Fig. 5.7 Line-spread functions.

5.5 Plane-Source Efficiency of a Collimated Detector

For a point-source viewed by an uncollimated detector at some distance, we can conveniently define a 'point-source efficiency' as the number of gamma-rays entering the detector divided by the number emitted by the source in the same time. This measure of efficiency is a pure number, and would be unaffected by the presence of a collimator so long as the source were within its visual field. To provide ourselves with a parameter which is a measure of the efficiency of the detector and which is indicative of the effect of a collimator, we consider instead a flat planar source, the activity being spread uniformly over an area at least as great as the circle of view. The plane-source efficiency is then defined as the number of gamma-rays entering the detector divided by the number of gamma-rays emitted *per unit area* of the source in the same time. It is clear from this definition that the plane-source efficiency has the dimensions of an area. Its physical significance is that it is that area of the source whose total gamma-rays emission (in all directions) is equal to the number of gamma-rays received by the detector. The plane-source efficiency can be calculated by considering the plane to be made up of a large number of elements $\Delta x \times \Delta y$, finding the point-source efficiency for each of them, and then performing a double integration over the whole (infinite) plane. The result is

$$E_g = \frac{\pi r^2 r'^2}{4t^2}$$

E_g being the symbol used to mean 'geometric plane-source efficiency'.

Some idea of the magnitude of E_g can be obtained from the simple geometric construction of Fig. 5.8. A point P is found on the detector axis at a

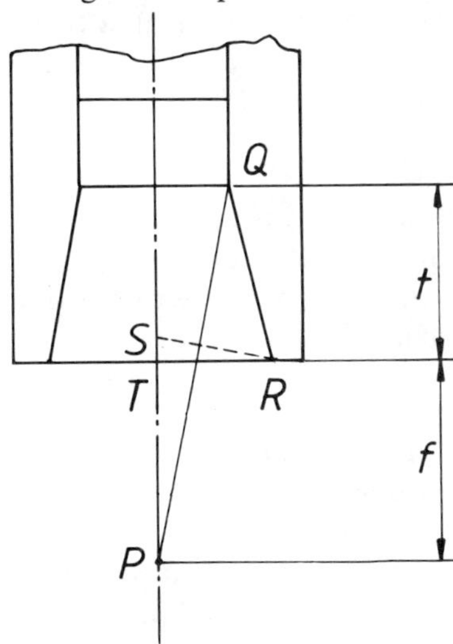

Fig. 5.8 Construction for plane-source efficiency $E_g = \pi s t^2$.

distance F below the end of the collimator. P is joined to Q on the edge of the crystal and a line RS is drawn from the inner edge of the collimator, at right angles to QP and intersecting the axis at S. T is the centre of the end-plane of the collimator. It is easy to prove that

$$\mathrm{ST} = \frac{rr'}{2t}$$

so that E_g is equal to the area of the circle with radius ST.

It should be noted that the expression for Eg involves all the essential dimensions of the collimator and is therefore a good parameter for assessing the relative performance of two or more collimators. On the other hand, it is quite independent of F, the distance between the collimator and the plane source. This means that for a given detector and collimator, the count-rate from a plane source is independent of the source distance, a result which appears at first sight to be at variance with the inverse-square law. Also the expression is independent of the relative magnitudes of r and r'; in the present case r' is greater than r but for the so-called focusing collimators (see Section 5.11) r' is less than r but the same formulae as those in this section apply.

The determination of E_g for a given collimated detector need not involve the preparation of a uniform plane source, for it may be obtained from the LSF. For if we have a uniform line source which emits N gamma rays per unit length during the time required to display the whole LSF, and the length of the line is Y, the total gamma emission is NY. Also if the line travels a distance X, the area swept out is XY. (It is assumed that X and Y are both greater than the diameter of the circle of view, $2R$.) Therefore the gamma ray emission per unit area swept out is $NY/XY = N/X$. Also if the total number of counts detected is c, the number of gamma rays entering the detector is c/ε, where ε is the intrinsic efficiency. It follows from the definition of the plane-source efficiency that

$$E_g = (c/\varepsilon) \div (N/X) = cX/\varepsilon N$$

so that E_g can easily be calculated from experimental observations.

Since in the above equation, E_g, X, ε and N are all independent of F, it follows that c is also independent. Now c is proportional to the area under the LSF curve, and we conclude that the area under all the LSF curves is the same, whatever the value of F.

5.6 The Effects of Absorption, Scattering and Incomplete Localisation of Radionuclide on Uptake Measurements

So far we have been considering measurements made only in air, but when uptake in body organs is to be determined the problem is more complicated. It is clear that absorption of radiation by tissue can only reduce the observed count-rate to less than it would have been if the source were entirely in air. On the other hand, scattering of radiation originating within the organ will

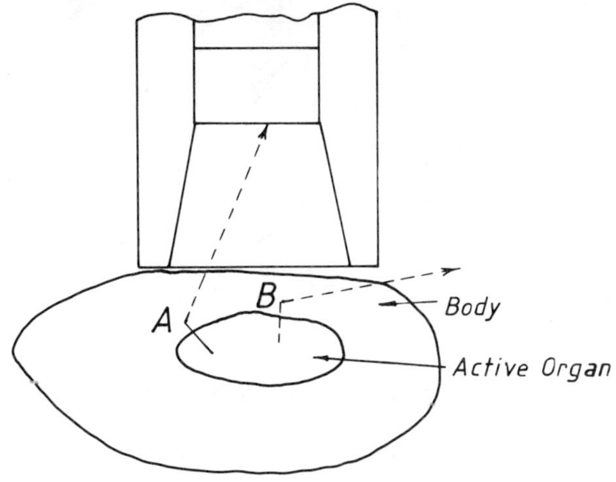

Fig. 5.9 Scattering effects in tissue.

probably have very little effect on the count-rate. That this is so is illustrated in Fig. 5.9; a gamma-ray A, which could not have been detected if the source were in air, might enter the detector if suitably scattered, whereas scattering affects gamma-ray B in the opposite sense. The position of the point of scatter could be within the organ itself, or at some other location in the body.

The combined effects of absorption and scattering can be investigated experimentally by using sources in tissue-like materials; ordinary water is generally good enough although some experimenters prefer wooden or Perspex blocks. Thus isocount lines are found to be more closely crowded together where tissue-like material is present (Fig. 5.10), and LSF curves are

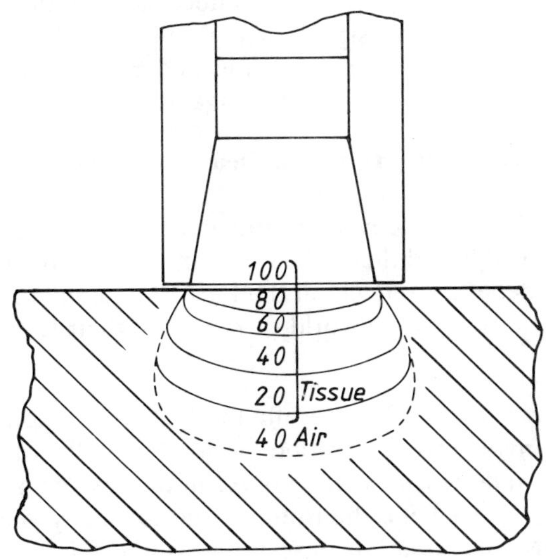

Fig. 5.10 Isocount lines in tissue.

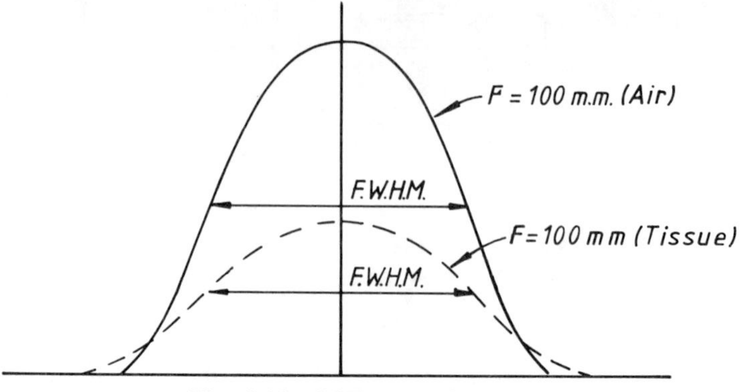

Fig. 5.11 LSF curves in tissue.

reduced in area as F is increased, while retaining general shape and FWHM values (Fig. 5.11). In any one subject the effects of scattering and absorption are likely to remain constant with time, so as far as serial counting on the same subject is concerned they can be ignored.

If the radionuclide administered to the subject fails to localise completely in the organ of interest, so that there is unwanted activity in other organs, such as the bloodstream, the count-rate associated with the organ will be too high. The extra count-rate arises in two ways. First, there is the fact that extraneous activity will be within the circle of view and can contribute directly. Secondly, radiation originating outside the circle of view has a certain probability of entering the circle and of being scattered towards the detector. The first of these effects can only be corrected for by taking counts over some other region of the body and subtracting a 'tissue background', due adjustments being made for the different volumes of tissue being counted in the two sites. To some extent the second effect can be reduced by employing energy discrimination to the signals being counted, on the principle that large-angle Compton scattering produces secondary gamma-rays of markedly lower energy which can be discriminated against.

5.7 Some Standard Uptake Tests and their Interpretation

We give here in a highly abbreviated form an account of some typical uptake tests which have established themselves over the years. Procedural details are omitted as these depend so much on the kind of equipment available; probably no two laboratories in the world carry out all their tests in precisely the same manner.

(*a*) *Thyroid Activity Tests.* Historically the first to be applied on a general scale, these tests involve administration, preferably orally, of 0.2–0.4 MBq of an iodine isotope and measuring the uptake in the thyroid gland for a period of up to 4 days. ^{131}I has been the isotope used for many years, although by modern standards the radiation dose to the subject is so high that many laboratories are now discarding it for this reason, among others. ^{123}I gives a much lower radiation dose and is always to be preferred in the cases of young people and adolescents. Uptake in normal (euthyroid) subjects attains a value of about 50% of the administered dose within 24 hours, after which no change except for physical decay is detectable. In myxoedematous subjects the uptake is slower and may not reach 10%, whereas hypothyroidism is characterised by more rapid uptake of as much as 90% of the dose administered followed by a fall-off perhaps to only 30% after 4 days. The reduction of activity in the thyroid gland in hyperthyroidism is accompanied by the appearance of iodine bound to protein in the bloodstream, which is easily tested for in blood-samples taken from the subject after 48 hours. The early part of the uptake, i.e. in the first hour or so, seems to be governed by a purely 'trapping' mechanism as it is inhibited by the presence in the blood of ions such as perchlorate which are also rapidly absorbed by the gland but not converted to compounds analogous to thyroxin. Information about the trapping mechanism can be obtained in studies of uptake of ^{99m}Tc as pertechnetate ion, which is a chemical analogue to perchlorate. Uptake at later stages must involve organic synthesis in addition, as it is inhibited by various drugs such as thiouracil which is known to slow down hormone synthesis. Low uptake in normal subjects occurs if they have large amounts of stable iodine in the diet. Some important *in vitro* tests of thyroid function are discussed in (d) and (e).

(*b*) *Renal Activity Tests.* Uptake tests in which activity over the kidneys is displayed for a period of some 30 min following intravenous injection of ^{131}I-labelled hippuran have been used for many years as an indication of renal function. When single-collimated detectors are used over each kidney, a plot of uptake versus time is known as the renogram. In the most sophisticated version of the test, the curves are corrected for activity still remaining in the blood bathing each kidney, the correction factor to be applied being derived

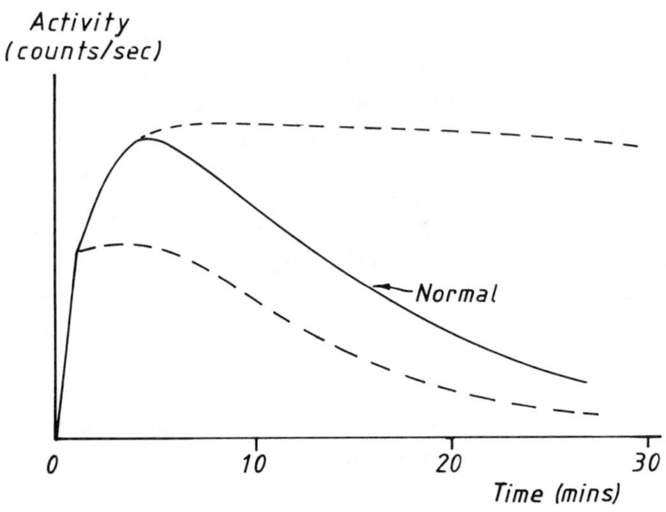

Fig. 5.12 Typical renograms.

from counts taken simultaneously from a third detector placed over the heart. The corrected curves are supposed to show the activity present in the kidney tissue alone and therefore to be of better diagnostic value, but many investigators find the uncorrected curves adequate diagnostically. Typical uncorrected renogram curves are shown in Fig. 5.12. Three distinct 'phases' are shown in renograms from normal subjects; these are (a) a very sharp rise consequent on the appearance of activity in the blood, (b) a more gentle rise lasting some 3–4 min showing an overall concentration of the tracer in the kidney, and (c) a final decrease with a half-period of some 15 min as a nett loss of activity in the kidney supervenes due to the urine flow. Inability to absorb the hippuran, due to any reason, is shown by absence of phase (b) in the curve, whereas an abnormally slow fall in phase (c) may be interpreted as diminished flow in the ureter.

(c) *Haematological Function Tests.* Much useful information on a subject's haematological status can be gained by labelling a sample of his red blood cells with ^{51}Cr, re-injecting them and following the appearance of activity in various organs. Labelling is fairly effectively done by incubating the whole blood with ^{51}Cr as chromate ion for 30 min at 37 °C. The labelled red blood cells must be thoroughly washed with acid–citrate–dextrose to remove unreacted ^{51}Cr, and then re-suspended in normal saline before re-injection. It is usual to inject just 50% of the preparation, retaining the remainder to be used as a counting standard. A convenient standard can be made up by adding heparin to haemolyse the cells and then making up to a volume of 500 ml with acidified saline and storing in a plastic bottle.

Measurement of activity in a known volume of blood withdrawn from the subject after sufficient time has elapsed for the labelled cells to mix with the rest of the blood in the body enables one to calculate the total mass of red blood cells. The rather simple calculation employed here is an example of the so-called principle of isotope dilution. Measurement of activity in the faeces can be used to detect blood loss by intestinal haemorrhage; faecal loss in normal humans is 1.5 ± 0.5 ml of whole blood per day.

Measurement of activity in red blood cells in samples withdrawn from the subject over several 24 hour intervals show a decrease which is due to two main causes; (a) the 'elution loss', which amounts to about 1% of the existing activity per day as ^{51}Cr atoms become detached from the labelled cells, and (b) the actual destruction of the red blood cells. In normal subjects the two effects give an apparent half-period of 30 ± 5 d.

Measurements of the activity in various body organs, as shown by external collimated detectors, can be helpful diagnostically. Usually one and the same detector is used in reproducible positions over the heart (precordium), liver, spleen and sacrum. The subject lies prone for the latter count. Counts are taken at intervals of an hour or two on the first day and once daily thereafter. The precordial counts show essentially the activity levels in the blood and generally fall steadily over the whole period of the investigation. Sacral counts should remain quite constant. Count-rates over liver and spleen usually rise slightly and in normal subjects a ratio of liver/spleen counts is maintained at about 1.2. Subjects suffering from haemolytic anaemia may be recognised by large increases in the liver count-rate (60% or more over a two-week period). Hypersplenism (i.e. the abnormally rapid breakdown of red blood cells in the spleen, where the ^{51}Cr atoms remain) is indicated by excessive spleen count-rates. Various parameters have been suggested as quantitative measures of hypersplenism; one such is the sum of the percentage increase in spleen count-rate and the percentage decrease in heart count-rate at any time, as compared with the initial values. The normal range of this parameter is 0–30%; values in abnormal cases may reach 100%.

No account, however brief, of these techniques would be complete without a mention of the fact that in many subjects there seem to be difficulties in establishing reliable values for count-rates 'at time zero'. Often there are rapid changes in count-rate over the organs, both in the upward and downward direction in the first few hours of counting. This makes the estimation of the proper values of initial count-rates almost a matter of guess-work. In such cases many workers prefer to relate all count-rates obtained subsequently, to their values at 24 hours instead of the time of injection. These remarks apply also to the activity found in blood samples, where again the first one or two values often should be ignored. The cause of the difficulty seems to be that in some subjects the red blood cells are liable to suffer damage of some kind during the labelling process, however much care is taken to control the conditions.

Other valuable information of haematological importance can be obtained by studies with ^{59}Fe (which are often carried out concurrently with the ^{51}Cr tests). This radionuclide, injected intravenously as ferrous citrate, is removed from the plasma with a half-period, in normal subjects, of 1.5 ± 0.3 h. Measurements are best made by withdrawing blood samples at about 30 min intervals, centrifuging, and counting standard volumes of the supernatant plasma. Very much shorter half-periods (as low as 0.5 h) are shown by subjects with iron deficiency anaemia, polycythaemia vera, and some forms of infective and haemolytic anaemias. Larger half-periods (up to 4 h) may indicate aplastic anaemia, acute viral hepatitis, and other conditions.

Organ counting, as for the chromium studies, shows in normal subjects a steady increase, by a factor of 5, in count-rate over the sacrum as the iron is incorporated into bone marrow. Thereafter, sacral count-rates increase more slowly for a day or two and then decrease with a half-period of about 2 d. Meanwhile heart, liver and spleen count-rates rise slowly; some $90 \pm 5\%$ of the injected activity normally appears in circulating red blood cells in 7–10 days. In abnormal subjects increased erythropoietic activity is shown by an even faster rise in sacral count-rates, whereas a slower rise may indicate aplastic bone marrow or erythropoiesis in abnormal sites. Abnormal splenic activity leads to rising splenic count-rates.

It will be realised that the combined ^{51}Cr–^{59}Fe study uses the dual isotope technique of counting, with appropriate channels set for the 320 keV gamma ray of ^{51}Cr and the 1.10 and 1.29 MeV gamma rays of ^{59}Fe. This means that all counts in the 320 keV channel have to be corrected for contribution from the ^{59}Fe. It is a good plan, whenever a combined study is contemplated, to begin with the ^{59}Fe measurements on the first day, completing the plasma-clearance investigation before preparing for the ^{51}Cr work. The ^{51}Cr measurements should be started on the following day. The plan has the advantage that blood samples taken on the second day have only a low count-rate in the ^{59}Fe channel, so that the ^{51}Cr can be counted more accurately.

(*d*) *T-3* (*Resin Uptake*) *Tests*. It is convenient to discuss these here, although they are *in vitro* rather than *in vivo* tests, because they involve the same kind of manipulations of blood and other fluid samples as do the haematological tests and depend on radioactivity measurements carried out with the same apparatus. This is most often a sodium iodide well-crystal; most laboratories find they soon have a workload justifying the use of an automatic sample-changer which will deal with a few hundred samples. There are several ways of conducting the tests, using materials supplied by different manufacturers; they all are based on the principle of isotope dilution.

The T-3 test is performed to get a measure of the capacity of the proteins in the blood plasma to take up tri-iodothyronine (T-3). It was originally done by adding T-3 labelled with radioiodine to a sample of whole blood and finding the resultant equilibrium distribution of activity as between the plasma and the red blood cells. However, the effective capacity of the latter varies somewhat from subject to subject and much more meaningful results are obtained by taking a sample of serum and adding to it a fixed amount of an anion exchange resin such a Amberlite IRA-400. On adding labelled T-3 and mixing well to ensure equilibrium, it is found that the activity divides itself into protein-bound and resin-bound portions, according to the availability of binding sites on the proteins. The ratio of activities in the two fractions can be estimated if we know the initial activity per millilitre of the mixture and measure either the activity captured by the resin or the activity per millilitre of the supernatant (protein fraction) after centrifuging. The latter technique is the one adopted by The Radiochemical Centre (Amersham) and available under the trade-name Thyopac-3. As in most of these tests, the iodine isotope used is ^{125}I. The results can be expressed in standard form if tests are carried out on unknown samples and also on standards prepared from mixtures of blood samples from known normal donors. Then if C_u and C_s are observed count-rates of unknown and standard sera we have

$$\text{(T-3 value of unknown)} = (C_u/C_s) \times \text{(T-3 value of standard)}$$

With the standard taken arbitrarily as 100%, values for the unknown usually lie between about 92 and 118 (normal subjects). Hypothyroid subjects give higher values, because more protein-binding sites are available due to the low concentration of thyroxine (T-4) in the blood of these subjects. High values are also found for pregnant women and on women taking oral contraceptives, not only because of changes in their T-4 concentrations but also because they tend to have increased numbers of binding sites on the plasma proteins. Conversely, low T-3 values may indicate an excess of circulating T-4, which occurs in cases of hyperthyroidism and occupies binding sites otherwise available for T-3. Many drugs, besides oral contraceptives, can affect the T-3 values.

(*e*) *Radioimmunoassay Tests.* The problem here is to measure the concentration of a hormone in the blood-serum, the general technique being known as 'saturation analysis'. We illustrate it by the example of thyroxine (T-4). Typical methods begin with the extraction of T-4 from a fixed volume of serum with ethanol, and then mixing the extract with radioiodine-labelled T-4. To the mixture is added a sample (of fixed weight) of human thyroxine-binding globulin (TBG) a protein which in a short time combines with the T-4 present, both active and inactive, until is is saturated. If it so happens that the total T-4 in the free state is exactly equivalent to the available binding sites in the TBG, then all the activity will also become protein-bound. But if the total T-4 is in excess, only a fraction of it can become bound, the remainder continuing to exist in the free state. Consequently, the activity will be distributed between the bound and free states, with a distribution ratio which is a linear function of the amount of excess T-4. In some techniques, the activity in the bound fraction is measured by adding a specific antigen to the mixture which causes precipitation of a TBG-antigen complex which can be centrifuged off and counted. The higher the count-rate, the smaller the T-4 excess. Alternatively a count can be taken on the supernatant. In either case, by using sera containing known concentrations of T-4 it is possible to construct a calibration curve (which is nearly linear) from which the observed count-rate obtained from an unknown serum can be related at once to its T-4 concentration.

Other methods avoid the somewhat unreliable use of an antigen to precipitate the TBG from the original mixture. If instead, a fixed amount of Amberlite IRA-400 resin is added, a fixed fraction of the free T-4 will be absorbed by it and can be separated from the rest of the mixture by centrifugation. The activity of the resin is now a linear function of the concentration of free T-4, so that again by using standard samples it is possible to construct a calibration curve of resin activity versus T-4 concentration in the serum. In the procedure worked out at The Radiochemical Centre (Amersham) the activity measurement is taken on a sample of the supernatant, rather than the resin. The kits supplied from TRC under the trade-name Thyopac-4 provide two standard samples, one containing about 2 μg/100 ml and the other about 16 μg/100 ml (referred to as the low standard and high standard respectively). Detailed analysis shows that the reciprocal of the count-rate of the final supernatant is nearly linearly related to the T-4 concentration (see Fig. 5.13). For further details, reference should be made to the TRC publication from which the figure is taken.

Normal values of T-4 lie between 3.5 and 12.5 μg/100ml, the limits varying somewhat with the population studied. In SI units we may quote limits of

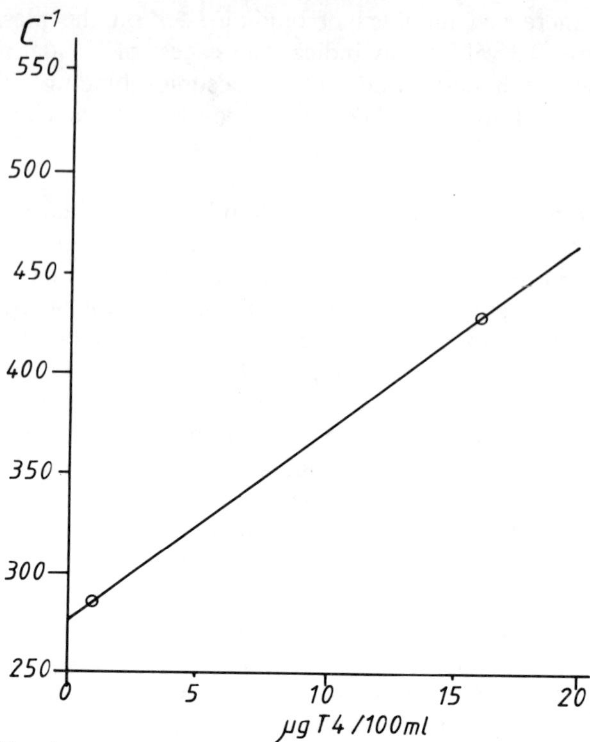

Fig. 5.13 Calibration curve for T-4 concentrations. (Reproduced from Thyopac-4 leaflet, P15/76, by permission of The Radiochemical Centre, Amersham.)

50 to 150 nanomoles T-4 per litre in round figures. Lower values indicate low thyroid activity, and vice versa. For pregnant women and those on oral contraceptives, the upper half of this range is considered normal (6.1–13.7 µg/100 ml or 80–160 nm/l). A diagnostically useful parameter is obtained by evaluating the index (T-4 value) × 100/(T-3 uptake), which may be called the free thyroxine index, or free thyopac index (FTI), as appropriate. This has the same numerical range of normal values as the T-4 concentration itself. Pregnant women and those on oral contraceptives also give FTI values more nearly within these limits because their T-3 uptakes and T-4 values are affected to roughly the same degree. T-3 uptakes and T-4 values may also be modified if the subject is taking certain drugs.

The field of radioimmunoassay is expanding rapidly and procedures are well established for determination of serum concentrations of human placental lactogen, oestriol, vitamin B_{12}, folic acid, cortisol, insulin, digoxin and others. Basically these involve no different principles from the above and we must refer the reader to other texts for their technical details and clinical interpretation.

(*f*) *Schilling Test.* This test as devised by Schilling is no longer performed in the original way as technical advances have improved its reliability considerably. It is concerned with assessing the utilisation within the body of vitamin B_{12} as ingested in the diet. In normal subjects an enzyme called intrinsic factor (IF) is secreted in the stomach lining and has the property of combining with vitamin B_{12} to form a complex which can pass through the intestinal wall and enter the bloodstream, from whence the vitamin is extracted and stored by the liver. In subjects with pernicious anaemia (PA), the IF is not produced by the stomach; in others failure to absorb and store the vitamin is due to a basic inability of the intestinal wall to admit the complexed vitamin (true malabsorption). Anaemic subjects can be investigated by a developed form of Schilling's test, using material obtainable from The Radiochemical Centre (Amersham) under the trade-name Dicopac. In this procedure the fasting subject is given, orally, capsules containing both vitamin B_{12} labelled with ^{58}Co, and a complex of ^{57}Co-labelled vitamin and intrinsic factor. A short time after administration the subject is injected intramuscularly with 1000 units of (inactive) vitamin B_{12}. This constitutes a body burden far in excess of the normal and as it is slowly released from the muscle it causes almost complete elimination of the vitamin from the vascular system into the urine in the next 24–48 hours. Urine collection is in fact performed, as a rule, over the first 24 hours only as this gives sufficiently reliable clinical data. The ^{57}Co and ^{58}Co content of the sample is measured using the dual isotope technique (Section 3.8). Measurements are also carried out on a standard sample prepared from capsules identical with those administered, and dissolved in water containing a little inactive cobalt carrier made up to the same volume as the urine.

Experience shows that with normal subjects about 25% of both ^{57}Co and ^{58}Co administered is collected in the 24 hour urine. Subjects with PA can absorb the complexed vitamin and therefore excrete some 25% of the ^{57}Co; however they cannot absorb the uncomplexed vitamin, although a certain amount of exchange of IF in the stomach undoubtedly occurs, and their excretion of ^{58}Co is low—5% or even less. This means that the excretion ratio ^{57}Co/^{58}Co in their urine is much greater than 1. The critical ratio is 1.3 ± 0.1; subjects with a ratio 1.0 ± 0.3 are classified as normal. Subjects who have a true malabsorption may have a normal ^{57}Co/^{58}Co ratio but the fraction of each isotope excreted will be less than about 10%. As with all these tests, each laboratory should establish its own considered ranges of normality, which will depend on clinical judgment and on the characteristics of the population under study.

IMAGING

5.8 General Principles of Radionuclide Imaging

While uptake measurements aim to assess the activity of a radionuclide in an internal organ taken as a whole, imaging procedures are designed to display the spatial distribution of activity within the organ. We are faced immediately with the difficulty, as in radiography, that practical displays are two-dimensional (images) whereas the organs themselves are three-dimensional (objects). One of the first essentials of imaging devices is therefore that they should be convenient for viewing the subject from more than one aspect, for instance at right angles. The situation is further complicated by the question of what we may call, by analogy with ordinary photography, the depth of focus of the system used. Thus a gamma camera produces an image with least blurring of those parts of an object which are closest to its detector head, more distant parts being progressively less well focused. On the other hand scanners have a rather narrow depth of focus, there being a focal plane at a distance of some 50 or 100 mm from the operating head; parts of the object near this plane are well focused but parts nearer to or farther away from the head give blurred images. We now describe briefly the configuration of typical examples of both these imaging devices before discussing the physics of their focusing properties.

(*a*) *Gamma Cameras.* The most widely used kind of gamma camera features a large (up to 400 mm diameter) sodium iodide crystal which receives radiation from the subject through a multi-hole collimator, electronic circuitry being provided not only for detection but also for positional location of the point of origin of the scintillations within the crystal. The collimator holes are (usually) parallel to each other and at right angles to the plane of the crystal; ideally therefore if a gamma ray originates from a point with coordinates (x, y) in any plane parallel to the crystal it may pass through the collimator and produce a scintillation at a poiint also characterised by coordinates (x, y). It is helpful therefore to visualise a 'scintillation image' within the crystal corresponding to the radioactive object. In some applications the gamma camera may be fitted with a collimator whose holes are divergent; this produces a scintillation image which is smaller than the object. Alternatively a single pin-hole collimator may be used to provide a larger, but inverted, image. These possibilities are illustrated in Fig. 5.14.

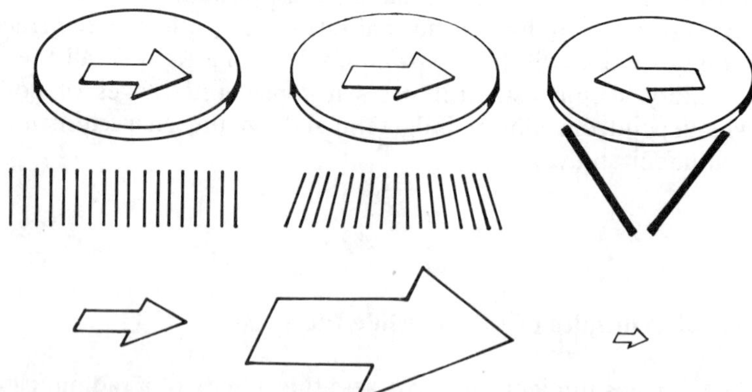

Fig. 5.14 Scintillation images in gamma cameras (the distance between collimators and crystal is exaggerated for clarity).

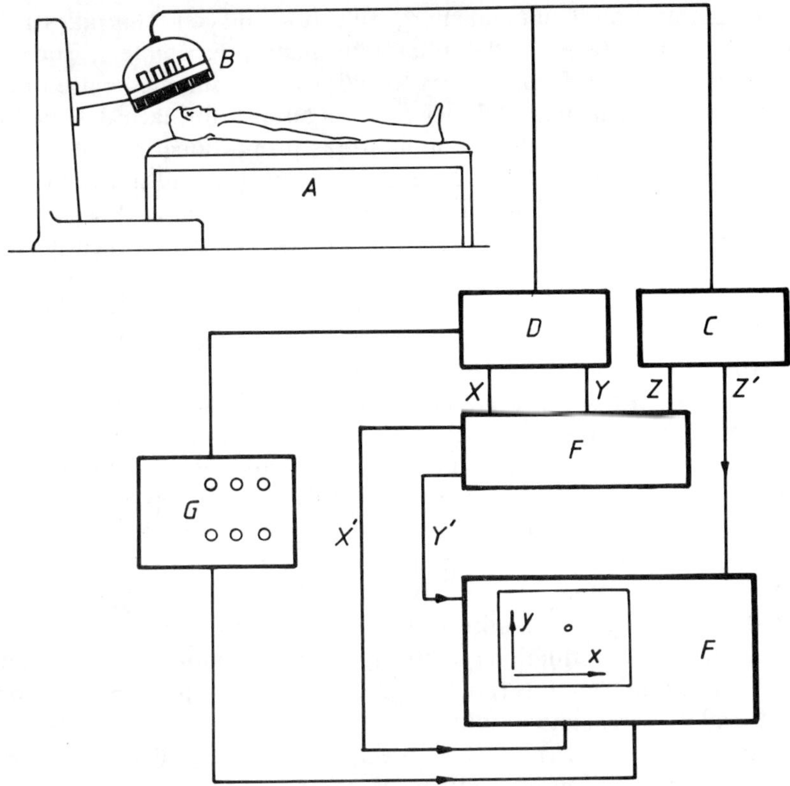

Fig. 5.15 Essential features of a gamma camera.

The layout of the rest of the gamma camera system is illustrated schematically in Fig. 5.15, the essential features being as follows:

1. Examination couch, or other support for the subject (A). Each gamma-camera picture may require an exposure time of 2–5 min so that the subject may need to be supported comfortably in order to restrain involuntary movement. The gamma-camera head (B) can be moved upwards and downwards and also tilted at an angle, so that is posssible to have the subject sitting or even standing; alternatives which are sometimes useful.
2. Detector head (B). This contains the collimated sodium iodide detector and its shielding. Unlike the simpler systems considered earlier, the crystal is optically coupled to a matrix of photomultiplier tubes, of which there may be 19 or 37, or in recent systems 61, in a close-packed hexagonal array. Each photomultiplier tube has its own adjustable H.T. supply and preamplifier and produces a voltage pulse whose amplitude is proportional to the amount of light received, from each scintillation, at its face. The pulse will be larger if the scintillation occurs near the photomultiplier tube and smaller if it occurs at a more distant site.

3. A summing amplifier and energy selection unit (C). This takes as input the signals from all the photomultiplier preamplifiers, adds them together and performs energy selection on the summed pulse just as in a conventional single-tube circuit. Its output, for each detected gamma-ray, is a pulse whose amplitude is proportional to the total light generated by the absorption event (except that it is affected by statistical fluctuations.) This pulse we refer to as a Z pulse. A second pulse, which we will refer to as the Z' pulse is also generated, slightly delayed in time behind the Z pulse, and of a form suitable for application to a storage oscilloscope in order to cause momentary brightening of the display. This means that a bright spot will appear on the oscilloscope whenever a gamma-ray between specified energy limits is detected.
4. A pair of summing amplifiers, called X and Y summers (D). Each of these takes as input the preamplifier pulses and combines these, each multiplied by a predetermined gain factor. For each scintillation detected, the X summer produces a pulse whose amplitude is approximately proportional to the x coordinate of the scintillation, and similarly with the Y summer. The X and Y pulses are, like the Z pulse, affected by statistical fluctuations in the amount of light constituting the scintillation. In addition, the X and Y pulses are affected by the fact that the scintillation quanta are emitted randomly in all directions.
5. A pair of ratio circuits (E). These compare the X and Z pulses, and the Y and Z pulses, and provide new output pulses X' and Y'. These have amplitudes which are more precisely proportional to the x and y coordinates, because the effect of statistical variations in the number of light quanta has been removed. The X' and Y' pulses are 'flat-topped' and are applied to the X and Y input terminals of the shortage oscilloscope (F).
6. A storage oscilloscope (F). This receives the X' and Y' pulses, and during the times they are applied the Z' (bright-up) pulse also arrives, thus creating a bright spot on the screen with the correct (x, y) coordinates. The process is repeated for each successive gamma-ray, the oscilloscope beam shifting to a new position just before the arrival of a bright-up signal. In this way a picture of many stationary dots is built up; its overall density is controlled by the number of dots processed and by their individual brightness. The subject may be positioned in front of the detector with the system running continuously but with frequent cancellation of the record, and the relative positions of subject and camera adjusted until it is seen that the organ of interest is within the field of view. Then a permanent record may be obtained by exposing a Polaroid film to the oscilloscope either for a predetermined time or until a total of say 2×10^5 dots have been recorded. With more sophisticated systems the x and y coordinates of each dot may be encoded and impressed on magnetic tape or disc. Various operations (background subtraction, smoothing and corrections for some forms of

image distortion) can then be carried out under computer control and displays finally produced on video-display units.

7. Control console (G). This unit enables the operator to switch C and D on and off, set the limits on the energy selection circuits, control the overall intensity on the display, and choose time or count limits for the recording. Although the gamma camera is a complex piece of equipment, its control console is usually rather easy to become familiar with and is simpler than that of a scanner. A useful accessory is an anatomical marker, which enables intense additional dots to be impressed on the record to mark the position of reference points on the subject, e.g. the umbilicus, etc., and also to indicate top and bottom, and left and right, where confusion could arise.

(*b*) *Scanners.* In contrast to gamma cameras in which the whole of a body organ is imaged at one time, scanners work on the principle of recording activity serially in time from small regions of the body which are selected according to the pattern of motion of the detecting head. Positional information may be said to be obtained by mechanical rather than by electronic means. The principal features of a typical scanner are illustrated in Fig. 5.16

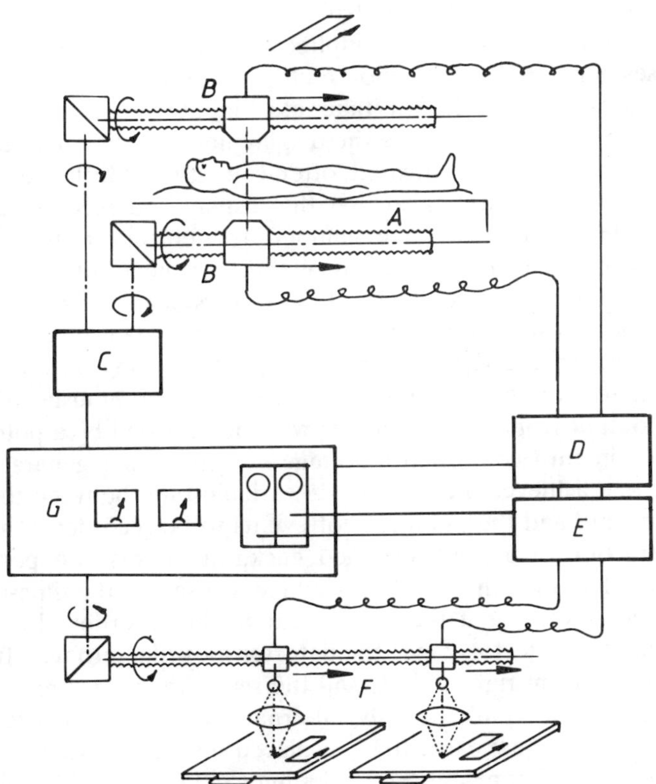

Fig. 5.16 Essential features of a scanner.

and are as follows:

1. Examination couch (A). This is a rather more important item in a scanner than in a gamma camera since the subject has to remain still for rather longer times—typically 15–20 min. Often it forms an integral part of the system, one of the detectors being mounted on it and travelling beneath the subject. It may be fitted with webbing and specially shaped head-rests to restrain involuntary movement of the subject during examination.
2. One or more detectors (B). These consist of sodium iodide crystals each with multi-hole collimators whose holes converge so that radiation from only a small region of tissue is received. If two detectors are used they usually face each other on the same axis, with the subject between them.
3. A system of motor drives (C). These move the detectors on a raster pattern over the area of interest. In some scanners the plane of movement is always horizontal but in others it may be tilted to some extent. In a recent development two detectors may be rotated about an axis lying along the length of the subject. The length of the lines traversed by the detectors, their spacing, and the speed of motion are all under the operator's control.
4. An electronic system for amplification and energy selection of the pulses generated by the photomultipliers in the detector heads (D).
5. A processing unit (E). This takes as its input the pulses from D and has an output suitable to drive the display unit (F). In early scanners this unit was very unsophisticated, often consisting of little more than a set of scale-of-two circuits so that the display unit received a pulse for every 1, or 2, or 4, . . . etc. gamma-rays detected. In this way control could be exerted in a crude way over the frequency of events which the display unit had to cope with. In later versions the problem of adjusting the density of the final record to the density of information gathered by the detectors was attacked by using an analogue rate-meter to monitor the input pulse rate; the continuously variable voltage output from this rate-meter was then divided by a potentiometer which in turn controlled a voltage-sensitive pulse generator. In this way one achieved a continuously variable ratio between the detected count-rate and the pulse-rate of events to be recorded. The effective subtraction of a predetermined background was also possible. The system had the inherent disadvantage, caused by the time-constant of the rate-meter, that the pulses sent to the recorder always 'lagged' behind the detector. Since the detector moved alternately from left to right and from right to left, and the recording device reproduced the same motion simultaneously, alternate lines in the photographic or printed record appeared to be displaced sideways, giving a characteristic 'scalloped' appearance. Scalloping can be reduced by using a rate-meter with a short enough time-constant, but if the time-constant is too short the behaviour of the pulse-generator becomes erratic and the

display unit tends to operate in short 'bursts' which occur sporadically and confuse the record. The modern tendency is to use digital logic circuits instead of analogue systems and to operate on fixed time intervals. This means that in a short interval, say 0.1 s, the number of events detected by the crystal, c_α, is stored and a coded form of this number is passed to the processor. Here a new number, c_p, is calculated and this number of new pulses is sent, equally spaced in time, to the recording unit during the *next* 0.1 s interval. c_p and c_d are of course whole numbers, and the operator may choose to make c_p the nearest whole number to some specified fraction of c_d. But there are other possibilities; for instance he may choose to subtract a constant number from c_d before forming the fraction. In this way the system effectively performs 'background erase'. Also the processor may be set to increase c_p more than proportionately for large values of c_d, which results in 'contrast enhancement' because dense parts of the record will appear even denser than they would otherwise do. The two facilities can be combined, as is illustrated in the graph of Fig. 5.17. With this digital system, scalloping is entirely eliminated because the motor drives to the recorder are arranged to start and to stop just 0.1 s later than those on the detector heads. This exactly compensates for the delay in processing. The use of stepping motors makes precise control of the mechanical movements possible; the system may also include a simple scaling circuit which will cause the motion of the recorder to be only one-half, or one-fifth, of that of the detector. The recorded image will therefore be full size or have a 'minification' factor of two or five.

A display unit (F). This provides the two-dimensional image for visual examination. It may take several forms, some or all of which

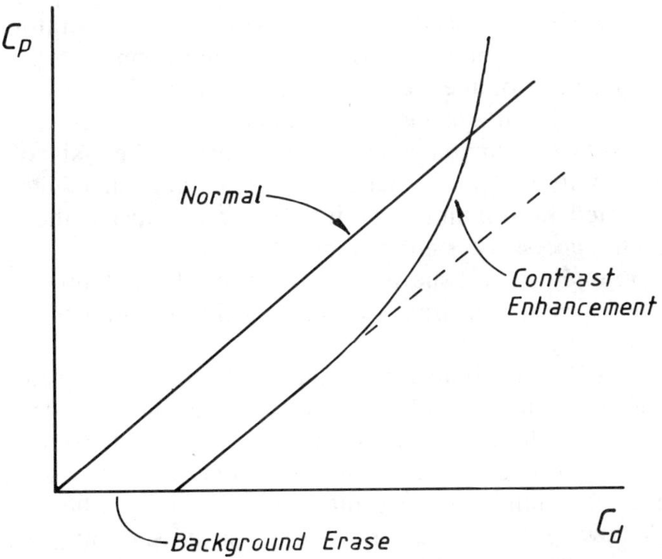

Fig. 5.17 Background erase and contrast enhancement for a scanner.

might operate in parallel on the one machine. The simplest is possibly a persistence oscilloscope which, while not usually giving good enough definition to be clinically useful, is nevertheless invaluable to the operator in ensuring that the whole of the region of interest has been explored. Permanent and higher quality records ('scans') are produced by two main methods. In the first of these a recording head with a built-in source of light is driven over a raster pattern corresponding to the detector-head motion under control of the processing unit. Light from the source is focussed optically and is recorded on standard X-ray film. The finished radiograph is full-size or minified depending on the motor drive speed ratios, with the advantage that a whole-body scan may be accommodated on an ordinary film. The operator has the facility to move the recording head independently of the detector head, so that it is possible to start a record at any point on the film. Thus four views of say the liver (anterior-posterior, posterior-anterior, and left and right laterals at half natural size can be fitted on one film. This not only economises on film but is quite helpful to the viewer in assessing any clinical information present. In the second system a printing head is used and is driven across thin paper sheets with carbon paper for copying. In some of these 'dot scan' devices the printing ribbon is multicoloured, the colour in use at any instant being controlled by an auxillary rate-meter. Thus areas of low count-rate may be printed in blue and areas of high count-rate in red on the top copy. Information is thus conveyed by colour coding as well as the spatial frequency of printed dots. Coloured dot-scans can certainly appeal to the untrained eye but many radiologists seem to prefer 'grey' dot-scans or X-ray films, possibly because of their familiarity with ordinary X-radiographs.

6. A control console (G). From the console the operator can select windows for the photoelectric peak of the radionuclide and set speed of movement of the detector heads in accordance with count-rate observed (on an auxiliary rate-meter) with the detector aligned with the point of maximum activity in the subject. The limits of traverse of the head are under continuous control, so that valuable time need not be wasted in scanning over inactive areas outside the subject. An essential accessory is an anatomical marker.

The diagram (Fig. 5.16) shows the basic layout for a double-headed scanner with photographic recording, with persistence oscilloscopes mounted in the control console.

More sophisticated scanners have facilities for recording scans on computer discs and/or magnetic tapes so that data processing post-scan is possible, and the effects of different background erase and contrast enhancement can be examined on one set of data. Some scanners can record scans from two radionuclides simultaneously present in the subject, together with 'subtraction scans' showing the difference in the two activities. This technique is of value in scanning the pancreas, advantage being taken of the fact that selenomethionine labelled with ^{75}Se concentrates in the liver and also in the pancreas, whereas sulphur colloid labelled with ^{99m}Tc concentrates only in the liver.

5.9 Geometrical Factors in the Design of Gamma-Camera Collimators

As has already been mentioned, gamma-camera collimators are most often parallel-hole structures and we will consider them first. The holes may be cylindrical in section, or square or even hexagonal, depending on the method of manufacture; for simplicity we will consider only circular holes but our general conclusions will apply with only minor modifications to the other shapes. On this basis we may regard the collimator, having N holes, as a collection of N single-hole collimators as previously discussed in Section 5.3, and with $r = r'$. Again taking the depth of the collimator as t, we can gauge some of its elementary properties when viewing a plane source in air at a distance F from Fig. 5.18. It is easy to prove that each collimator hole will

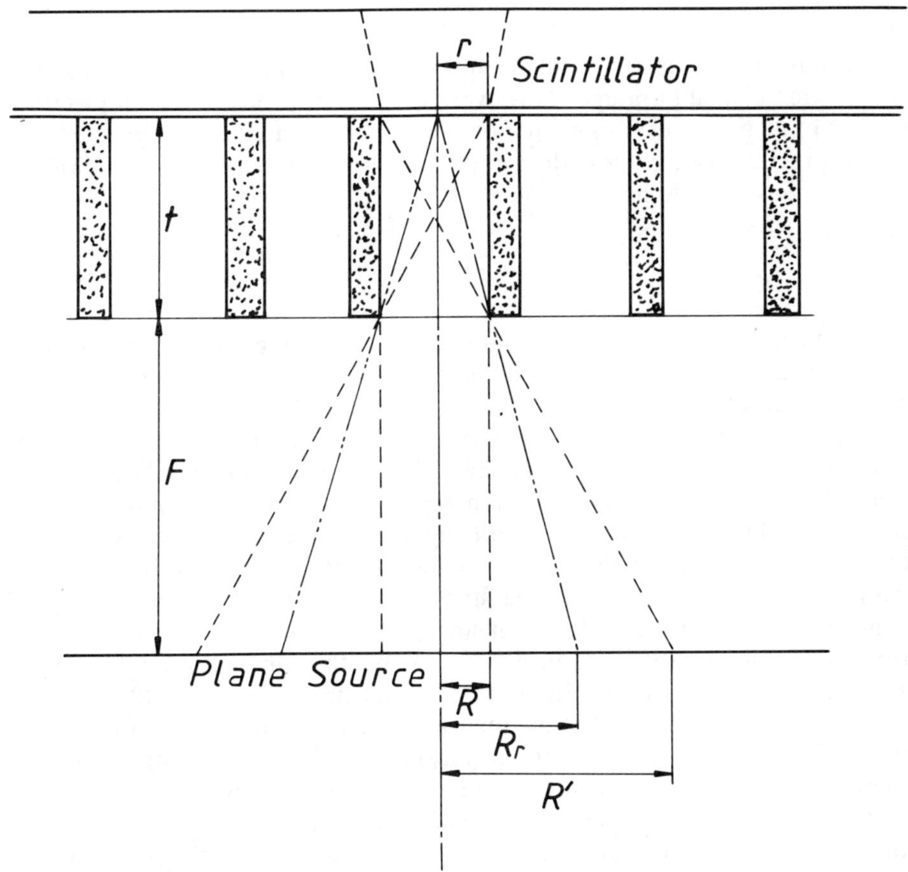

Fig. 5.18 Gamma camera viewing a plane source in air.

have a visual field of radius $R' = r$, a circle of view of radius $R = r + 2Fr/t$ and a resolution radius $R_r = r(t + F)/t$. If we consider any one hole it is clear that a gamma-ray originating within the visual field is almost equally likely to produce a scintillation anywhere within a portion of the detector, which is a slightly divergent cone with base radius r, immediately above that hole. If the coordinates (in the XY plane) of the scintillation are to bear a close relationship with the coordinates of the point of origin of the gamma-ray, it is obvious that r must be as small as possible. Moreover some areas of the plane source within the circle of resolution of one hole may also be within the circle of resolution of a neighbouring hole, or even within its visual field. The circle of resolution must therefore be made as small as possible, which for any fixed value of T means that t must be made large. A practical limit to t is set by the size and therefore weight of the complete collimator; it is commonly some 70–80 mm.

A second consideration is the efficiency of the system. If there are N holes, we shall have a plane-source efficiency of

$$E_g = \frac{N\pi R^4}{4t^2}$$

For a collimator of fixed overall diameter N could be made indefinitely large if R were made small enough. However $N\pi R^2$ is just the total area of cross-section of all the holes; this cannot be as much as the area of the detector face but in practice is rarely less than one-third of it. If then we regard $N\pi R^2$ as roughly constant, we have

$$E_g \propto \frac{R^2}{t^2}$$

from which it is immediately obvious that the requirement for good resolution, of small R and large t, is in direct opposition to the requirement of high plane-source efficiency which calls for large R and small t. In this respect collimator design has to be based on compromise. Another important consideration is that of the thickness of the walls (septum thickness) between the holes. When discussing uptake collimators, we did not lay down any criteria for their wall thickness, as there is no upper limit except for the consideration that the total weight of the system should not be inconveniently large. But with a multi-hole collimator one has to bear in mind the possibility of gamma-rays entering a hole at an angle, penetrating a wall, and passing through a neighbouring hole into the detector, thus giving false information about their point of origin. The problem is illustrated in Fig. 5.19

In case (a) the path of the gamma-ray is in a randomly chosen direction, not necessarily in the plane of the paper, i.e. the plane containing the axes of the neighbouring holes. In this general case the path-length through the wall material is a very complicated function of the point of origin with respect to the pattern of holes and to the angles involved. However, if the path happens to be in the plane defined by neighbouring hole axes, it can be seen that there

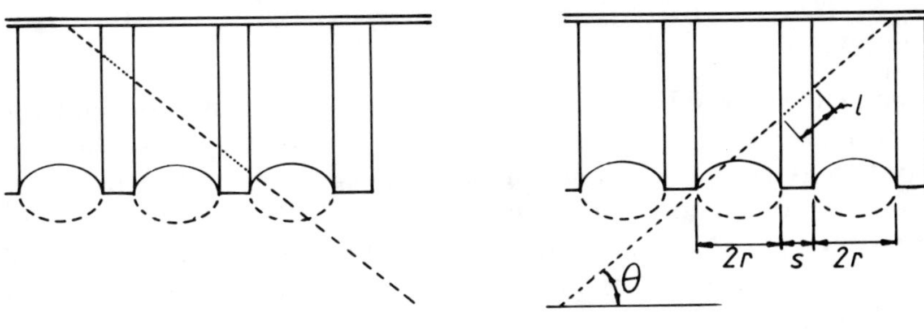

(a) (b)

Fig. 5.19 Septal penetration: (a) general case, (b) least-path case.

is a certain path-length through the wall, illustrated at (b), which is shorter than any other possibility. It can be seen from the geometry of the figure that this minimum length l is related to the septum thickness s by

$$\frac{l}{s} = \operatorname{cosec}(\theta)$$

$$= \frac{1}{\sin(\theta)}$$

$$= \frac{\sqrt{[t^2 + (s + 4r)^2]}}{s + 4r}$$

i.e.

$$l = \frac{s}{s + 4r}\sqrt{[t^2 + (s + 4r)^2]}$$

$$= \frac{st}{s + 4r}\left\{1 + \left(\frac{s + 4r}{t}\right)^2\right\}^{1/2}$$

$$= \frac{st}{s + 4r}\left\{1 + \frac{1}{2}\left(\frac{s + 4r}{t}\right)^2 + \cdots\right\}$$

where we have expanded the expression raised to the power $\frac{1}{2}$ by the binomial theorem, writing only the first two terms.

Quite often in practice t is several times as big as $s + 4r$ so that the second and all later terms in the expansion are small compared to unity and to a fair approximation

$$l \approx \frac{st}{s + 4r}$$

whence

$$s \approx \frac{4lr}{t - l}$$

It is instructive now to insert some numercial values and find what dimensions these formulae imply. It is clearly wise to specify l so large that septal penetration is acceptably small; a practical figure for the penetration of gamma-rays in the geometry (b) may be taken as 2%. This is of course the most favourable geometry for penetration, so that the average overall gamma-rays will be much less than this figure. We will assume exponential absorption, i.e.

$$\frac{I}{I_0} = \exp(-\mu l)$$

where μ is the linear absorption coefficient (see Section 3.11). If we design a collimator for use with ^{113m}In, for which the gamma-ray energy is 390 keV, we find μ in lead to be about 0.025 mm^{-1}. If then we choose l so that $\mu l = 4$, because $\exp(-4) = 0.018 = 1.8\%$, we find $l = 16$ mm. It is clear from the geometry that t has to be much greater than this; taking a typical value $t = 70$ mm and inserting these values for l and t in the formula for s we find $s \approx 1.2r$. Thus, for some arbitrary value of r, say 5 mm, we estimate the required septal thickness to be 6 mm. These measurements will obviously limit the number of holes we can accommodate on a crystal of, say, 300 mm diameter. The number, n, of holes on a diameter must in fact be equal to or less than $300/(2r + s) = 300/16$, i.e. 17 (the largest possible odd number). The total number N is then

$$N = \frac{3n^2 + 1}{4} = 217$$

using a formula for the number of holes in a close-packed array within a circle having n holes on a diameter. As an illustration of a close-packed array for $n = 7$, see Fig. 5.20.

We are now able to calculate the plane-source efficiency E_g for any required value r, and assuming that the maximum number of holes has been used. The results are shown graphically in Fig. 5.21. The graph also shows E_g values calculated for ^{99m}Tc gamma-rays (140 keV). For these, the absorption coefficient in lead is 2.5 mm^{-1} which on the same assumption of $\mu l = 4$ gives $l = 1.6$ mm and $s \approx 0.094r$. We may conclude from the graph that for any given hole size a collimator for ^{99m}Tc can have at least twice the plane-source efficiency as a collimator for ^{113m}In, assuming the same septal penetration factor, because of the larger number of holes allowable. Alternatively one may say that for collimators with equal plane-source efficiency a ^{99m}Tc collimator can have holes with a diameter of only 0.6 times that of the holes in a ^{113m}In collimator, which implies that much better performance in terms of

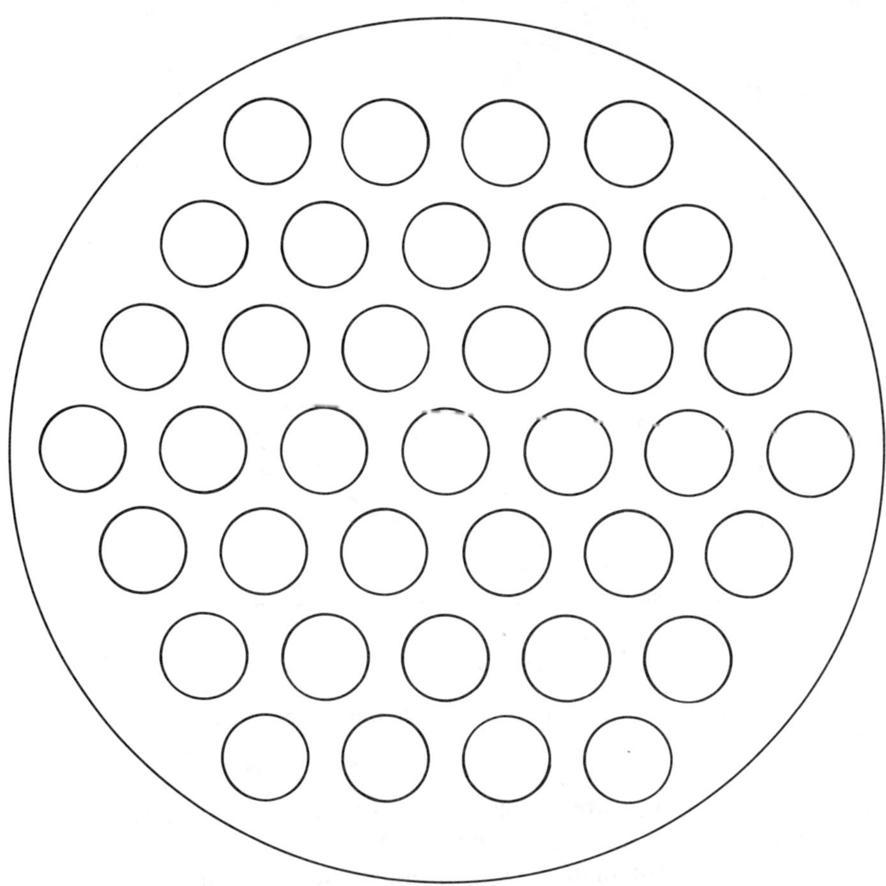

Fig. 5.20 Close-packed array for $n = 7$ ($N = 37$).

resolution in the image. In this connection it should also be borne in mind that ^{99m}Tc has the advantage of giving relatively little radiation dose, per megabecquerel injected, to the subject. Therefore quite low E_g values are acceptable and, indeed, commercially available ^{99m}Tc collimators may have r as small as about 2 mm. For these, the calculated septum thickness becomes so small that manufacture to this specification becomes impossible—the collimator would not be self-supporting. One popular way of forming the structure is to press a set of parallel crimps in sheets of 0.5 mm thick lead, then to bind sets of sheets together so that the crimps form roughly triangular channels. The reproduction of fine detail in the object then becomes more a matter of how well the electronic part of the system can determine the X and Y coordinates, than of the collimator performance.

On the other hand, a collimator for ^{113m}In, as we have seen, might well have $r = 5$ mm—rather large holes in order to get a high enough plane-source efficiency—and a septal thickness of 6 mm, which is so high that the pattern of collimator holes can be seen superimposed on the image. The

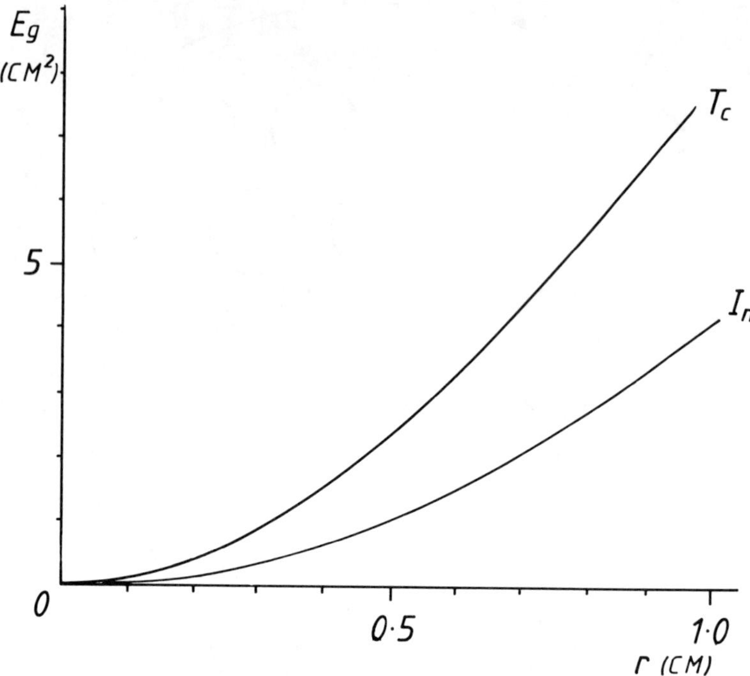

Fig. 5.21 Plane efficiency as a function of hole radius, for radiation from ^{113m}In and ^{99m}Tc.

situation for positron emitters such as ^{18}F is, of course, even worse because of the annihilation radiation energy of 510 keV.

Gamma-camera collimators with divergent holes are useful in enabling images of large organs to be formed. Their geometry is rather more complicated than the above simple treatment but it is evident that good-quality imaging, as far as resolution is concerned, will only be possible with low-energy gamma emitters. The images obtained are sometimes a little hard to interpret since parts of the object nearest to the camera are magnified to a lesser extent than are distant parts.

5.10 The Localisation of Scintillations in Gamma-Camera Crystals

The collimator of a gamma camera ensures that the coordinates in the XY plane of the point of origin of a gamma-ray correspond to the coordinates of its point of entry into the crystal, although with an uncertainty which is of the order of the hole size. We now turn to an elementary consideration of the uncertainties inherent in the localisation of the scintillation caused by the ray.

The first possible source of uncertainty lies in the fact that the light quanta comprising the scintillation do not originate at a point but along the track of the photoelectrons and/or Compton scattered electrons. (We are here neglecting the ranges of the low-energy secondary electrons and the distances between them and the luminous centres in the crystal; but all these dimensions are only one or two orders of magnitude greater than atomic spacings and are negligible in the present context.) If the gamma-ray were from ^{99m}Tc most of the photoelectrons would have energies of 140 keV, which would give them ranges (in sodium iodide) of the order of 0.01 mm. Even for annihilation radiation of 510 keV, the photoelectrons would have ranges of only some 10 times this. If all the interactions were by photoelectric absorption, one could conclude that the positional uncertainty of the scintillation due to the non-zero range of the photoelectrons is quite negligible. However a fraction of the interactions—depending on the gamma-ray energy—involve Compton scattering so that the light quanta are generated along the tracks of two or more primary electrons. For instance, for 140 keV gamma-rays, Fig. 3.28 shows that the photoelectric attenuation coefficient in sodium iodide is of the order 10 times that for Compton scattering, which means that the relative frequency of Compton events is about 10% of the whole. Now a Compton gamma-ray scattered at 90° will carry an energy of 100 keV and will have a half-thickness for absorption in sodium iodide of some 4 kg m^{-2}, i.e. about 1 mm. However, by no means all the scattering occurs at nearly 90° and in a good many cases the scattered photon will diverge from the primary track more nearly in the forward or backward direction. It is therefore reasonable to expect most of the energy absorption within 1 mm from the primary track. Taking the photoelectric and Compton events together, it seems highly probable that the fraction of the light emitted from sites further than 1 mm from the original gamma-ray track is less than 1%. In the present application, this is probably negligible.

On the other hand for, say, annihilation radiation (510 keV), Compton events* are more than three times as frequent as photoelectric absorption events; also Compton gamma-rays scattered at 90° carry 350 keV of energy giving them a half-thickness value for absorption in sodium iodide of 50 kg m^{-2} or about 14 mm. Again the Compton scattered gamma-rays are emitted with a distribution of all angles but it is clear that a not-inconsiderable fraction of the total light generated by them when they are absorbed will originate at sites as much as 10 mm from the original gamma-ray track. The process is illustrated in Fig. 5.22.

A second kind of consideration is concerned with the statistical fluctuations in the numbers of scintillation quanta available at the photomultiplier faces. For gamma-rays from ^{99m}Tc there are about 5000 quanta per scintillation on average and these are 'shared out' among several photomultiplier tubes. Any tube may, therefore, receive a much smaller number, so that for a series of scintillations all at one fixed site, the pulses it produces will be small and will have a relatively large spread in amplitude. The following mathematical

* The Compton scattering coefficient is three times the photoelectric coefficient. This means that Compton events dissipate three times as much energy as photoelectric events; since less energy is deposited in a Compton event than in a photoelectric event, the actual ratio of *events* is considerably more than three.

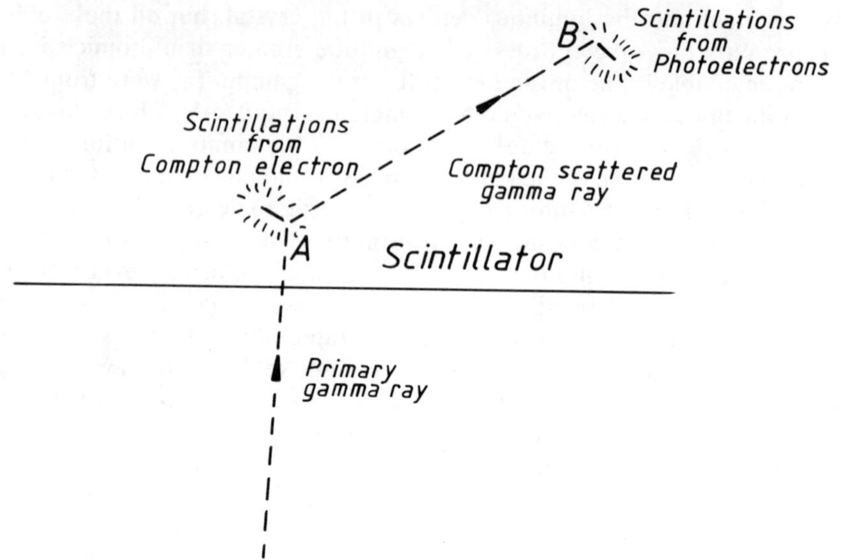

Fig. 5.22 Origins of light quanta in scintillations.

treatment does not go very far but supplies a semi-quantitative idea of the magnitude of this effect.

Suppose we consider the face of a photomultiplier tube, of radius b, in contact with a scintillator; a scintillation occurs at a point P which is at a depth a and a horizontal distance p from the tube axis (Fig. 5.23). We take the origin of coordinates at the centre of the photomultiplier face, so that P lies along the negative direction parallel to the X axis. We choose a point C within the photomultiplier tube face, with coordinates x and y, and draw a small element of the area Δx by Δy centred at C. The normal to the XY plane at C makes an angle θ with the line PC. This means that, viewed from P, the

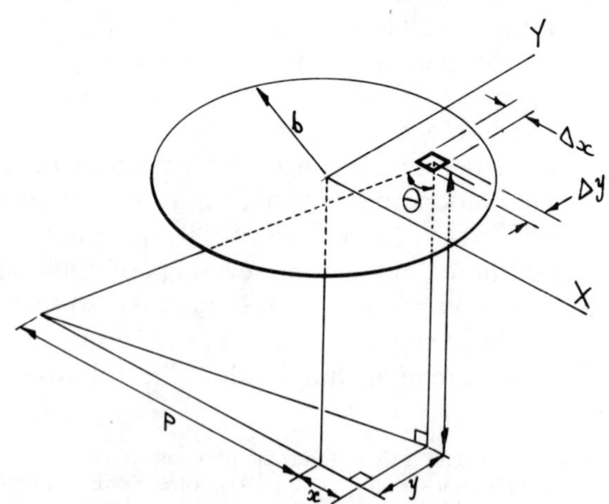

Fig. 5.23 Geometry of scintillations in a gamma camera.

element is tilted through an angle θ and the projection of its area at right angles to PC is $\Delta x \Delta y \cos(\theta)$. The solid angle subtended at P is therefore

$$\Delta\Omega = \frac{\Delta x \Delta y \cos(\theta)}{(\mathrm{PC})^2}$$

$$= \frac{\Delta x \Delta y a}{(\mathrm{PC})^3}$$

$$= \frac{\Delta x \Delta y a}{\{(p + x)^2 + y^2 + a^2\}^{3/2}}$$

by applying Pythagoras' theorem twice to the right-angled triangles marked in the figure.

Now we can find the whole solid angle subtended by the photomultiplier face by adding together the elementary angles for all possible positions of C within the circle. In the limiting case of Δx and Δy indefinitely small, this is expressed as a double integral

$$\Omega = \iint \frac{a dx dy}{\{(p + x)^2 + y^2 + a^2\}^{3/2}}$$

where the integration limits are set for x between $-b$ and $+b$ and for y between $-\surd(b^2 - x^2)$ and $+\surd(b^2 - x^2)$. This double integration can be evaluated analytically only in the special case of $p = 0$, but for non-zero values of p we have recourse to numerical methods which will give Ω as a function of p to any required accuracy, for fixed values of a and b. Using these and taking the typical values $a = 10$ mm and $b = 25$ mm, we get results presented graphically in Fig. 5.24. The figure shows that if a scintillation occurs at $p = 0$ (i.e. on the axis of the tube), the solid angle is almost exactly 4 steradians, so that if the light from the point scintillation is uniformly directed in space, with no allowance for reflections, the fraction which would enter the tube is $4/4\pi = 0.32$ or 32%. Sideways displacement of P by as much as 10 mm would have rather little effect. However if P were immediately below the edge of the tube ($p = 25$ mm), while the solid angle is only 2 steradians, the slope of the curve is at its greatest. Thus the fraction of light received is 16% but its rate of change with distance is about 0.16 steradians per millimetre, or about 8% of its value per millimetre.

In Section 4.19 we gave a formula for the FWHM of a photoelectric peak in a sodium iodide crystal, the spread being due to the statistical fluctuations in the number of light quanta. This reads

$$\mathrm{FWHM} = 2.36\sqrt{\left[\frac{E}{33\bar{f}\bar{p}} \cdot \left(\frac{n}{n - 1}\right)\right]}$$

where $\bar{f}$ is the fraction of light collected and was taken to be approximately 1 for a crystal encapsulated in reflecting material. For the present case this fraction is however $2/4\pi$, and with $n = 3$ and $p = 0.1$ as explained before we

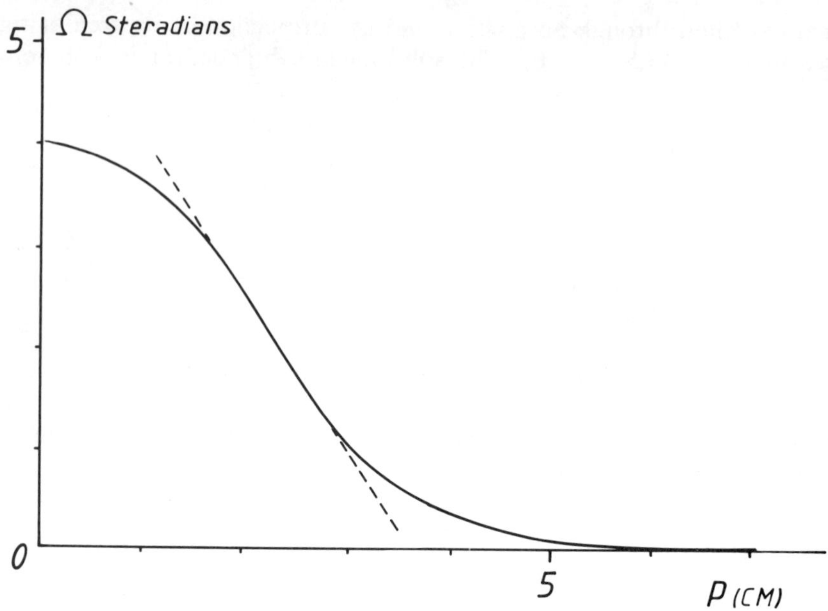

Fig. 5.24 Solid angle Ω as a function of p.

find

$$\text{FWHM} = 2.66\sqrt{E}$$

Thus for ^{99m}Tc gamma-rays we find a FWHM of 32 keV, or a relative FWHM of 32/140 = 0.23 or 23%. Comparing this with the effect of lateral displacement of the point of scintillation, we see it corresponds to a displacement of 23/8 = 3 mm approximately. What this means is that when the tube generates a pulse of such a size that a scintillation appears to have originated at $p = 25$ mm, this distance is in fact uncertain to the extent of 3 mm in FWHM, which corresponds to a standard deviation of about 1.3 mm.

If a scintillation actually occurs in a gamma-camera crystal close to the edge of a photomultiplier tube, it may well be located somewhat more precisely than our estimate because of additional information from a neighbouring tube. On the other hand, if it occurs much nearer to the axis of one tube it will not be located so precisely by that tube and the information available from other tubes will also be meagre because their light collection will be low. The performance of the whole system will obviously be quite complicated, depending on the spacing between photomultipliers and their pattern of arrangement as well as the site of the scintillation. Actual measurements with very narrowly collimated sources seem to show that there is an 'intrinsic' resolution in the latest designs of gamma camera of 3–4 mm for the gamma-rays from ^{99m}Tc. For gamma rays from ^{113m}In, a similar calculation suggests a resolution of 2–3 mm—it is smaller because of the larger light output per scintillation. From the point of view of statistics, therefore, the gamma camera's resolution is best for high-energy sources. Our previous discussion showed that low-energy sources would be preferable because the generation of the scintillation photons is more closely confined to the path of the gamma-rays in the crystal. There is an optimum value of the gamma-ray energy required for best resolution; this is generally thought to be about 200 keV, but will obviously depend to some extent on the design parameters of the gamma camera.

5.11 Geometrical Factors in the Design of Scanner Collimators

In a radionuclide scanner the purpose of the collimator is to delineate a small volume within an extended radioactive source; gamma radiation from the chosen volume is to be detected by a sodium iodide crystal immediately behind the collimator, radiation from points outside the volume being rejected. In its simplest but perhaps most effective and certainly most generally used form, the collimator comprises a matrix of conical holes, hexagonally close-spaced, in a cylindrical lead block. The axes of the holes meet at a common point P which is also the apex of all the cones (see Fig. 5.25). In what follows we will assume this geometry, with holes circular or nearly so in

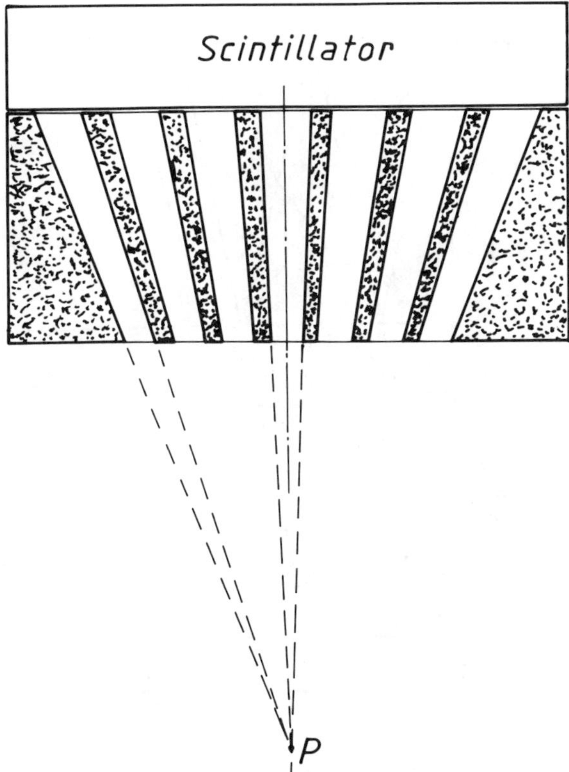

Fig. 5.25 Section of scanner collimator.

section, although some manufacturers produce collimators with hexagonal-shaped holes which give somewhat higher sensitivity. Also for some special purposes it is useful to have a collimator whose inner rings of holes are directed to a focal point closer to the collimator face than that for the outer holes. However, consideration of the simpler forms gives insight into all the major design features.

For all practical collimators it is possible to define a point on the axis such that a 'point source' placed there, in air, gives a higher count-rate than it would do at all other points. This we naturally call the *focus*; for simple collimators it is for practical purposes identical with the apical point P defined geometrically above. The plane through the focus and perpendicular to the axis is the focal plane; its distance from the front face of the collimator is the focal length f. Note the distinction between the focal length f and the distance we have denoted previously by F, which is the distance between the collimator face and *any* chosen plane. We may now redraw Fig. 5.4 with reference to the central (axial) hole of the collimator and the focal plane, and obtain Fig. 5.26. (We have very much exaggerated the hole radius in comparison with the thickness t in order to make the geometry clear.)

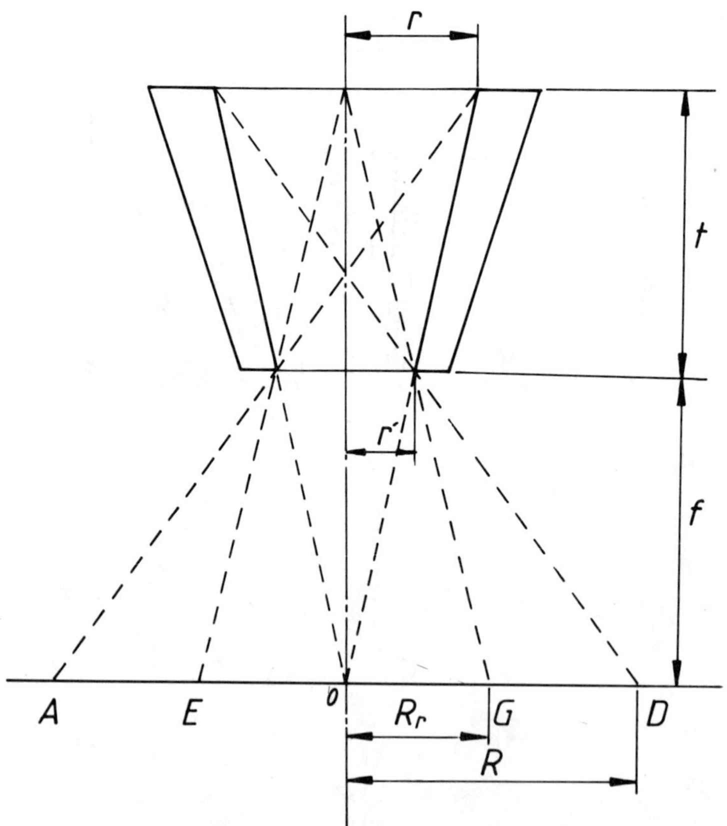

Fig. 5.26 Geometry of central hole in scanner collimator.

In the focal plane we see that the visual field has shrunk to the focal point P itself and that the circle of resolution EG has exactly half the diameter of the field of view AD. For the resolution radius EG/2 we have

$$R_r = \frac{r'(t + f)}{t}$$

or alternatively

$$R_r = \frac{rf}{t}$$

It will be noticed that for a focused collimator, which has a unique focal length f, the resolution radius can be understood as a parameter of the collimator, the implication being that the circle of resolution is that in the focal plane. (For a non-focusing collimator, for which $r' \geq r$, one always has to *specify* the distance F at which the circle of resolution is to be drawn.)

When we come to consider the other holes in a collimator, we find the geometry is complicated by the fact that they are inclined at an angle to the axis. Thus in Fig. 5.27, the central hole and one in the outer ring are shown in section. In the diagram the dimension D is the outer diameter of the collimator; the crystal diameter is a little larger. We have adopteed the convention—which is done to simplify some of our calculations—that the off-axis hole has

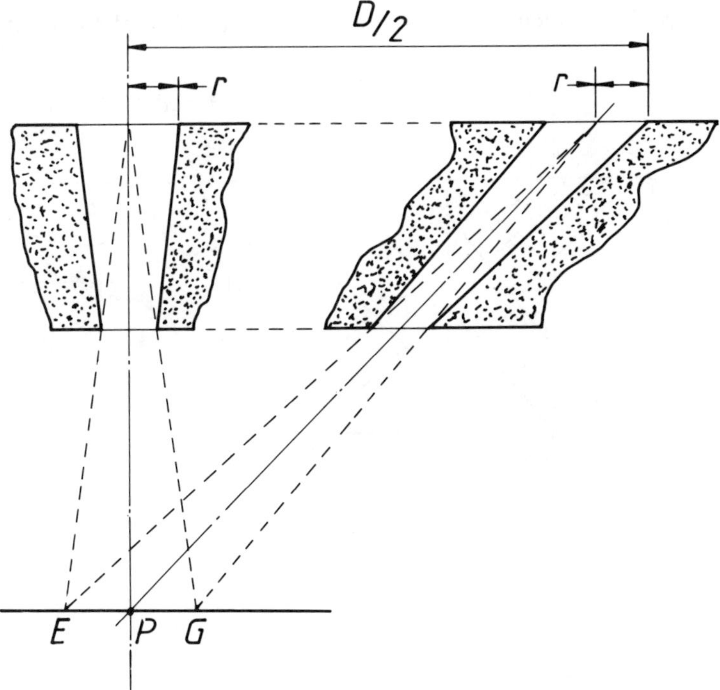

Fig. 5.27 Central and angled hole at edge of a focusing collimator.

a 'width' of $2r$ at the top and $2r'$ at the bottom, as for the central hole. This convention ensures that all holes view the same resolution circle with diameter EG. But these widths are not perpendicular to the axes of the holes, and as drawn the holes would be elliptical in cross-section. Real collimators are however usually made with all the holes circular in cross-section because of greater ease of construction. The edge of the off-axis hole as seen at the collimator face is elliptical, as is illustrated in Fig. 5.28.

The length of the semi-minor axis of the holes as seen at the face is still r but the semi-major axis is $r/\cos(\theta)$ or $r \sec(\theta)$, where θ is the angulation of the hole axis. In practice θ is rarely greater than 30° even for the outermost ring of holes, so that the semi-major axis is usually less than or equal to $1.15r$. Clearly, angulated holes will have an ellipse of resolution in the focal plane, rather than a true circle, and its semi-major axis will be as much as 15% larger than R_r. This means that the effective resolution of the collimator, taking all the holes into account, will be a few per cent worse than it would be if all the holes were strictly equivalent to the central one.

Now let us consider the plane-source efficiency. For the central hole we have

$$E_g = \frac{\pi r^2 r'^2}{4t^2}$$

For tilted holes we should replace t by $t' = t \sec(\theta)$, which will make E_g smaller by a factor $\sec^2(\theta)$. On the other hand, the tilted holes view an ellipse, in reality, which has an area greater than that of the resolution circle

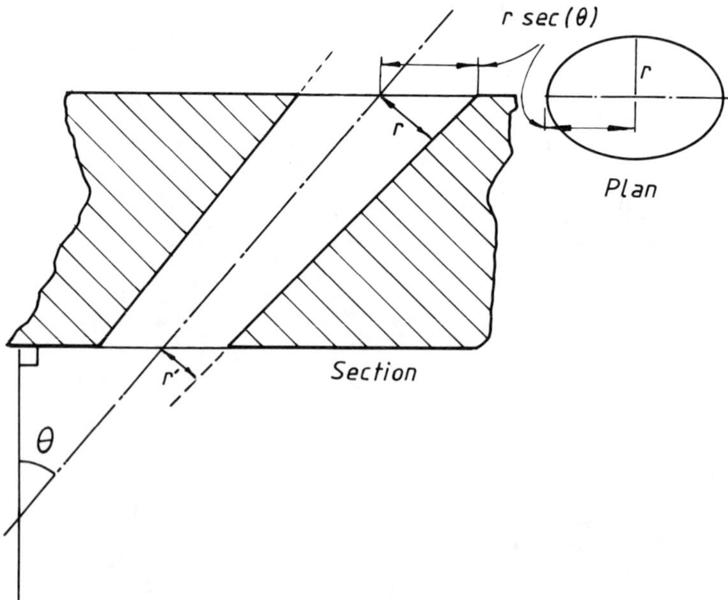

Fig. 5.28 Geometry of off-axis hole.

by a factor sec (θ). For the outer holes, therefore, the correct plane-source efficiency is

$$E_g = \frac{\pi r^2 r'^2}{4t^2} \cdot \cos(\theta)$$

Again if in practice θ is always less than about 30° the error in replacing cos (θ) by 1 is less than 15%. Taking all the holes into consideration we conclude that we shall be in error by less than about 10% if we neglect the effect of the angulation, and that for N holes in all the total plane-source efficiency is

$$E_g = \frac{N\pi r^2 r'^2}{4t^2}$$

We can usefully substitute for r and r' in this formula; for by simple geometry

$$\frac{r'}{r} = \frac{f}{t + f}$$

and with

$$R_r = \frac{rf}{t}$$

we find easily

$$E_g = \frac{N\pi R_r^4 t^2}{4f^2(t + f)^2}$$

An alternative form, written in terms of the radius of the circle of view R is

$$E_g = \frac{N\pi R^4 t^2}{64f^2(t + f)^2}$$

a formula sometimes called the Beck formula after its originator.

As with gamma-camera collimators, an important parameter is the septal thickness s, which we here take to mean the thickness of lead wall between neighbouring holes as seen at the collimator face. The septa are of course slightly tapered and a somewhat better formula for the minimum path-length l (see Section 5.9) than that given for parallel holes is

$$l = \frac{st}{s + 4r}\left\{\frac{4f^2 + 4ft}{4f^2 + 4ft + t^2}\right\}$$

where l is a mean value for the minimum path-lengths taken over all the septa. As will be seen from Fig. 5.29, the minimum path-lengths depend on the tilt of the septa and on whether the gamma-ray is entering an inner hole and penetrating into an outer hole, or vice versa. The extra multiplying factor in this formula is always less than 1 and is generally greater than 0.9 because t is generally less than f in practice. Nevertheless, it may be taken as 0.9 to give

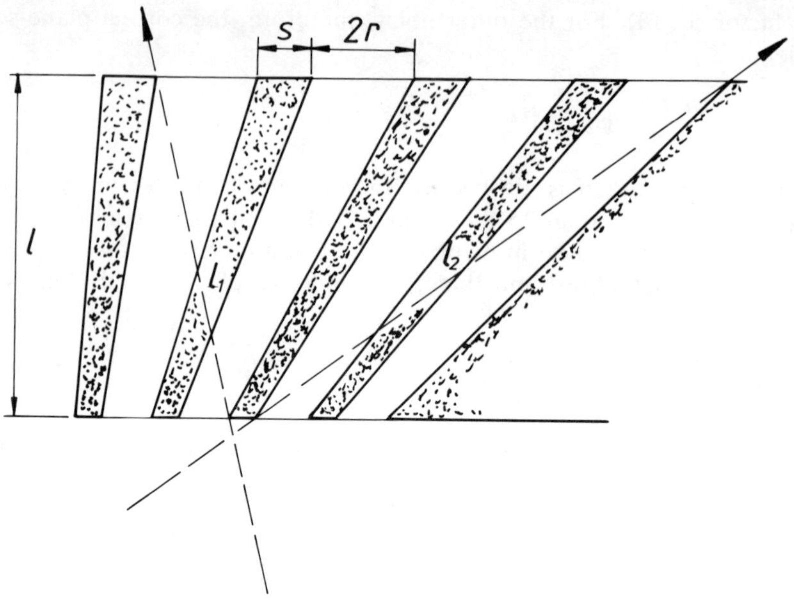

Fig. 5.29 Septal penetration in a focusing collimator: l_1 is less than l_2, l_2 is slightly more than it would be in a parallel-hole collimator with same values for r, s and t.

a conservative estimate for s, namely

$$s = \frac{4lr}{0.9t - l}$$

In working out the septal thickness required for a collimator for ^{99m}Tc, it is immaterial whether this formula is used or the simpler one for parallel holes. But for gamma-rays from ^{113m}In, using the dimensions $r = 5$ mm and $t = 70$ mm, as we did for a gamma-camera collimator, we find $s = 7$ mm—a millimetre greater. The necessity for slightly greater septal thickness, together with the fact that the holes in the outer rings are elliptical, means that rather fewer holes can be accommodated in the collimator than would otherwise be possible. Another point of interest here is that radiation passing through the outer ring of holes will enter the very edge of the detecting crystal and is likely to be detected with diminished efficiency, especially if it is of high energy. As is the case with the gamma camera, there are therefore a number of reasons for preferring ^{99m}Tc to ^{113m}In and other higher-energy isotopes for imaging.

A very important feature of scanner collimators is the so-called depth of focus. This, by analogy with the optical case, means the distance, both above and below the focal plane, within which spatial variations in radionuclide concentration will be acceptably well reproduced in the scan image. To an extent the concept is subjective, but it can be placed on a semiquantitative basis by the following argument. Referring to Fig. 5.30, we show a collimator for which we have drawn the axes of the central hole and of two holes in the outer ring, meeting at the focal point P. The circle of resolution is EG as

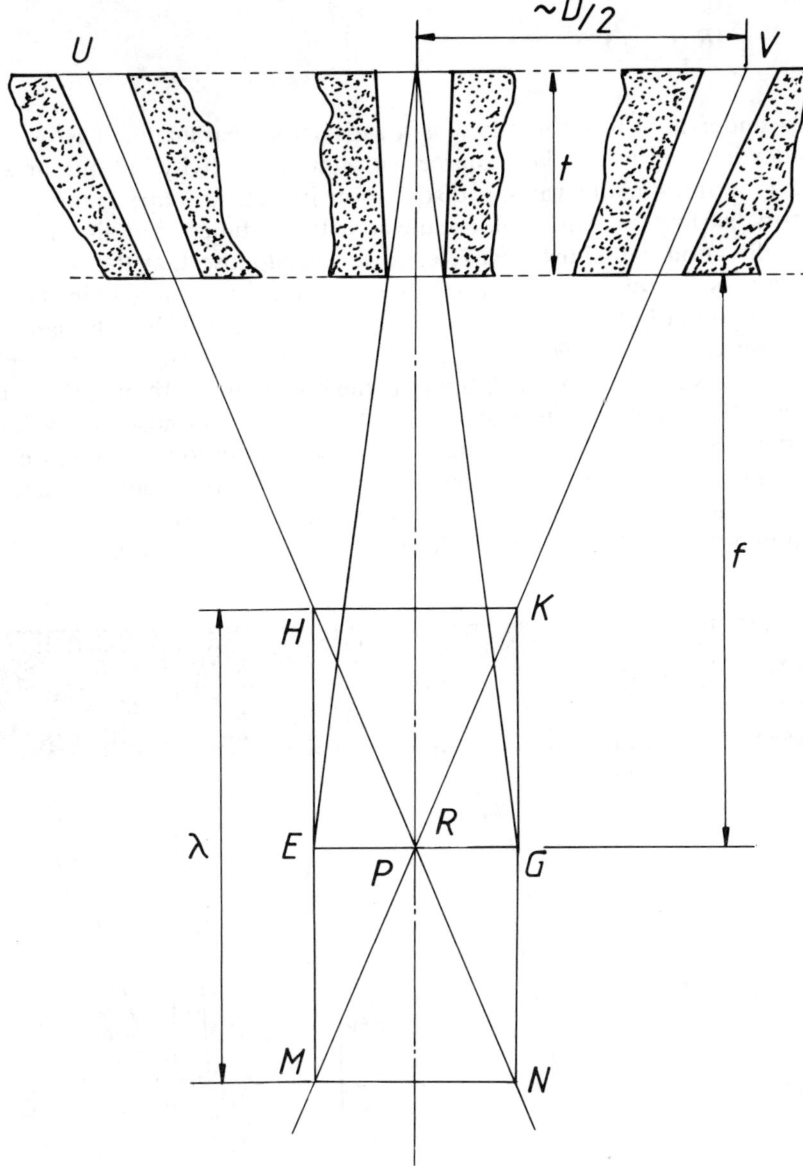

Fig. 5.30 Definition of depth of focus.

before and we have completed a rectangle HKNM, where H, K, N and M are points on the axes and HK = EG = MN. Then HM = KN = λ is defined to be the depth of focus. To a good enough approximation the distance UV in the figure is equal to the collimator diameter D, and we easily have from geometry that

$$\frac{\lambda}{\mathrm{HK}} = \frac{t + f}{D/2}$$

or

$$\lambda = \frac{4R_r(t + f)}{D}$$

To understand in what way λ is a satisfactory parameter it is useful to consider imaging not a plane source with uniform activity but one in which there is a sinusoidally varying distribution of activity superimposed on a uniform distribution. Thus a plot of the activity as a function of distance in say the X direction will show a waveform; let us suppose that the length of a half-wave is just equal to the resolution. Then if the source is in the focal plane, the outer holes and the central hole will all be 'viewing' the same part of the source—say a peak in the waveform, as in Fig. 5.31a. But if the plane source is moved a distance λ/2 towards the collimator, although the central hole will still respond to the peak in the activity, the outermost holes will view almost exactly as much of a 'trough' as the peak. Moreover, if the source is moved sideways in the X direction, as a peak enters the resolution circle of one outer hole, a trough will enter the circle of the other one. A similar situation occurs if the source is displaced a distance λ/2 below the focal plane.

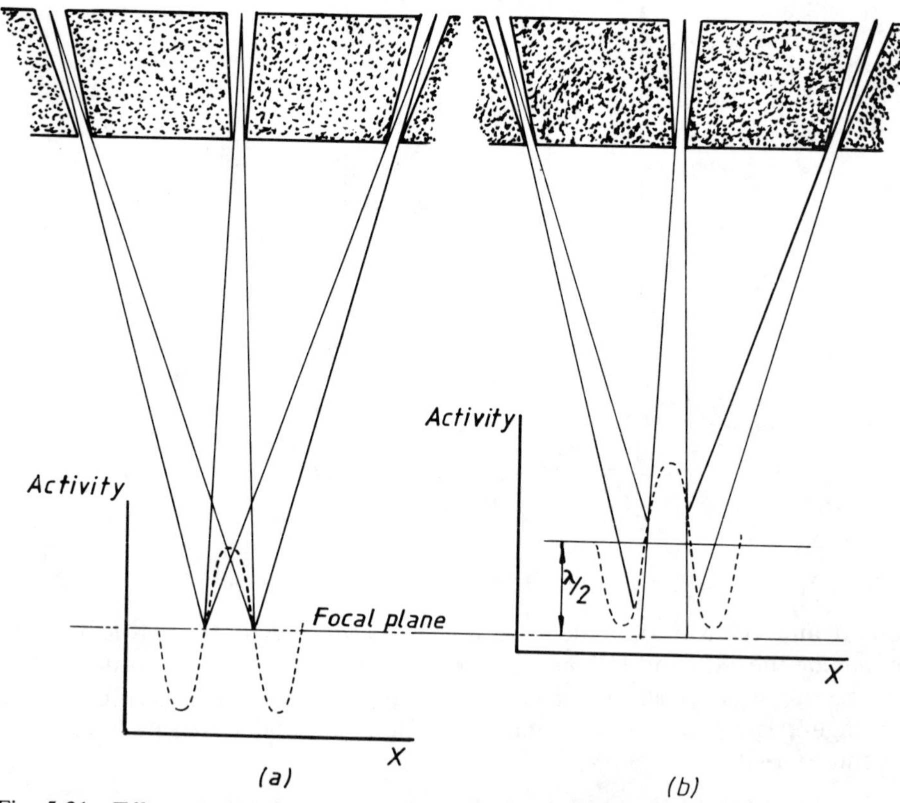

Fig. 5.31 Effect of raising a plane source, with sinusoidal activity distribution from the focal plane (a) to a position λ/2 above the focal plane (b).

With either of these displacements, therefore, the outermost pair of holes is failing to give us any information about the structure of the source; as far as they are concerned the source is indistinguishable from a uniform distribution.

The same argument of course applies in the Y direction, and we can see that if we had a source consisting of small 'islands' of high activity surrounded by regions of low activity, the whole outer ring of collimator holes will add nothing to the quality of the image when the source is outside the depth of focus, but only add to the general blurring. It is clear too that intermediate holes are affected to a lesser extent, and in clinical practice one gets the general impression that the focusing in a scanner is rather 'sharp'. This has advantages; thus in lateral views of the brain it is helpful to observe whether a lesion appears well defined in the view taken from the left side but is diffuse on the other view, because this would indicate abnormality in the left hemisphere. If the two views are nearly equivalent there may be an abnormality nearer the mid-line. For other organs, e.g. the liver, wherein it is more important to demonstrate whether an abnormality exists or not than to show exactly where it is, a long focal depth is preferable. A scanner with a large (125 mm) crystal might then be fitted with a collimator whose outer rings of holes have a longer focal distance than the inner rings. Alternatively, of course, a gamma camera should be used, as this in the present context has an infinite depth of focus.

5.12 Effects of Absorption and Scattering on Gamma-Camera and Scanner Images

So far in our discussions of the efficiency and resolution of radionuclide imaging systems, we have deliberately omitted the effects of absorption and scattering and assumed throughout that the sources were in air (strictly, a vacuum). The effect of tissue or indeed any material between the source and detector is to reduce the detection efficiency, the relevant mass-attenuation coefficients for 140 keV gamma-rays being about 0.015 m^2 kg^{-1} and for 390 keV gamma-rays about 0.010 m^2 kg^{-1} for soft tissue (approximately the same as for water). Thus the E_g values for plane sources of ^{99m}Tc and ^{113m}In respectively should be multiplied by the factors given in Table 5.1, for the thicknesses of interposed tissue listed. The factors are only approximate, as they are calculated on the basis of *detection* of 140 keV, or 390 keV, gamma-rays respectively; in practice with a scintillation detector one usually sets an energy window-width of some 10–20% of the peak energy, with the result that gamma-rays Compton scattered through small angles are also recorded. A given thickness of tissue interposed between a source and the detector therefore reduces the count-rate rather less than would be expected. The calculated attenuation factors therefore tend to be too small, perhaps by as much as a factor of two for a tissue thickness of 200 mm.

Even when allowances are made for this effect, it is still clear that the imaging devices will be more sensitive to activity near the body surface than at

Table 5.1 Attenuation Factors in Soft Tissue for 140 keV and 390 keV Gamma-Rays

Thickness (mm)	Attenuation factor	
	140 keV	390 keV
0	1.000	1.000
20	0.74	0.82
40	0.55	0.67
60	0.41	0.55
80	0.30	0.45
100	0.22	0.38
120	0.165	0.30
140	0.123	0.247
160	0.091	0.20
180	0.067	0.165
200	0.050	0.135

a depth, and by quite a considerable factor in many practical situations. A noticeable result of this is that gamma-camera pictures of the lateral aspect of the brain, using ^{99m}Tc as pertechnetate ion as tracer in the bloodstream, often show diffuse activity in the middle of the image which is due to radionuclide in the musculature at the side of the skull. The scanner on the other hand with its focusing property gives pictures in which the peripheral activity is so blurred out as to be much less perceptible. Moreover, it can be removed altogether from the image by setting a degree of background subtraction.

The attenuation coefficient for 140 keV gamma-rays in bone is somewhat higher than it is in soft tissue. (For 390 keV gamma-rays the difference is negligible.) One result of absorption in bone is sometimes seen on gamma-camera pictures of the lungs, when very faint shadows appear which are due to the ribs. The effect does not seem to occur with scanner images because the ribs are so far from the focal plane that their shielding efect is not localised. Both scanner and gamma-camera images from obese female subjects sometimes show diminished intensity at the lung bases owing to attenuation in breast tissue.

The main effect of scattering, in any system, is the degradation of image quality, which may be expressed quantitatively as an increase in effective resolution radius. The blurring due to scattering may be reduced only by the use of a narrow energy window; on the other hand, the use of too narrow a window can result in an unacceptable loss in overall count-rate. Obviously a compromise has to be reached. It should be noted that the position of the focal plane of a scanner is unaffected by scattering and absorption, in the sense that the circles of view of all the holes still coincide here, although it

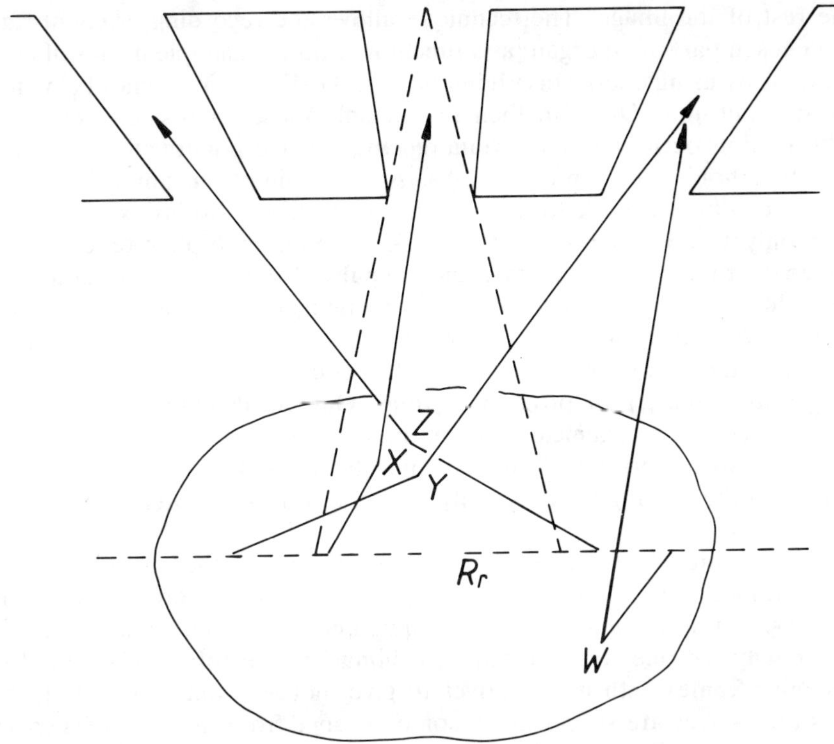

Fig. 5.32 Effect of scattering on resolution. Gamma-rays originating in the focal plane but outside the circle of resolution might be scattered at points such as W, X, Y and contribute to the observed count-rate. Large-angle scatters (point W) can be discriminated against.

may be that a point source will give the greatest count-rate when placed slightly nearer to the collimator (see Fig. 5.32).

5.13 Special Features of Radionuclide Imaging Systems

Gamma cameras have a unique advantage over scanners in that they view the whole of the organ being examined at every instant of time. Moreover because of their much larger detector area and hence plane-source efficiency they obtain information at a much faster rate. These two factors in combination enable dynamic studies to be carried out on quite short timescales. Thus the initial stages of clearance of ^{131}I-hippuran from the blood by the kidneys can be visualised by taking exposures at intervals of a few seconds. The flow of blood in the carotid arteries can be visualised by injecting a bolus of ^{99m}Tc as pertechnetate ion and taking exposures at about one-second intervals. Fast dynamic studies are of course much facilitated by storing the signals as produced on a computer disc or magnetic tape and playing them back to some form of visual display unit at leisure.

A second important technique for which the gamma camera is suited is that of selection of particular areas of the image for data processing independently of the rest of the image. The technique allows the recording of count-rate over a chosen part of an organ, as a function of time. A simple means of area selection is by using energy discrimination on the X' and Y' signals being fed to the display unit. One can then record only those counts for which the 'bright-up' dots occur within a certain rectangle, corresponding to the chosen energy thresholds and window widths. Another simple technique involves cutting a cardboard mask to fit over the oscilloscope display, so that dots appear only through a hole in the mask. A photomultiplier tube is then arranged over the oscilloscope face and the pulses from it recorded serially. It is useful to record the original data on magnetic tape and to play it back with differently shaped masks, if necessary. Interfacing the gamma camera to a small computer makes area selection and of course more sophisticated processing only a matter of program writing. Gamma cameras are available which incorporate mechanical scanning motion, so that whole-body images are possible with a crystal of only 300 mm diameter. Here again storage of signals from the moving head can only be done with a computer memory and processing facility.

An interesting development of scanner technology is that of radionuclide scanning tomography. Although ideas for carrying this out were current in the early 1960s, it is only recently that equipment has been made available commercially. In one form, a pair of collimated detectors are mounted in rectangular frames with motor drives to give motion from end to end; the frames themselves are supported at some distance from the ends of a cross-member which can be rotated about a horizontal axis (see Fig. 5.33).

The subject lies on a narrow couch between the detectors in such a way that the axis of rotation passes through the organ of interest (e.g. the liver, or

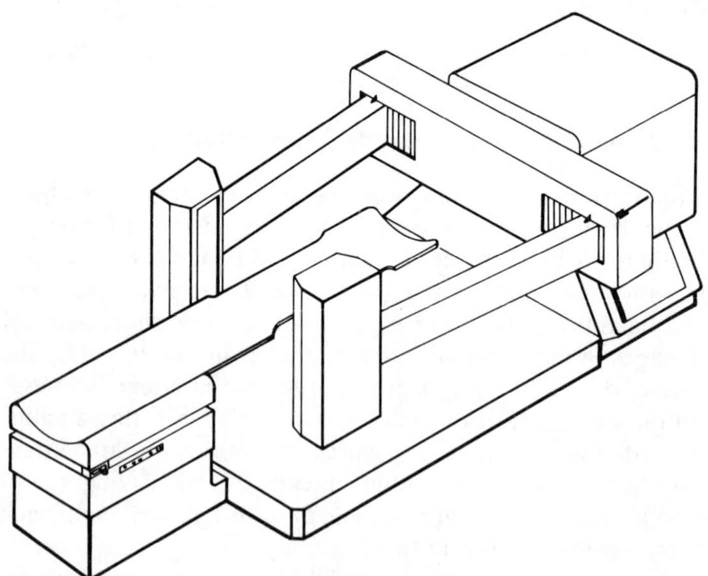

Fig. 5.33 Perspective view of a tomographic scanner (third-angle projection).

brain). A single line of scanning is performed with the detectors moving in synchronism along their frame supports. Outputs from the detectors are summed and, for brain images, counts in 80 intervals each of 3.4 mm are stored sequentially in a small computer (e.g. PDP-8 or PDP-11). The scanner frames are then indexed round by 6°, under computer control, and a second line of data collected. The process is repeated for a total of 30 traverses. Somewhat larger cell sizes, e.g. using counting intervals of 6 mm, may be chosen for large organs such as livers. The object in the scanning plane is not of course infinitely thin, the resolution in the direction of the rotation axis being typically about 15 mm (FWHM). The final visual image can be formed on ordinary X-ray film or as a dot scan. The plane of scanning is easily changed by moving the subject along the scanner axis.

Other tomographic instruments are being developed in which a large number (typically 100) of detectors are fixed in a ring in a vertical plane and face each other across a space in which the subject is positioned. The radionuclide used must be a position emitter, e.g. ^{13}N, ^{15}O, giving annihilation radiation which is detected by coincident pulses in opposite pairs of detectors, 'single' pulses being ignored. This technique obviates the need for detector collimation, although some degree of shielding is required.

Radionuclide tomographic images can never show the good resolution which is achieved by computerised axial tomography using X-rays (CAT scanning) because of the far larger number of photons used per exposure in the latter technique. However, radionuclide tomography has the advantage inherent in all radionuclide imaging of presenting functional as well as structural information. There seems no doubt but that several imaging systems beside those mentioned await exploitation, a development made possible largely by the advent of small high-speed computers.

5.14 Assessment of the Performance of Imaging Equipment

The ultimate test of the performance of imaging equipment is its usefulness as a diagnostic tool; the clinician wants to know whether the images he sees are compatible with a state of normality in the subject and whether, if an abnormality exists, he can identify its nature. To him a good scanner or gamma camera is one which will enable him to make a high proportion of accurate diagnoses, particularly perhaps for cases in which other diagnostic aids are unreliable. Also, from a clinical point of view some diagnoses carry greater value than others. For instance a brain scan of a subject who has suffered a small cerebrovascular accident (CVA), which is essentially a haemorrhage in the brain, may be scarcely distinguishable from a 'normal' scan; and in fact it is of little consequence to the subject himself whether the scan is reported as normal or not since his hospital treatment is the same. On the other hand, it might be vitally important to be sure of the presence of an operable tumour. To the physical scientist, many of the judgments the clinician has to make are somewhat subjective, and hardly susceptible to proper statistical analysis because of ethical considerations. One should not subject hospital patients to tests merely to settle a point of instrumental capability. What the physical scientist can do, however, is to devise purely physical tests on the basis of which he can advise the clinician about the performance he can expect; for example, the clinician can be safely deterred from seeking a pair of tiny lesions only 5 mm apart if the resolution radius of the scanner is known to be 10 mm. We here discuss some of these tests which are in common use.

On a gamma camera a simple but effective test of overall performance is that of image uniformity. A flat cylindrical vessel, about 10 mm deep and 300 mm diameter, is filled with an aqueous solution of a suitable radionuclide, e.g. ^{99m}Tc as $NaTcO_4$. This then constitutes a test object which should have a uniform distribution of activity over a diameter as large as the gamma-camera detector. Images of this 'uniformity phantom' should themselves be almost perfectly uniform, although a very faint pattern corresponding to the positions of the photomultiplier tubes is sometimes distinguishable. The test may be made more objective using area selection techniques and indeed corrections for non-uniformity can be applied by computer methods. Another good visual test employs the 'Williams' phantom (developed by Professor E. Williams of the Middlesex Hospital), which is a shallow perspex vessel, filled with radioactive solution, whose depth varies from point to point in the way illustrated in Fig. 5.34.

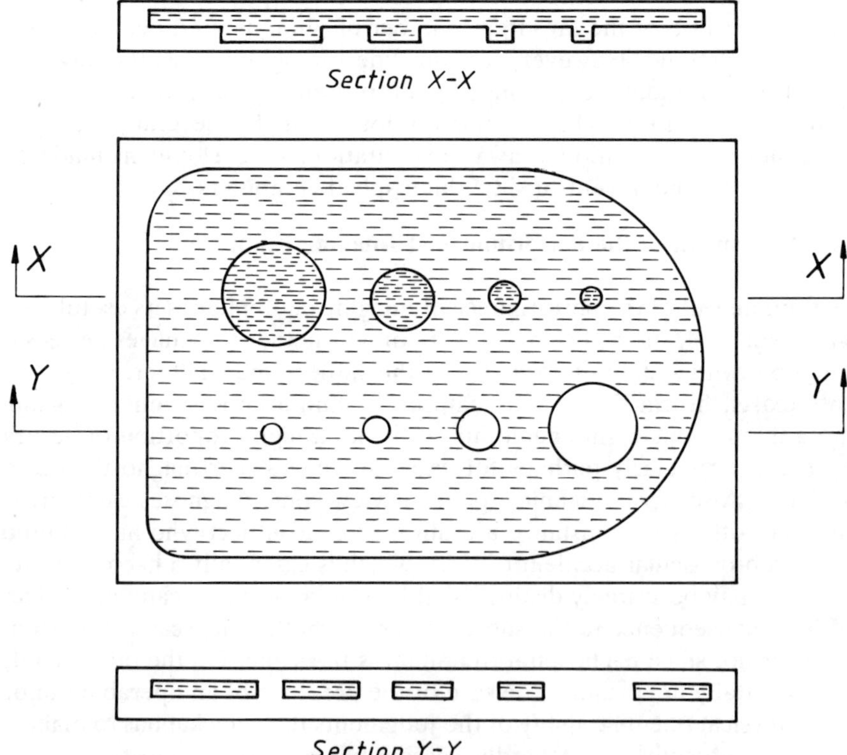

Fig. 5.34 The Williams phantom (plan view).

Other simple but more quantitative tests involve the use of line-sources. If a line-source is presented to the gamma camera in a direction parallel to say the X-axis and the Y' signals are displayed on a multichannel analyser, one sees the LSF in the Y direction, and vice versa. With the line-source in the same (X) direction a display of the X' pulses should be flat-topped; this forms a check on the uniformity in that direction. It is obviously very important to

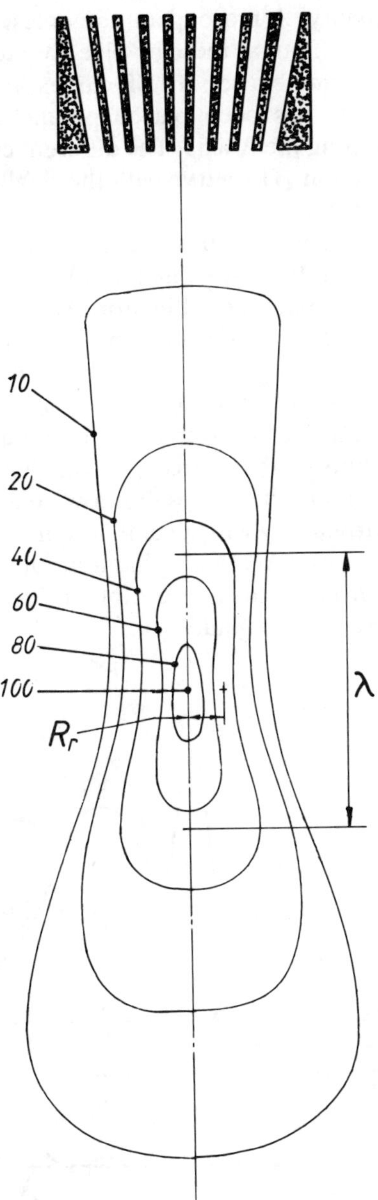

Fig. 5.35 Isocount line for a scanner collimator (in air).

set the line-source accurately along the correct direction, and it is instructive to take photographs of the image with the line-source placed in several positions all parallel to one another. Curvature of the lines on the image, especially near the edges, is often noticeable ('barrel' distortion, or 'pin-cushion' distortion, depending on the curvature of the lines).

Tests on a scanner's performance as an imaging device scarcely need include a uniformity test, but the Williams phantom gives a very good visual impression of image quality. A more quantitative test would be the mapping of isocount lines (Fig. 5.35) but as these involve very tedious experiments it is much more usual to determine a set of LSF curves. A typical set is shown in Fig. 5.36; the line-source has been placed parallel to the Y axis and its direction of travel was along the X axis. The different curves apply to different values of f in the Z direction. The curve with the smallest FWHM is obtained when $f = F$, the focal distance.

Tests with line-sources and Williams phantoms can be made both with the sources in air and also with tissue-like material (e.g. Perspex) between them and the gamma-camera or scanner collimator. These experiments show in a very practical way the effects of absorption and of scattering on image quality. Still another simple test, although one not easily quantified, uses a uniformity phantom, or flood source, filled with a solution containing a low-energy gamma-ray emitter such as ^{99m}Tc, over which is laid a pattern of lead strips. Over one of the quadrants of the source one may have parallel strips 3 mm wide and 3 mm apart, on another strips 4 mm wide and 4 mm apart, and so on. It is a good plan, particularly with a gamma camera, to record an image of such a 'bar phantom', or of a Williams phantom, at daily intervals. If these tests show deterioration in performance other more rigorous tests may be needed while readjustments are made.

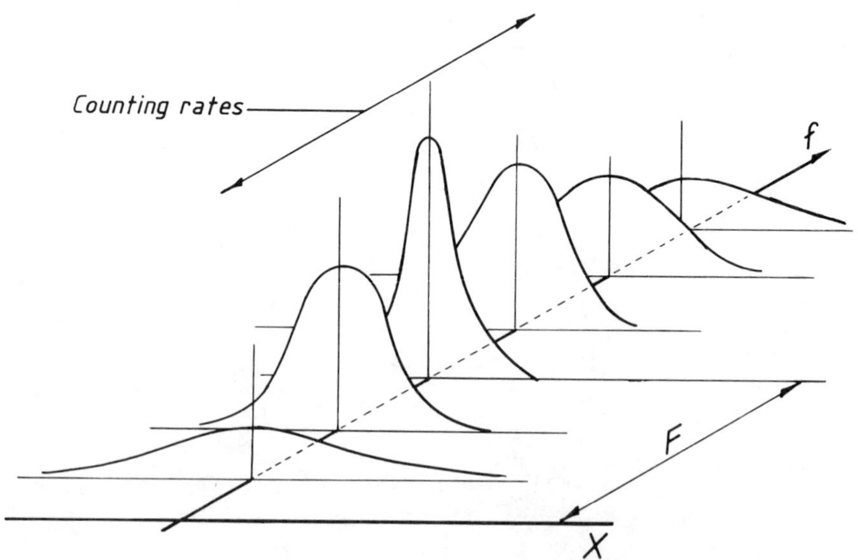

Fig. 5.36 LSFs for a scanner collimator.

5.15 Some Typical Radionuclide Images and their Interpretation

The number of radionuclide images in existence in the world must now run into many millions and it is impossible in a book of this size to begin to give a representative example of each of the categories into which they fall. In the following pages, we reproduce an image for each kind of scan commonly met with (viz. brain, liver, lung, placenta, skeleton) and one or two others, for which a clinical report would be unequivocally normal. Also we have selected some striking examples wherein the image shows well recognised abnormality. All too often in clinical practice one finds images whose interpretations, taken by themselves, are open to doubt although they may be helpful in conjunction with other medical evidence; we omit examples of these because they deserve a book to themselves directed to the medical specialist. Our selection is, therefore, highly biased.

(*a*) *Phantoms.* Pictures 5.37a are of a Williams phantom and a bar phantom, and picture 5.37b is of a uniformity phantom, taken on a gamma camera. The radionuclide used was ^{99m}Tc. Even the smallest hot-spots and cold-spots of the Williams phantom are clearly discernible. The bar widths are 3, 5, 5 and 5 mm and the spacings 3, 5, 7 and 10 mm respectively in the bar phantom; it is only just possible to discern a pattern in the 3 mm quadrant. The uniformity in Fig. 5.37b is slightly imperfect but is good enough for most clinical purposes.

Picture 5.37c is of the same Williams phantom taken in the focal plane of a scanner. The definition is not quite so good as on the gamma camera and there is an artifact due to faulty tracking of the light source on the fourth line from the bottom of the scan.

(*b*) *Brain Scans.* The radiopharmaceutical used in these scans is sodium pertechnetate (^{99m}Tc). An intravenous injection of 400 MBq is made one hour before scanning. The activity remains in the bloodstream so that the scans show essentially the vascular areas in the skull. Scans 5.38a and b are of a normal subject. The outlines are due to blood in the scalp and beneath the bony floor (sella) of the brain. Within the outline there is relatively faint activity due to vascularity in the lateral sinus and the two transverse sinuses. Scans 5.38c and d are of an abnormal subject with a large area of increased activity peripherally on the left side. This appearance is typical of a subdural haematoma.

Scans 5.39a and b show a large ovoid lesion in the left hemisphere. From the rounded shape and extremely high count density, one can infer that this is a very vascular tumour, probably a meningioma.

Scans 5.39c and d show a widely dispersed abnormal area on the right side. The A–P and P–A views show that it is largely peripheral, and the right lateral view shows that it follows the course of the mid-cerebral artery (left of this view) and the anterior cerebral artery (right half of this view). It clearly represents haemorrhage along the course of these vessels, i.e. a CVA (cerebrovascular accident).

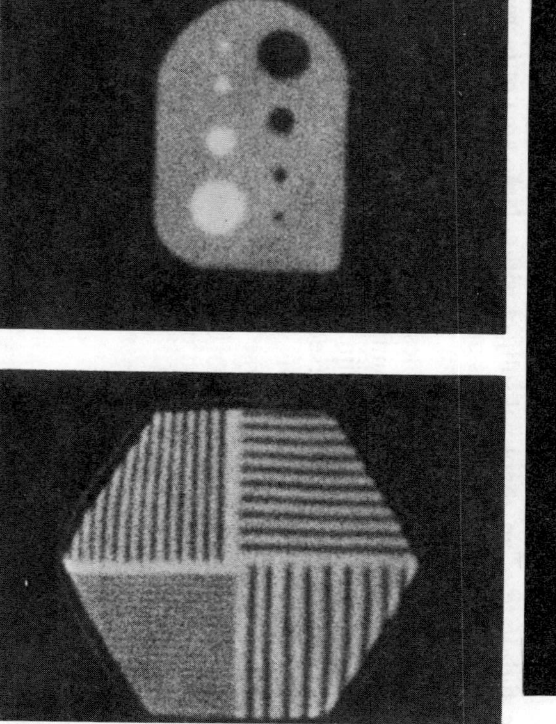

Fig. 5.37 Phantoms.

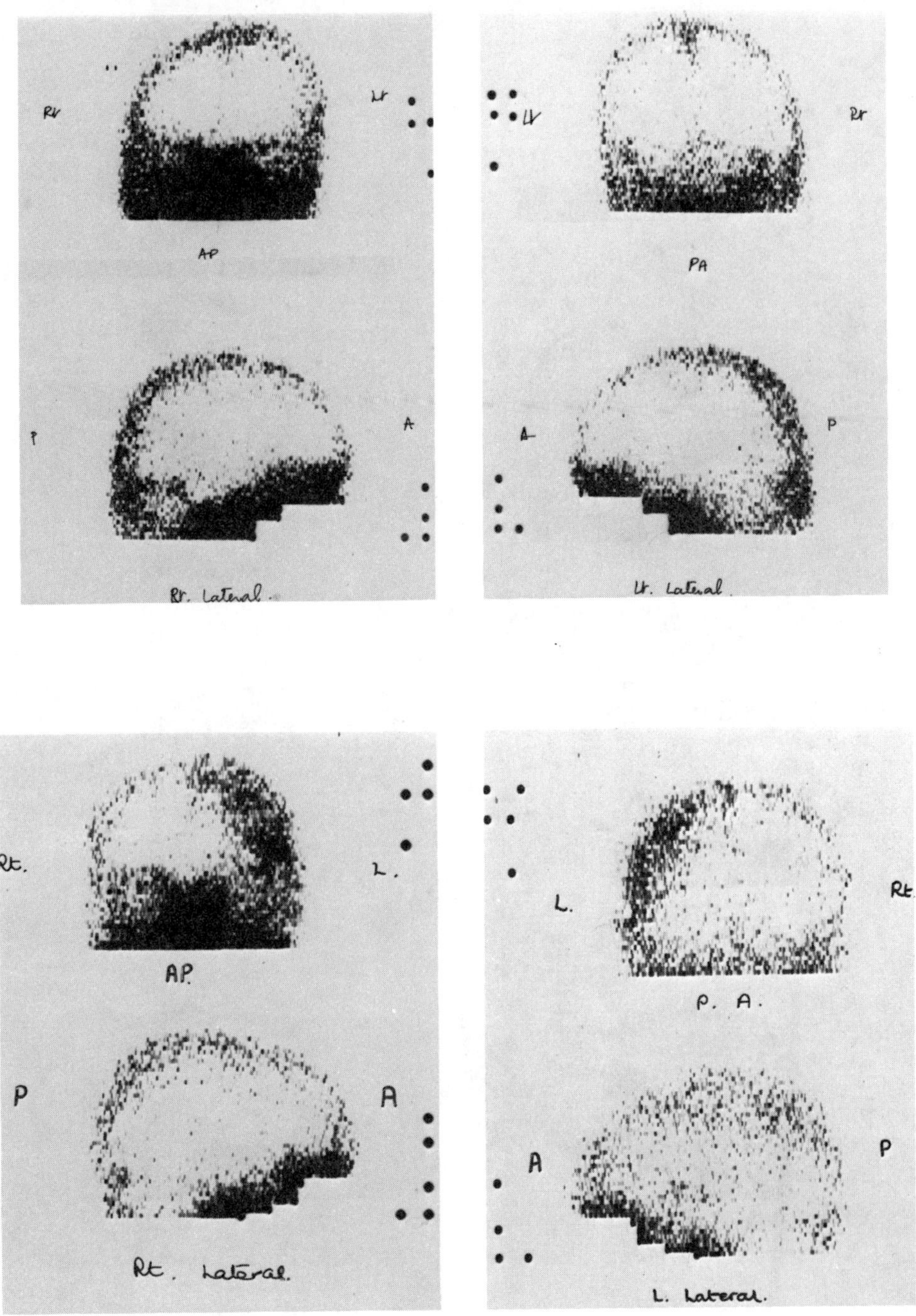

Fig. 5.38 Brain scans.

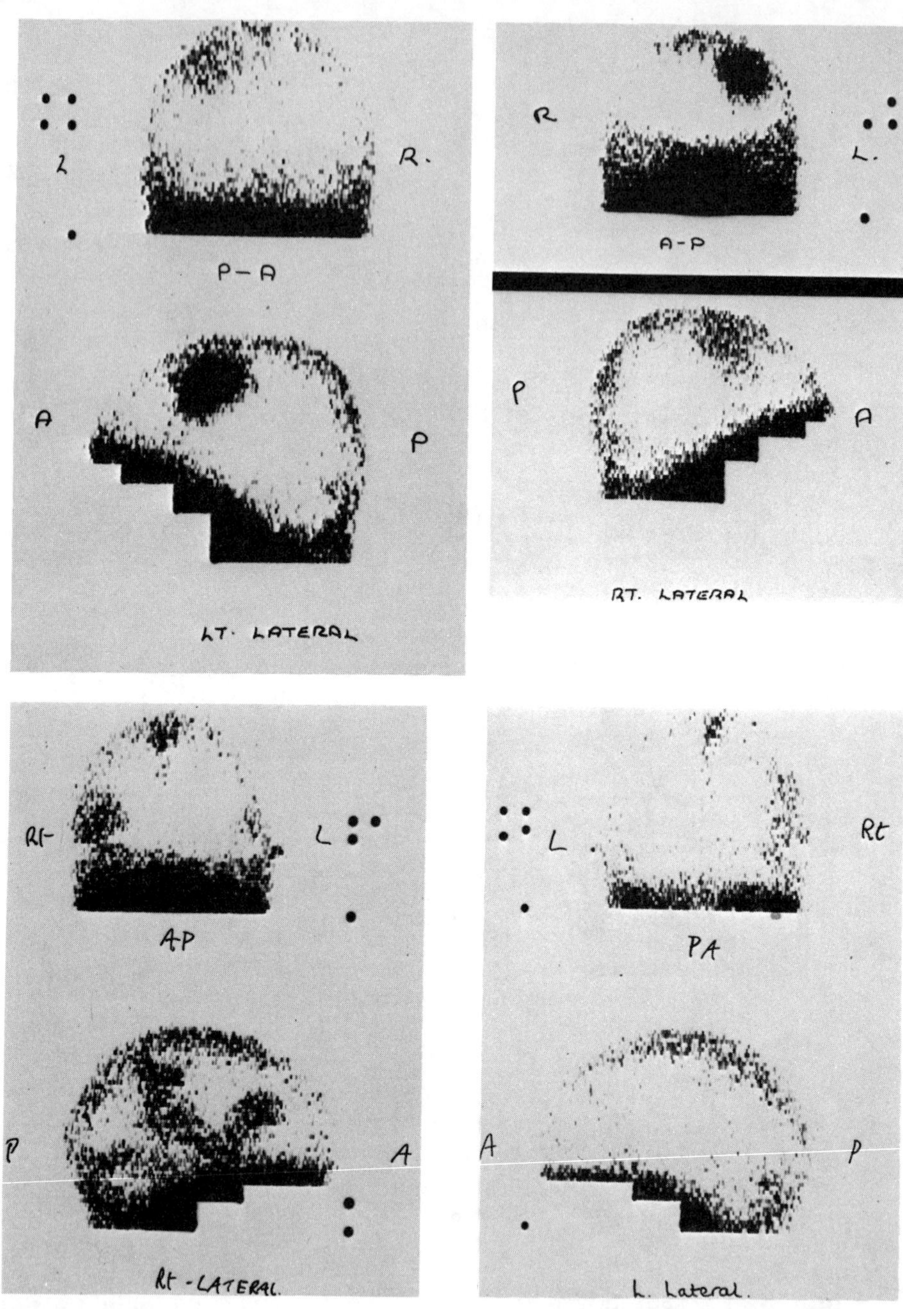

Fig. 5.39 Brain scans.

(*c*) *Liver Scans.* Scans 5.40a and b show four views of a liver, using sulphur colloid labelled with ^{99m}Tc as the radiopharmaceutical (dose used 100 MBq). The triangular shape of the liver on the left side of the A–P view is very typical of a normal organ, although quite wide variations in this shape are met with. The small area of uptake to the right of the liver is the spleen; this appears larger and denser on the P–A view, in conformity with normal anatomy. The right lateral view shows the liver alone, but the left lateral view shows both liver and spleen, the latter somewhat overlying the former. The extent to which the organs overlap varies a good deal from subject to subject and depends also very critically on how the subject is disposed to the plane of scanning.

Scans 5.40c and d show a liver in which almost the whole of the organ has failed to take up the colloid. There are many rounded defective regions and the scan is typical of multiple secondary carcinoma.

The line marked in on the A–P views represents the lower edge of the rib cage (costal margin) which gives a good anatomical reference. Notice the slightly irregular appearance of the upper border of the liver in scans 5.40a and b, which is caused by the subject's breathing.

Scans 5.41a and b are of a liver showing a single very large defective area in the lower half of the organ. There is no way of diagnosing from the scan alone the cause of this; it was in fact an abscess. After drainage, the scan returned to within normal limits.

Cirrhosis of the liver leads to a generalised functional failure of the organ, usually with some regeneration which may show as a liver enlargement, particularly of the left lobe. As the disease progresses, uptake of colloid in the liver decreases. However, cells in other organs, particularly in the spleen and bone marrow, appear to take over the function of capturing colloidal particles. Scans 5.41c show this effect to a remarkable degree, there being virtually no sign of the existence of the liver whereas the spleen is greatly enlarged and has high uptake. The spine is clearly delineated and even the ribs betray activity in the marrow.

(*d*) *Lung Scans.* Scans 5.42a and b are A–P and P–A views of lungs taken after intravenous injection of 80 MBq of macroaggregated human serum albumin (MAA) labelled with ^{99m}Tc. The aggregated particles have been trapped in the alveolar capillaries. The density of uptake is reasonably uniform and the lung outlines were found to correspond well with the lung spaces shown on the X-ray. (Note that the position of the sternal notch, SN, has been marked in as an anatomical reference.) The right-left asymmetry on the A–P view is, of course, due to the heart.

In scans 5.42c and d there is obvious lack of perfusion in the apex and at the base of the right lung. The apparent defect in the middle of the medial border of the left lung, seen on the P–A view, is due to the heart, and is of no consequence. The X-ray appearance of the lungs was normal so the perfusion defects are undoubtedly the result of blockage of major blood vessels supplying these regions (pulmonary emboli).

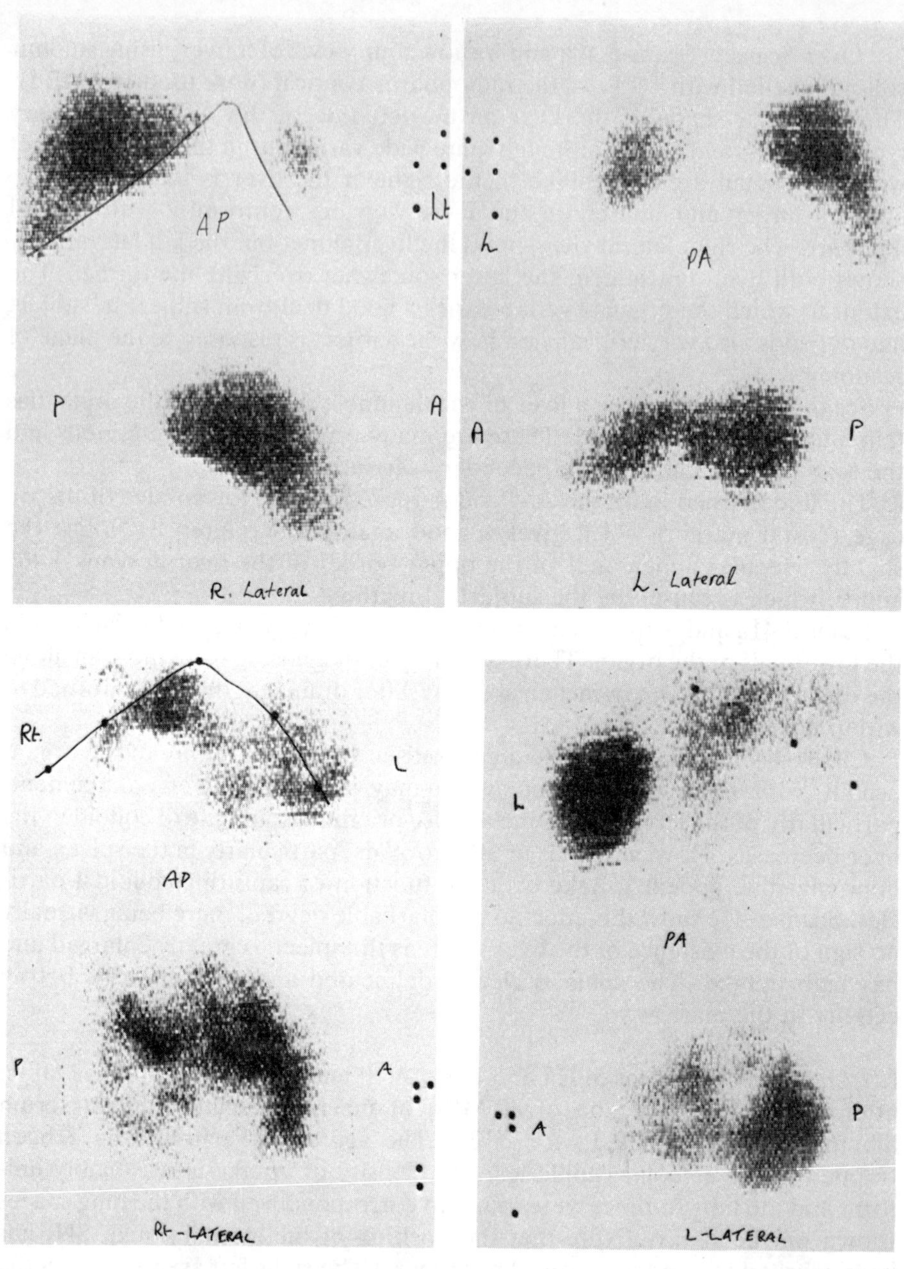

Fig. 5.40 Liver scans.

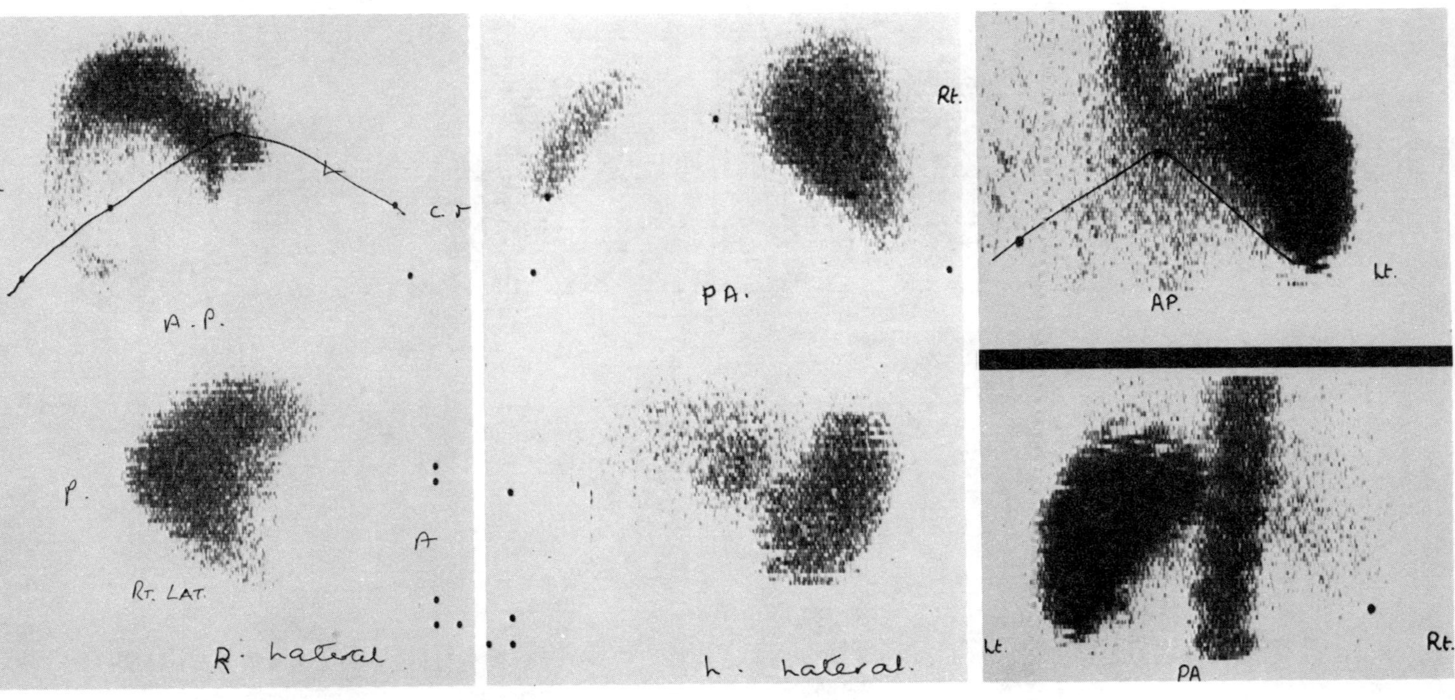

Fig. 5.41 Liver scans.

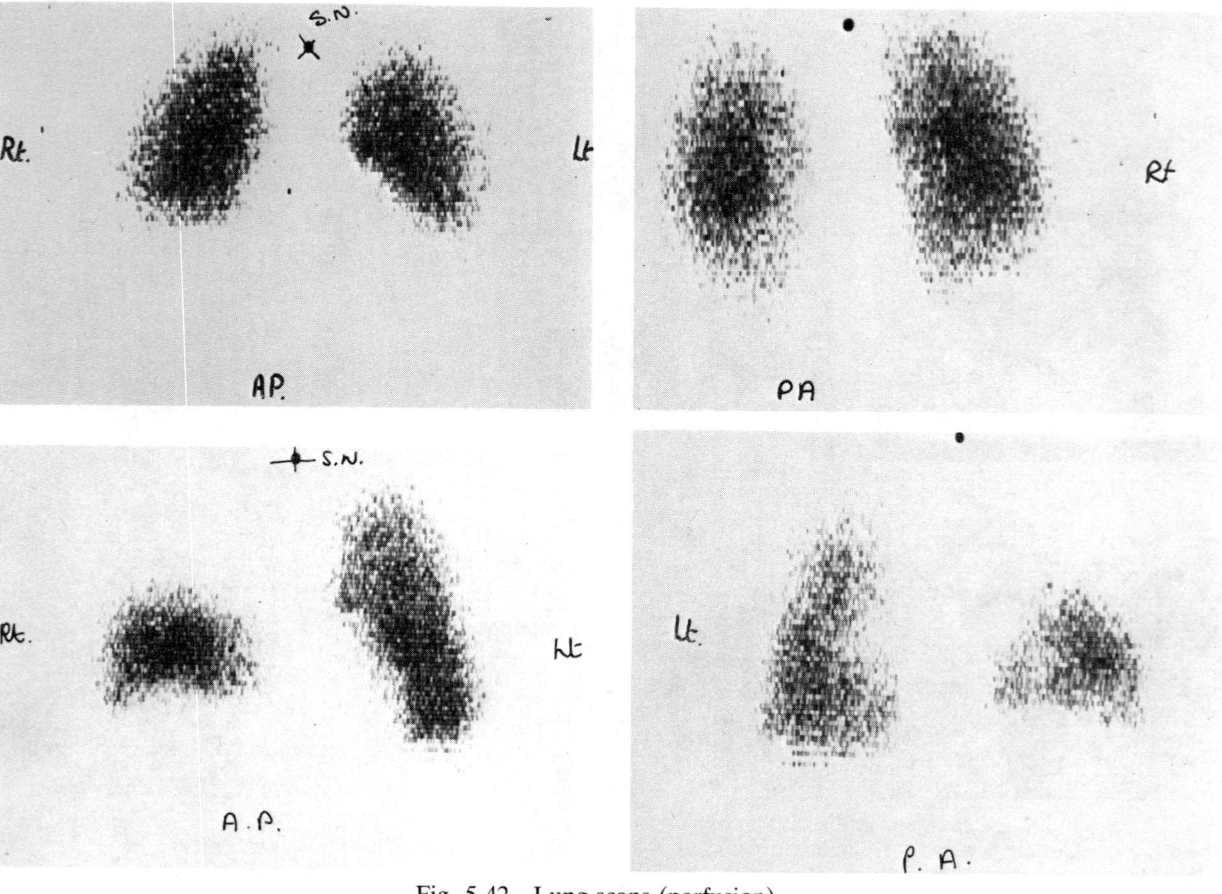

Fig. 5.42 Lung scans (perfusion).

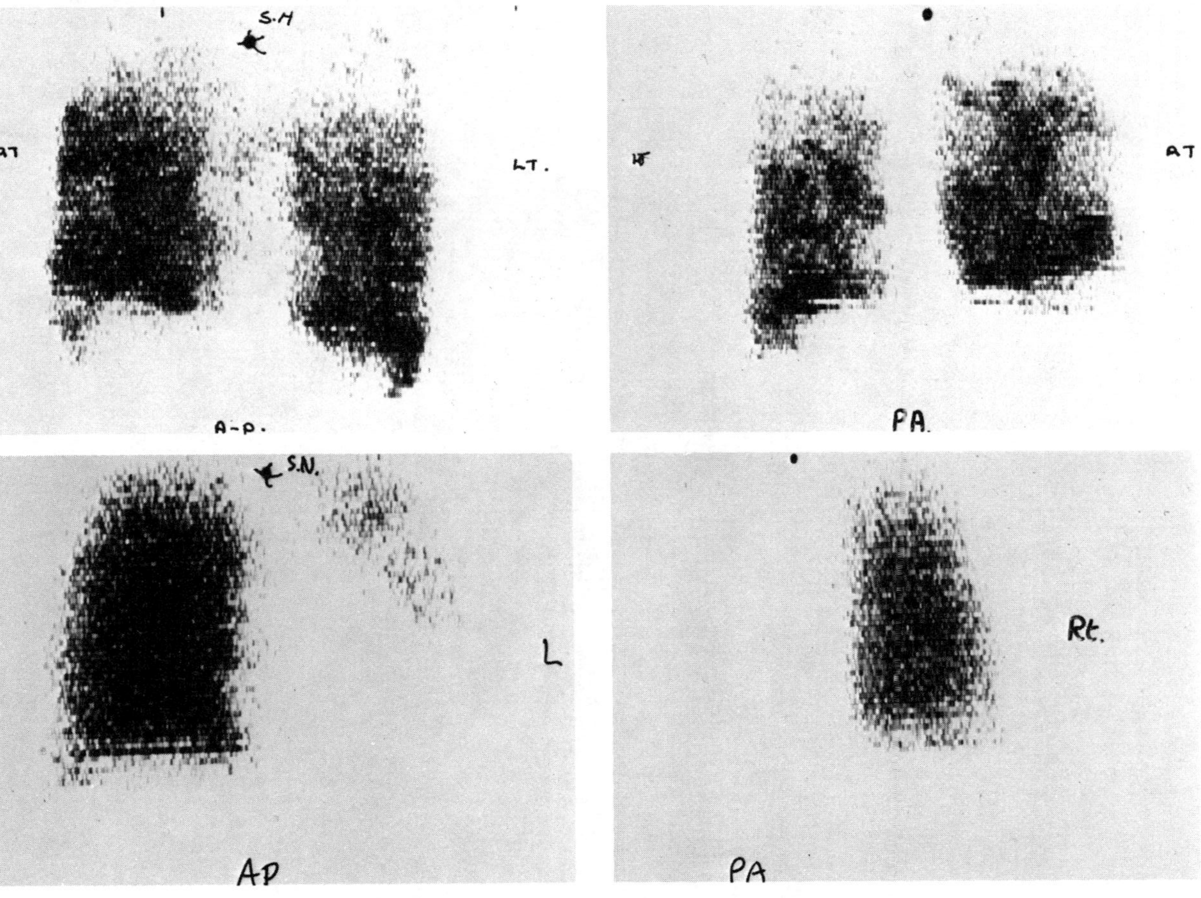

Fig. 5.43 Lung scans (perfusion).

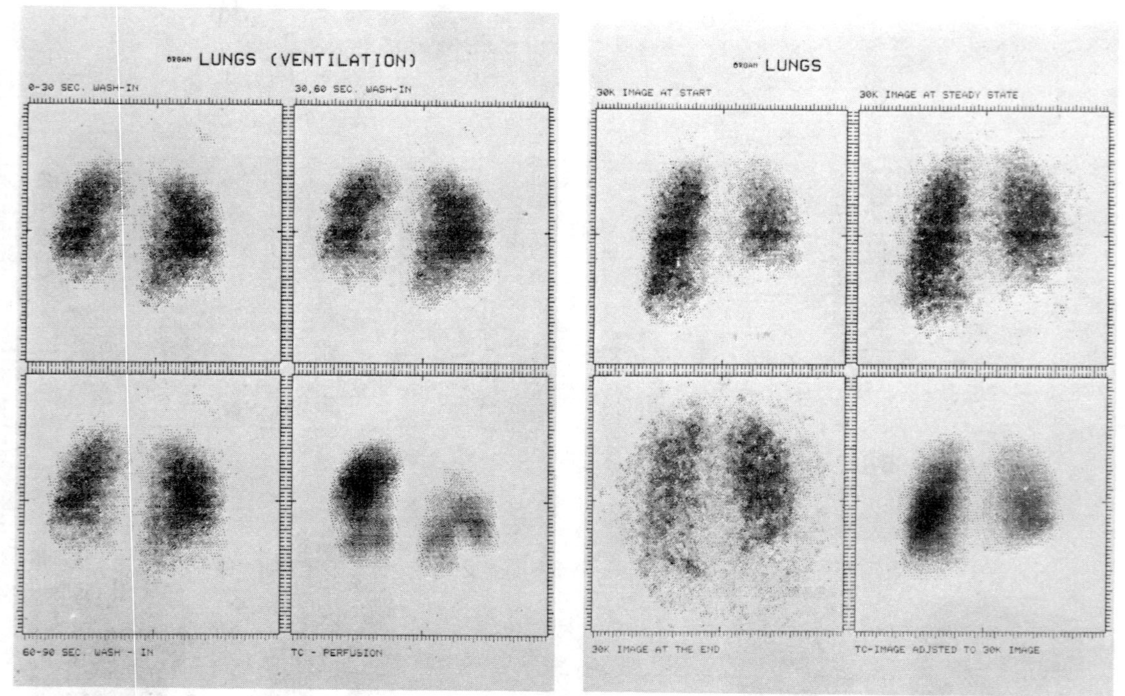
ORGAN LUNGS (VENTILATION)
0-30 SEC. WASH-IN
30,60 SEC. WASH-IN
60-90 SEC. WASH - IN
TC - PERFUSION
ORGAN LUNGS
30K IMAGE AT START
30K IMAGE AT STEADY STATE
30K IMAGE AT THE END
TC-IMAGE ADJSTED TO 30K IMAGE

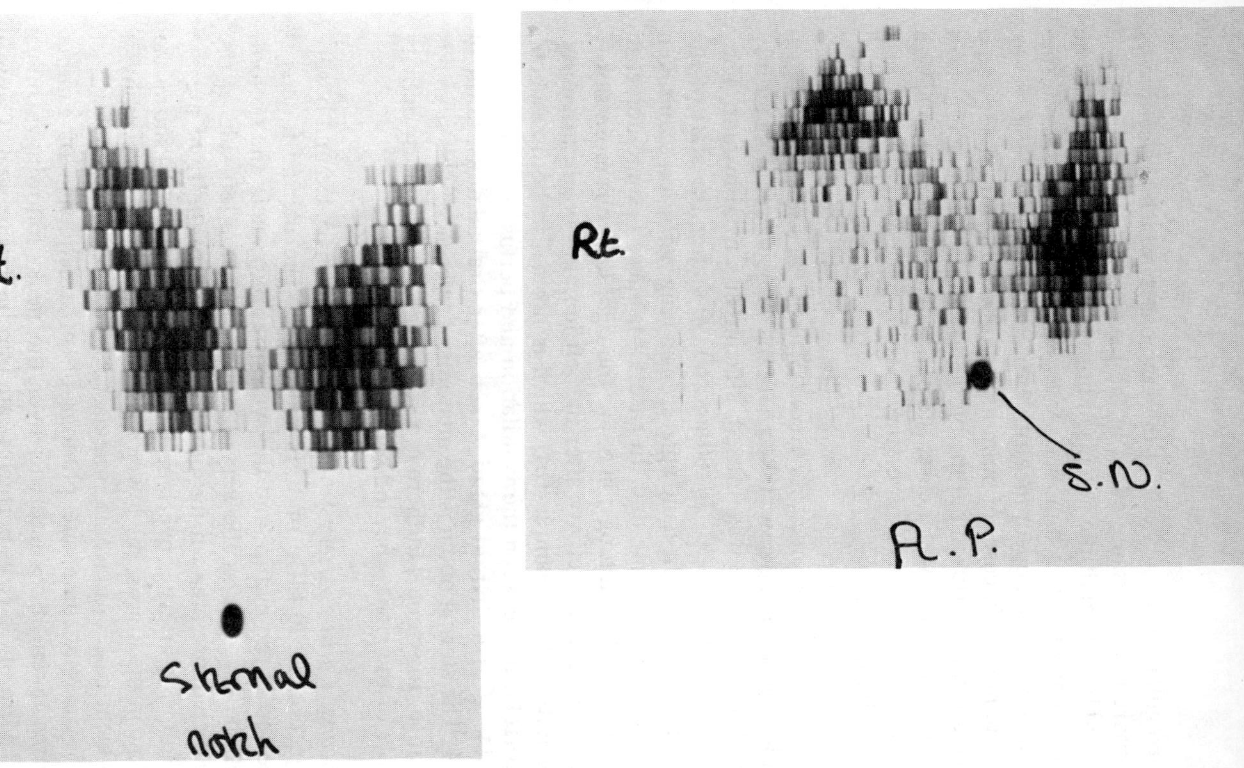

Fig. 5.44 Lung scans (ventilation/perfusion; thyroid scans).

Notice the irregularities observable at the lung bases, which are caused by breathing; compare with the same effect in the liver scans.

Scans 5.43a and b show multiple defects in both lung fields. These could be due to pulmonary emboli, or to tumours. A decision between these alternatives could be made on the appearance of ventilation scans (see Fig. 5.44), which can be quite normal in cases of pulmonary emboli. In scans 5.43c and d the right lung is normally perfused but the blood flow in the left lung is almost completely absent. In this particular case, the patient was presumed to have a massive pulmonary embolus, as neither the chest X-ray nor the ventilation scan showed any abnormality.

(*e*) *Ventilation/Perfusion Lung Scans; Thyroid Scans.* In scans 5.44a and b the first three pictures are P–A views taken with the subject breathing air into which a steady supply of ^{133}Xe was infused, starting at a fixed instant. They were obtained during times 0–30, 30–60 and 60–90 s after the start of the infusion. The fourth (lower right-hand) picture is a lung perfusion scan taken on the same subject on the same day. In case 5.44a there is some defect in ventilation at the base of both lungs; in 5.44b the right lung fills more slowly than the left and also retains activity after the left lung has begun to empty. Except for the left lung of example (b), which is normal in both respects, there is little correlation between the ventilation and perfusion.

The scan 5.44c is of a thyroid gland, taken after oral administration of 0.4 MBq of ^{131}I. The sternal notch is the anatomical reference. This scan has normal appearance. In scan 5.44d there is a very large area of diffuse activity at the lower pole of the right lobe, indicating a large goitre.

(*f*) *Placenta Scans; Carotid Artery Study.* Placenta scans 5.45a and b were taken after injection of 20 MBq of ^{113m}In as $InCl_3$. The uterine fundus is shown near the top of the picture and the symphysis pubis near the bottom. Scan (a) shows the placenta as the very dense area on the left side of the uterus; the smaller dense area on the other side is due to the right uterine artery. The left uterine artery is overlain by the placenta. Scan (b) shows the placenta very low in the uterus, the lower edge coinciding with the symphysis pubis; this is a clear case of placenta praevia.

Illustration 5.45c is a composite photograph of a sequence of gamma camera exposures taken at 5 s intervals over the neck, following a bolus injection of 300 MBq of ^{99m}Tc as pertechnetate into the basilic vein. This is a normal case showing equal flow on both sides of the neck.

(*g*) *Whole-Body Tumour Scans.* Scans 5.46a and b are A–P and P–A views of the whole body taken 3 days after the administration of 50 MBq of ^{67}Ga as gallium citrate. In subject (a) most of the activity has been excreted and the only organ in which there is appreciable uptake is the liver. This scan would be classified as normal. In case (b) there is widespread activity whose density equals or exceeds that in the liver. The abnormal areas indicate lesions most of which—but perhaps not all—appear to be associated with bone.

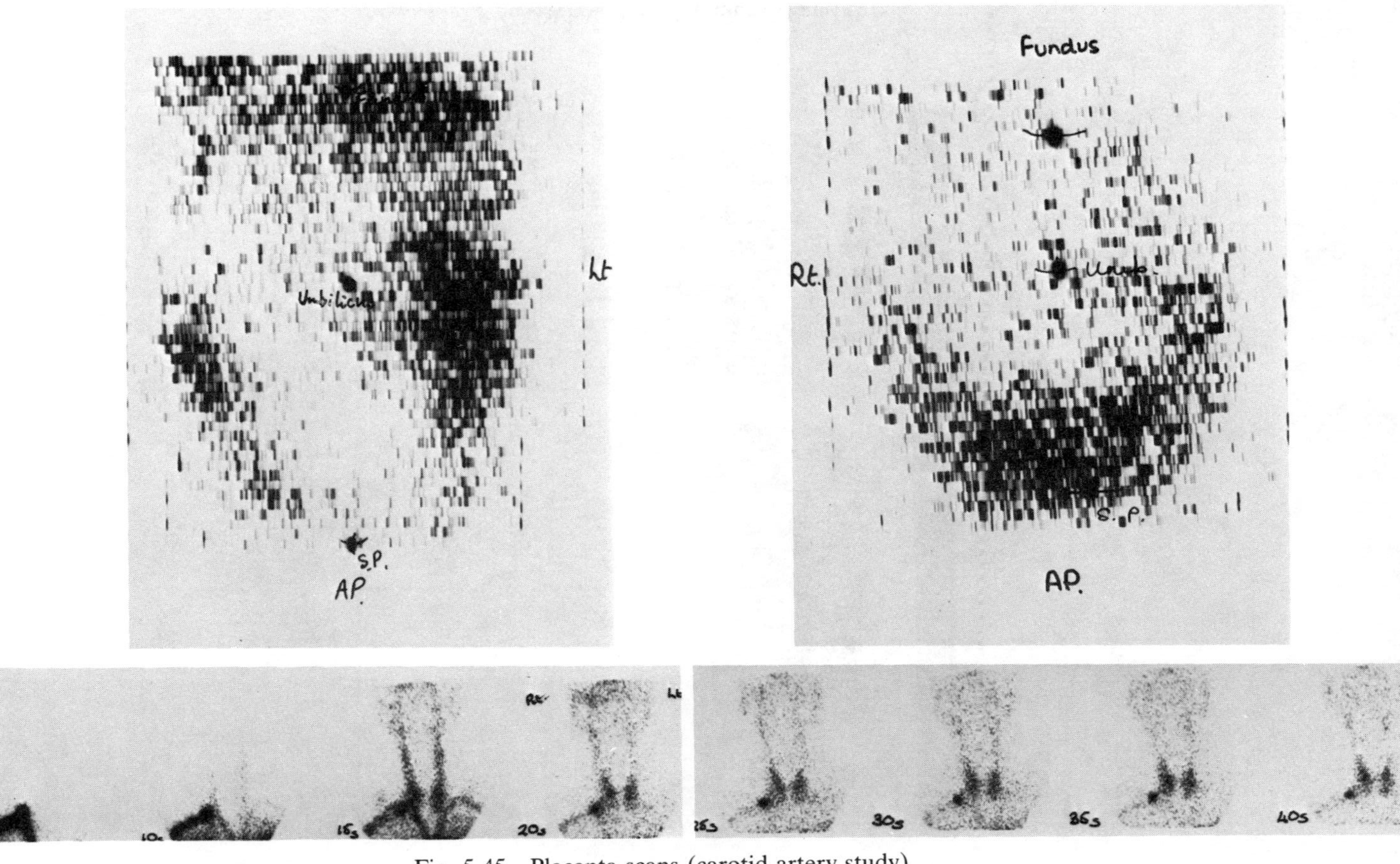

Fig. 5.45 Placenta scans (carotid artery study).

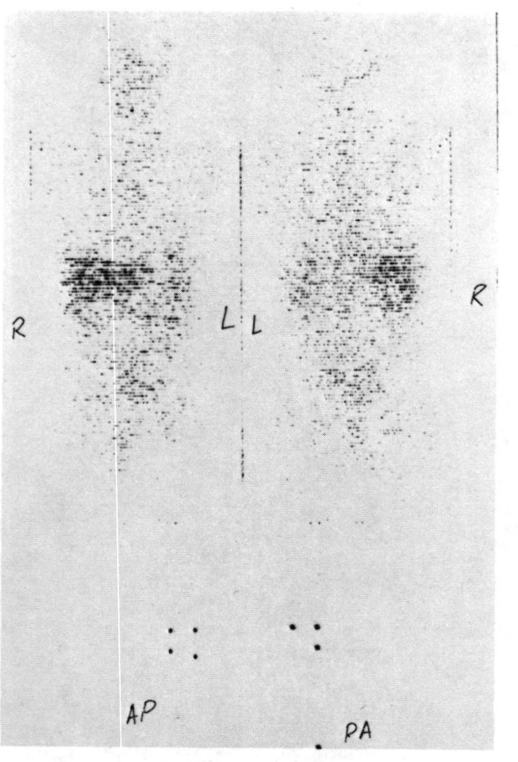

Fig. 5.46 Whole-body tumour (a) and (b).

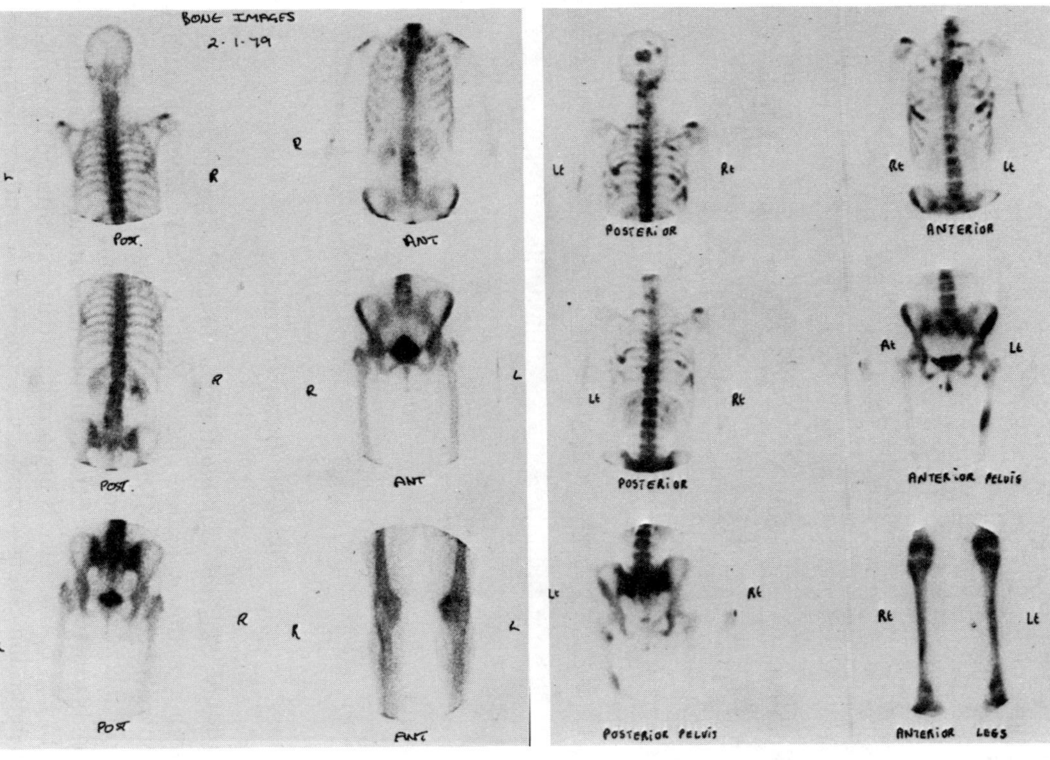

Fig. 5.47 Whole-body skeletal gamma-camera images.

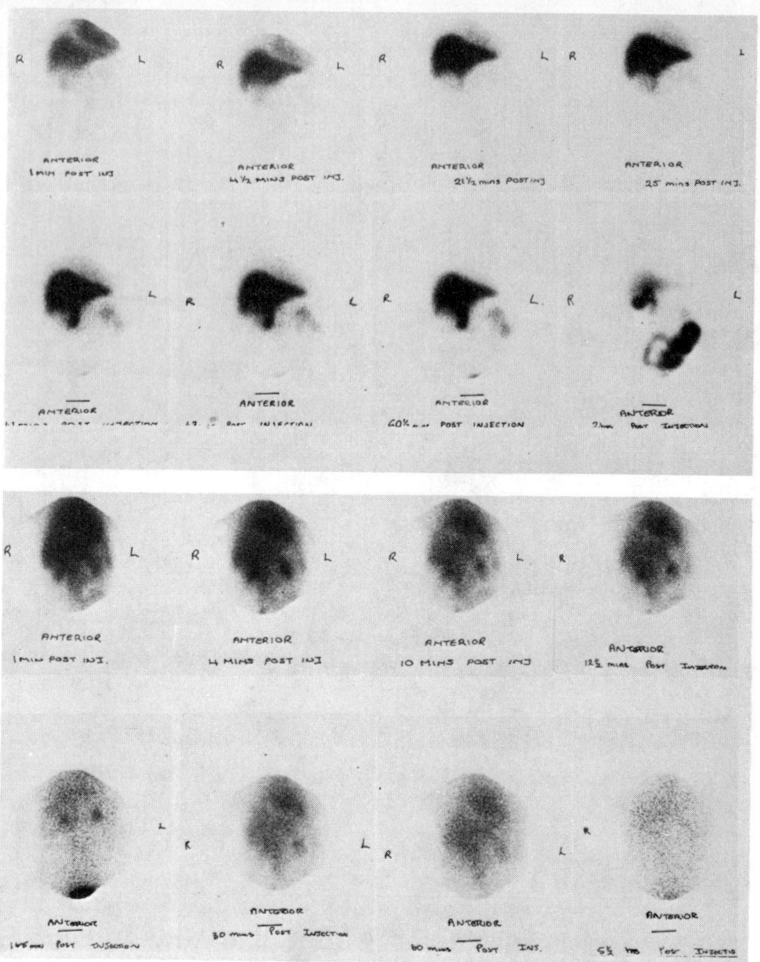

Fig. 5.48 Biliary tract study.

(*h*) *Whole-Body Skeletal Gamma-Camera Images*. The images were obtained two hours after injection of 500 MBq of ^{99}mTc-labelled methylene-diphosphonic acid complex (MDP). Fig. 5.47a shows the normal appearance of activity in the bony skeleton and in addition some activity in the kidneys and urinary bladder as part of the radiopharmaceutical is broken down and excreted. In Fig. 5.47b there is abnormal activity in scattered areas throughout the skeleton indicating widespread deposits of secondary carcinoma.

(*i*) *Biliary Tract Study*. The illustrations are gamma-camera images taken at intervals of a few minutes after injection of 100 MBq of ^{99m}Tc-HIDA complex (*N*-(2,6-diethylacetanilido)-imino-diacetic acid). Concentration of the activity in the liver is evident within a few minutes. In a normal case (Fig. 5.48a), its appearance in the bile duct is shown as early as 20 min post-injection and at the end of two hours it has passed almost completely into the intestine. In the case of an obstructed bile duct (Fig. 5.48b), the only excretory pathway for the activity is via the kidneys.

5.16 Radiation Dose to the Body from Ingested Radionuclides

The risk of injury from absorbed radiation in any person being subjected to a diagnostic or therapeutic procedure with radionuclides must of course be carefully assessed. Present-day codes of practice recommend a maximum permissible absorbed dose-rate (for persons over 18 years of age) of 5 rem, or 0.05 Sv, per annum, over the whole body. Particular organs, e.g. the extremities, are allowed up to 75 rem, or 0.75 Sv, per annum. The dangers of exceeding these dose-rates by appreciable factors are well documented and it must be left to clinical judgment as to whether in a particular hospital patient the hazard from a diagnostic or therapeutic measure is outweighed by the benefit gained by more certain diagnosis, or by destruction of malignant cells, as the case may be. There is no way of directly measuring the absorbed dose of radiation in a human being but in this section we outline the methods of estimating this dose on the basis of the underlying physics. In the nature of the case, the estimates can only be rather rough because of biological and geometrical variations among the subjects, but they can be made conservatively and are certainly accurate enough for practical purposes.

The important factors that have to be taken into account are:

(a) The total energy radiated, per unit time, by the radionuclide. The gamma-ray energy for all commonly used radionuclides is well documented; for radionuclides emitting beta particles the beta particle energy as given in tables is generally the maximum E_{max} and one can assume that the energy absorbed in tissue is $E_{max}/3$. Energy values given for conversion electrons must not be divided by 3 as these are monoenergetic.

(b) The absorption effects, for beta particles and electrons. The ranges are of the order of only one millimetre in tissue so that all their available energy is absorbed within the organ containing the administered radionuclide. For gamma-rays with their nearly exponential attenuation in matter, only a fraction of their energy will be absorbed in the organ, another fraction will be absorbed elsewhere in the body, and the remainder will escape altogether. The exact evaluation of these fractions is quite difficult.

(c) The distribution of radionuclide within the body. It may be sufficiently accurate to assume complete localisation of administered radionuclide in one or perhaps two particular organs, but localisation to the extent of just 100% is not always achieved, resulting in a diminished dose in the organs themselves but a raised dose in the rest of the body.

(d) The time-scale of the radiation exposure. It is usually assumed that localisation of administered radionuclide occurs quite quickly (if incompletely) and that thereafter the dosage is controlled by the diminution of the activity with a known effective half-life (Section 4.10). It is then important to know where the radionuclide relocates after leaving the organ in question; obvious sites are the excretory systems, and it may be necessary to estimate the radiation absorption by the kidneys and bladder.

With these considerations in mind we now calculate the absorbed dose to organs on the assumption of a given initial content of radioactive material. The notation used in standard texts before the introduction of SI units employed the letter C to denote the amount of radionuclide present, in microcuries, per gram of the organ. Here we will assume that the initial activity is A Bq in an organ of mass m, so that the specific activity is $A/m = A_s \text{Bq kg}^{-1}$. For simplicity we assume one beta particle emitted per disintegration, and with a mean energy $\bar{E}$ in MeV. For the gamma-rays we take the exposure rate from a point source of 1 Bq at a distance of 1 m to be as calculated in Section 3.10. For commonly used radionuclides this means using the data in Table 3.1 of that section; at the time of writing there is no international agreement on the symbol to be used here and we propose to quote the exposure rate at 1 m as γ R $\text{Bq}^{-1}\ \text{s}^{-1}$. Using the f factors given in Table 3.2 of Section 3.11, the absorbed dose rate is γf Gy $\text{Bq}^{-1}\ \text{s}^{-1}$.

(*a*) *The Beta-Particle Dose.* The initial rate of energy radiated in the beta-particle emission is clearly $A_s E \times 10^6$ eV $\text{kg}^{-1}\ \text{s}^{-1}$. If the activity A decays exponentially with effective decay constant λ_e, the rate at time t is then $10^6 \times A_s\bar{E} \exp(-\lambda_e t)$ eV $\text{kg}^{-1}\ \text{s}^{-1}$ and assuming complete decay of the radionuclide we find the total energy radiated to be

$$\begin{aligned}\text{total energy per kilogram} &= 10^6 A_s\bar{E}\int_0^\infty \exp(-\lambda_e t)\text{d}t\\ &= 10^6 A_s\bar{E}[\exp(-\lambda_e t)/(-\lambda_e]_0^\infty\\ &= 10^6 A_s\bar{E}/\lambda_e \quad \text{eV kg}^{-1}\end{aligned}$$

In view of the short range of the beta particles in tissue, we assume that all the energy is absorbed in the organ in question, and writing the expression in terms of the more familiar half-life $(t_{1/2})_e$ (in days) by replacing λ_e by $0.6932/((t_{1/2})_e \times 60 \times 60 \times 24)$ we have for the total beta-particle dose to the organ

$$\begin{aligned}D_\beta &= \frac{10^6 \times 60 \times 60 \times 24}{0.6932} A_s\bar{E}(t_{1/2})_e \quad \text{eV kg}^{-1}\\ &= \frac{10^6 \times 60 \times 60 \times 24}{0.6932} A_s\bar{E}(t_{1/2})_e \quad \text{eV kg}^{-1}\\ &= 1.994 \times 10^{-8} A_s\bar{E}(t_{1/2})_e \quad \text{Gy}\end{aligned}$$

An easily remembered rule is, therefore, that when the specific activity of the radionuclide is A_s GBq kg^{-1}, or A_s MBq g^{-1}, the total beta-particle dose is $20 A_s \bar{E}(t_{1/2})_e$ Gy. For example, for a 50 g thyroid gland containing 200 MBq of ^{131}I and subject to an effective half-life of 3 d, we have (since the mean beta-particle energy E is 0.178 MeV) that

$$D_\beta = 20 \times \frac{200}{50} \times 0.178 \times 3$$

$$= 43 \text{ Gy}$$

This would be considered by many physicians as a rather low therapeutic dose in a case of hyperthyroidism.

(*b*) *The Gamma-Ray Dose.* From the absorbed the dose rate γf Gy Bq^{-1} s^{-1} given when a point-source is at a distance of 1 m, as explained above, we can easily get the total dose assuming complete decay with an effective half-life $(t_{1/2})_e$ (days); the result is

$$\text{total dose (at 1 m)} = \frac{60 \times 60 \times 24}{0.6932} \gamma f (t_{1/2})_e$$

$$= 1.246 \times 10^5 \, \gamma f (t_{1/2})_e \quad \text{Gy Bq}^{-1}$$

We must consider the absorbed dose at a point O within the body due to the activity in a tissue element with sides Δx, Δy, Δz at a point (x,y,z) (Fig. 5.49).

Assuming a tissue density ρ kg m^{-3} and a specific activity of A_s Bq kg^{-1}, the element approximates to a point source of activity $A\rho\Delta x\Delta y\Delta z$ Bq at a distance $R = \sqrt{(x^2 + y^2 + z^2)}$ from the origin. The exposure at 0 is governed by the inverse-square law and is also modified by the attenuation of the gamma radiation in the medium. The element therefore contributes an

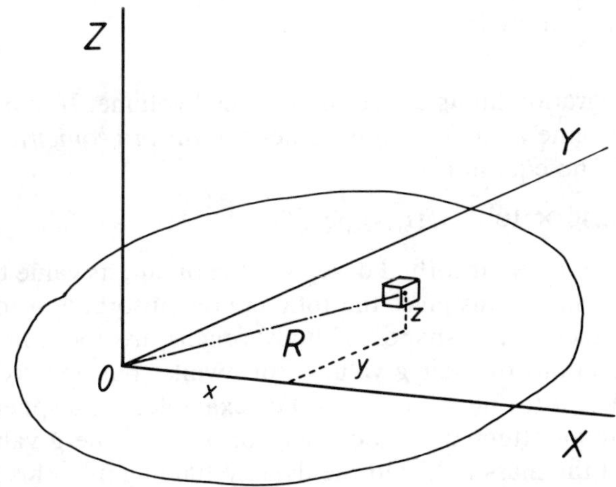

Fig. 5.49 An element $\Delta x\Delta y\Delta z$ at a distance R from the origin of coordinates in a body organ.

absorbed dose at 0 of

$$1.246 \times 10^5 A_s \gamma f (t_{1/2})_e \rho \frac{\exp(-\mu R)}{R^2} \Delta x \Delta y \Delta z \quad \text{Gy}$$

Summing over all elements and proceeding to the limit of indefinitely small elements, we have

$$D_\gamma = 1.246 \times 10^5 A_s \gamma f (t_{1/2})_e \rho \iiint \frac{\exp(-\mu R)}{R^2} \, \mathrm{d}x \mathrm{d}y \mathrm{d}z \quad \text{Gy}$$

where the integration limits are set by the physical extent of the active organ. The integral can be evaluated analytically only for some special cases; for instance when $\mu = 0$ and the origin is at the centre of a sphere of radius R its value is $4\pi R$; when the origin is on the surface of the sphere it is $2\pi R$. However, for tissue μ is about 0.3 m^{-1} and the integral has to be evaluated by numerical methods. It is generally known as the *geometrical factor* for a point with respect to the particular shape of the organ being considered and is denoted by g. The point may lie inside the organ or outside, of course. The basic equation then becomes

$$D_\gamma = 1.246 \times 10^5 A_s \gamma f (t_{1/2})_e \rho g \quad \text{Gy}$$

It is important to realise that the equation gives us the absorbed dose—that is to say the amount of energy absorbed per kilogram of the medium—at one particular point. What is often much more important is the absorbed dose in some specified volume in space, for example the active organ itself. To find this, we may generalise the above argument by finding the absorbed dose not merely at the origin but at any point in space. The geometric factor g then becomes a function of the spatial coordinates and we can obtain its mean value over a volume V by evaluating the integral

$$\bar{g} = \iiint \frac{g}{V} \, \mathrm{d}x \mathrm{d}y \mathrm{d}z$$

where the integration limits define the specified volume. If in particular these limits are the same as before, $\bar{g}$ becomes the *mean geometric factor* for the active organ. The equation

$$D_\gamma = 1.246 \times 10^5 A_s \gamma f (t_{1/2})_e \rho \bar{g} \quad \text{Gy}$$

then gives us the *mean* absorbed dose over the organ; its value times the mass of the organ in kilograms gives the total energy absorbed in joules.

In actual practice the shapes of body organs are too complex to make detailed calculations of their $\bar{g}$ values worthwhile, but approximate calculations are sufficient for most purposes. For example, for a sphere of radius R and possessing an attenuation coefficient of 0.3 m^{-1} the $\bar{g}$ value is approximately $2\pi R$; if the mass is M and the density 1000 kg m^{-3} this is numerically about 0.4 M. (The dimension of g and $\bar{g}$ is of course that of length.) Extensive compilations have been published for other geometrical shapes, in particular

Table 5.2 Mean Geometric Factor $\bar{g}$ (M) for Cylinders

(Reproduced from G. Hine and G. Brownell, *Radiation Dosimetry*, by permission of Academic Press.)

	Radius (m)				
Length (m)	0.03	0.05	0.1	0.2	0.3
0.02	0.17	0.22	0.30	0.36	0.39
0.05	0.22	0.32	0.48	0.62	0.68
0.1	0.25	0.38	0.61	0.87	0.98
0.2	0.26	0.41	0.69	1.05	1.26
0.4	0.26	0.41	0.72	1.16	1.43
0.6	0.26	0.42	0.73	1.18	1.48
0.8	0.26	0.42	0.73	1.19	1.50
1.0	0.26	0.42	0.73	1.19	1.50

the cylinder. For radiation protection purposes, it is sufficient to simulate the human body by a collection of spheres and cylinders of suitable dimensions and to estimate a $\bar{g}$ value for the whole body by calculating weighted means over the collection. Table 5.2 and 5.3 show estimates of $\bar{g}$ for cylinders and for human bodies of different dimensions; they are abridged from the more extensive data given in *Radiation Dosimetry* by Hine and Brownell.

As an example we will calculate the gamma-ray dose to the whole body for a subject weighing 80 kg and whose height is 1.8 m, consequent upon an injection of 400 MBq of ^{99m}Tc as sodium pertechnetate. As the activity remains largely in the bloodstream, we will assume uniform distribution over the whole body; allowing for excretion we will take the effective half-life as 4 h. As far as whole-body dose, these are probably conservative estimates although concentration of the radionuclide in the urinary bladder will give a

Table 5.3 Mean Geometric Factor for Whole Human Body

(Reproduced from G. Hine and G. Brownell, *Radiation Dosimetry*, by permission of Academic Press.)

	Height (m)			
Weight (kg)	1.4	1.6	1.8	2.0
40	1.10	1.08	1.05	1.02
60	1.28	1.22	1.19	1.17
80	1.48	1.43	1.38	1.34
100	1.54	1.47	1.42	1.38

higher dose in that organ. The factors we need are then:

(a) the specific activity A_s = 400/80 = 5 MBq kg^{-1} = 5 × 10^6 s^{-1} kg^{-1}
(b) the γ factor, 0.056 × 10^{-14} R Bq^{-1} s^{-1}
(c) the f factor, approximately 0.01 Gy R^{-1}
(d) the effective half-life in days = 0.167 d
(e) the tissue density ρ, approximately 1000 kg m^{-3}
(f) the $\bar{g}$ factor, which is estimated in Table 5.3 as 1.38 m.

Substituting in the formula

$$D_\gamma = 1.246 \times 10^5 A_s \gamma f (t_{1/2})_e \rho \bar{g}$$

which already contains the factor converting seconds to days, we have

$$D_\gamma = 8.0 \times 10^{-4} \text{ Gy}$$

The calculation of absorbed dose in any body organ due to its radionuclide content proceeds in the same way. It is a much more difficult problem to estimate the absorbed dose in one organ A, due to radionuclide present in some other organ B. Fortunately no such calculation needs to be done at all accurately, and it is usually sufficient to assume that all the activity in the organ B can be replaced for calculation purposes by a point source of appropriate strength at its centre of gravity. The absorbed dose to some representative small portions of the organ A can be found and the total calculated. Sometimes in this connection it is helpful to employ the reciprocal dose theorem, which states that if two organs A and B contain radionuclide at equal specific activity, the absorbed dose in A due to activity in B is equal to the absorbed dose in B due to the activity in A. Here again the assumption is made that the radionuclide distribution is uniform within the two organs.

Finally, let us consider the radiation dose to an organ from a radionuclide which is not assumed to be present initially but which accumulates in it over an appreciable time. For example, a radiopharmaceutical may be injected quite rapidly into the bloodstream and we may wish to calculate the absorbed dose to the liver, or some other organ which takes up the activity comparatively slowly. Let us assume an injection of $N_1(0)$ radioactive nuclei, all at time zero, into the bloodstream. For any nucleus the probability per unit time of radioactive decay is, of course, the decay constant λ_p; let us assume a probability per unit time of transfer to the organ in question of λ_s. Then at any time t the rate of change of $N_1(t)$, the number of radioactive nuclei in the blood, is

$$\frac{dN_1(t)}{dt} = -(\lambda_p + \lambda_s)N_1(t)$$

whose solution is

$$N_1(t) = N_1(0) \exp\{-(\lambda_p + \lambda_s)t\}$$

If, as before, we take the effective decay constant of the activity in the organ as λ_e, we can express the rate of change of the number of radioactive nuclei in the organ, $N_2(t)$, as

$$\frac{dN_2(t)}{dt} = \lambda_s N_1(t) - \lambda_e N_2(t)$$

whose solution is

$$N_2(t) = \frac{\lambda_s N_1(0)}{(\lambda_e - \lambda_p - \lambda_s)}[\exp\{-(\lambda_p + \lambda_s)t\} - \exp(-\lambda_e t)]$$

since we are assuming $N_2(t) = 0$. (Compare the similar problem of Section 4.7.) The activity in the organ at time t is then

$$A = \lambda_p N_2(t) \quad \text{Bq}$$

where λ_p is the physical decay constant, and the specific activity is

$$A_s = \frac{A}{m} = \frac{\lambda_p N_2(t)}{m} \quad \text{Bq kg}^{-1}$$

The total energy, per kilogram, absorbed by the organ in respect of the beta particles as the decay proceeds is

$$10^6 \bar{E} \int_0^\infty A_s dt = 10^6 \bar{E} \frac{\lambda_p N_1(0)}{m} \cdot \frac{\lambda_s}{(\lambda_e - \lambda_p - \lambda_s)} \int_0^\infty [\exp\{-(\lambda_p + \lambda_s)t\} - \exp(-\lambda_e t)]dt$$

$$= 10^6 \bar{E} \frac{\lambda_p N_1(0)}{m} \cdot \frac{1}{\lambda_e} \cdot \frac{\lambda_s}{\lambda_p + \lambda_s} \quad \text{eV kg}^{-1}$$

Now $\lambda_p N_1(0)/m = A_s'$ is just the specific activity which the radionuclide would have had if it were present in the organ at time zero. Also if we let

$$\lambda_u = \lambda_p + \lambda_s$$

so that the effective half-life for removal of activity from the blood is

$$0.6932/\lambda_u = (t_{1/2})_u$$

we have

$$\frac{\lambda_s}{\lambda_p + \lambda_s} = \frac{\lambda_u - \lambda_p}{\lambda_u}$$

$$= 1 - \frac{\lambda_p}{\lambda_u}$$

$$= 1 - \frac{(t_{1/2})_u}{(t_{1/2})_p}$$

With the above relationships we can then derive the absorbed dose to the

Table 5.4 Absorbed Dose from Internally Administered Radionuclides

Doses are in units of 10^{-9} Gy Bq^{-1} (mGy MBq^{-1}) except where indicated. (Reproduced, slightly abridged and converted to SI units, from E.L. Saenger, *Recent Advances in Nuclear Medicine. Vol. 3* (Ed. J.H. Lawrence), 1971 by permission of Grune & Stratton.)

Radio-nuclide	Radiopharma-ceutical	Admin. route	Total Body	Blood	Bone	Bone marrow	Brain	Gonads	Kidneys	Liver	Lungs	Spleen	Thyroid	Other tissues
^{3}H	Water	Oral	0.03–0.06											
	Water	IV	0.03–0.06											
^{11}C	Monoxide	Inhalation (1)	2	3.5				1			2–3.5	2		Foetus 0.0003
^{18}F	Sodium flucride	IV	0.01–0.03		0.03–0.1									Bladder 0.5–1.5
^{22}Na	Chloride	IV	6		0.1	30								
^{24}Na	Chloride	Oral	0.6											Foetus 0.4–0.6
	Chloride	IV	0.3–0.6											
^{32}P	Sodium phosphate	IV	2–3	1.5	3–15	5–10	1			5–10		10		
	DFP	IV	0.3	10–15										
^{42}K	Chloride	IV	0.3–0.6				0.5			0.2		0.4		Muscle 0.4
^{45}Ca	Chloride	Oral	3		20									
	Chloride	IV	4		15–40									

^{47}Ca	Chloride	Oral	1		6	4						
	Chloride	IV	1–2		5–20							
^{51}Cr	Labelled RBC	IV	0.03–0.6	0.3–1.5			0.03–0.08		0.5–0.8		0.8–1	Foetus 0.3–0.6
	Labelled RBC (modified)	IV	0.06–0.1				0.03–0.1		0.6		5–15	
	Sodium chromate	IV	0.03–0.08	0.1–0.2				0.5–1.5	2–10		3–7	
	EDTA	IV					0.003	0.01	0.02			
^{55}Fe	Chloride	Oral	0.1–0.15		0.15				0.4		0.7	
	Chloride	IV	1–1.5	10–15	1.5	5			4		7	
^{59}Fe	Chloride	Oral	1		0.4							
	Chloride	IV	5–10	25	4	20			8		8	
	Citrate	Oral	1		2		1	1	1		0.6	
	Citrate	IV	10		40		40	25	40		60	
^{57}Co	Vitamin B_{12}	Oral	0.5–1.5				20–40		10–40			
^{58}Co	Vitamin B_{12}	Oral	5–40				40		100–150			
^{60}Co	Vitamin B_{12}	Oral	70–170				150		300–1200			
^{67}Ga	Citrate	IV	0.06	0.06	0.2	0.1–0.25	0.2	0.1	0.1	0.06	0.1	Pancreas 0.06
^{68}Ga	Citrate	IV	0.01		0.1			0.06				

Table 5.4 (*Continued*)

Radio-nuclide	Radiopharma-ceutical	Admin. route	Total Body	Blood	Bone	Bone marrow	Brain	Gonads	Kidneys	Liver	Lungs	Spleen	Thyroid	Other tissues
^{75}Se	L-seleno-methionine	IV	2–2.5	2–3				0.1–0.2	3	8		4–6	1.5–2	Pancreas 3
^{81m}Kr	Gas	Inhalation (2)									0.2–0.4			
^{85}Kr	Gas	Inhalation (1)						0.015			0.7			Trachea 2–3
	Solution in saline	IV				3		0.000 03			0.0015			Trachea 0.004
^{85}Sr	Nitrate	IV	1.5–6		8–15									
	Chloride	IV	1.5–6		8–15									
^{87m}Sr	Nitrate	IV	0.003			0.03								
	Chloride	IV	0.003		0.03–0.1									
^{99m}Tc	Pertechnetate	IV	0.003–0.005				0.002	0.003–0.01	0.03	0.01–0.02			0.03–0.15	Stomach 0.01–0.05
	Albumin	IV	0.003–0.005	0.13				0.01						
	Sulphur colloid	IV	0.003–0.005	0.005		0.005–0.008		0.003–0.005		0.05–0.1		0.05–0.15		
	Gluco-heptonate	IV	0.003					0.06	0.08					Bladder 0.07

	DTPA	IV	0.005	0.01			0.003–0.005	0.03					Bladder 0.15
	DMSA	IV	0.004			0.004		0.4	0.015				Ovaries 0.003
	HIDA	IV	0.008					0.1	0.01				Upper large intestine 0.20
	MAA	IV	0.003					0.02	0.01	0.07	0.05	0.02	Bladder 0.04
	MDP	IV	0.002					0.01					Bladder wall 0.12
^{113m}In	Ferric hydroxide	IV	0.004							0.15–0.25			
	Colloid	IV	0.004	0.003–0.02	0.006		0.006		0.015–0.15	0.1–0.15	0.02		
	DTPA	IV	0.003–0.006	0.01			0.006		0.1	0.003–0.015	0.02–0.03		Bladder 0.1
^{123}I	Sodium iodide	IV	0.01									5	
^{125}I	Sodium iodide	Oral	0.1									100–400	
	Albumin	IV	0.15–0.2	0.8									
	Hippuran	IV	0.003–0.01				0.005–0.08	0.06					
	Maa	IV	0.003						0.06				
^{131}I	Sodium iodide	Oral	0.15–1	0.2–0.6			0.6–0.8		0.3			300–600	

Table 5.4 (*Continued*)

Radio-nuclide	Radiopharma-ceutical	Admin. route	Total Body	Blood	Bone	Bone marrow	Brain	Gonads	Kidneys	Liver	Lungs	Spleen	Thyroid	Other tissues
	Albumin	IV	0.3–0.8	1.5–6				0.6–2.5		0.3			6–15	
	MAA	IV	0.03–0.1	0.15–0.4				0.1–0.35	1.6	0.15–0.6	1–1.5	0.1–0.8	20–60	
	Rose Bengal	IV	0.1–0.3							0.2–0.8				
	Hippuran	IV	0.01–0.06					0.015–0.06	0.03–0.3	0.02			40	Bladder 0.1–2
	PVP	IV	0.3–0.03 0.8				0.04	3	3.5	0.15–2.5		3	10	
	Oleic acid	Oral	0.2											Intestines 6
^{132}I	Sodium iodide	Oral	0.03–0.06	0.06										Intestines 6
^{125}Xe	Gas	Inhalation (3)		1				0.2			1.5		40*	Major airways mucosa 7
^{127}Xe	Gas	Inhalation (3)		1				0.3			1.5			Major airways mucosa 5
^{133}Xe	Gas	Inhalation (3)		3				0.2			3			Major airways mucosa 35

	Solution in saline	IV	0.0001					0.00005–0.01			0.001		Fat 0.0005 Trachea 0.006
^{131}Cs	Chloride	IV	0.05–0.1		0.05		0.003	0.3	0.6	0.1	0.01	0.06	Pancreas 0.02
^{198}Au	Colloid	IV	0.15–0.6	0.15		0.6–2		0.1–0.15	2	6–10		2–10	

* Radiation dose from daughter ^{125}I. *Inhalation doses in milligrays:*

(1) One minute breathing air containing 100 MBq l^{-1}.

(2) Dose absorbed during collection of 2×10^6 counts, sufficient for 6 gamma-camera views.

(3) Dose absorbed during 3 min breathing of recirculated air containing 100 MBq l^{-1}. (An equivalent technique using a single-breath inhalation at a higher concentration followed by free breathing gives about the same absorbed dose.)

organ from the beta particles as

$$D_\beta = 1.994 \times 10^{-8} A_s' \bar{E} (t_{1/2})_e \left\{ 1 - \frac{(t_{1/2})_u}{(t_{1/2})_p} \right\} \text{ Gy}$$

The same argument applies of course to the gamma-ray dose and we may write for the gamma-ray dose to an organ, due to its radioactive content alone,

$$D_\gamma = 1.246 \times 10^5 A_s' \gamma f (t_{1/2})_e \rho \bar{g} \left\{ 1 - \frac{(t_{1/2})_u}{(t_{1/2})_p} \right\} \text{ Gy}$$

There will be an additional gamma-ray dose to the organ from radionuclide external to it, i.e. the radionuclide not taken up from the bloodstream.

It will be seen that the above formulae reduce to those given earlier, if the half-period for the clearance of activity from the blood is negligible compared with the physical half-life of the radionuclide. Otherwise the doses are reduced by the factor in the curly brackets; they are also reduced if mechanisms exist which enable a fraction of the radionuclide to be excreted from the system without passage through the particular organ. Dosages to the lining of the urinary bladder are obviously indeterminate to a limited extent, because activity is removed from this organ step-wise, not continuously. Some estimates of absorbed radiation doses are shown in Table 5.4.

5.17 Exercises

1. Write a computer program which will enable you to read in data for a line-spread function and calculate the FWHM.
2. Estimate the essential dimensions of a focusing collimator assuming a diameter D = 125 mm, giving a resolution of 7 mm at a focal distance of 150 mm, for use with (a) ^{99m}Tc and (b) ^{18}F. Compare the geometric plane-source efficiency of the two collimators.
3. Estimate the radiation absorbed dose to a normal thyroid gland of 25 g which has taken up 0.2 MBq of ^{131}I.
4. Estimate the radiation absorbed dose to a liver of 0.75 kg following intravenous injection of 400 MBq of ^{99m}Tc-sulphur colloid. Assume an effective half-life of 3 h for the activity in the liver and a half-life for clearance from the blood of 10 min.
5. Confirm the estimate of the radiation absorbed dose to the bladder consequent upon the administration of ^{99m}Tc-DTPA for a kidney scan, as given in Table 5.4.
6. Devise a proof of the reciprocal dose theorem.

6
System Analysis

BIOLOGICAL SYSTEMS

6.1 Introduction

The basic data obtained from *in vivo* and *in vitro* measurement are, in general, sets of numbers which represent count-rates of one sort or another. These numbers generally require only rather simple manipulation to give figures (e.g. percentage uptake, T3/T4 values) of diagnostic value to the clinician. Similarly, one of the functions of a scanner or a gamma camera is to 'process' information about the distribution of count-rates over the organ in question, and to present its results in a form (the scan or autoradiograph) which is supposed to contain diagnostic information. In both these instances we can discern three essential steps:

(a) the gathering of 'raw' data,
(b) data processing, and
(c) the presentation of processed data in an intelligible form.

In this chapter we are going to broaden our viewpoint very considerably by adopting a *systems analysis* approach. This will involve some references to the material in Chapter 1, but we shall not attempt mathematical rigour.

First we need some definitions. We regard a *system* as any collection of equipment, experimental technique, or whatever else may be required which has (i) a recognisable input ('raw data') and (ii) a recognisable *output* ('significant data'). It corresponds therefore to (*b*) in the three steps we listed earlier. The breadth of the concept of system is illustrated by the following examples. Thus, a desk calculator could be regarded as a 'system' with an input consisting of numbers keyed in, and an output in the form of other numbers in the display. On a very different level, a hospital is a system whose input is sick patients, and whose output is (hopefully) healthy people. Clearly complicated systems like hospitals are interacting groups of smaller simpler systems such as calculators and scanners. All systems however share some common principles, and when these are expressed mathematically it becomes easier to understand how a complex and unfamiliar system works, by analogy with a simple and familiar system.

6.2 Linearity

One important restriction we must acknowledge is that because of mathematical difficulties it is generally possible to study only linear systems. A formal definition of linearity is also follows. If an input X_1 gives an output Y_1 and an input X_2 gives an output Y_2 and also an input $(X_1 + X_2)$ gives an output $(Y_1 + Y_2)$, then the system is linear. For example, we may consider a system as comprising a radiation detector, a scaler–timer, and a calculator which automatically subtracts background counts. The input may be radiation from sources of different strengths and the output is a list of count-rates.

If we find

$$\text{input} = X_1 = 3\ \text{MBq}; \quad \text{output} = Y_1 = 6000\ \text{s}^{-1}$$
$$\text{input} = X_2 = 4\ \text{MBq}; \quad \text{output} = Y_2 = 8000\ \text{s}^{-1}$$
$$\text{input} = X_3 = 7\ \text{MBq}; \quad \text{output} = Y_3 = 14\,000\ \text{s}^{-1}$$

then we may conclude that we have a linear system, because the output from a 7 MBq source is equal to the sum of the outputs from a 3 MBq source and a 4 MBq source taken separately.

Note that in this example the background subtraction is an essential feature of the system. If we had had a background count of 1000 s^{-1}, the 'gross' count-rates would have been

$$Y_1' = 7000\ \text{s}^{-1}$$
$$Y_2' = 9000\ \text{s}^{-1}$$
$$Y_3' = 15\,000\ \text{s}^{-1}$$

so that $Y_3' \neq Y_1' + Y_2'$, and the system *without* background subtraction would not be linear. It is important to realise that we can often describe a system in a number of slightly different ways, depending on the exact definitions of what we are taking as input and/or output; in general we should try to choose the input and output parameters so as to make the system linear.

Linearity cannot always be achieved, however. If we have the radiation detection equipment just mentioned, we might wish to study the output (nett) count-rate for a constant fixed source but as a function of the high voltage supplied to the detector. Then although the physical apparatus is still the same, we are considering the 'H.T.' as being the input to the system, and will find that the output is very far from being a linear function of the input. Moreover there is no way of modifying our basic choice of input and output which will result in a linear relationship.

Sometimes, it is necessary to consider a system with more than one input (or more than one output). Thus, the result of an investigation might depend on the values of two independent (nett) count-rates; an example is the dual isotope counting technique discussed in Section 3.28. For a system with two inputs, we extend the criterion of linearity as follows. Suppose the inputs are variable quantities X and X', and the output is some function F of these, so that we may write for the output, Y, that $Y = F(X,X')$. Then any particular input values X_1, X_1' will give an output Y_1 say, while another input pair X_2, X_2', will give an output Y_2, where

$$Y_1 = F(X_1, X_1')$$
$$Y_2 = F(X_2, X_2')$$

Then we will define our system as being linear if the pair of inputs $(X_1 + X_2)$ and $(X_1' + X_2')$ gives rise to the output $Y_1 + Y_2$; i.e. if

$$Y_1 + Y_2 = F((X_1 + X_2), (X_1' + X_2'))$$

It is easy to see that the system is linear if the function F just means a summation, for then we have

$$Y_1 = X_1 + X_1'$$
$$Y_2 = X_2 + X_2'$$

and the condition for linearity is

$$Y_1 + Y_2 = (X_1 + X_2) + (X_1' + X_2')$$

which is obviously true. But if the function F is a multiplication so that

$$Y_1 = X_1 X_1'$$
$$Y_2 = X_2 X_2'$$

the system is not linear because

$$Y_1 + Y_2 \neq (X_1 + X_2).(X_1' + X_2')$$

The criterion we have been using to test for linearity is sometimes called the *principle of superposition*.

6.3 Static and Dynamic Systems

A radiation detection system as outlined above may be used in a *static* mode— i.e. to give a measure of the strength of a single fixed source—or it may be used to give an output which varies with time as the source itself is allowed to vary with time. For instance, if a patient is injected with ^{51}Cr-labelled red blood cells and counts are taken at intervals over the liver, we could regard the system (patient + uptake equipment) as a *dynamic* system, the interest lying in the variation with time of the counts from the liver rather than their absolute magnitude. A scanner is a detector working in a dynamic mode in the sense that it receives radiation sequentially from different regions of the organ being examined; the scanner, however, generally presents its output as a *static* image in the sense that all the information appears at one time on the photoscan. The distinction between static and dynamic modes is therefore somewhat arbitrary, and the real crux of the matter is that we should not limit ourselves to considering systems where input/output is just a single fixed quantity; inputs and outputs in general can be quantities which vary with time (e.g. uptake measurements in a single organ), or with spatial coordinates (scans), or even with both time and space coordinates (sequential pictures from a gamma camera).

Mathematically, input and output are to be represented as functional quantities $I(x,y,t)$ and $O(x,y,t)$. We shall not attempt such generality in what follows because we can illustrate the power of our methods using only a single variable—that is to say, we will consider inputs and outputs which depend on only one spatial coordinate, or only on the time. But much of the mathematical formulation we are going to employ could in fact be replaced by a matrix notation and would then apply to much more complicated situations.

6.4 Modelling

A central feature of systems analysis is that the real system we are trying to study is often so complicated and appears to have so many, perhaps irrelevant, side-issues that we find it impossibly hard to understand. But we would hope, by a mixture of luck, experience and intuition to construct a simpler system, called a model, which would have the properties:

(a) its inputs and outputs would be analogous to those of the real system in some recognisable way, and

(b) its mode of operation would be known to us; we would 'know how the model worked' and could therefore infer how the real system works.

The idea is an extension of that of a radioactive phantom described earlier—which is really a 'model' of a scanner input, not a true *system* model. A better analogy to a system model be a toy locomotive which might reproduce on a small scale the behaviour of a 'real' locomotive. But models need not bear any physical resemblance to the systems being modelled. For example, the flow of blood in an arterial system might be modelled by the flow of water in a plastic pipe, although such a model is hardly likely to be very satisfactory because water and plastic have many physical properties (e.g. viscosity, elasticity) which are rather dissimilar from those of blood and arterial walls respectively. Our model would have to simulate variations in diameter of the lumen—again a matter of some practical difficulty. A more convenient model might be a flow of electric charge through a circuit consisting of a network of components such as resistors, capacitors and inductors. There would be almost no physical resemblance to the 'real' system but the model would be more flexible in use because it would be so much easier to modify any of its parameters (the actual values of the components and their arrangement in the network). And the argument does not stop at this point: while it seems natural to represent a fluid flow by an electric current, it is no less *effective* to represent the numerical value of any physical variable (e.g. concentration of a blood constituent, composition of expired air) by the numerical value of an electrical quantity (e.g. current, voltage). We are, therefore, led to the concept of an electrical model as a four-terminal network: two terminals being used for input, viz. an electrical signal such as a current or voltage whose magnitude is proportional to the magnitude of some kind of stimulus being applied to a biological system; and two terminals being used as output, viz. an electrical signal which simulates the effect of the stimulus. What we are saying is that if we can find a model which correctly *represents* the real biological system, we can expect the workings of the model to tell us something about the workings of the real system, however tenuous the physical resemblance between the two.

One advantage of inventing electrical models is that their behaviour can very often be predicted from a straightforward application of physical principles. This prediction of behaviour is often termed 'solution of the model', because what happens in practice is that the model is never physically constructed at all; instead mathematical relationships are written down which describe the currents, voltages, etc., in the network. The relationships are, in general, differential equations whose solutions are the output y (say as a function of t the time), which is to be expected from a given input x (also say a function of the time). For our present purposes, it is sufficient to consider single inputs and outputs only, but the concepts are of very general nature and their application to a wide range of problems is limited only by mathematical difficulties.

6.5 Compartment Analysis: Basic Concept

A very powerful model of a biological system is that of a series of compartments through which materials are flowing at a constant rate. In the absence of evidence to the contrary, each compartment is assumed to be *homogeneous*, that is to say it has the same properties, such as composition, at all points within it. Compartments may be defined arbitrarily to suit the problem in hand; they may be physically definable spaces within the body, for example the bloodstream or skeleton, or definable constituents of such spaces. Thus it is often convenient to define the red blood cells (RBC) and the blood plasma as two separate compartments, although they occupy the same physical space.

In such a case, the flow we are concerned with might be, for example, that of iron (bound to transferrin) in the plasma; flow not in a mechanical sense but in the sense that a 'flow-sheet' express in a diagrammatic way the movement of iron atoms from the plasma compartment, via erythropoietic tissues into the red blood cell compartment. Such flows are said to be *steady*; there may be slight variations over short time intervals, but on a reasonably long-term basis it must be that the rate of ingress of iron (from some previous compartments in a chain) is equal to the rate of outflow, otherwise there would be a continuous increase (or decrease) in the iron content of the particular compartment, e.g. plasma. Of course, as a disease or malfunctioning of the system evolves, the iron content of the plasma may slowly change from values which lie between acceptable limits of normality to some other value, but we generally assume a steady value for such parameters, at any rate over the time needed to carry out an experimental measurement. It should be noted that a mere measurement of the concentration, or total amount, of iron in the plasma tells us nothing about the flow. However, we may define the *mean transit time* as the time spent on the average by the individual iron–transferrin molecules in the plasma, and the following relationship must hold

$$\text{Flow} = \frac{\text{Mass of material in compartment}}{\text{Mean transit time (days)}} \qquad \text{kg d}^{-1}$$

—a relationship which we have put in general terms as it is applicable to any constituent of any compartment. Other units of mass and time may be more suitable in particular cases, of course.

To express this mathematically we may let M be the mass of any constituent of a compartment at time t; then in an element of time Δt a mass ΔM enters the compartment from a previous one in the chain and an equal amount leaves it in the same time. The flow *through* the compartment is then $\Delta M/\Delta t$ or $\mathrm{d}M/\mathrm{d}t$ in the limit. Our basic relationship is then that this flow is numerically equal to M/T, T being the mean transit time. It is sometimes convenient to work in terms of the reciprocal of T, i.e. $\lambda = 1/T$, so that the fundamental equation becomes

$$\frac{\mathrm{d}M}{\mathrm{d}t} = \frac{m}{T} = \lambda M$$

We may refer to λ as the *rate-constant*, or *transfer-constant*. It should be emphasised that this equation has a special meaning as $\mathrm{d}M/\mathrm{d}t$ represents the flow *through* the compartment. Some care is therefore needed in interpretation when the equation is integrated. The *nett* effect on M of the flow into the compartment (equal to $\mathrm{d}M/\mathrm{d}t$) and the outflow (also equal to $\mathrm{d}M/\mathrm{d}t$ but in the reverse direction) is of course zero. If we were to interpret $\mathrm{d}M/\mathrm{d}t$ to mean the *nett* rate of increase in M, we would have

$$\frac{\mathrm{d}M}{\mathrm{d}t} = 0$$

which on integration means that M is a constant, which of course we are assuming to be true.

It is also generally assumed that for a given system the mean transit time, or T, is also constant with time. This implies that if we artificially increase the amount of material in the compartment (e.g. by injecting an extra amount of iron–transferrin in the above example), then the flow-rate will also increase. What we always observe is that the amount of material in the compartment decreases with time until the original mass is more or less closely re-established. It seems reasonable to suppose that the rate of outflow is increased, and to postulate that the increase in flow-rate is proportional to the additional mass of the particular constituent. It is not, however, very clear that the constant of proportionality will be exactly the same as that implied in our basic equation, viz. $1/T$ or λ. It could be unchanged in value, if exactly the same mechanisms for regulating the flows of material into and out of the compartment were to operate both before and after the change (addition of new material) which we imposed upon the system. However, it is always possible that an imposed change could act as a trigger to initiate new mechanisms, besides speeding up already existing ones. The overall effect would still be to re-establish the original state of the system, but it might be impossible to tell experimentally whether new mechanisms were operating or not. Clearly then, in order to obtain accurate numerical values of the mean transit time, we need to make measurements of the flow, as well as the amount of material in the compartment, under conditions which do *not* change either of these quantities significantly, that is to say, 'without disturbing the system'.

It is in the great power of the radioactive tracer technique that this condition can be met. We can indeed add to the compartment in question a small sample, suitably labelled with radionuclide, of the material being studied and then follow its transfer to other compartments without appreciably disturbing the system. There is one troublesome practical difficulty, which is that radioactively labelled material can usually be injected at only one physical point in the compartment, and we must be able to allow sufficient time for it to become mixed with the substance being traced throughout the compartment. Ideally, the tracer molecules are chemically identical with those of the substance being traced, although this is not always achieved in practice. For instance, albumin labelled with ^{131}I, i.e. albumin molecules chemically combined with ^{131}I atoms, it is believed, will satisfactorily trace ordinary albumin which has no stable iodine in its composition. The investigation of tracer behaviour can be carried out in a number of ways, and in principle either by measuring its rate of removal from the compartment into which it has been placed, or by measuring its rate of arrival in other compartments.

The success of the radioactive tracer technique depends on two important principles. The first is that the mass of added tracer material is so small that it does not significantly affect the processes regulating the flow, and, therefore, the flow itself is unchanged. The second is that the mass of tracer injected into the system, m, obeys the basic equation in the sense that it corresponds to the outflow only (since no radioactive material is being transferred into the compartment from a previous one). Then in a time Δt the amount of tracer leaving the compartment is Δm, say; the *nett* rate of increase in m is $-\Delta m/\Delta t$, or $-\mathrm{d}m/\mathrm{d}t$ in the limit. We then have that

$$-\frac{\mathrm{d}m}{\mathrm{d}t} = \frac{m}{T} = \lambda m$$

This equation although superficially resembling the previous one for M has however quite a different meaning, because $-\mathrm{d}m/\mathrm{d}t$ represents the *outflow* of activity, there being no inflow; also m is now a variable (function of time). In particular it can be integrated to read

$$m = m_0 \exp(-\lambda t)$$

Also since the activity A, and also the count-rate C observed under fixed geometrical conditions, due to a radioactive sample is always proportional to its mass (after making corrections for decay),

$$A = A_0 \exp(-\lambda t)$$

and

$$C = C_0 \exp(-\lambda t)$$

In these equations A_0 and C_0 of course mean the activity and count-rate of the tracer at a specified time (usually taken as the time at which the injection took place). For the rest of this chapter, we will assume that these quantities have been corrected for physical decay.

Our model then leads us to the prediction that corrected count-rates as a function of time should show an exponential decrease, as occurs with ordinary physical decay, except that the rate-constant λ is determined by the biological process of transfer of material from compartment to compartment. There exists a biological half-life, during which an initial count-rate decays to half its value, and numerically equal to $0.6932/\lambda$. This time is indeed more often quoted in the literature than the more fundamental mean transit time T, which is $1/\lambda$ or $1.443 t_{1/2}$.

It is of interest to see that if C_0 is regarded as the input to the model and C as the output, the model is linear in the sense we have already defined. For at any fixed time t the term $\exp(-\lambda t)$ is a constant, and it is easy to show that a system in which the output is proportional to the input is linear, by applying the principle of superposition. What this means is that a tracer experiment will give the same result for the value of T or λ independently of the absolute amount of tracer used, just as the value of a half-life in a purely physical decay process is independent of the initial amount of radioactive sample.

We turn now to some other models, which are progressively less and less similar to biological systems, showing however that they are all equivalent to the above because they all lead to the same mathematical formulation.

6.6 A Single-Compartment Model: The Continuously Stirred Tank Reactor or CSTR

This model is taken from a standard piece of chemical engineering equipment and consists of a tank of fluid of fixed volume which is continuously stirred so as to be always thoroughly mixed. Fluid flows continuously into the tank at a constant rate and there is an outflow of mixed fluid which keeps the volume of fluid in the tank steady. We now investigate the behaviour of this system by injecting a tracer activity, of negligibly small volume, at time zero. The activity remaining in the tank is measured as time proceeds (Fig. 6.1).

Suppose we have a unit flow rate, that is to say the fraction of the fluid leaving the tank in Δt seconds is just $\Delta t \times (1/V)$. Then if at any time t the activity remaining in the tank is A, we must have

$$-\Delta A = \frac{\Delta t}{V} A$$

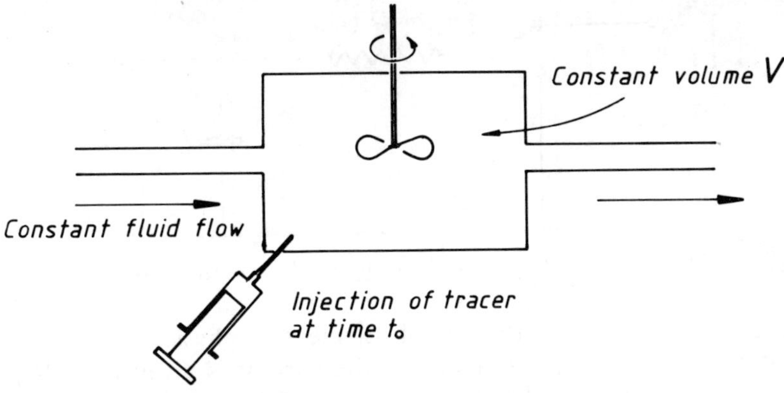

Fig. 6.1 The continuously stirred tank reactor (CSTR).

or

$$\frac{dA}{dt} = -\frac{A}{V} = -\lambda A$$

where

$$\lambda = \frac{1}{V}$$

This model therefore, leads to just the same equations as before, both for the activity in the tank and for the count-rate observed by a radiation detector in a fixed position nearby. One must be careful not to confuse the parameters of the model with those of the original biological system. For instance, a biological system may have the bloodstream of a patient as a compartment and this bloodstream will of course have its own 'volume' and 'flow'. There will also be, for example, the 'flow' of iron atoms from plasma to RBC; it is the latter 'flow' which is modelled by a CSTR whose own (unit) flow and volume are such that

$$\frac{\text{flow}}{V} = \lambda$$

simulates the iron 'flow' as expressed at the fraction of iron molecules transferred per second out of the plasma.

6.7 A Single-Compartment Model: The Charged Capacitor

Consider the four-terminal network of Fig. 6.2. Suppose a voltage V_0 is applied to the input terminals for a long time extending up to the instant $t = 0$. During this time, with the switch in the position shown, the capacitor will become fully charged, and the output voltage will also be V_0. Now at the instant $t = 0$, let the switch be suddenly reversed, so that the input voltage is reduced abruptly to zero. The capacitor will discharge through the resistor R

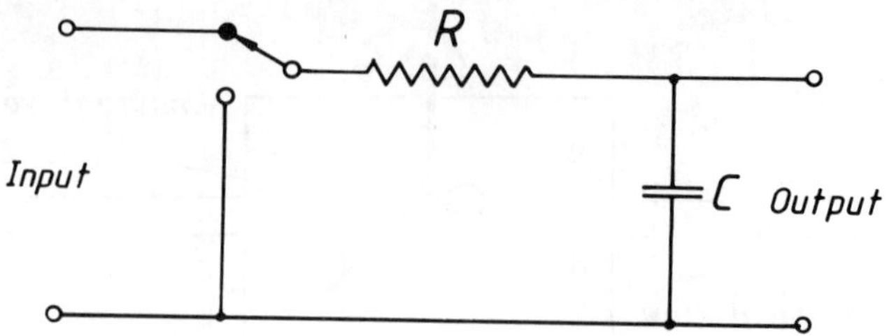

Fig. 6.2 Single-compartment model: the charged capacitor.

and the output voltage, $V(t)$, will fall. Elementary physics tells us that if the capacitor holds a charge Q at time t, the current is given by $-\mathrm{d}Q/\mathrm{d}t$, or $-C\,\mathrm{d}V/\mathrm{d}t$, and also by V/R. Therefore

$$C\frac{\mathrm{d}V}{\mathrm{d}t} + \frac{V}{R} = 0$$

or

$$\frac{\mathrm{d}V}{\mathrm{d}t} + \frac{1}{RC} \cdot V = 0$$

This equation has an identical form to that of the previous section with the voltage replacing the count-rate and with $1/RC$ for λ. The solution is of course

$$V = V_0 \exp(-\lambda t) = V_0 \exp(-t/RC)$$

We can say that the charged-capacitor model with the 'step' input (input V_0 changing to zero at $t = 0$) exactly simulates the biological system; the voltage $V/(t)$ is the *analogue* of the count-rate $C(t)$.

An important variation of the above is obtained by using a digital voltmeter (DVM) to indicate the instantaneous voltage across the capacitor at instants which are equally spaced in time. A record of DVM readings will then constitute a *sampled* output of the system. If the readings are taken at intervals of T, it is easily shown that the kth reading ($k \geq 0$) will be

$$V_k = V_0 \exp(-k\lambda T)$$

6.8 A Single-Compartment Model: The Discrete-Time Analogue

This is a conceptual model in which we imagine a system which can accept as input a stream of separate signals of magnitude $x_1, x_2, x_3, \ldots$ which are applied at equally spaced intervals of time. Correspondingly the output produced by the system is a stream of signals of magnitude $y_1, y_2, y_3, \ldots$ at equal intervals. The process of converting the input into the output is specified by the flow diagram of Fig. 6.3.

The essential features of the model are: (1) the delay unit, which can accept any signal and pass it on to the next component after a delay of one unit of time; (2) an amplifier which multiplies the magnitude of any incoming

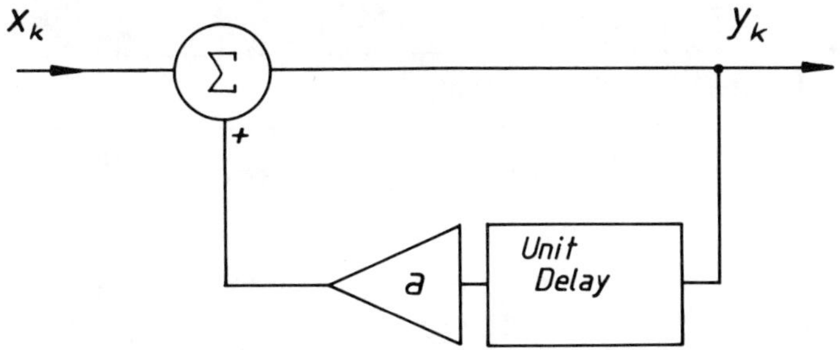

Fig. 6.3 Unit-delay discrete-time model.

signal by a factor a; and (3) a summing unit which accepts any pair of signals which arrive simultaneously and passes their sum onwards. To see how the model works, let us consider a simple case. This is when we have $x_k = 0$ for all negative values of the subscript k. Then it is obvious that all the outputs y_k will be zero for negative k. However we arrange for x_0 to have a non-zero value, say V_0; this will pass through the system and give rise to the output y_0 also equal to V_0. In addition, a signal of magnitude V_0 is also passed to the delay unit, from which it emerges one unit of time later. It is multiplied by a by the amplifier so that a signal V_0a arrives at the summer. Its arrival coincides with that of the next signal input, i.e. x_1. The summer therefore produces the output $y_1 = x_1 + V_0a$. In particular, if x_1 happens to be zero, the output $y_1 = V_0a$. To find y_2, we must take into account a signal V_0a being passed through the delay unit and amplifier, so that a new signal V_0a^2 must be added to x_2. Again if $x_2 = 0$ we find

$$y_2 = 0 + V_0a^2 = V_0a^2$$

Thus it is clear that we can summarise the behaviour of the model by listing the input and output sequences as follows.

$$\begin{array}{ll}
x_{-\infty} = 0 & y_{-\infty} = 0 \\
\vdots & \vdots \\
x_{-1} = 0 & y_{-1} = 0 \\
x_0 = V_0 & y_0 = V_0 \\
x_1 = 0 & y_1 = V_0a \\
x_2 = 0 & y_2 = V_0a^2 \\
\vdots & \vdots \\
x_k = 0 & y_k = V_0a^k
\end{array}$$

We find therefore that a *pulsed* input $x_0 = V_0$ (all other values of x_k zero) to this model gives an output sequence of the same form as the sampled capacitor-discharge model, and the models are functionally equivalent if $V_k = V_0 \exp(-k\lambda T) \equiv V_0a^k$, i.e. if the amplification factor $a = \exp(-\lambda T)$.

We are not much concerned here with a practical realisation of this model. It might be an electronic system in which the input is a series of voltage pulses of different amplitudes. Perhaps more usefully it could be realised in the form

Table 6.1

Time	Input	Output
$t_{-\infty}$	$x = x_{-\infty} = 0$	$y_{\infty} = 0$
$\vdots$	$\vdots$	$\vdots$
t_0	$x = x_0$	$y = y_0 = x_0 + y_{-1}a = x_0$
t_1	$x = x_1$	$y = y_1 = x_1 + y_0a = x_1 + x_0a$
t_2	$x = x_2$	$y = y_2 = x_2 + y_1a = x_2 + (x_1 + x_0a)a$
t_3	$x = x_3$	$y = y_3 = x_3 + y_2a = x_3 + \{x_2 + (x_1 + x_0a)a\}a$

of a computer program, or even a program for a programmable desk calculator. Such programs can easily be written so as to accept any arbitrary input, with non-zero values of $x_0, x_1, x_2, \ldots$ as may be required. Such programs can then simulate the behaviour of biological systems in which the tracer is not necessarily all injected at zero time, but comes into the system more or less gradually. We shall find this of importance when we study the transfer of materials in many-compartment systems. At the moment, we will just give two representations of the input–output relations for an arbitrary input $x_0, x_1, x_2, \ldots$. These we classify as (a) algebraic and (b) programmatic.

(*a*) *Algebraic.* We may make a list of input and output values at different times as shown in Table 6.1. We can see from this tabulation that at time t_{k+1}, x will be x_{k+1}, and y will be y_{k+1} and equal to $x_{k+1} + y_ka$. The general equation

$$y_{k+1} = x_{k+1} + y_ka$$

is called a *difference* equation, and is completely analogous to a differential equation for a continuous system. The analogy appears even closer by writing it as

$$y_{k+1} - ay_k = x_{k+1}$$

If we now define the *shift operator* S as an operator which replaces y_k by y_{k+1}, i.e. by the identity

$$S(y_k) \equiv y_{k+1}$$

the difference equation may be written

$$(S - a)\,(y_k) = x_{k+1}$$

We know that the solution to this equation, for all inputs (except x_0) equal to zero, is $y_k = y_0a^k$; so that the parallel with the differential equation $(D + \lambda)y = 0$ whose solution is $y = y_0 \exp(-\lambda t)$ is exact. There is in fact a whole calculus of difference equations in which the shift operator S plays just the same role as the differential operator D in differential calculus.

(*b*) Programmatic. The following FORTRAN segment suggests how to read in values for a rate-constant λ and input activities X at a succession of times, and to print out the corresponding values of Y. NN non-zero input values are allowed for and NNN output values.

```
      READ 1 LAMDA, NN, NNN
      A = EXP(-LAMDA)
      READ 2 (X(N), N = 1, NN)
      NN1 = NN + 1
      DO 3 N = NN1, NNN
      X(N) = 0.
    3 CONTINUE
      Y(1) = X(1)
      DO 10 N = 2, NNN
    5 M = N - 1
    6 Y(N) = X(N) + A * Y(M)
   10 CONTINUE
      PRINT 20 (Y(N), N = 1, NNN)
```

It will be seen how in this program the statement labelled 5 corresponds to the required unit delay, so that in the 'summation' statement 6, the Nth input is added to A times the *previous* output. Alternatively statement 5 can be regarded as a shift operator since its function is to change a subscripted value by one unit.

6.9 Single-Compartment System with Sampled Output

We have introduced the idea of sampling in our models, and of course the same concept applies to the system itself. In practice, we rather rarely measure count-rates continuously; rather, individual samples are obtained, say at regular intervals T'. For example, activity in the blood plasma might be measured on samples taken at daily intervals. If the activity does follow an exponential law, we could simulate the biological system by a sampled charged-condenser model, so that the kth count-rate $C_k = C_0 \exp(-k\lambda T')$ would correspond with the kth voltage $V_R = V_0 \exp(-k\lambda T)$. It should be noted that the timescale of the biological system and of the model can be very different; thus T' might be an interval of 1 d and T an interval of 1 s. All this means is that the constant λ in the biological system will be in units of d^{-1} whereas in the model the units will be s^{-1}, but the numerical value will be the same. Thus, while it may take many days to acquire data on the actual system, the data may be simulated on the model, each simulation taking just so many seconds. Furthermore it is a simple matter to construct a model in which the values of R and C can be varied at will, so that the behavior of the model can be made to 'fit' the experimental data from the actual system.

6.10 Two-Compartment Models

The simplest possible two-compartment system would be one in which the transfer of material from one compartment to a second is followed by transfer

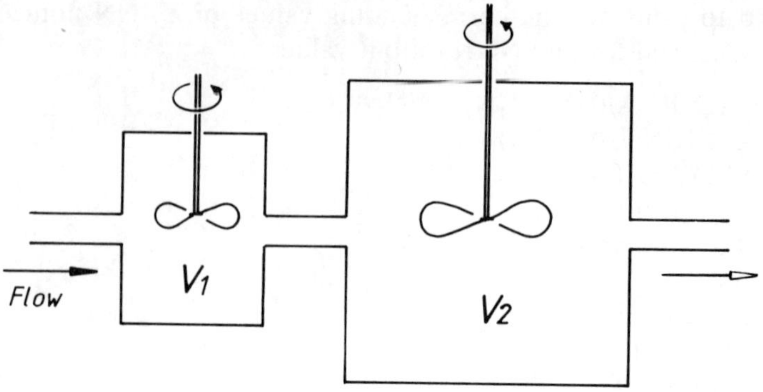

Fig. 6.4 Two CSTRs as a model of a two-compartment system.

from the second to a third, the transfers being characterised by parameters λ_1 and λ_2 respectively. A dynamic model would be, for example, a pair of CSTRs in tandem as shown in Fig. 6.4. The fluid flow is the same in both tanks, but as they are of different volumes V_1 and V_2, the fractional flows $1/V_1$ and $1/V_2$ correspond to the constants λ_1 and λ_2 respectively.

Again it should be emphasised that the *volumes* of the CSTRs do not necessarily simulate the compartmental volumes; V_1/V_2 is *not* the same as the ratio of the volumes of the compartments in the biological system.

If an injection of tracer is made into the first CSTR in single-exponential fall in activity in it can be observed as before. For the second CSTR, using C_2 to mean the activity at time t, we have

$$\frac{\mathrm{d}C_2}{\mathrm{d}t} + \lambda_2 C_2 = \lambda_1 C_1$$

the term on the right-handside representing the influx of activity from the first CSTR. This equation can be solved by substituting for C_1;

$$\frac{\mathrm{d}C_2}{\mathrm{d}t} + \lambda_2 C_2 = \lambda_1[(C_1)_0 \exp(-\lambda_1 t)]$$

Then the solution for C_2 is

$$C_2 = \frac{\lambda_1 (C_1)_0}{\lambda_1 - \lambda_2}[\exp(-\lambda_2 t) - \exp(-\lambda_1 t)]$$

6.11 Discrete-time Analogues for Two-Compartment Systems

As shown earlier, the behavior of a single-compartment system can be modelled by a device comprising a summer, delay unit and amplifier. One easy way to extend this device to model a second compartment is to feed its output sequence into a second unit which will work in the same way but will have an amplifying factor say b, where as before $a = \exp(-\lambda_1 T)$ and now also $b = \exp(-\lambda_2 T)$, as shown in Fig. 6.5.

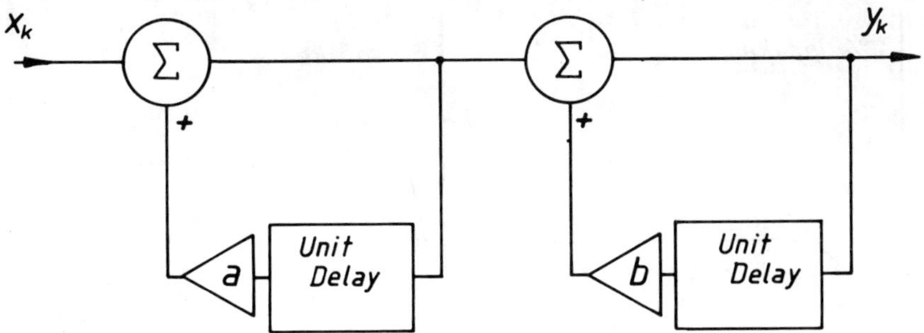

Fig. 6.5 A two-compartment discrete-time model.

The theoretical justification for this procedure is given by the following argument. By taking the differential equation for C_2 and differentiating again we easily have (compare Section 1.12)

$$\frac{d^2C_2}{dt^2} + (\lambda_1 + \lambda_2)\frac{dC_2}{dt} + \lambda_1\lambda_2C_2 = 0$$

or in D notation

$$D^2C_2 + (\lambda_1 + \lambda_2)DC_2 + \lambda_1\lambda_2C_2 = 0$$

Factorising,

$$(D + \lambda_1)(D + \lambda_2)C_2 = 0$$

The discrete-time analogy is

$$(S - a)(S - b)y_k = x_{k+2}$$

showing that two successive shift-and-multiply operations on a sequence y_k will give the required output sequence. Another technique uses the analogue

$$\{S^2 - (a + b)S + ab\}y_2 = x_{k+2}$$

directly, giving rise to the model shown in Fig. 6.6 which has the advantage of having only one summing unit.

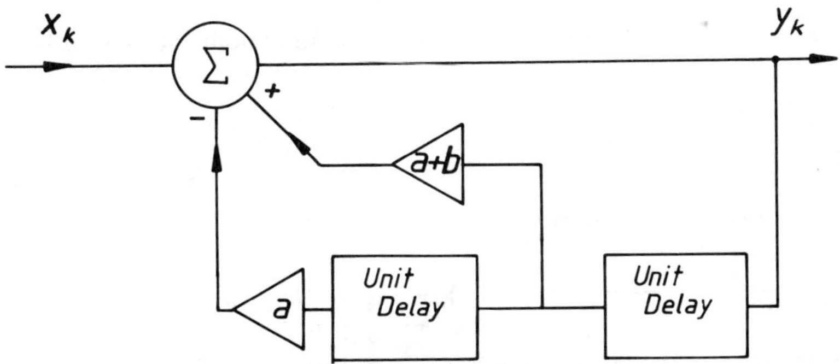

Fig. 6.6 Alternative two-compartment discrete-time model.

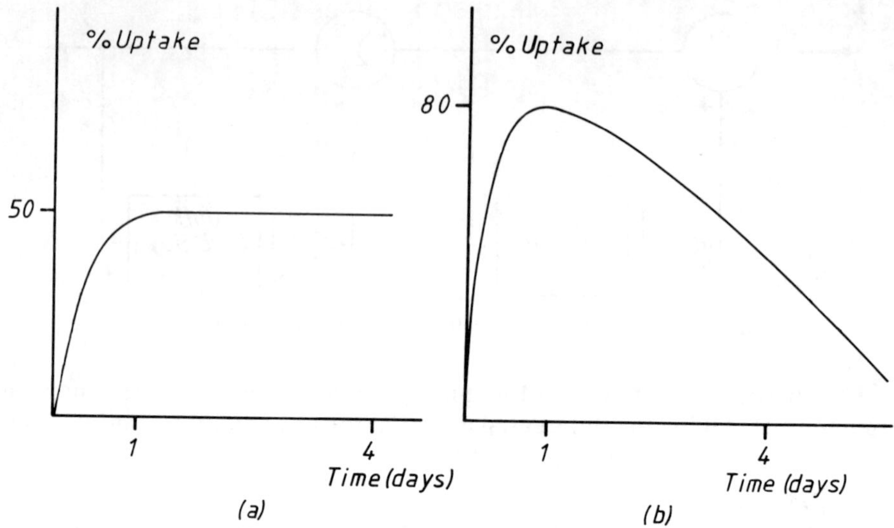

Fig. 6.7 Uptake of ^{131}I in thyroid: (a) normal subject, (b) hyperthyroid subject.

6.12 A Model for Uptake of Iodine in the Thyroid

A model of the fate of ^{131}I as iodide in the bloodstream supposes that there is a fractional rate of uptake into the thyroid gland represented by a constant λ_1; another fraction is excreted via the kidneys and we will assume a corresponding rate constant λ_2. Also the thyroid gland itself releases ^{131}I as protein-bound iodide into the bloodstream, governed by a rate constant λ_3. Curves of uptake in the thyroid as functions of time are shown in Fig. 6.7 for typical normal and hyperthyroid subjects it being assumed that an activity $A_1(0)$ is injected into the bloodstream at time t_0.

Denoting the activity of ^{131}I in the bloodstream at time t by A_1, we have

$$\frac{dA_1}{dt} = -(\lambda_1 + \lambda_2)A_1$$

whence

$$A_1 = (A_1) \exp \{-(\lambda_1 + \lambda_2)t\}$$

If A_2 is the activity in the thyroid at time t (of course being zero at time $t = 0$), we find

$$\frac{dA_2}{dt} = \lambda_1 A_1 - \lambda_3 A_2$$

whose solution is easily found to be

$$A_2 = \frac{\lambda_1 (A_1)_0}{\lambda_1 + \lambda_2 - \lambda_3} [\exp(-\lambda_3 t) - \exp\{-(\lambda_1 - \lambda_2)t\}]$$

The significance of the terms in this last equation is illustrated in the graph of Fig. 6.8.

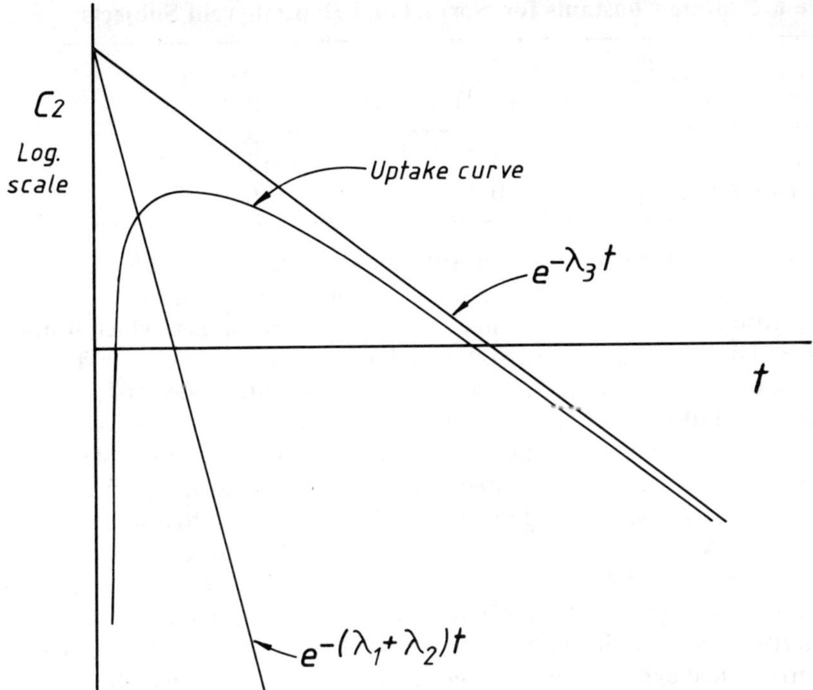

Fig. 6.8 Theoretical uptake curve for ^{131}I in thyroid.

Clearly the exp $\{-(\lambda_1 + \lambda_2)t\}$ term largely governs the shape of the rising part of the uptake curve. For both normal and hyperthyroid subjects the initial rate of uptake is quite fast, the upward slope of the curves corresponding to a half-life of roughly 0.1 d. We therefore have

$$\lambda_1 + \lambda_2 = 0.6932/0.1 \approx 7\ \mathrm{d}^{-1}$$

On the other hand, the term exp $(-\lambda_3 t)$ determines the final slope of the falling part of the uptake curve. For normal subjects the final slope is very small, the apparent half-life being at least 100 d. For these subjects λ_3 is therefore of the order 0.007 d^{-1}. Also for normal subjects we notice that the uptake at $t = 1$ d is approximately 0.5. Putting $t = 1$ and $A_2/A_1(0) = 0.5$ in the equation for A_2 we find

$$0.5 = \frac{\lambda_1}{7 - 0.007}[\exp(-0.007) - \exp(-7)]$$

$$\approx \frac{\lambda_1}{7}$$

since the first exponential term is nearly 1 and the second is negligibly small. The value is λ_1 is thus approximately 3.5, and this means that λ_2 also is approximately 3.5.

For hyperthyroid subjects, however, the final slope of the uptake curve is quite large, showing an apparent half-life of as little as 3 d. The uptake at

Table 6.2 Rate Constants for Normal and Hyperthyroid Subjects

Subject	λ_1 (d^{-1})	λ_2 (d^{-1})	λ_3 (d^{-1})
Normal	3.5	3.5	0.007
Hyperthyroid	0.2	6.8	0.23

$t = 1$ d is often as high as 0.8. Substitution of these figures into the equations enables us to calculate rough values λ_1, λ_2 and λ_3 as before.

For ease of reference, the values of these rate-constants, which it must be emphasised are rough representative values only, are collected in Table 6.2.

For any particular subject, of course, more accurate values of λ_1, λ_2 and λ_3 could be found by trial and error to produce a theoretical curve fitting the experimental points as closely as possible. Alternatively a simple electrical model could be used; for example, consider the circuit of Fig. 6.9.

In the model, a step voltage rising suddenly from zero to a steady value V_{in} would represent the appearance of activity in the bloodstream. If for the moment we think of R_2 as extremely high and the two capacitors C_1 and C_2 replaced by an equivalent $C_3 = C_1C_2/(C_1 + C_2)$, we see that the potential at X will rise exponentially with a time constant $R_1C_3 = 1/(\lambda_1 + \lambda_2)$. Assuming no current leakage through R_2 the output voltage V_{out} will be a fraction $C_1/(C_1 + C_2)$ of that at X. However if R_2 is not infinitely high, V_{out} will decrease with a time-constant $R_2C_2 = 1/\lambda_3$. The variation of the output voltage with time is therefore

$$V_{out} = V_{in}\left(\frac{C_1}{C_1 + C_2}\right)[\exp(-t/R_2C_2) - \exp(-t/R_1C_3)]$$

The normal thyroid case could then be modelled with $C_1 = C_2 =$ say 1 μF, so that $C_3 = 0.5$ μF and $V_{out} \approx 0.5V_{in}$ when the first exponential term in the bracket is nearly 1 and the second is negligibly small. As is suggested in Section 6.9, we can have the model working on a timescale in which 1 s corresponds to 1 d in the biological case. Then, for the rising part of the curve, we would have in units of s^{-1},

$$\lambda_1 + \lambda_2 = 7 = 1/R_1C_3 = 1/R_1 \times 0.5 \times 10^{-6}$$

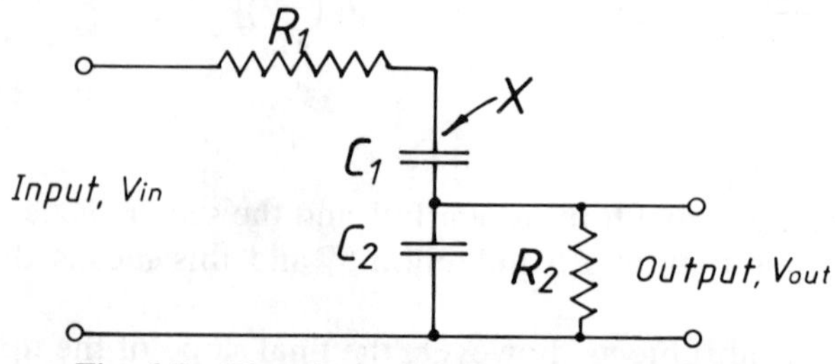

Fig. 6.9 Electrical model of thyroid uptake.

therefore

$$R_1 = 1/7 \times 0.5 \times 10^{-6}\ \Omega \approx 300\ \text{k}\Omega$$

also

$$\lambda_3 = 0.007 = 1/R_2C_2 = 1/R_2 \times 10^{-6}$$

and therefore

$$R_2 = 1/0.007 \times 10^{-6}\ \Omega \approx 140\ M\Omega$$

This value for R_2 justifies our original assumption that this would be a very high resistance and indeed it would be more convenient experimentally to make $R_1 = 300\ \Omega$ and $R_2 = 140\ \text{k}\Omega$. This scaling-down by a factor of 1000 would mean that the model would work in units of 1 ms representing 1 d of the biological system. This would be very convenient for oscilloscope display, using a square-wave input voltage from a pulse generator with a repetition frequency of a few hertz. The use of variable resistors for R_1 and R_2, with values smaller than those just worked out, would then give displays which could be easily adjusted to fit data from abnormal subjects.

6.13 Extension to a Biological System with Variable Volumes: The Cardiovascular System

It might be thought that models based on CSTRs with fixed volumes are of restricted utility for the simulation of biological systems in which some or all of the compartments have variable volumes, but this is not so, as we will show by using a simple example. The diagram (Fig. 6.10) represents a conceptual model of blood flow originating from the left ventricle, passing through the arterial tree, the venous tree, and entering the right atrium. The symbols in the diagram have the following significance:

- Q_1 Flow-rate of blood leaving left ventricle.
- Q_2 Flow-rate of blood between arterial and venous pathways.

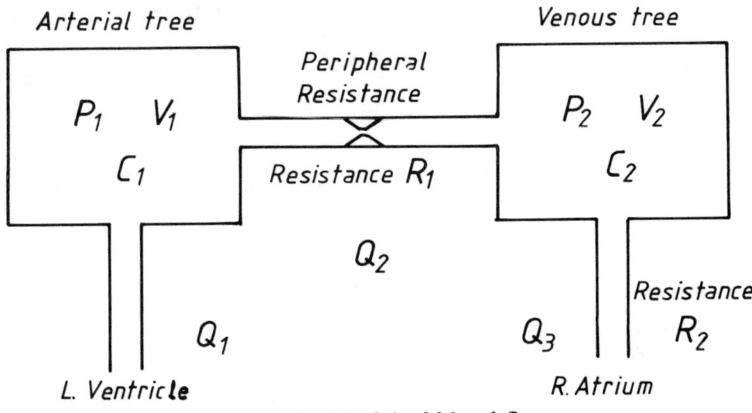

Fig. 6.10 Model of blood flow.

Q_3 Flow-rate of blood entering right atrium.
P_1 Pressure of blood in arterial system.
P_2 Pressure of blood in venous system.
(We assume zero pressure in the right atrium.)
V_1 volume of arterial blood.
V_2 Volume of venous blood.
(Note that P_1, P_2, V_1 and V_2 are not constants.)
C_1 Compliance of arterial system $= \mathrm{d}V_1/\mathrm{d}P_1$.
C_2 Compliance of venous system $= \mathrm{d}V_2/\mathrm{d}P_2$.
(For perfectly elastic arterial and venous systems, the compliances will be constants.)
R_1 Peripheral resistance, i.e. the resistance to flow which the blood experiences in making the transition from the arterial to the venous trees.
R_2 Resistance to the flow of blood from the venous system into the right atrium. (An electrical analogy would be the internal resistance of a battery.)

We may now ask, what is the relationship between Q_1 and Q_3? To find this we can write down three kinds of equation which relate the various parameters. These are:

(1) *Flow equations.* These express the fact that the *nett* flow into a compartment must equal its rate of change of volume. Therefore

$$\frac{\mathrm{d}V}{\mathrm{d}t} = Q_1 - Q_2 \quad \text{and} \quad \frac{\mathrm{d}V_2}{\mathrm{d}t} = Q_2 - Q_3$$

(2) *Compliance equations.* It follows from the definition of the compliances that

$$\frac{\mathrm{d}V_1}{\mathrm{d}t} = C_1\frac{\mathrm{d}P_1}{\mathrm{d}t} \quad \text{and} \quad \frac{\mathrm{d}V_2}{\mathrm{d}t} = C_2\frac{\mathrm{d}P_2}{\mathrm{d}t}$$

(3) *Resistance equations.* It follows from the definition of a resistance that

$$Q_2 = (P_1 - P_2)/R_1 \quad \text{and} \quad Q_3 = P_2/R_2$$

It is then quite easy to show that

$$R_1C_1.R_2C_2\frac{\mathrm{d}^2Q_3}{\mathrm{d}t^2} + (R_1C_1 + R_2C_2 + R_2C_1)\frac{\mathrm{d}Q_3}{\mathrm{d}t} + Q_3 = Q_1$$

or

$$\frac{\mathrm{d}^2Q_3}{\mathrm{d}t^2} + \left(\frac{R_1C_1 + R_2C_2 + R_2C_1}{R_1C_1.R_2C_2}\right)\frac{\mathrm{d}Q_3}{\mathrm{d}t} + \frac{1}{R_1C_1.R_2C_2}\cdot Q_3 = \frac{Q_1}{R_1C_1\cdot R_2C_2}$$

This is a second-order differential equation whose coefficients are reasonably near to being constants, if the subject is in a 'steady' state. (The peripheral resistance will take on different values according to whether the subject is resting at low temperature, or active when warmer.) In particular, if Q_1 is taken to be a single sharp pulse of flow, so that its Laplace transform is just unity, it is easy to show that the flow Q_3 is modelled by a two-compartment system in which the constants λ_1 and λ_2 are merely functions of R_1, R_2, C_1 and C_2—they have no physical relationship with the volumes of the venous or arterial trees at all, but take on different values if the resistances change. It is in fact generally true that the actual parameters (volumes, flow-rate) in a physically realisable model (e.g. a chian of CSTRs) may have very little apparent relationship to the biological system being modelled. The physical model serves merely as a means of visualising the significance of the underlying differential equation and only in this sense gives insight into the biological system.

6.14 A Generalisation for Complex Dynamic Systems

Real biological systems are often so complex that the construction of useful physical models becomes impracticable. However, a 'working' model is really only of academic interest, for a sketch of any proposed compartmental system enables us to write down the mathematical model, i.e. a set of differential equations. (It may be that in the future the concept of compartments will be replaced by some other which will facilitate the derivation of more appropriate equations.) Then, as these equations are linear, it is clear that the general method of Section 1.15 will provide solutions. To make this point clear we may consider any linear system whatever which has an input $x(t)$ and an output $y(t)$, as in Fig. 6.11.

Now if $x(t)$ were merely a unit impulse at time zero, we would get an output function $y(t)$ which we will call $h(t)$, which is just characteristic of the system itself. But in fact, $x(t)$ can be thought of as a series of impulses, of different magnitudes, contiguous to one another. For a typical one occurring at time t_0, i.e. $x(t_0)$, there will be a contribution to the output $y(t)$ at time t.

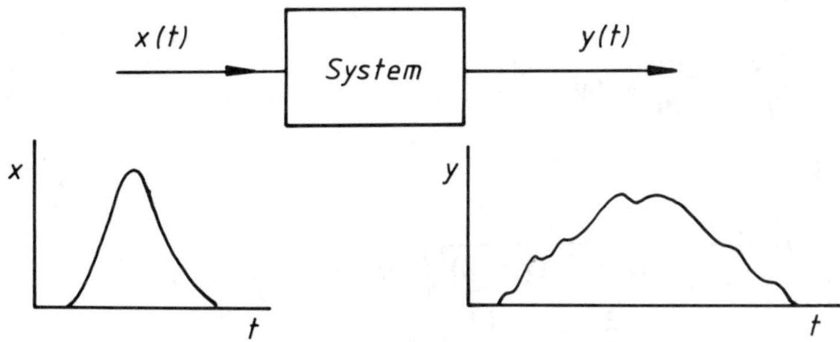

Fig. 6.11 The general linear system.

The fraction of $x(t_0)$ which appears as output at time t is clearly that value of h appropriate to the intervening time interval $(t - t_0)$, viz. $h(t_0 - t)$; the contribution to $y(t)$ will be $x(t_0)\,h(t - t_0)$. Summing all contribution for all values of t_0 we get

$$y(t) = \int_0^t x(t_0)h(t - t_0)\mathrm{d}t_0$$

This is immediately recognisable as the convolution integral so that Sections 1.18 and 1.19 are relevant. In particular the function $h(t)$ corresponds to a transfer function $H(s)$ and we find that the solution for $y(t)$ for unit impulse input into a system described by linear differential equations (derived on the basis of compartmental analysis) is expressible as the sum of a number of exponential terms.

Example 1
If the system equation is

$$\frac{\mathrm{d}^2y}{\mathrm{d}t^2} + 5\frac{\mathrm{d}y}{\mathrm{d}t} + 6y = \frac{\mathrm{d}x}{\mathrm{d}t} + 3x$$

i.e.

$$(\mathrm{D}^2 + 5\mathrm{D} + 6)y = (\mathrm{D} + 3)x$$

we find

$$S^2Y + 5Ys + 6Y = sX + 3X$$

so

$$H = \frac{Y}{X} = \frac{s + 3}{s^2 + 5s + 6}$$

$$= \frac{s + 3}{(s + 3)(s + 2)} = \frac{1}{s + 2}$$

and so

$$h(t) = \exp(-2t)$$

Example 2
If

$$\frac{\mathrm{d}^2y}{\mathrm{d}t^2} + 5\frac{\mathrm{d}y}{\mathrm{d}t} + 6y = \frac{\mathrm{d}x}{\mathrm{d}t} + x$$

we get

$$H = \frac{Y}{X} = \frac{s + 1}{(s + 3)(s + 2)}$$

$$= \frac{2}{s + 3} - \frac{1}{s + 2}$$

and so

$$h(t) = 2 \exp(-3t) - \exp(-2t)$$

Example 3
Reverting to the cardiovascular problem of the previous section, and substituting

$$R_1C_1 + R_2C_2 + R_2C_1 = k_1 \qquad \frac{1}{R_1C_1.R_2C_2} = k_2$$

we get

$$D^2Q_3 + k_1k_2DQ_3 + k_2Q_3 = k_2Q_1$$

from which

$$H(s) = \frac{k_2}{s^2 + k_1k_2s + k_2}$$

$$= \frac{k_2}{(s + p_1)(s + p_2)}$$

where

$$p_1 + p_2 = k_1k_2 \qquad \text{and} \qquad p_1\,p_2 = k_2$$

i.e.

$$p_1 = \frac{k_1k_2 + \sqrt{(k_1k_2 - 4k_2)}}{2}$$

and

$$p_2 = \frac{k_1k_2 - \sqrt{(k_1k_2 - 4k_2)}}{2}$$

Then

$$H(s) = \frac{k_2}{p_2 - p_1}\left[\frac{1}{s + p_1} - \frac{1}{s + p_2}\right]$$

and

$$h(t) = \frac{k_2}{p_2 - p_1}\left[\exp(-p_1t) - \exp(-p_2t)\right]$$

which shows that even in this complicated case, the impulse response of the system can be expressed as the sum of exponential terms, the coefficients in which are related, although in a not very simple way, to well defined parameters of the system.

The mathematical treatment given here should be compared with that of the *RLC* circuit of Section 1.13. It is theoretically possible for the above solution for $h(t)$ to be oscillatory instead of purely exponential.

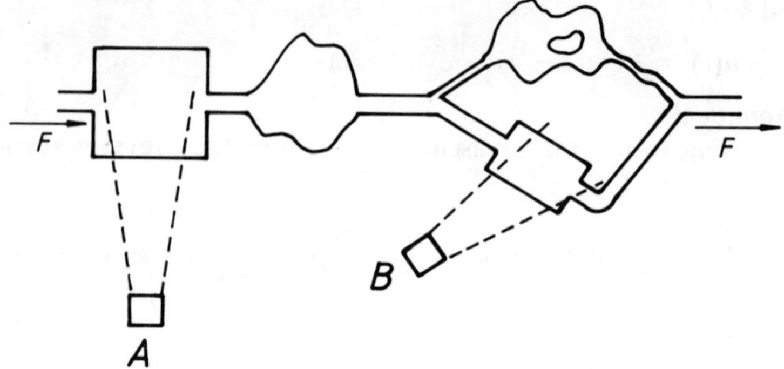

Fig. 6.12 A constant-flow system with detectors A and B.

6.15 The Occupancy Theorem

The occupancy theorem is a generalisation applicable to dynamic systems which provides an important complement to the foregoing. Suppose we consider any system through which there is a constant flow F of any particular material (Fig. 6.12). The system may have branches so that the flow is divided at some points and re-united at others. We may consider any portion of the system, which might consist of a compartment, or a part of a compartment in the sense we have already used. Within this portion A, there will be a mass of material, m_A say, which is steady in time, equal amounts of material entering and leaving A per unit time. We may inject a tracer for this material, ideally without disturbing the flow, and then measure the fraction $f_A(t)$ of the activity appearing in A as a function of time. If external detectors are used as in Fig. 6.12, $f_A(t)$ is just the count-rate $c(t)$ observed divided by the count-rate which would be produced if all the activity were in the part A. Clearly $f_A(t)$ will be a function which has zero value initially, rises to positive values, and then diminishes to zero if we count over long enough time. The time integral of $f_A(t)$ over all time is $\int_0^\infty f_A(t)\mathrm{d}t$, and has the dimension of time. It is called the *occupancy* of the compartment A. The occupancy theorem states that, for all possible portions A, B, C, . . . of the system,

$$\frac{\int_0^\infty f_A \mathrm{d}t}{m_A} = \frac{\int_0^\infty f_B \mathrm{d}t}{m_B} = \cdots = \frac{1}{F}$$

It is not easy to give a rigorous proof of this theorem, but that it is reasonable is shown by two simple considerations.

(a) If A is a large part of the system, m_A will be large but the fraction of tracer which passes through the part in unit time will be small. The time taken for all the tracer to move through the part will therefore be large and a large value of $\int_0^\infty f_A \mathrm{d}t$ can be expected. It is not unreasonable to find that $\int_0^\infty f_A \mathrm{d}t$ is in fact proportional to m_A, with $1/F$ as the constant of proportionality.

(b) Where the system branches, the fraction of material going into each branch must be equal to the fraction of tracer going into that branch. For two branches A and B, we therefore expect

$$\frac{\int_0^\infty f_A \mathrm{d}t}{m_A} = \frac{\int_0^\infty f_B \mathrm{d}t}{m_B}$$

In a sense the occupancy theorem is an extension, to dynamic systems, of the principle of 'isotope dilution', exemplified in Section 5.7 (measurement of red cell mass).

6.16 Application to Protein Turnover Studies

We sketch here some elementary ideas about the fate of a typical protein in the body. For details the original literature should be examined (see, for example, the work of Vitek, Wraight, Orr and others in the list of references). A very much simplified representation of the biological system is shown in Fig. 6.13. Proteins are synthesised, primarily in the liver, and we assume these to be transferred at a constant rate into the first compartment of the system, namely the blood plasma. This compartment is in dynamic equilibrium with a second, whose extent is ill-defined, except in the negative sense that it excludes the plasma; it is referred to as the extra-vascular space. In this space proteins are constantly being broken down (catabolism), the degradation products (amino acids) being returned to the plasma, from which they are rapidly excreted via the kidneys and appear in the urine. As a tracer it has been common practice to inject albumin labelled with ^{131}I into the plasma. Measurements of activity in the plasma as a function of time are carried out by withdrawing samples of blood at intervals. When catabolism takes place, the activity may be present as free iodide or in forms bound to the amino acids; in either case, if the thyroid gland is blocked by administration of stable iodide, the activity will be carried into the urine. Measurement of urine activity will therefore be an indication of the catabolic process. We have, therefore, shown two extra compartments in Fig. 6.13, just as if the protein were transferred as such (not as its degradation products) into the plasma and then the urine.

One of the first investigations of this system in experimental animals was made by Vitek, whose notation we will use as it has been adopted by later authors (although it is not consistent with our own so far). Thus we denote by X_p and X_u the count-rates obtained from samples (with standard volumes) of plasma and urine respectively, each normalised in a particular way. These quantities are of course functions of time. The count-rates $X_p(t)$ are normalised by taking $X_p(0) = 1$, that is to say all count-rates obtained after $t = 0$ are divided by the initial count-rate. The urine activity is regarded as cumulative, so that $X_u(t)$ represents the total count-rate obtained on all urine sample up to time t, normalised in such a way that $X_u(\infty) = 1$. These conventions mean that $X_p(t)$ represents the *fraction* of the labelled protein present in the plasma

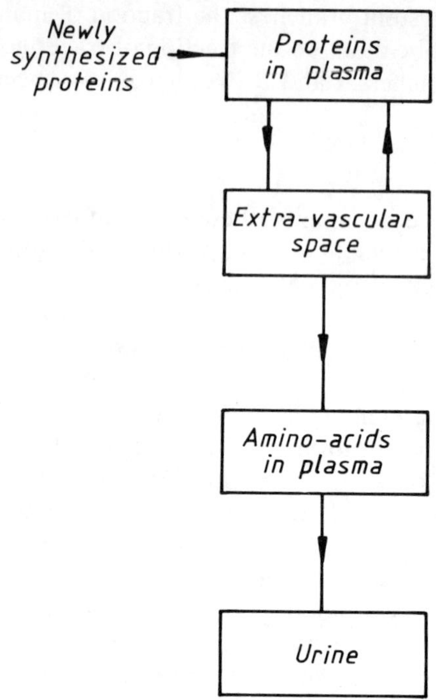

Fig. 6.13 Compartments in protein turnover.

at time t, and $X_u(t)$ represent the *fraction* of the activity which has accumulated in the urine during the time t. The fraction of the activity which is in neither plasma nor urine at time must of course be in the extravascular space; we denote this fraction $X_{ex}(t)$ where

$$X_{ex}(t) = 1 - X_p(t) - X_u(t)$$

It should be noted that $X_{ex}(t)$ cannot be measured directly, since we have no practicable way of taking samples of the extra vascular space.

We now introduce a function $G_{p,ex}$, which is also a function of time. With reference to Fig. 6.14, this function is used to specify that fraction of the activity present in the plasma at time τ which appears in the extra-vascular space at the later time t. Clearly this fraction is given by $G_{p,ex}(t - \tau)$, and just as in Section 6.14, we find that $X_{ex}(t)$ is given by the integral

$$X_{ex}(t) = \int_0^t X_p(\tau)G_{p,ex}(t - \tau)d\tau$$

In a similar way we can denote by $G_{p,u}(t - \tau)$ that fraction of the plasma activity at time τ, which contributes to the urine activity accumulated at time t, and write

$$X_u(t) = \int_0^t X_p(\tau)G_{p,u}(t - \tau)d\tau$$

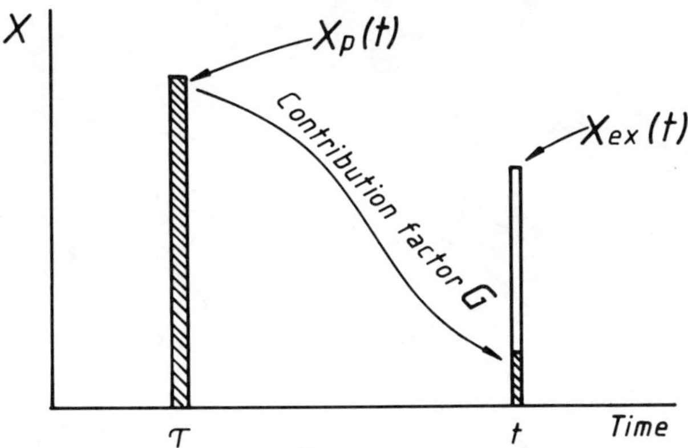

Fig. 6.14 Relation between $X_p(t)$ and $X_{ex}(t)$.

These two equations are clearly convolutions; we can write them in terms of the Laplace transforms: $x_{ex}(s) \leftrightarrow X_{ex}(t)$, $x_p(s) \leftrightarrow X_p(t)$, $x_u(s) \leftrightarrow X_u(t)$, $g_{p,ex}(s) \leftrightarrow G_{p,ex}$ and $g_{p,u}(s) \leftrightarrow G_{p,u}(t)$, so that

$$x_{ex}(s) = x_p(s) g_{p,ex}(s)$$
$$x_u(s) = x_p(s) g_{p,u}(s)$$

or

$$g_{p,ex}(s) = x_{ex}(s)/x_p(s)$$

and

$$g_{p,u}(s) = x_u(s)/x_p(s)$$

We see therefore that if the functions $X_p(t)$, $X_{ex}(t)$ and $X_u(t)$ are established from experimental measurements, we need only to find their Laplace transforms to be able to calculate the transfer functions $g_{p,ex}(s)$ and $g_{p,u}(s)$; the inverse transforms of the latter are then $G_{p,ex}(t)$ and $G_{p,u}(t)$.

A straightforward, although lengthy, procedure for carrying out the transforms was developed by Wraight (see references) who fitted experimental curves of X_p, X_{ex}, X_u (Fig. 6.15) to sums of exponential terms, for example,

$$X_p(t) = A \exp(-at) + B \exp(-bt) + C \exp(-ct)$$

the constants A, B, C, a, b, c being chosen to give the best least-squares fit.

It is important to realise in this that no physical significance need be assigned to these constants. Indeed, any functional form for X_p, X_{ex}, X_u could be assumed, the only necessity being a good fit to experimental values. The transforms x_p, x_{ex} and x_u are thus easily evaluated; the inverse transforms of $g_{p,ex}$ and $g_{p,u}$ are only a little more difficult.

From the functions $G_{p,ex}(t)$ and $G_{p,u}(t)$ at least two important quantities can be calculated:

(1) $G_{p,ex}(0)$. This represents the *initial* value of the fraction of activity in the plasma which is being transferred to the extra-vascular space, per unit time, and is equal to the fractional flow of albumin in this direction. (At later

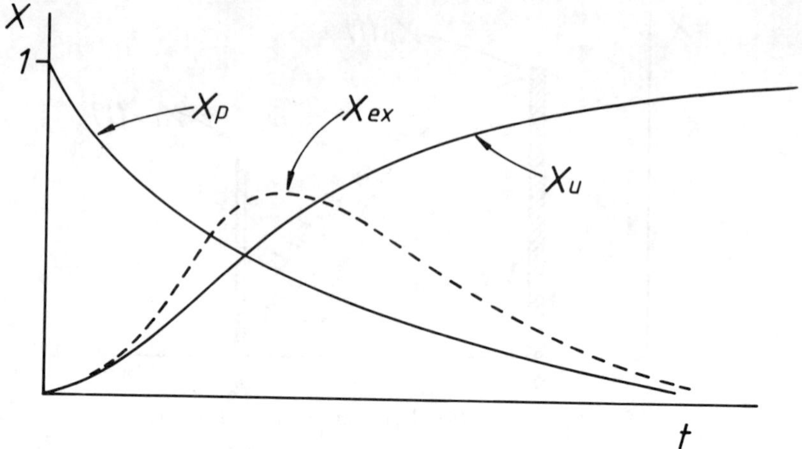

Fig. 6.15 Typical curves for X_p, X_{ex}, X_u. (Reproduced from F. Vitek *et al.*, *J. Nuc. Biol. Med.*, **10**, 124 (1966).)

times the fractional flow of *activity* in this direction will be *less* than that of the albumin, as the activity begins to trace the *reverse* flow.) Further, if we measure the absolute amount N_p of albumin in the plasma, the absolute flow $P_{p,ex}$ is given by

$$P_{p,ex} = G_{p,ex}(0)N_p$$

This is the total area under the curve for $G_{p,ex}(t)$. The importance of this quantity is seen as follows. If we denote by N_{ex} and N_p the absolute amount of

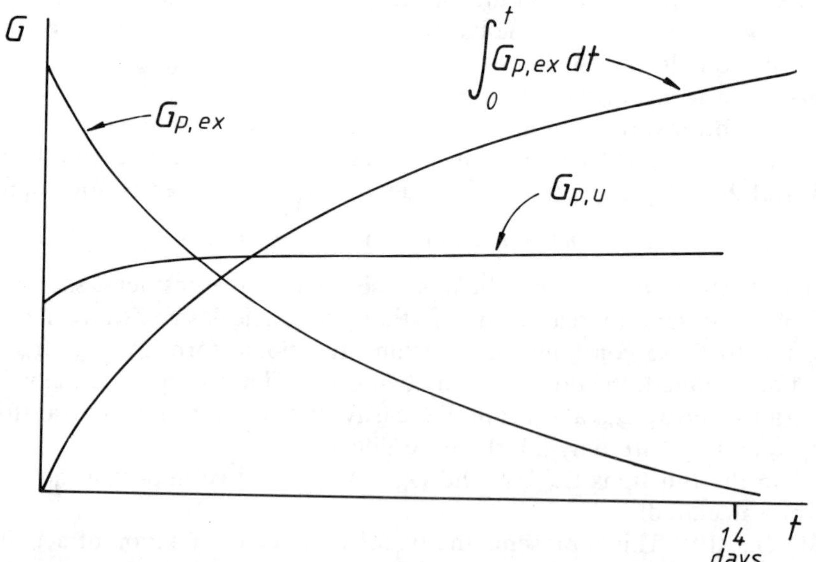

Fig. 6.16 The functions $G_{p,ex}$, $G_{p,u}$. (Reproduced from F. Vitek *et al.*, *J. Nuc. Biol. Med.*, **10** 124, (1966).)

albumin in the extra-vascular and plasma compartments respectively, we have from the occupancy theorem

$$\frac{\int_0^\infty X_{ex}(t)\mathrm{d}t}{N_{ex}} = \frac{\int_0^\infty X_p(t)\mathrm{d}t}{N_p}$$

We can then denote by R the ratio of these quantities, i.e.

$$R = \frac{N_{ex}}{N_p} = \frac{\int_0^\infty X_{ex}(t)\mathrm{d}t}{\int_0^\infty X_p(t)\mathrm{d}t}$$

and as Vitek showed, this may be reduced to

$$R = \frac{N_{ex}}{N_p} = \int_0^\infty G_{p,ex}\mathrm{d}t$$

The amount of albumin in the plasma N_p is experimentally accessible, so this means we have a way of estimating the extra-vascular albumin, which is not directly measurable because we have no clear way of defining the extent of the extra-vascular space. The general features of the $G_{p,ex}$ and $G_{p,u}$ functions are shown in Fig. 6.16.

(2) In addition, Orr (see references) has pointed out that the R given above may be quite accurately estimated from the integrals $\int_0^\infty X_{ex}(t)\mathrm{d}t$ and $\int_0^\infty X_p(t)\mathrm{d}t$ on the assumption—which seems to be borne out well enough in practice—that these functions show 'straight-line' decays for large t when plotted semi-logarithmically (Fig. 6.17).

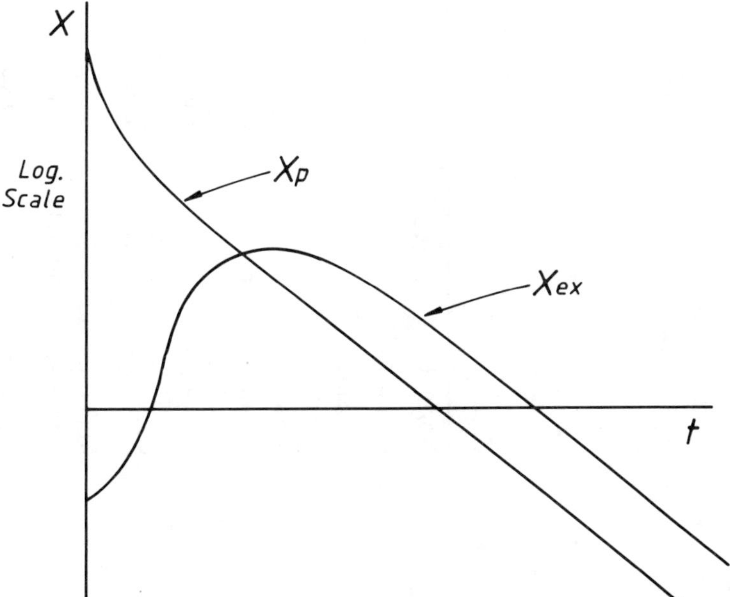

Fig. 6.17 X_{ex} and X_p plotted semi-logarithmically.

Table 6.3 Albumin Turnover Data in Humans

(Reproduced, in abridged form, from E.H. Belcher and H. Vetter, *Radioisotopes in Medical Diagnosis*, by permission.)

Case	Serum concentration (g/100 ml)	Fractional turnover (%/day)	Absolute turnover		$D = N_p/(N_p + N_{ex})$
			(g/day)	(mg/kg body wt per day)	
Normal	3.9	8.0	8.9	158	0.41
Protein losing enteropathies:					
Stomach	2.4	24.5	13.8	243	0.56
Small intestine	0.8	42.2	6.8	151	0.69
Colon	1.8	33.8	17.0	265	0.30
Diffuse	1.9	17.0	9.2	188	0.53
Nephrosis	0.96	48.7	9.4	187	0.49
Malabsorption (post-gastrectomy)	1.6	2.9	1.7	32	0.57
Cirrhosis	2.65	7.7	6.7	113	0.41

Moreover, the curves become parallel to one another at large t and it is claimed that a 25% error in the estimation of their final slope can make only a 2% error in the values of the integral. Orr prefers to record the urine count as the *daily* excretion rather than the cumulative excretions; with this change in notation a semi-logarithmic plot of X_u also tends to the same slope as X_p and we find that the daily catabolic rate F is given by

$$F = \frac{\int_0^\infty X_u(t)dt}{\int_0^\infty X_p(t)dt}$$

Clinical studies on humans have been carried out over many years, and for some other proteins besides albumin. A summary of results for normal patients and those in various disease states, is given in Chapter 19 of *Radioisotopes in Medical Diagnosis*, (Belcher and Vetter). Table 6.3 is an abbreviated version of this summary, giving data for albumin only and with a minor recalculation (D in Belcher and Vetter's table is related to our R by $R = (1 - D)/D$.) The table includes data for some patients who have protein-losing gastroenteropathies, which lead to loss of activity in the faeces, not allowed for in our treatment.

Although in all disease states the protein concentration in the plasma is diminished below the normal level, the fractional and absolute protein turnover rates tend to be at least as high as normal, and in many cases are much higher than normal, except for the gastrectomy cases. There are wide variations in the ratio of protein in the extra-vascular space to that in the plasma.

6.17 Application to Renography

Over the years many attempts have been made to improve the purely empirical approach to the curves obtained by the conventional hippuran renography test outlined in Section 5.7(b). While the shape of these curves, unmodified by any mathematical treatment, is diagnostically significant in any particular case, their interpretation in a fundamental sense is beset by two main difficulties. First, it is clear, from the physiology of the kidney, that its operation as a body compartment is wholly different from what we have hitherto assumed. Material extracted from the previous compartment (the blood plasma) is not mixed with an already existing reservoir of material and immediately made available for excretion; instead, after glomerular filtration it flows steadily through the nephrons before being discharged via the ureters. This means that if the extracted material carries a 'spike' of radioactive tracer, the activity in the kidney will remain constant during the time taken to traverse the nephrons, eventually falling off as each nephron discharges its contents into the urine. The nephrons are not all of exactly the same length, nor is the flow-rate exactly equal in all of them, so that the fall in activity in the kidney as a whole will not be instantaneous. In order to characterise the system, we have to find the form of its impulse response function, it being recognised from the outset that the models we have used so far are not appropriate and in particular that we cannot expect the function to be expressible as a sum of exponential terms. The renogram curves are obviously only distantly related to the impulse response function, although it is a fair guess that the time taken for the maximum count-rate to appear is about the same as the average time spent by material in the nephrons.

Secondly, there is the very important point that the detectors placed over each kidney respond to hippuran activity not only in the material extracted by the kidney but also to that remaining in the blood that is present in the field of view. In principle, the latter activity can be allowed for by subtraction. The activity in the blood is measured, as a function of time, by using a third detector over the praecordium, or other region of the body well perfused with blood. If a gamma camera is used for the study, it is possible to define an area close to the kidney to obtain the 'blood-background' counts, and these are probably a more accurate reflection of the strength of the activity reaching the kidney than are the praecordial counts. The blood activity at time t is then multiplied by a suitable normalisation factor and subtracted from the renal activity at that time. The 'blood-background-subtracted' curves so obtained should then contain information relevant only to the kidney function itself. Parameters easily derivable form the corrected curves, e.g. the time taken for the peak activity to appear, and the time taken after the peak for the activity to decrease to half its value, should then be of more certain diagnostic significance than the parameters of the original curves. They are, however, only one step towards the derivation of the impulse response curves.

It is important to realise that the kidneys work completely independently of each other, each, in the normal case, removing about 90% of the ^{131}I-hippuran content of the blood flowing to it (i.e. a flow of some 600 ml min^{-1}, or 10–12% of the cardiac output). However, in an abnormal case, in which one kidney is failing to clear hippuran from its blood supply, the activity reaching the other must be correspondingly higher. The shape of the (uncorrected) renogram curve given by the normal kidney will therefore be modified by the malfunction of the abnormal kidney. Blood-background subtraction should clearly mitigate this effect. The practical difficulty in applying the technique is that one has no *a priori* knowledge of the relative amounts of blood in the fields of view of the renal and extra-renal detectors. There is therefore some doubt about the magnitude of the normalisation factor to be applied; moreover it is a different factor for each kidney. Some rather intricate computational techniques, which really amount to trial-and-error solutions, have been tried but do not seem to have been universally accepted. Another way out of the difficulty is to take additional measurements using a second tracer which indicates the blood activity in all three detectors but which is not extracted by the kidneys. This theoretically ideal solution is however beset with practical difficulties; for example if one wishes to inject the second tracer simultaneously with the hippuran, the radioactive label cannot be ^{131}I, or the tracers will be indistinguishable; yet the two radionuclides must have rather similar energy spectra to ensure the same ratio of counting efficiencies for all the detectors. The alternative is to use ^{131}I-labelled tracers for both sets of measurements, which then have to be made at times separated by some hours, with all the attendant difficulties of accurate re-positioning of the subject with respect to the detectors. Finally, there is the question of the chemical nature of the tracer itself; clearly it must be such as not to concentrate in any other organ of the body, or to undergo glomerular filtration in the kidney, or to suffer decomposition yielding active products with properties different from its own. It is at least doubtful whether such a substance has been found.

A quite different approach is to carry out a transformation of the renogram curves, a technique which gives us an improved way of expressing the fundamental mode of action of the kidney and a particularly simple means of eliminating the effect of blood background. We assume that each kidney possesses an impulse response function $h(t)$ such that a unit pulse of activity arriving at the kidney at time zero produces a count-rate $h(t)$ at time t in the detector over that kidney, *from the tracer present within the kidney tissue.* The count-rate actually observed (allowing for ordinary background) will include the blood-ground; we denote this by $h'(t)$. If then the activity actually in the plasma at a time τ, following a 'spike' injection, is expressed as the function $p(\tau)$, we find that the activity $p(\tau)$ makes a contribution to the kidney-tissue count-rate at time t of $p(\tau).h(t - \tau)$. The contribution to the observed count, including blood-background, is $p(\tau).h'(t - \tau)$. If, therefore, it were possible to count the kidney tissue alone, its count-rate as a function of time would be

$$r(t) = \int_0^t p(\tau).h(t - \tau)\mathrm{d}\tau$$

The renogram is, however, a record of the observed count-rate

$$r'(t) = \int_0^t p(\tau).h'(t - \tau)\mathrm{d}\tau$$

In addition we are making the assumption that the renogram counts are the sum of those due to activity in the kidney itself and those due to an unknown amount of blood, the latter being at any time a constant fraction of the extra-renal count.

We may express this by the equation

$$r'(t) = r(t) + Cp(t)$$

C being a constant. From these equations we may easily find the relation between h and h'. For, by making the transformations $r(t) \leftrightarrow R(s)$, $r'(t) \leftrightarrow R'(s)$, $h(t) \leftrightarrow H(s)$, $h'(t) \leftrightarrow H'(s)$ and $p(t) \leftrightarrow P(s)$, we have

$$R(s) = P(s).H(s)$$
$$R'(s) = P(s).H'(s)$$
$$R'(s) = R(s) + CP(s)$$

Therefore

$$H'(s) = \frac{R'(s)}{P(s)}$$

$$= \frac{R(s) + CP(s)}{P(s)}$$

$$= H(s) + C$$

Now taking the inverse Laplace transforms we have

$$h'(t) = h(t) + C\delta$$

where we have written $C\delta$ to mean an impulse of magnitude C occurring at time zero. It follows that if we can use the *observed* renogram count-rates to find the function $h'(t)$, we can obtain the impulse response function $h(t)$ from this merely by omitting a 'spike' appearing at time zero. This simple step is equivalent to 'correcting' the renogram for blood background. Furthermore it is clear that $h'(t)$ is given by

$$h'(t) = \mathcal{L}^{-1}(R'(s)/P(s))$$

$\mathcal{L}^{-1}$ being the inverse Laplace transform operator.

It has recently been shown that the activity in the plasma can be resolved into two, or at most three, exponential components. It is convenient to normalise the extra-renal count-rate to unity at time zero, so that we have

$$p(t) = p_1 \exp(-k_1 t) + p_2 \exp(-k_2 t) + p_3 \exp(-k_3 t)$$

where $p_1 + p_2 + p_3 = 1$, and the values of $p_1, p_2, p_3, k_1, k_2, k_3$, are found by resolution of the count-rate curve recorded by the extra-renal detector. These constants may have significant relationships with the transfer constants of a postulated dynamic model of the distribution of activity between blood and some other physiological compartments, but the nature of these is irrelevant to the present argument. In practice, all that is necessary is that values of the constants listed above be obtained such that calculated values of $p(t)$ agree with experimental ones within reasonable error. Transforming the equation for $p(t)$ we have

$$P(s) = \frac{p_1}{s + k_1} + \frac{p_2}{s + k_2} + \frac{p_3}{s + k_3}$$

and

$$H'(s) = R'(s) \Bigg/ \left\{\frac{p_1}{s + k_1} + \frac{p_2}{s + k_2} + \frac{p_3}{s + k_3}\right\}$$

This last equation can be written in the form

$$H'(s) = \frac{s^3 + q_1 s^2 + q_2 s + q_3}{s^2 + r_1 s + r_2} R(s)$$

if we define the coefficients of s as

$$q_1 = k_1 + k_2 + k_3$$
$$q_2 = k_1k_2 + k_2k_3 + k_3k_1$$
$$q_3 = k_1k_2k_3$$
$$r_1 = (k_2 + k_3)p_1 + (k_3 + k_1)p_2 + (k_1 + k_2)p_3$$
$$r_2 = p_1k_2k_3 + p_2k_3k_1 + p_3k_1k_2$$

The numerical values of all these coefficients are thus easily obtained. The expression for $H'(s)$ can now be written

$$H'(s) = \left\{s + l_1 + \frac{l_2s + l_3}{s^2 + r_1s + r_2}\right\}R(s)$$

where the new coefficients are

$$l_1 = q_1 - r_1$$
$$l_2 = q_2 - r_2 - r_1(q_1 - r_1)$$
$$l_3 = q_3 - r_2(q_1 - r_1)$$

Finally we may write

$$H'(s) = \left\{s + l_1 + \frac{u_1}{s + v_1} + \frac{u_2}{s + v_2}\right\}R'(s)$$

where

$$v_1 = [r_1 - \sqrt{(r_1^2 - 4r_2)}]/2$$
$$v_2 = [r_1 + \sqrt{(r_1^2 - 4r_2)}]/2$$
$$u_1 = (l_3 - l_2v_1)/(v_2 - v_1)$$
$$u_2 = (l_3 - l_2v_2)/(v_1 - v_2)$$

The last equation for $H(s)$ is now clearly the Laplace transform of

$$h'(t) = \frac{\mathrm{d}}{\mathrm{d}t}\{r'(t)\} + l_1r'(t) + u_1\{\exp(-v_1t) * r'(t)\} + u_2\{\exp(-v_2t) * r'(t)\}$$
$$\equiv \frac{\mathrm{d}}{\mathrm{d}t}\{r'(t)\} + l_1r'(t) + u_1\int_0^t \exp(-v_1\tau).r'(t-\tau)\mathrm{d}\tau + u_2\int_0^t \exp(-v_2\tau).r'(t-\tau)\mathrm{d}\tau$$

We therefore have that the impulse response function $h'(t)$ is given by the sum of two convolution terms, *plus* a constant times the actual count-rate at time t, *plus* the slope of the renogram curve at that time. The last term may not be very easy to evaluate accurately in practice because of statistical fluctuations in the observed count-rate. The difficulty is avoided if we define a new function $g'(t)$ such that

$$g'(t) = \int_0^t h'(\tau)\mathrm{d}\tau$$

and by integration of the expression for $h'(t)$ obtain

$$g'(t) = r'(t) + \left(l_4 + \frac{u_1}{v_1} + \frac{u_2}{v_2}\right) \int_0^t r'(\tau)\mathrm{d}\tau$$
$$-\frac{u_1}{v_1} \int_0^t \exp(-v_1\tau).r'(t - \tau)\mathrm{d}\tau - \frac{u_2}{v_2} \int_0^t \exp(-v_2\tau).r'(t - \tau)\mathrm{d}\tau$$

where $l_4 = l_1 + u_1/v_1 + u_2/v_2$. An alternative, and simpler, way of obtaining this relation is to start with the expression for $H(s)$, divide it throughout by s to obtain

$$\frac{H(s)}{s} = \left\{1 + \frac{l_1}{s} + \frac{u_1}{s(s + v_1)} + \frac{u_2}{s(s + v_2)}\right\} R(s)$$

and then to form the inverse transform, term by term. The computation of the convolution terms still needs to be done, but there is now no differential term. The evaluation of the right-hand side can in fact be done with reasonable accuracy and in a few minutes with a programmable calculator (HP-25). Incidentially it is worth noting that the method of derivation of the expressions for $h'(t)$ and $g'(t)$ makes it clear that additional components in the blood-activity curve would merely result in additional convolution terms. Thus we may write for the general case of an input function being expressible as a summed exponential, that

$$g'(t) = r'(t) + l_4 \int_0^t r'(\tau)\mathrm{d}\tau + \sum_1^n w_n \exp(-v_n t) * r'(t)$$

where the constants l_4 w_n, v_n are simple functions of the coefficients appearing in the expression for $p(t)$, and the number n of convolutions is one less than the number of exponential terms in $p(t)$.

Corresponding to the function $h(t)$ we may also define $g(t)$ by the equation

$$g(t) = \int_0^t h(\tau)\mathrm{d}\tau = g'(t) - \mathrm{C}$$

Except at $t = 0$, the differential coefficients of $g(t)$ and $g'(t)$ are the same, and equal to $h(t)$. The form of the functions $r'(t)$, $g(t)$, $h(t)$ and $h'(t)$ for a typical normal kidney is shown in Fig. 6.18.

For a normal kidney the upward slope of $g(t)$ is constant for $0 < t < 3$ min, so that $h(t)$ is constant over this interval, and equal to $h(0)$. After this time $g(t)$ tends asymptotically to a constant value, as $h(t)$ falls monotonically to zero. We may define a mean transit time T_m for hippuran

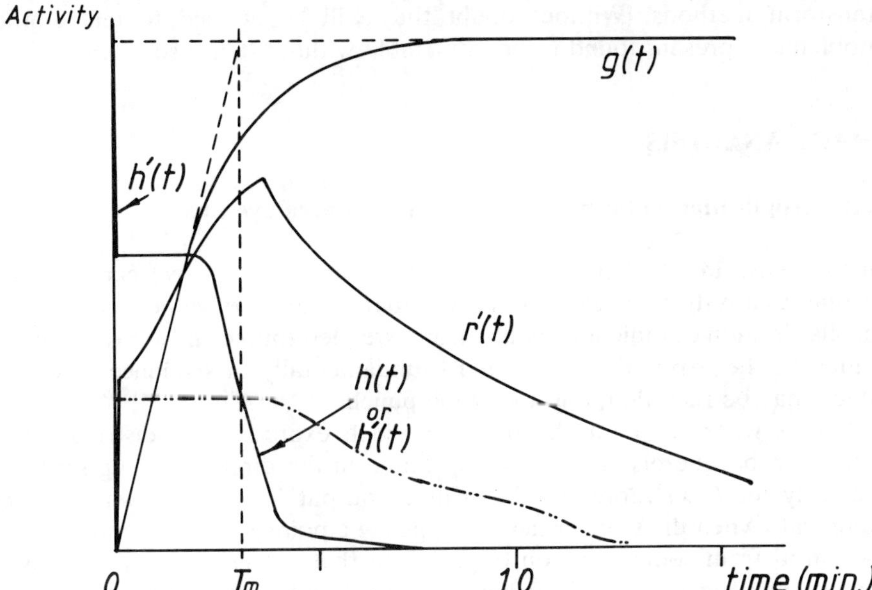

Fig. 6.18 Renogram $r'(t)$ and the functions $g(t)$ and $h(t)$ for a normal kidney.

molecules in the kidney, such that

$$h(0).T_{\mathrm{m}} = \int_0^\infty h(t)\mathrm{d}t$$

The integral on the right-hand side is just the limiting value of $g(t)$ when t becomes infinite, i.e. $g(\infty)$. Also $h(0)$ is the initial slope of $g(t)$, or $(\mathrm{d}g/\mathrm{d}t)_0$ we conclude that

$$T_{\mathrm{m}} = g(\infty)/(\mathrm{d}g/\mathrm{d}t)_0$$

T_{m} is then given by the simple graphical construction shown in the figure. It is not necessarily the time at which the renogram curve reaches its maximum value, although in many cases the latter time is not a bad approximation to T_{m}. In the figure the curve for $h(t)$, after the initial nearly horizontal part, falls towards zero quite precipitously, becoming almost indistinguishable from zero in about a further 4 min. Cases on which the urine flow is obstructed show a very much slower approach to zero (shown by the dotted curve in the figure). The horizontal scale of the figure is somewhat dependent, of course, on the rate of urine flow, and it may be important in comparing one subject with another that their renograms are obtained under conditions of approximately equal urine floow. As between the right and left kidney in any one subject, the relative heights of the impulse response curves is a measure of the relative effective renal plasma flow.

The investigation of this system has been described in some detail, not only because of its currrent interest but as a good example of the power of transform methods. Without doubt they will be applied to many other problems at present found to be intractable without their use.

IMAGE ANALYSIS

6.18 Application to Scanner and Gamma-Camera Systems

In the sense in which we have been discussing systems, we can regard a scanner as a system whose quantitative input is a series of values representing the distribution of radioactivity along an extended source and whose output is a line on the exposed film or, more fundamentally, a sequence of counts which may be recorded, e.g. by a tape punch.

Let us write $I(x_0)$ for the function which expresses the distribution of activity in bacquerels at points along a line in the direction being scanned. Similarly let $O(x)$ represent the scanner output in terms of an observed count-rate when the scanner head is opposite a point x. Both x_0 and x will be measured from some common origin; note that the notation enables us to discuss the count-rate when the head is at position x, due to activity present at the point x_0, which may be a different point. If the measured line-spread function is expressed as $L(x_0)$, then it is clear from Section 1.16 that we may write the convolution

$$O(x) = kI(x_0) * L(x_0) = k \int_{-\infty}^{+\infty} I(x - x_0)\, L(x_0) \mathrm{d}x_0$$

where we will use $-\infty$ as the lower limit of integration because it is convenient to express the line-spread function symmetrically about the origin of coordinates. The line-spread function is assumed normalised to 1, so that the constant k represents the ratio of the count-rate (over the whole line-spread) to the activity constituting the line-source; in other words the counting efficiency of the detector.

An important special case occurs if we suppose $I(x_0)$ to be an exponential function, of the form

$$I(x_0) = I_0 \exp{(sx_0)}$$

where I_0 and s are constants. Then we can write

$$I(x - x_0) = I_0 \exp{(sx - sx_0)}$$

and obtain

$$O(x) = k \int_{-\infty}^{+\infty} I_0 \exp{(sx - sx_0)} L(x_0) \mathrm{d}x_0$$

We need to insert the limits $+\infty$ and $-\infty$ if $I(x_0)$ is allowed to extend over all distances.

Since x is a constant as far as the integral is concerned, this reduces to

$$O(x) = kI_0 \exp(sx) \int_{-\infty}^{+\infty} L(x_0) \exp(-sx_0)\mathrm{d}x_0$$

The integral can be recognised as the (double-sided) Laplace transform of the line-spread function. It is, therefore, a constant for any fixed value of s. We conclude that the output function $O(x)$ shows an exponential variation of form

$$O(x) = kI_0\mathcal{L}(L(x_0)).\exp(sx)$$

This means that a source having an exponential variation with distance will give rise to an output from the scanner which varies with distance in exactly the same way, and this will be true whatever the form of the line-spread function. Thus in Fig. 6.19 the ratio BC/AC is constant and equal to $k\mathcal{L}(L(x_0))$.

Now let us replace s by the complex number $2\pi\nu\mathrm{i}$. We then have

$$I(x_0) = \exp(2\pi\nu x_0\mathrm{i}) = \cos(2\pi\nu x_0) + \mathrm{i}\sin(2\pi\nu x_0)$$

The corresponding $O(x)$ will also have both real and imaginary parts, but equating the real parts we easily get

$$O(x) = kI_0\left[\cos(2\pi\nu x)\int_{-\infty}^{\infty} L(x_0)\cos(2\pi\nu x_0)\mathrm{d}x_0 - \sin(2\pi\nu x)\int_{-\infty}^{\infty} L(x_0)\sin(2\pi\nu x_0)\mathrm{d}x_0\right]$$

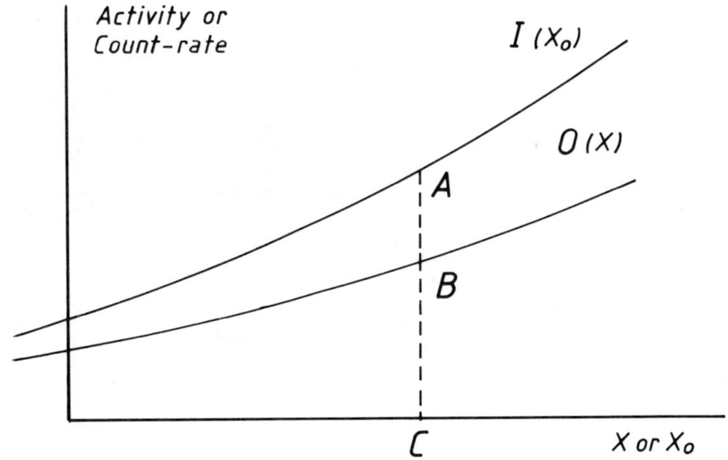

Fig. 6.19 Exponential variation of activity and count-rate. The ratio BC/AC = $k\mathcal{L}(L(x_0))$.

This result can be simplified if $L(x_0)$ is symmetrical about the vertical axis, for then

$$\int_{-\infty}^{\infty} L(x_0) \sin (2\pi \nu x_0) dx_0 = 0$$

because for every value of x_0, contributing a value of $L(x_0) \sin (2\pi\nu x_0)$ to the integral, there will be a value $-x_0$ giving $L(-x_0) \sin (-2\pi\nu x_0) = -L(x_0) \sin (2\pi\nu x_0)$. We find then that

$$O(x) = kI_0 \cos (2\pi\nu x) \int_{-\infty}^{+\infty} L(x_0) \cos (2\pi\nu x_0) dx_0$$

which means that a source whose activity varies with distance according to a cosine law should produce a count-rate, when the scanner head is at any distance x, just proportional to the activity at that point (Fig. 6.20).

The integral

$$\int_{-\infty}^{+\infty} L(x_0) \cos (2\pi\nu x_0) dx_0 \equiv \mathcal{F}(L(x_0))$$

is the *Fourier* transform of the line-spread function. (A more general treatment involves also a sine term, when one needs to transform functions not symmetrical about the origin, but the simple form above is adequate for our present purpose.) Note that the activity repeats itself when $2\pi\nu x = 0, 2\pi, 4\pi, \ldots$. ν is therefore a *spatial* frequency, whose units are mm^{-1}. For

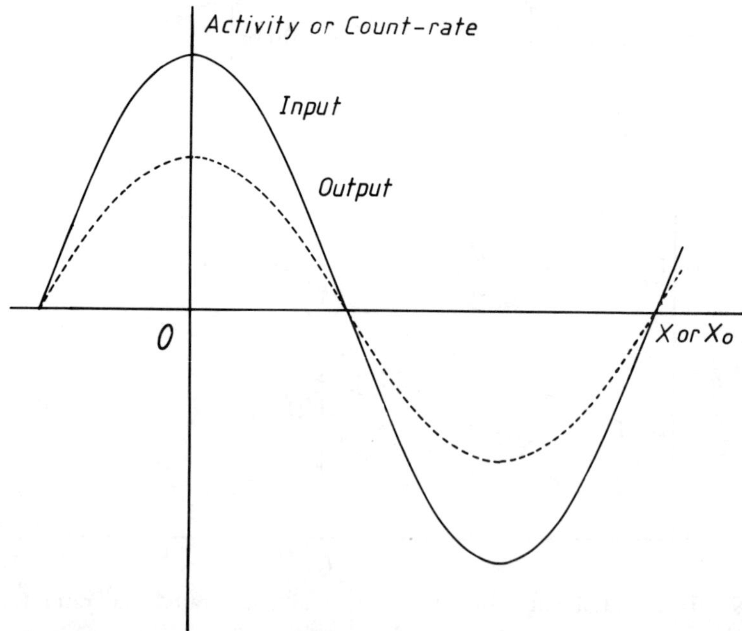

Fig. 6.20 Input cosine function giving an output cosine function.

example, a value of $\nu = 0.2\ \text{mm}^{-1}$ signifies a sinusoidally varying source with a spacing of 5 mm from peak to peak.

An objection may be raised that we cannot have a negative-going source strength. However, this difficulty is avoided by considering a sinusoidal source variation superimposed on a constant level of activity. Thus, for the special case $\nu = 0$, i.e. an input $(I_0)_1$, say, showing no spatial variation, the output $O_1(x)$ is

$$O_1(x) = k(I_0)_1 \int_{-\infty}^{+\infty} L(x_0)\mathrm{d}x_0$$

$$= k(I_0)_1$$

since the LSF is normalised to 1. Superimposing a sinusoidally varying input $(I_0)_2 \cos(2\pi\nu x_0)$ whose theoretical output is

$$O_2(x) = k(I_0)_2 \cos(2\pi\nu x)\mathcal{F}(L(x_0))$$

the combined source will have an output given by

$$O(x) = k\{(I_0)_1 + (I_0)_2 \cos(2\pi\nu x)\mathcal{F}(L(x_0))\}$$

The input and output functions are plotted schematically in Fig. 6.21. The summed inputs and outputs do not become negative, and so are physically realisable, if the value of $(I_0)_1$ is at least as large as that of $(I_0)_2$. A useful parameter to describe these curves is the *modulation* defined as the amplitude (peak-to-peak) of the sinusoid divided by twice the mean value of the curve. We get for the input modulation

$$m_\mathrm{I} = \frac{2A_\mathrm{I}B_\mathrm{I}}{2B_\mathrm{I}C_\mathrm{I}} = \frac{(I_0)_2}{(I_0)_1}$$

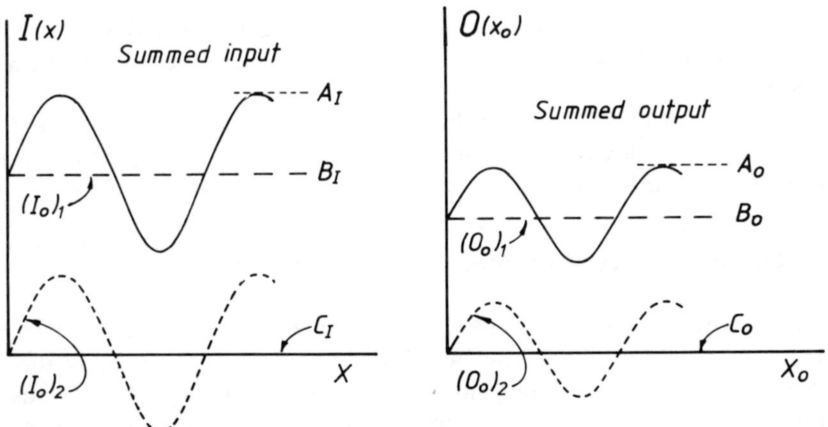

Fig. 6.21 Sinusoidally varying inputs and outputs superimposed on constant inputs and outputs.

and for the output

$$m_O = \frac{2A_OB_O}{2B_OC_O} = \frac{2k(I_0)_2\mathcal{F}(L(x_0))}{2k(I_0)_1} = \frac{(I_0)_2}{(I_0)_1} \cdot \mathcal{F}(L(x_0))$$

Combining these we easily have

$$m_O/m_I = \mathcal{F}(L(x_0))$$

This means that the ratio of output to input modulation is equal to the Fourier transform of the line-spread function. We must emphasise that the Fourier transform can be evaluated for any desired value of the spatial frequency ν; the ratio m_0/m_I is thus a *function* of ν and is called the modulation transfer function (MTF).

Although we have been treating the line-spread function as being continuous, the experimental data, as discussed in Section 1.14 take the form of count-rates at equal spatial intervals Δx_0, forming a sequence L_{-n}, $L_{-n+1}, \ldots, L_{-1}, L_0, L_1, \ldots, L_n$ where L_0 is the count-rate at the peak of the function. If now we choose any arbitrary value of ν and calculate the values of the terms in the sequence

$$\cos(2\pi n\nu\Delta x_0), \cos\{2\pi(n-1)\nu\Delta x_0\}, \ldots \cos(0), \ldots \cos(2\pi n\nu\Delta x_0)$$

we easily have

$$\text{MTF}(\nu) = \frac{L_{-n}\cos(2\pi n\nu\Delta x_0) + \cdots + L_0 + \cdots + L_n\cos(2\pi n\Delta x_0)}{L_{-n} + \cdots + L_0 + \cdots + L_n}$$

$$= \frac{\sum_{j=-n}^{n} L_j \cos(2\pi j\nu\Delta x_0)}{\sum_{j=-n}^{n} L_j}$$

The calculation is quite straightforward on a programmable calculator; only half-a-dozen well chosen ν values will give a good idea of the form of the MTF function.

The extension of these principles to the measurement of the MTF for a gamma camera is obvious. The only technical difference lies in the method used to obtain the line-spread function.

6.19 The Significance of the Modulation Transfer Function

There is a very general theorem (Fourier's theorem) which states that any distribution function of the kind we have been discussing, namely the distribution of radioactive strength in a spatial dimension, is expressible as the sum of a series of sine and cosine terms with suitable multiplying factors, the series has a fundamental term which contains the longest wavelength neces-

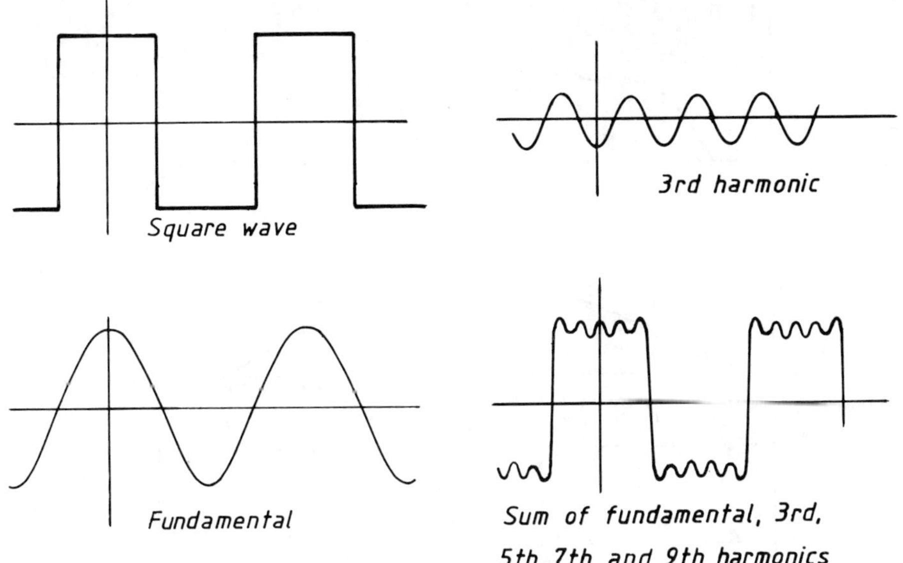

Fig. 6.22 Fundamental and harmonics in a square wave.

sary to describe the distribution and a number of harmonics of shorter wavelength. Thus a Fourier series of cosine terms which would approximate to a square-wave distribution centred at the origin are sketched (Fig. 6.22).

It is clear that a very large number of terms in a Fourier series might be necessary to get an acceptable approximation to any desired function which has sharp discontinuities. In biological systems these will occur, for example, at the edges of organs and sometimes at the edges of discrete lesions. In radionuclide scanning, however, we are hardly in a position to require very accurate delineation of these discontinuities and often are quite content to establish the presence or absence of a lesion.

If a lesion is, for the sake of argument, an object of 10 mm diameter, we would expect that a Fourier series expressing the distribution of activity in the organ to need terms with a wavelength at least as short as 10 mm (Fig. 6.23). In this figure the position of the lesion corresponds to the 'valley' in the sine wave of highest spatial frequency; other waves of longer wavelength will be needed to interfere with this wave at all points on either side of the lesion.

It is intuitively clear that a wavelength of $\lambda = 10$ mm is necessary even to determine whether or not a lesion is present, and if its shape is to be defined with any accuracy even shorter wavelengths will be needed. Now, all this refers entirely to the *input* to the scanner or gamma camera. It is immediately apparent that the *output* of the imaging device can be an accurate reproduction of the input only if the MTF is unity over the whole range of wavelengths considered. Even a poor-quality imaging device might be able to reproduce gross features of the source, but the reproduction of fine detail will be less

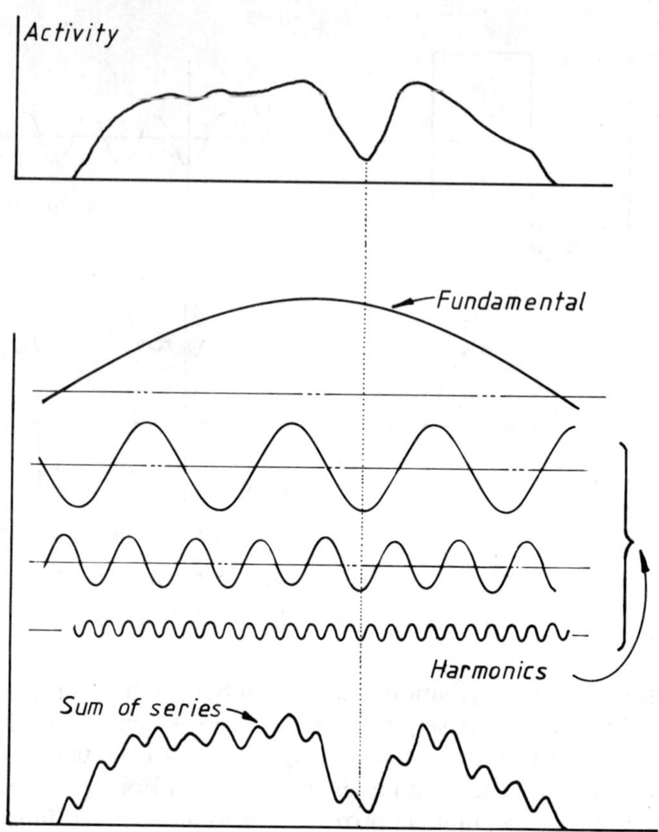

Fig. 6.23 A set of harmonics to represent an organ having a small lesion.

than perfect insofar as the MTF will be less than unity for shorter wavelengths. This is illustrated schematically in Fig. 6.24.

This diagram makes it clear that an imaging device might give unequivocal evidence of the existence of a lesion of a certain size (or of course a larger lesion) while others of a smaller size might be reproduced in the scanner or gamma-camera output with such a small amplitude as to be scarcely noticeable. The value of a MTF determination is then that it tells us, as a result of only one experimental measurement, how we may expect the imaging device to respond to a variety of activity distributions. For a complete assessment of the performance of an imaging device, one would of course wish to obtain MTFs under a whole range of conditions, such as distance between the line-source and the detector, and the presence or absence of scattering material. Such data facilitate the intercomparison of performance as between two machines in a quantitative way. This assessment of performance does not, however, take into account the effects of the particular display system incorporated in the device (colour or dot display) which are to some extent subjective.

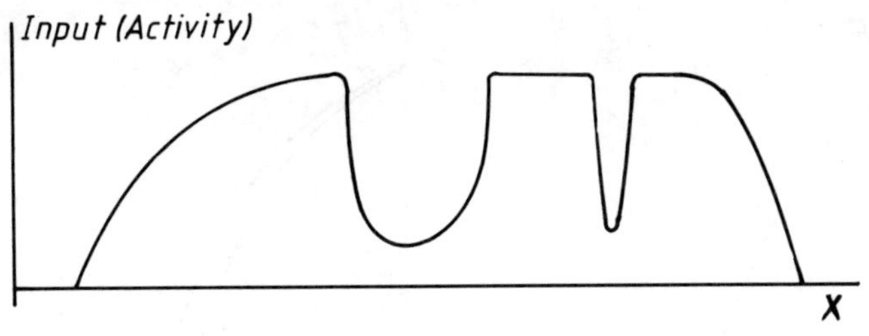

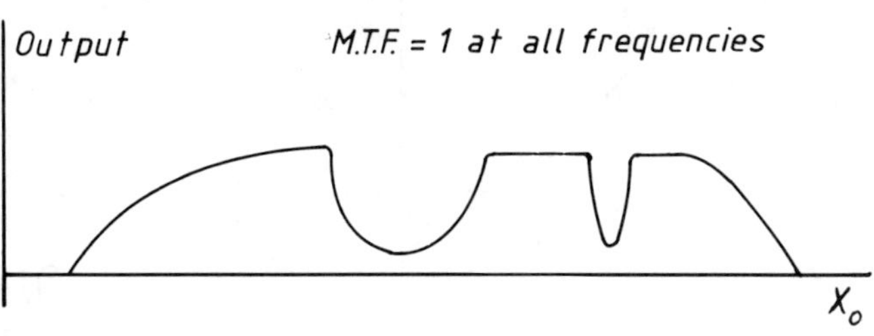

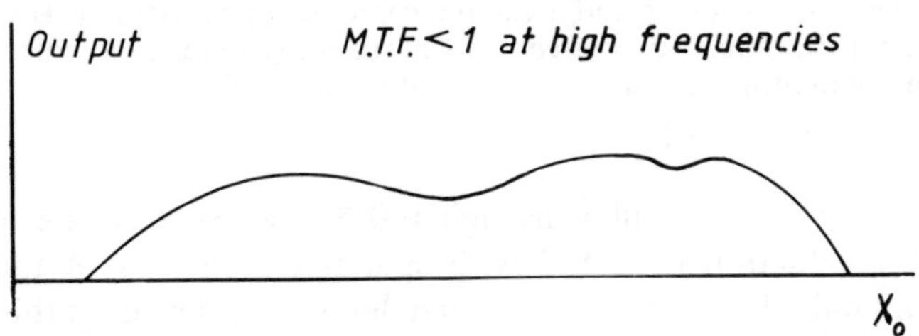

Fig. 6.24 Input–output relationships with different MTFs.

6.20 Some Results of MTF Calculations

In Fig. 6.25, we reproduce some typical results for MTF calculations for sources in air, the transfer function being plotted against ν values. The curves are shown in each case for sources at different distances from the detector. The following points should be noted.

(a) At low values of ν—up to say 0.02 mm^{-1} the MTF value approaches 1, so that there is almost as much modulation in the scanner or camera output as in the inputs. Therefore, sources which could be adequately described by sinusoids with a peak-to-peak spacing of 50 mm or more (1/0.02) will be imaged well enough.

(b) MTF values for objects which are well out of the focal plane of scanners fall very rapidly above $\nu = 0.02$ mm^{-1}; typically the MTF is 0.1 or less at $\nu = 0.05$ mm^{-1}. However, in the focal plane (curve 2 in

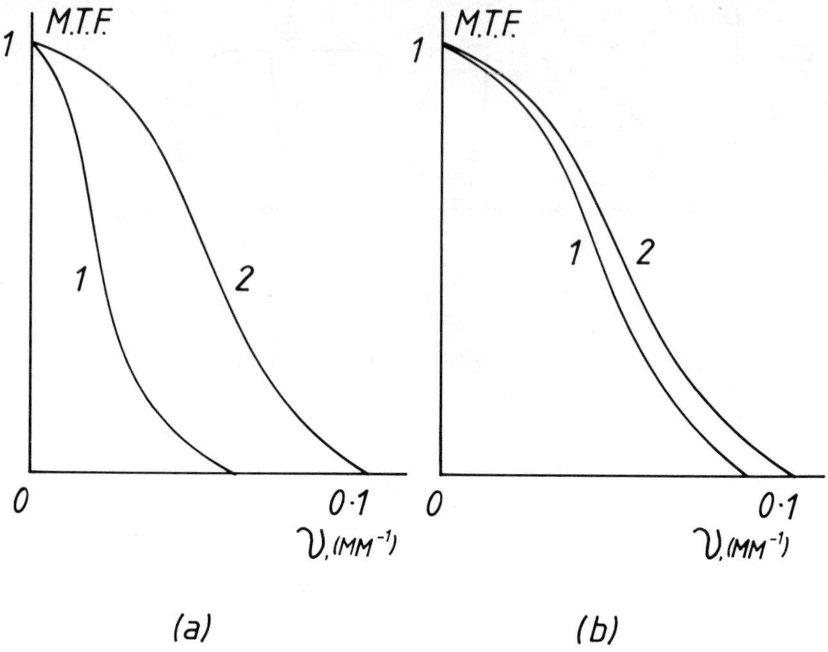

Fig. 6.25 MTFs for a scanner and a gamma camera. (a) Scanner: 1, source 100 mm above or below focal plane; 2, source in focal plane. (b) Gamma camera: 1, source 200 mm from collimator; 2, source on collimator face.

(*a*)), the MTF may still be as high as 0.6 or so. So a scanner will show a fair reproduction of detail with spatial dimensions of the order of 20 mm in the focal plane while such detail very far out of this plane will be lost. MTF curves for sources only 10–20 mm above, or below, the focal plane are usually found to be quite close to that for a source in the plane, the good agreement being consistent with a fair depth of focus.

(c) MTF curves for a gamma camera are much the same shape as for a scanner in the focal plane, but of course, do not show the same variation with focal distance. Indeed, a gamma camera tends to give clearer images of structures closest to it, but the depth dependence is generally very small (in air).

The effect of scattering material (e.g. tissue) between the line-source and the imaging device is to degrade the MTF in a fashion displayed in Fig. 6.26. The graph shows the effect of interposing a thickness of 150 mm tissue-equivalent material between a line-source and the face of a gamma camera. Dpending on the exact parameters of the imaging device and the amount of scattering material, there may be considerable loss of transfer function for $\nu \geq 0.02\ \text{mm}^{-1}$ (wavelength $\leq$ 50 mm). A similar distortion of the MTF curve is also seen for collimators whose septal penetration is too high for the radionuclide being used.

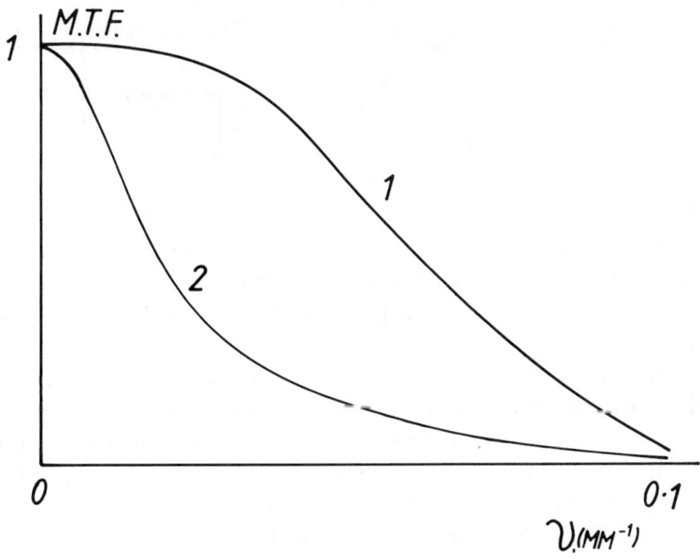

Fig. 6.26 MTF: 1, without, 2, with scatterer.

6.21 MTF and Statistical Considerations

An order-of-magnitude calculation based on the number of 'counts' recorded by a scanner or gamma camera in routine imaging is instructive. Over the whole imaging field of say 10^5 mm^2, the total number of recorded counts will usually be between 10^5 and 10^6; let us calculate on the basis of 10^6 counts. This means that if the object is uniform, some 10^3 counts will be observed within a typical resolution circle of 100 mm^2. But the figure of 10^3 counts is subject to statistical variations which are of the order of its square root, i.e. 30. The relative variations are, therefore, of the order of 3%. It follows that neighbouring areas in the image, which should just be resolved, will show 'natural' variations of this magnitude, even if there is no 'true' spatial variation in the source. Consequently if the source is imaged with a MTF of only 0.03 at a spatial frequency of $\nu = 0.1$ mm^{-1}, no true variations in the source can be distinguished from statistical variations. The situation is of course worse in regions of lower count density. It seems a fairly good rule of thumb that radionuclide imaging is ineffective in practice if the MTF is less than 0.05. The resolution of a pair of small lesions with a centre-to-centre spacing of 10 mm is unlikely to be successful unless the MTF is at least 0.1.

6.22 Exercises

1. Investigate a two-compartment model in which the compartments are in dynamic equilibrium with one another: that is to say, the flow of material from compartment 1 is exactly balanced by the reverse flow. Let $C_1(t)$ and $C_2(t)$ be the count-rates given by tracer in the two compartments and

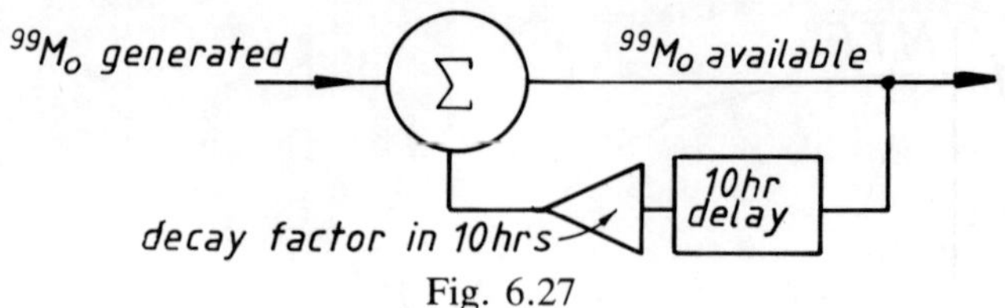

Fig. 6.27

take $C_1(0) = C_{1,0}$ and $C_2(0) = 0$. Write simultaneous differential equations for DC_1 and DC_2. Solve these for C_1 and C_2 (*a*) by direct elimination, and (*b*) by taking Laplace transforms before eliminating C_2 and C_1.

2. Refine the model of Section 6.12 to take into account the fact that uptake of iodine into the thyroid by hyperthyroid patients may be up to twice as rapid as for normals. Also estimate suitable values for the resistances and capacitances in the electrical model to that it would simulate a hyperthyroid case, assume uptake rising with a half-life 0.2 d to a steady value of 15% of the dose administered.
3. ^{99}Mo ($t_{1/2} = 70$ h) is produced by a reactor: in successive 10 h periods the activity available was 250, 600, 650, 700 and 200 TBq. Calculate the decay factor for a 10 h period (i.e. the fraction of ^{99}Mo which survives for 10 h). Calculate the *total* amount of ^{99}Mo available at the end of these, and of some successive, 10 h periods. Use the model illustrated in Fig. 6.27.
   Show that the output from the model is a convolution.

Table 6.4

Lateral displacement of scanner axis from the line-source (mm)	Count-rate (above background) (s^{-1})
0	10 100
1	9 265
2	7 800
3	6 200
4	4 850
5	3 825
6	2 750
7	1 950
8	1 350
9	900
10	610
11	450
12	300
13	220
14	175
15	130

4. The figures shown in Table 6.14 were obtained for the line-spread function of a scanner. Derive the modulation transfer function of the system. Would the function be appreciably different if the data for displacements greater than 10 mm had been ignored? Is it likely that the function could be more accurately determined by using data for displacements greater than 15 mm?
5. Write a program for the deconvolution of renograms.

   Arrange to read in 25 values of a variable H(I) and a variable R(I) representing nett counts obtained at one minute intervals from an extra-renal and a renal detector respectively.

   Write an iterative program to find the best values of constants P1, P2, P3, K1, K2, K3 which will minimise the sum of squares

$$\begin{aligned}&(\mathrm{H(I)} - \mathrm{P1} * \mathrm{EXP}(-\mathrm{K1} * \mathrm{T}) - \mathrm{P2} * \mathrm{EXP}(-\mathrm{K2} * \mathrm{T}) \\ &\quad - \mathrm{P3} * \mathrm{EXP}(-\mathrm{K3} * \mathrm{T}))**2\end{aligned}$$

   Write a program to evaluate G(I) corresponding to the function $g'(t)$ defined on p.372 Evaluate the mean transit time TM.
6. Starting from the equation for $H'(s)$ on p. 370 obtain the equation for $g'(t)$ on p.371. Use the result that if the Laplace transform of a function $f(\tau)$ is $F(s)$, then

$$\mathcal{L}\int_0^t f(\tau)\mathrm{d}\tau = F(s)/s$$

6.23 Bibliography

GABEL R.A. and ROBERTS R.A. *Signals and Linear Systems* (John Wiley)

BELCHER E.H. and VETTER H. *Radioisotopes in Medical Diagnosis* (Butterworth)

Various authors: *Dynamic Studies with Radioisotopes in Medicine, 1974, Proceedings of a Symposium* (IAEA, Vienna)

Appendix 1
Numerical Constants and Definitions

Numerical Constants

e = base of natural logarithms = 2.718 28
e^{-1} = 0.367 88
ln 2 = 0.693 15
ln (X) = 2.302 59 log (X)

Physical Constants

	Symbol
Avogadro's constant = 6.022×10^{23} mol^{-1}	N
Charge on one electron = 1.602×10^{-19} C	
Mass of one unified mass unit = (mass of ^{12}C atom)/12 = 1.6604×10^{-27} kg	u
Mass of one electron = 9.1096×10^{-31} kg	m_e
Mass of one hydrogen atom = 1.6734×10^{-27} kg	m_H
Mass of one neutron = 1.6748×10^{-27} kg	m_n
Energy of one electronvolt = 1.602×10^{-19} J	
Energy equivalent of one electron mass = 0.511 MeV	
Energy equivalent of one proton = 938.3 MeV	
Energy equivalent of one neutron = 939.6 MeV	
Number of seconds in one day = 8.64×10^4	
Number of seconds in one year = 3.1588×10^7	
Velocity of light = 2.9979×10^8 m s^{-1}	c
Planck's constant = 6.6626×10^{-34} J s	h

Product of photon energy and wavelength = hc J m
energy (keV) = 1.238/λ (nm)
energy (keV) = 12.38/λ (Å)
(1 Å = 1 ångström unit = 10^{-10} m)

Compton wavelength shift = change in wavelength associated with photon scatter through 90°

$$= h/m_e c^2 = 0.002\ 42 \text{ nm}$$

Definitions

Symbol

Unit of radioactivity: The unit is the becquerel, a rate of nuclear transformation of one per second. (Previous unit, the curie (Ci), or 3.7×10^{10} Bq; this is approximately the rate of disintegration of 1 g of ^{226}Ra.) Bq

Unit of radiation exposure: The quotient of ΔQ by Δm where ΔQ is the sum of all the electrical charges of one sign produced in air, when all the electrons (negative electrons and positrons) liberated in a volume of air whose mass is Δm are completely stopped in air:

$$X = \Delta Q/\Delta m$$

The special unit of exposure is the roentgen R

$$1 \text{ R} = 2.58 \times 10^{-4} \text{ C kg}^{-1}$$

Unit of absorbed dose: The quotient of ΔE_d by Δm where ΔE_d is the energy imparted by ionising radiation to the matter in a volume element, and Δm is the mass of matter in the volume element:

$$D = \Delta E_d/\Delta m$$

The special unit of absorbed dose is the gray Gy

$$1 \text{ Gy} = 1 \text{ J kg}^{-1}$$

(Previous unit, the rad, equal to 100 erg g^{-1} or 0.01 J kg^{-1}. 1 rad = 0.01 Gy).

Unit of dose equivalent: The unit is the sievert. This has the same biological effect as 1 Gy when absorbed from 'hard' X-rays. (Previous unit, the rem, was 0.01 Sv.) Sv

Appendix 2
Energy of an Accelerated Particle

Classical Mechanics

From Newton's third law a force F will produce an acceleration a in a particle of mass m, where

$$F = ma = m\frac{\mathrm{d}v}{\mathrm{d}t} = m\frac{\mathrm{d}x^2}{\mathrm{d}t^2}$$

The work done by the force in acting over the distance from the origin to the point $x = a$ is then

$$\int_0^a F\mathrm{d}x = \int m\frac{\mathrm{d}x}{\mathrm{d}t} \cdot \mathrm{d}v = \int mv\mathrm{d}v$$

with suitable limits on v. If $v = 0$ initially and $v = V$ when $x = a$, with the proviso that m is a constant, the work done by the force is

$$\int_0^V mv\mathrm{d}v = \frac{1}{2}mV^2$$

This is then the kinetic energy of the particle,

$$E_{\mathrm{K}} = \frac{1}{2}mV^2$$

Relativistic Mechanics

In this case the mass of the particle is *not* constant but varies with v according to the relation

$$m = m_0/[\sqrt{(1 - \beta^2)}]$$

where m_0 is the (constant) rest-mass and $\beta = v/c$, c being the velocity of light, 3×10^8 m s^{-1}. Relativistic mechanics applies strictly to all particles, but reduces to the classical case when β is negligibly small. However when β is not negligible, we have

$$\text{work done} = \text{kinetic energy} = \int_0^V m_0 \frac{v\,dv}{\sqrt{[1 - (v^2/c^2)]}}$$

$$= m_0c^2[1/\sqrt{(1 - \beta^2)} - 1]$$

i.e.

$$E_K = mc^2 - m_0c^2$$

Einstein recognised m_0c^2 as the energy of the particle *in virtue of its mass alone*. The *total* energy E_T is the sum of the *rest energy* m_0c^2 and the kinetic energy:

$$E_T = m_0c^2 + E_K = mc^2$$

Tables A2.1 and A2.2 give some useful numerical values for the mass and velocity of electrons and of protons as a function of energy. It is useful to note that the rest-mass of an electron, in energy units, is 511 keV ($1\ \text{eV} = 1.602 \times 10^{-19}$ J).

Table A2.1 Mass and Velocity of Electrons

Energy (keV)		Mass (kg)			Velocity (m s^{-1})
Kinetic	Total	$\times 10^{-13}$	m/m_0	β	$\times 10^{-8}$
0	511	9.1096	1.00	0.0	0.0
50	561	10.00	1.10	0.413	1.24
100	611	10.90	1.20	0.548	1.64
500	1111	18.01	1.98	0.863	2.58
1000	1511	26.9	2.96	0.941	2.82
5000	5511	98.1	10.77	0.996	2.98

Table A2.2 Mass and Velocity of Protons

Energy (MeV)		Mass (kg)			Velocity (m s^{-1})
Kinetic	Total	$\times 10^{27}$	m/m_0	β	$\times 10^{-8}$
0	938	1.6734	1.00	0.0	0.0
10	948	1.69	1.011	0.144	0.43
20	958	1.71	1.021	0.203	0.61
50	988	1.76	1.053	0.314	0.94
100	1038	1.85	1.106	0.428	1.28
1000	1938	3.45	2.064	0.875	2.62

Bibliography and References

The following abbreviations are used:

Brit. J. Radiol.: British Journal of Radiology
Circ. Res.: Circulation Research
Eur. J. Nuc. Med.: European Journal of Nuclear Medicine
I.E.E.E. Nuc. Sci.: Institute of Electrical and Electronic Engineers, Nuclear Science
I.J.A.R.I.: International Journal of Applied Radiation and Isotopes
I.J. Nuc. Med. Biol.: International Journal of Nuclear Medicine and Biology
J. Nuc. Med.: Journal of Nuclear Medicine
Med. Biol. Eng.: Medical and Biological Engineering
Nuc. Medizin: Nuclear Medizin
Pharmacol. Rev.: Pharmacological Reviews
P.M.B.: Physics in Medicine and Biology
Sem. Nuc. Med.: Seminars in Nuclear Medicine

An asterisk* denotes a review article.

Activation Analysis

AL-HITI, K., THOMAS, B.J. *et al.* Spinal calcium; its *in vivo* measurement in man. *I.J.A.R.I.* **27**, 97 (1976)

BELL, C.M.J., LEACH, M.O., *et al.* Problems in the interpretation of the *in vivo* measurement on calcium by the ^{37}Ar method. *J. Nuc. Med.* **19**, 54 (1978)

BIGLER, R.E., LAUGHLIN, J.S., *et al.* Estimation of skeletal calcium in humans by exhaled ^{37}Ar measurement: evaluation of fast neutron requirements. *I.J. Nuc. Med. Biol.* **3**, 105 (1976)

BODDY, K., HOLLOWAY, I., *et al.* A simple facility for total body *in vivo* activation analysis. *I.J.A.R.I.* **24**, 428 (1973)

BUSH, D. Measurement and control of dose equivalent to patients undergoing whole-body irradiation for *in vivo* neutron activation analysis. *P.M.B.* **17**, 32 (1972)

CHAMBERLAIN, M.J., FREMLIN, J.H., *et al.* Use of the cyclotron for whole-body neutron activation analysis; theoretical and practical considerations. *I.J.A.R.I.* **21**, 725 (1971)

COHN, S.H., DOMBROWSKI, C.S. *et al.* *In vivo* neutron activation analysis in man. *I.J.A.R.I.* **21**, 127 (1970)

DUBI, A., HOROWITZ., Y.S. Simultaneous two-sample activation measurement immune to detector intercalibration. *I.J.A.R.I.* **28**, 291 (1977)

ELLIOTT, A., HOLLOWAY, I. *et al.* Neutron uniformity studies related to total body *in vivo* neutron activation analysis. *P.M.B.* **23**, 269 (1978)

LEACH, M.O., BELL, C.M.J. *et al.* *In vivo* measurement of calcium by the ^{37}Ar method: a study of the effect of recirculating breath collection systems on the exhalation rate. *P.M.B.* **23**, 282 (1978)

LEACH, M.O., THOMAS, B.J. *et al.* Total body nitrogen by the N(n,2n)^{13}N method; a study of interfering reactions and variation of spatial sensitivity with depth. *I.J.A.R.I.* **28**, 263 (1977)

LEWELLEN, T.K., CHESUNT, C.H. *et al.* Preliminary observations on the excretion of ^{37}Ar from man following whole-body neutron activation. *J. Nuc. Med.* **16**, 672 (1975)

MCNEILL, K.G., THOMAS, B.J. *et al.* *In vivo* neutron activation analysis for calcium in man. *J. Nuc. Med.* **14**, 502 (1973)

OZBAS, E., CHETTLE, D.R. *et al.* *In vivo* neutron activation analysis measurements of calcium using the ^{41}Ca(n,γ)^{37}Ar reaction. *I.J.A.R.I.* **27**, 227 (1976)

PALMER H.E. Feasibility of determining total-body calcium in animals and humans by measuring ^{37}Ar in expired air after neutron irradiation. *J. Nuc. Med.* **14**, 522 (1973)

PALMER, H.E., NELP, W.B. *et al.* The feasibility of *in vivo* neutron activation analysis of total body calcium and other elements of body composition. *P.M.B.* **13**, 269 (1968)

VESELSKY, J.C., NEDBALEK, M. The determination of protein-bound iodine by neutron activation analysis. *I.J.A.R.I.* **21**, 225 (1974)

Assessment of Performance of Radionuclide Imaging Systems

ARONOW, S. Performance analysis of imaging systems. *Sem. Nuc. Med.* **3**, 239 (1973)

BAKER, R.G., SCRIMGER, J.W. Investigation of parameters of scintillation camera design. *P.M.B.* **12**, 51 (1967)

BARBER, D.C., MALLARD, J.R. Conditions for high resolution radioisotope scanning. *P.M.B.* **14**, 675 (1969)

BARBER, D.C. Comments on the use of intrinsic plane source efficiency. *I.J.A.R.I.* **24**, 61 (1973)

BELL, T.K., CRANLEY, K. Rapid investigation of the response of focussing and non-focussing collimators. *Brit. J. Radiol.* **48**, 759 (1975)

BROOKEMAN, V.A. Component resolution indices for scintillation camera systems. *J. Nuc. Med.* **16**, 228 (1975)

CRADDUCK, T.D., FEDORUK, S.O. *et al.* A new method for assessing the performance of scintillation cameras and scanners. *P.M.B.* **11**, 423 (1966)

CAUSER, D.A., MALLARD, J.R. Measurement of the sensitivity of gamma ray imaging systems. *I.J.A.R.I.* **25**, 119 (1974)

COHEN, G., KEREIAKES, J.G. *et al.* Quantitative assessment of field uniformity for gamma cameras. *Radiology* **118**, 197 (1976)

COX, N.J., DIFFEY, B.L. A numerical index of gamma camera uniformity. *Brit. J. Radiol.* **49**, 734 (1976)

DAY, M.J. Specification and additivity of unsharpness in diagnostic radiology. *P.M.B.* **21**, 399 (1976)

DOWDEY, J.E., MURRY, R.C. *et al.* Effect of collimator motion on image quality in nuclear medicine. *Radiology* **114**, 411 (1975)

EHRHARDT, J.C., OBERLEY, L.W. *et al.* Effect of a scattering medium on gamma-ray imaging. *J. Nuc. Med.* **15**, 943 (1974)

GALVIN, J.M., SIMMONS, G.H. *et al.* A comparison of radio-isotope imaging devices using plane source distributions. *P.M.B.* **15**, 735 (1970)

GILLESPIE, F.C., WYPER, D.S. Conditions for high resolution radio-isotope scanning. *P.M.B.* **14**, 673 (1969)

GOODWIN, P.N., HIMELSTEIN, E. Methods for comparing the performance of different gamma cameras. *Sem. Nuc. Med.* **7**, 299 (1977)

HUDSON, F.R., DAVIS, J.B. A phantom for simulating fast dynamic studies on a gamma camera. *P.M.B.* **22**, 87 (1977)

JAHNS, E., HINE, G.J. A line-source phantom for testing the performance of scintillation cameras. *J. Nuc. Med.* **8**, 829 (1967)

JASZCZAK, R.J. Line-spread response for a scintillation camera and movable filter-plate system. *P.M.B.* **19**, 362 (1974)

JONES, J.P., BRILL, A.B. The validity of an equivalent point source assumption made in quantitative scanning. *P.M.B.* **20**, 455 (1975)

KENNY, P.J. Spatial resolution and count-rate capacity of a positron camera. *I.J.A.R.I.* **22**, 21 (1977)

KEYES, W.I., GILLESPIE, P.J. Rational selection of parameters in ratemeter controlled rectilinear scanning. *I.J.A.R.I.* **21**, 649 (1970)

MACINTYRE, W.S., ALFIDI, R.J. *et al.* Comparative modulation transfer functions for the EMI and Delta scanners. *Radiology* **120**, 189 (1976)

MCCULLOUGH, E.C., PAYNE, J.T. *et al.* Performance evaluation and quality assurance of computerised tomography scans. *Radiology* **120**, 173 (1976)

MALLARD, J.R., MYERS, M.J. The performance of a gamma camera for visualisation of radio-isotopes *in vivo*. *P.M.B.* **8**, 165 (1963)

MORRISON, L.M., BRUNO, F.P. *et al.* Sources of gamma-camera inequalities. *J. Nuc. Med.* **12**, 785 (1971)

MUEHLLEHNER, G., BUCHIN, N.P. *et al.* Performance parameters of the positron imaging camera. *I.E.E.E. Nuc. Sci.* **NS23**, 528 (1976)

NUSYNOWITZ, M.L., BENEDETTO, A.R. Simplified method for determining the modulation transfer function for the scintillation camera. *J. Nuc. Med.* **16**, 1200 (1975)

OKUMURA, Y. The application of the modulation transfer function to reduce unsharpness in scintigrams. *I.J.A.R.I.* **22**, 49 (1971)

PARKER, R.P. Degradation of spatial resolution in gamma cameras employing NaI or Ge detectors. *P.M.B.* **15**, 493 (1970)

PAYNE, J.T., WILLIAMS, L.E. *et al.* Comparison and performance of Anger cameras. *Radiology* **109**, 381 (1973)

PAYNE, J.T., LOKEN, M., *et al.* Limitations of available gamma camera oscilloscopes for whole-body image or minified multi-image display. Radiology **113**, 730 (1974)

PEŘINOVA, V., HUŠAK, V. Remark on the concept 'resolution' in scintillation scanning. *P.M.B.* **12**, 333 (1967)

RAO, G.U.V. Contrast perception in imaging systems. *I.J.A.R.I.* **21**, 571 (1970)

ROLLO, F.D. An index to compare the performance of scintigraphic imaging systems. *J. Nuc. Med.* **15**, 757 (1974)

SANDERSON, G.R. Erroneous perturbation of the modulation transfer function derived from the line-spread function. *P.M.B.* **13**, 661 (1968)

SHARMA, R.R., FOWLER, J.F. Conditions for high-resolution radio-isotope scanning. *P.M.B.* **14**, 670 (1969)

SHARMA, R.R., FOWLER, J.F. Threshold detection tests in radio-isotope scanning. *P.M.B.* **15**, 289 (1970)

SHARMA, R.R. The 'working region' of distance-independence for point sensitivity in air of moving and stationary scanners. *P.M.B.* **16**, 257 (1971)

SHARMA, R.R. Intrinsic plane sensitivity: a new parameter. *I.J.A.R.I.* **23**, 121 (1972)

SHARMA, R.R. Intrinsic plane sensitivity: a reply. *I.J.A.R.I.* **24**, 62 (1973) (See BARBER, D.C. *ibid.* p. 61)

SHARMA, R.R. Intrinsic plane sensitivity: still the most suitable for comparing moving and/or stationary scanners. *I.J.A.R.I.* **26**, 225 (1975) (See CAUSER, D.A. *ibid.* p. 226)

SHARP, P., MALLARD, J.R. A proposed model for the visual detection of signals in radio-isotope display images. *P.M.B.* **19**, 348 (1974)

SIMONS, H.A.B., BAILY, J.M. An investigation into the usefulness of a 'figure of merit' as a criterion of a collimating system. *P.M.B.* **12**, 29 (1967)

SPECTOR, J.J., BROOKEMAN, V.A. *et al.* Analysis and correction of spatial distortions produced by the gamma camera. *J. Nuc. Med.* **13**, 307 (1972)

SVEDBERG, J.B. Image quality of a gamma camera system. *P.M.B.* **13**, 597 (1968)

SVEDBERG, J.B. Computed intrinsic efficiency and modulation transfer function for gamma cameras. *P.M.B.* **18**, 657 (1973)

WHITTINGHAM, T.A., MALLARD, J.R. Theoretical considerations for high-resolution radio-isotope scanning. *P.M.B.* **13**, 657 (1968)

ZIMMERMAN R.E., HOLMAN, B.L. Modulation transfer function for the Pho/Gamma 3 and Pho/Gamma HP scintillation cameras using ^{99m}Tc and ^{133}Xe (abstract only). *J. Nuc. Med.* **13**, 481 (1972)

Autoradiography

CROSS, S.A.M., GROVES, A.D. *et al.* A quantitative method for measuring radioactivity in tissues sectioned for whole-body autoradiography. *I.J.A.R.I.* **25**, 381 (1974)

GORE, D.J., THORNE, M.C. *et al.* The visualisation of fissionable radionuclides in rat lung using neutron-induced autoradiography. *P.M.B.* **23**, 149 (1978)

JACKSON, D.D., KAHN, M. Autoradiographic detection of ^{131}I using polaroid films. *I.J.A.R.I.* **20**, 742 (1969)

RONAI, P. High resolution autoradiography with ^{51}Cr. *I.J.A.R.I.* **20**, 471 (1969)

SMITH, J.W., GLORIOSO, J.C. A simple method for measuring radioactivity in autoradiograms of polyacrimide gels. *I.J.A.R.I.* **28**, 693 (1977) *van Rooijen, N. The separate detection of two radioactively labelled compounds in tissue sections or cell smears on a light-microscopic level: a review on double radionuclide autoradiography. *I.J.A.R.I.* **27**, 547 (1976)

Books

Defares, J.G., Sneddon, I.N. *An Introduction to the Mathematics of Medicine and Biology.* (Year Book Medical Publishers, Chicago, Ill.)

Dillman L.T., von der Lage, F.C. *Radionuclide Decay Schemes and Nuclear Parameters for Use in Radiation-Dose Estimation.* Pamphlet No 10. (Medical Internal Radiation Dose Committee of the Society of Nuclear Medicine, Maryville, Tennessee)

Gusev, N.G., Dimitriev, L.L. *Quantum Radiation of Radioactive Nuclides.* (Moscow)

Hennig, K., Woller, P. *Nuclear Medicine—Brief and to the Point.* (Theodor Steinkopf, Dresden)

Hine, G.J., Sorenson, J.A. *Instrumentation in Nuclear Medicine,* Vol 2. (Academic Press)

Neame, K.D., Homewood, C.A. *Liquid Scintillation Counting.* (Halstead Press, New York)

Perkins, W.J. *Biomedical Computing.* (Pitman)

Quinn, E. *Year Books of Nuclear Medicine.* (Year Book Medical Publishers, Chicago, Ill.)

Raynaud, C., Todd-Pokropek, E.E. *Information Processing in Scintigraphy.* (Commisariat a l'energie atomique, Dept de biologie, Service Hospitalier F. Joliot, Orsay, France)

Razzak, M.A., Sodee, D.B. *Textbook of Nuclear Medicine Technology.* (C.V. Mosby, St Louis, USA)

Ter-Poggosian, M.M., Phelps, M.E. *et al.* *Reconstruction Tomography in Diagnostic Radiology and Nuclear Medicine.* (University Park, Baltimore, USA)

Wang, Y. *CRC Handbook of Radioactive Nuclides.* (Chemical Rubber Company)

Calorimetry

Bewley, D.K., McCullough, E.C. *et al.* Heat defect in tissue-equivalent radiation calorimeters. *P.M.B.* **17**, 95 (1972)

Ditmars, D.A. Measurement of the average total decay power of two plutonium sources in ã Bunsen ice calorimeter. *I.J.A.R.I.* **27**, 469 (1976)

Collimator Design

Barber, D.C. Direct calculation of the collimator transfer function for single bore and multichannel focussing collimators. *I.J.A.R.I.* **25**, 193 (1974)

Beattie, J.W. Optimisation of parallel-hole collimators for gamma cameras. (abstract only). *J. Nuc. Med.* **13**, 411 (1972)

Bell, K.T., Johnston, A.R. Penetration effects in wide-angle collimators. *P.M.B.* **13**, 401 (1968)

BELL, K.T., SPIERS, E.W. *et al.* Calculation of penetration factors for point gamma-ray sources off the axis of cylindrical-hole collimators. *P.M.B.* **15**, 47 (1970)

BELL, K.T., SPIERS, E.W. *et al.* Penetration factors for point gamma-ray sources off the axis of parallel-hole collimators. *P.M.B.* **16**, 152 (1971)

BOSNJAKOVIC, V., BENNETT, L.R. An automated method for the evaluation of non-focussed collimator performance in water medium. *J. Nuc. Med.* **15**, 679 (1974)

BROOKEMAN, V.A., BAUER, T.J. Collimator performance for scintillation camera systems. *J. Nuc. Me.* **14**, 21 (1973)

CAUSER, D.A. The design of parallel hole gamma camera collimators. *I.J.A.R.I.* **26**, 355 (1975) (See also a correction, *ibid.* **27**, 186 (1976))

CAUSER, D.A., TAYLOR, C.G. A high-sensitivity high-resolution diverging collimator for use with low-energy isotopes. *P.M.B.* **20**, 318 (1975)

CLARKE, L.P., DUFFEY, G.J. *et al.* An improved uptake probe designed for a large crystal rectilinear scanners. *P.M.B.* **23**, 118 (1978)

CRADDUCK, T.D. Distortion produced by pinhole collimators. *J. Nuc. Med.* **13**, 778 (1972)

FOLLETT, D.H. Collimators for radio-isotope scanning. *P.M.B.* **14**, 667 (1969)

HUŠAK, V., PEŘINOVA, V. Design of collimators for radio-isotope scanning. *P.M.B.* **14**, 233 (1969)

HUŠAK, V. Collimators for radio-isotope scanning. *P.M.B.* **14**, 669 (1969)

*KENNY, P.J. Collimation for rectilinear scanners and gamma camera imaging equipment. *Sem. Nuc. Med.* **3**, 259 (1973)

KIRCOS, L.T., Leonard, P.F. *et al.* An optimised collimator for single-photon computed tomography with a scintillation camera. *J. Nuc. Med.* **19**, 322 (1978)

LAKSHMANAN, A.V., CAUSER, D.A. *et al.* A 'depth-independent' collimator for use with ^{99m}Tc in both conventional and transverse section scanning. *I.J. Nuc. Med. Biol.* **2**, 123 (1975)

MCKEIGHEN, R.E., MUEHLLEHNER, G. *et al.* Gamma camera collimator considerations for imaging ^{123}I. *J. Nuc. Med.* **15**, 328, (1974)

METZ, C.E. TSUI, M.W. *et al.* Theoretical prediction of the geometric transfer function for focussed collimators. *J. Nuc. Med.* **15**, 1078 (1974)

MOYER, R.A. A low-energy multihole converging collimator compared with a pin-hole collimator. *J. Nuc. Med.* **15**, 59 (1974)

MUEHLLEHNER, G. Septal penetration in scintillation camera collimators. *P.M.B.* **18**, 855 (1973)

MURPHY, P.H., BURDINE, J.A. *et al.* Clinical appraisal of an experimental converging collimator. *J. Nuc. Med.* **15**, 291 (1974)

NEILL, G.D.S. Predicted and measured performance of a segmented-channel focussing collimator. *I.J.A.R.I.* **24**, 563 (1973)

OBERLEY, L.W., LENSINK, S.C. *et al.* An accurate system for evaluating gamma camera collimators. *P.M.B.* **19**, 36 (1974)

SHIMMINS, J.G., SUMNER, D.J. *et al.* Experimental study of profile scanning optimisation. *J. Nuc. Med.* **14**, 895 (1973)

SIMONS, H.A.B. A computer method for calculation of the point-source response of a focussing collimator. *P.M.B.* **15**, 57 (1970)

SIMONS, H.A.B., BELL, T.K. *et al.* Calculation of collimator penetration effects. *P.M.B.* **16**, 329 (1971)

SIMMONS, G.H., CHRISTENSON, J.M. *et al.* A new design technique for parallel-hole collimators. (abstract only). *J. Nuc. Med.* **13**, 467 (1972)

SIMMONS, G.H., CHRISTENSON, J.M. *et al.* A non-linear programming method for optimising parallel-hole collimator design. *P.M.B.* **20**, 771 (1975)

SWANN, S., PALMER, D. *et al.* Optimised collimators for scintillation cameras. *J. Nuc. Med.* **17**, 50 (1976)

TSIALAS, S.P., LEWIS, B. A computer program for evaluating the performance of focussed collimators. *I.J.A.R.I.* **21**, 375 (1970)

WYPER, D.J., GILLESPIE, F.C. A quantitative method of evaluating focussed collimators. *J. Nuc. Med.* **12**, 197 (1971)

WYPER, D.J., GILLESPIE, F.C. *et al.* Method for reducing the energy dependence of the resolution of focussed collimators. *J. Nuc. Med.* **13**, 19 (1972)

Computer Techniques

ABRAMS, M.E., CRAWLEY, J.C.W. *et al.* A comparative study of digital and analogue computer techniques for deconvolution procedures in clinical tracer studies. *P.M.B.* **14**, 225 (1969)

ALPERT, N.M., BURNHAM, A.B. *et al.* Numedics: a system for on-line data processing in nuclear medicine. *J. Nuc. Med.* **16**, 386 (1975)

BARBER, D.C., MALLARD, J.R. Data processing for radio-isotope images for optimum smoothing. *P.M.B.* **16**, 635 (1971)

BARBER, D.C. Digital computer processing of brain scans using principal components. *P.M.B.* **21**, 792 (1976)

BOTTOMLEY, P.A., HINSHAW, W.S. *et al.* A computer driven photoscanner for medical imaging. *P.M.B.* **23**, 309 (1978)

BUDINGER, T.F., MACDONALD, B. Reconstruction of the Fresnel-coded gamma camera images by digital computer. *J. Nuc. Med.* **16**, 309 (1975)

CHACKETT, K.F., WELBORN, J.B. Computer assisted evaluation of liver scans. *P.M.B.* **16**, 533 (1971)

*GLASS, H.I. The role of computers in the nuclear medicine laboratories. *Sem. Nuc. Med.* **3**, 303 (1973)

HOUSTON, A.S. Mathematical tumours and their use in assessing data processing techniques in radio-isotope scintigraphy. *P.M.B.* **19**, 631 (1974)

KEYES, W.I., BARBER, D.C. *et al.* Quantitative multilevel display of radioisotope scan data via a small digital computer. *P.M.B.* **18**, 133 (1973)

REINKE, D.B., DAMM, D.W. *et al.* Data storage and retrieval system for a nuclear medicine department. *J. Nuc. Med.* **16**, 275 (1975)

SCHMIDLIN, P. Development and comparison of computer methods for organ motion correction in scintigraphy. *P.M.B.* **20**, 465 (1975)

SIMON, W. A method of exponential separation applicable to small computers. *P.M.B.* **15**, 355 (1970)

SKRETTING, A. An iterative computer algorithm for optimisation of radionuclide subtraction studies. *P.M.B.* **20**, 578 (1975)

TSIALAS, S.P., LEWIS, B. A computer program for evaluating the performance of focussed collimators. *I.J.A.R.I.* **21**, 375 (1970)

VERNON, P., GLASS, H.I. An off-line digital system for use with a gamma camera. *P.M.B.* **16**, 405 (1971)

*WEBER, D.A. Computers in nuclear medicine: introductory concepts. *Sem. Nuc. Med.* **8**, 107 (1978)

Counting Corrections

ADAMS, R., ZIMMERMAN, D. Methods for calculating the dead-time of Anger camera systems. *J. Nuc. Med.* **14**, 496 (1973)

ARNOLD, J.E., JOHNSTON, A.S. *et al.* The influence of true counting rate and the photopeak fraction of detected events on Anger camera deadtime. *J. Nuc. Med.* **15**, 412 (1974)

BEN-PORATH, M. Camera deadtime: rate-limiting factor in quantitative dynamic studies. *I.J. Nuc. Med. Biol.* **2**, 107 (1975)

BRUNINX, E., PRINS, H.J. Accurate measurement of single-channel gamma-ray counters by the proportional source method. *I.J.A.R.I.* **25**, 483 (1974)

COOPER, P.H., LERNER, S.R. *et al.* Unexpected deadtime losses in a modified rectilinear scanning system. *J. Nuc. Med.* **14**, 828 (1973)

DAVIES, P.T. DETERDING, J.H. Optimisation of counting of samples with double radioactive labelling. *I.J.A.R.I.* **23**, 293 (1972)

DIETHORN, W.S. Counter deadtime: a question of poor text-book advice. *I.J.A.R.I.* **25**, 55 (1974)

FREEDMAN, G.S., KINESELLA, T. *et al.* A correction method for high-countrate quantitative radionuclide angiography. *Radiology*, **104**, 713 (1972)

GLASS, H.I., VERNON, P. A method of correcting for count-rate losses in dynamic gamma-camera studies. *P.M.B.* **17**, 843 (1972)

HUTTIG, M. Anger camera deadtime. *J. Nuc. Med.* **15**, 468 (1974)

MCTAGGERT, W.G., CARDUS, D. An analysis of errors in a technique for combined use of multiple isotopes. *I.J.A.R.I.* **20**, 429 (1969)

MORIN, P.P., MORIN, J.F., *et al.* Caracterisation pratique de la linearité de comptage d'une gamma caméra. *I.J. Nuc. Med. Biol.* **3**, 101 (1976)

MUEHLLEHNER, G., JASZCZAK, R.J. *et al.* The reduction of coincidence loss in radionuclide imaging cameras through the use of composite filters. *P.M.B.* **19**, 504 (1974)

PORGES, K.G.A., DE VOLPI, A. Deadtime corrections in counting rapidly decaying sources. *I.J.A.R.I.* **22**, 581 (1971)

SORENSON, J.A. Methods of correcting Anger camera deadtime losses. *J. Nuc. Med.* **17**, 137 (1976)

UJHELYI, C. A new simple method to determine accurately the resolving time correction by using two gamma sources. *I.J.A.R.I.* **27**, 143 (1976)

VAN DAMME, K.J. A disregarded geometrical effect in an elegant method to measure the activity of an isotope emitting two gamma rays in cascade. *I.J.A.R.I.* **28**, 675 (1977)

Counting Methods: Detector Sensitivity

BLOCH, P., SANDERS, T. Reduction of the effects of scattered radiation on a sodium iodide imaging system. *J. Nuc. Med.* **14**, 67 (1973)

BOJSEN, J., VADSTRUP, S. A portable external 2-channel radiotelemetric

GM detector unit for measurement of radionuclide tracers *in vivo*. *I.J.A.R.I.* **25**, 161 (1974)

BRINKMAN, G.A., ATEN, A.H.W. *et al.* Absolute standardisation with a NaI(Tl) crystal. *I.J.A.R.I.* **14**, 153 (1963)

BRINKMAN, G.A., LINDNER, L. Two-crystal counting of ^{123}I. *J. Nuc. Med.* **18**, 1046 (1977) (See also reply by MPANIAS, P., GOLLNICK, D.A. *et al.* *J. Nuc. Med.* **18**, 1046 (1977)

CAMPION, P.J. The standardisation of isotopes by beta–gamma coincidence using high-efficiency detectors. *I.J.A.R.I.* **4**, 232 (1959)

CAZZOLA, M., BAROSI, G. *et al.* A simple method for simultaneous liquid scintillation counting of ^{59}Fe and ^{51}Cr in blood. *I.J.A.R.I.* **27**, 660 (1976)

CHARLTON, D.E. Use of the term 'Half-value Layer'. *P.M.B.* **14**, 326 (1969)

CHO, Z.H., CHAN, J.K. *et al.* Positron ranges obtained from biomedically important positron-emitting radionuclides. *J. Nuc. Med.* **16**, 1174 (1975)

COULTER, B.S. Calculation of precision in isotope dilution experiments. *I.J.A.R.I.* **20**, 271 (1969)

*CRADDUCK, T.D. Fundamentals of scintillation counting. *Sem. Nuc. Med.* **3**, 205 (1973)

DUBUQUE, G.L., GRANT, R.J. *et al.* Measuring mixed energy gamma radionuclides using a radionuclide calibrator. *I.J. Nuc. Med. Biol.* **3**, 145 (1976)

ELSBORG, L. Gas proportional counting compared with scintillation counting for assaying tritium in biological materials. *J. Nuc. Med.* **15**, 115 (1974)

TEN HAAF, F.E.L., VERHEIJKE, M.L. An improved gamma well counter for radioactive tracer applications. *I.J.A.R.I.* **27**, 79 (1976)

HARE, D.L., HENDEE, W.R. *et al.* Accuracy of well ionisation chamber isotope calibrators. *J. Nuc. Med.* **15**, 1138 (1974)

HOLMBERG, P., RIEPPO, R. The effect of well dimensions on the efficiency values of well-type sodium iodide detectors. *I.J.A.R.I.* **28**, 469 (1977)

HORROCKS, D.L. Energy calibration of a liquid scintillation counter. *I.J.A.R.I.* **24**, 49 (1973)

JOHNSTON, A.R., EARNSHAW, J.C. *et al.* Further investigation into a method of background subtraction in the simultaneous use of ^{59}Fe and ^{51}Cr in a scattering medium. *P.M.B.* **19**, 105 (1974)

LOWENTHAL, G.C., WYLLIE, H.A. Thin 4π sources on electrosprayed ion exchange resin for radioactivity standardisations. *I.J.A.R.I.* **24**, 415 (1973)

LOWENTHAL, G.C., PAGE, V. A re-entrant well gas-flow proportional counter for radioactivity measurements of solutions containing ^{35}S and other radionuclides. *I.J.A.R.I.* **20**, 130 (1969)

*LOWENTHAL, G.C. Secondary standard instruments for the activity measurements of pure beta emitters. *I.J.A.R.I.* **20**, 559 (1969)

MEYER, E., DESAULNIERS, G. *et al.* The Cd–Te subminiature semiconductor detector. *I.J.A.R.I.* **26**, 697 (1975)

MORRIS, A.C., BARCLAY, T.R. *et al.* A miniaturised probe for detecting radioactivity at thyroid surgery. *P.M.B.* **16**, 397 (1971)

MORSEY, S.M., YOUSSEF, S.K. Calibration of gamma sources without standards. *I.J.A.R.I.* **27**, 343 (1976)

*PARKER, R.P. Semiconductor nuclear radiation detectors. *P.M.B.* **15**, 605 (1970)

*PEREZ-MENDEZ, V. Multiwire proportional chambers in nuclear medicine. *I.J. Nuc. Med. Biol.* **3**, 29 (1976)

PLCH, J., ZDEERADICKA, J. *et al.* A windowless 4π scintillation counter with NaI(Tl) crystals. *I.J.A.R.I.* **24**, 407 (1973)

RIEPPO, R., BLOMSTER, K. Calculated absolute photopeak efficiency values for 3 × 3 inch NaI detector with cylindrical source geometry. *I.J.A.R.I.* **27**, 365 (1976)

RIEPPO, R. Efficiency values of NaI well-type detectors in the gamma-ray region of 10–150 keV for point and needle-shaped sources. *I.J.A.R.I.* **27**, 453 (1976)

RIEPPO, R. Calculated 10–150 keV gamma-ray efficiency values of NaI detectors for a 4π detecting geometry. *I.J.A.R.I.* **27**, 457 (1976)

RIEPPO, R. Monte Carlo calculation of the self-absorption of gamma rays in a volume-shaped source connecting a face-type and a well-type NaI detector. *I.J.A.R.I.* **27**, 605 (1976)

RIEPPO, R. A relationship of the absolute photopeak efficiency values of NaI detectors between point- and extended-source geometries. *I.J.A.R.I.* **27**, 609 (1976)

RIEPPO, R. A method for the calculation of the absolute photopeak efficiency values of NaI detectors with water as the medium in the source. *I.J.A.R.I.* **29**, 65 (1978)

ROEDEL, W., LEIDNER, L. *et al.* Internal-gas radioactivity counting with pure hydrogen sulphide as counting gas. *I.J.A.R.I.* **28**, 713 (1977)

DE ROOST, E., FUNCK, E. *et al.* Improvements in 4π beta counting. *I.J.A.R.I.* **20**, 387 (1969)

SMITH, T. How much light from rectangular scintillation counters? *P.M.B.* **20**, 282 (1975)

SNIPES, M.B., LENGEMANN, F.W. A practical method for the resolution of two beta-emitting nuclides by liquid scintillation counting. *I.R.A.R.I.* **22**, 513 (1971)

TANAKA, E., IINUMA, T.A. *et al.* Optimum isotope ratios in double-tracer studies. *P.M.B.* **12**, 345 (1967)

*TER-POGOSSIAN, M.M., PHELPS, M.E. Semiconductor detector systems. *Sem. Nuc. Med.* **3**, 343 (1973)

VAN DAMME, K.J. Method to calculate activity of a source from counting rates in single and multiple photopeaks. *J. Nuc. Med.* **18**, 1043 (1977) (See also HARPER, P.V., LATHROP, K.A., *ibid.* p. 1044)

WENZEL, M., WAHID, A.A. Use of Auger and conversion electrons for the detection of gamma-emitting nuclides in chromatographic quality control of radionuclides. *I.J.A.R.I.* **26**, 119 (1975)

WRAIGHT, E.P. Adaptation of a well-counter for measurement of cardiac output. *J. Nuc. Med.* **13**, 707 (1972)

YOSHIDA, M., MIYAHARA, H. *et al.* A source preparation for 4π counting with an aluminium compound. *I.J.A.R.I.* **28**, 633 (1977)

Dynamic Analysis

ANDERSON, J., OSBORN, S.B. *et al.* Clearance curves for radioactive tracers—sums of exponentials or powers of time? *P.M.B.* **14**, 498 (1969)

BAJZER, A., NOSIL, J. A simple mathematical lung model for quantitative regional ventilation measurement using ^{81m}Kr. *P.M.B.* **22**, 975 (1977)

BECK, J.S., RESCIGNO, A. Calcium kinetics: the philosophy and practice of science. *P.M.B.* **15**, 566 (1970)

BROWN, M. Measurement of mass flow rates of water and steam in evaporator tubes on operating boilers using radiotracers. *I.J.A.R.I.* **25**, 289 (1974)

*COHEN, M.L. Radionuclide clearance techniques. *Sem. Nuc. Med.* **4**, 23 (1974)

*CRADDUCK, T.D., MACKINTYRE, W.J. Camera-computer systems for rapid dynamic imaging systems. *Sem. Nuc. Med.* **7**, 323 (1977)

DE GRAZIA, J.A., SCHEIBE, P.O. Clinical application of a kinetic model of hippurate distribution and renal clearance. *J. Nuc. Med.* **15**, 102 (1974)

DIFFEY, B.L., HALL, F.M. *et al.* The ^{99m}Tc-DTPPA dynamic renal scan with deconvolution analysis. *J. Nuc. Med.* **17**, 352 (1976)

EBERSTADT, P.L., COO, J.J. A two-compartment model and glomerular filtration rate. *I.J. Nuc. Med. Biol.* **2**, 99 (1975)

FARMELANT, M.H., BURROWS, B.A. The renogram: physiological basis and current clinical use. *Sem. Nuc. Med.* **4**, 61 (1974)

FLEMING, J.S., GODDARD, B.A. A technique for the deconvolution of the renogram. *P.M.B.* **19**, 546 (1974)

FLEMING, J.S. Measurement of hippuran plasma clearance using a gamma camera. *P.M.B.* **22**, 426 (1977)

GARDNER, R.D., DUNN, T.S. The development of radiotracer methods for laminar-flow measurements in small channels. *I.J.A.R.I.* **28**, 347 (1977) (See also pp. 355 and 369)

HALKO, A., BURKE, G. *et al.* Computer-aided statistical analysis of the scintillation camera ^{131}I-hippuran renogram. *J. Nuc. Med.* **14**, 252 (1973)

HILL, D.W., THOMPSON, F.D. The use of a compartmental hypothesis for the estimation of cardiac output from dye-dilution curves and the analysis of radiocardiograms. *Med. Biol. Eng.* **11**, 43 (1973)

HOLROYD, A.M., GHISHOLM, G.D. *et al.* Quantitative analysis of renograms using a gamma camera. *P.M.B.* **15**, 483 (1970)

HUDSON, F.R., DAVIS, J.B. *et al.* A phantom for simulating fast dynamic studies on a gamma camera. *P.M.B.* **22**, 87 (1977)

KENNY, R.W., ACKERY, D.M. *et al.* Deconvolution analysis of the scintillation camera renogram. *Brit. J. Radiol.* **48**, 481 (1975)

KETY, S.S. Theory and applications of the exchange of inert gases at the lungs and tissues. *Pharmacol. Rev.* **3**, 1 (1951)

KING, M.A., KILPPER, R.W. *et al.* A model for local accumulation of bone imaging radiopharmaceuticals. *J. Nuc. Med.* **18**, 1106 (1977)

KIRCH, D.L., BROWN, D.W. *et al.* Frequency domain techniques applied to dynamic scintigraphic studies. (abstract only). *J. Nuc. Med.* **13**, 440 (1970)

KRAL, M., KUBA, J. Determination of regional cerebral blood flow using a ^{133}Xe inhalation technique. *P.M.B.* **18**, 100 (1973)

KRISHNAMURTHY, G.T., THOMAS, P.B. *et al.* Comparison of ^{99m}Tc-polyphosphate and ^{18}F kinetics. *J. Nuc. Med.* **15**, 832 (1974)

MELDOLESI, U., RONCARI, G. *et al.* The renal plasma flow and glomerular filtration rate as estimated through double radiocompound renography. *Nuc. Medizin* **13**, 279 (1975)

METZ, C.E., KIRCH, D.L. *et al.* A mathematical model for the determination of cardiac regurgitant and ejection fractions from radio-isotope angiocardiograms. *P.M.B.* **20**, 531 (1975)
NIMMON, C.C., McALISTER, J. *et al.* The influence of geometrical factors in ^{131}I-hippuran renography. *P.M.B.* **20**, 67 (1975)
NOWOTNY, R., KLETTER, K. Comments on the validity of compartmental analysis for hippuran distribution. *P.M.B.* **21**, 653 (1976)
NICKLES, R.J., AU, Y.F. The oxygen clock—a dual tracer physiological timer. *P.M.B.* **20**, 54 (1975)
ORR, J.S., WHELDON, T.E. *et al.* Integral transforms for deconvolution of radio-isotope kinetic studies. *P.M.B.* **16**, 529(1971)
PARKEY, R.W., LEWIS, S.E. *et al.* Compartmental analysis of the ^{133}Xe regional myocardial blood-flow curve. *Radiology* **104**, 425 (1972)
PHELPS, M.E., EICHLING, J.O. A quick method for calculation of the vascular mean transit time. *J. Nuc. Med.* **15**, 814 (1974)
*RAYNAUD, C. A technique for the quantitative measurement of the function of each kidney. *Sem. Nuc. Med.* **4**, 51 (1974)
SCHEIBE, P.O., YOSHIKAWA, B. Inputs for dose calculations from compartmental models. *J. Nuc. Med.* **15**, 1025 (1974)
VAN STEKELENBERG, L.H.M., KOOMAN, A. *et al.* A three-compartment model for transport and distribution of hippuran. *P.M.B.* **21**, 74 (1976)
VAN STEKELENBERG, L.H.M., AL, N. *et al.* Comments on the validity of compartmental analysis for hippuran distribution. *P.M.B.* **21**, 656 (1976)
VAN STEKELENBERG, L.H.M. Hippuran transit times in the kidney—a new approach. *P.M.B.* **23**, 291 (1978)
SUBRAMANYAN, R., ALPERT, N. *et al.* A model for regional cerebral oxygen distribution during continuous inhalation of $^{15}O_2$, $C^{15}O$ and $C^{15}O_2$. *J. Nuc. Med.* **19**, 48 (1978)
THYN, J., HANSSON, L. The residence-time distribution in systems with recirculation on the outlet flow. *I.J.A.R.I.* **26**, 748 (1975)
TOTHILL, P., McLOUGHLIN, G.P. *et al.* Techniques and errors in scintigraphic measurements of gastric emptying. *J. Nuc. Med.* **19**, 256 (1978)
VITEK, F., BIANCHI, R. *et al.* The study of distribution and catabolism of labelled serum albumin by means of an analog computer technique. *J. Nuc. Biol. Med.* **10**, 121 (1966)
WELCH, M.J., ADATEPE, M. *et al.* An analysis of ^{99m}Tc kinetics—effect of perchlorate and iodide pretreatment. *I.J.A.R.I.* **20**, 437 (1969)
WRAIGHT, E.P. The place of deconvolution analysis in plasma turnover studies. *P.M.B.* **14**, 463 (1969)
YASHUSHI, I., KAWAMURA, *et al.* Functional imaging of intrarenal blood flow using scintillation camera and computer. *J. Nuc. Med.* **16**, 899 (1975)
YANO, Y., McRAE, J. *et al.* Lung function studies using ^{81m}Kr and the scintillation camera. *J. Nuc. Med.* **11**, 674 (1970)
ZIERLER, K.I. Equations for measuring blood flow by external monitoring of radio-isotopes. *Circ. Res.* **16**, 309 (1965)

Imaging Techniques

ANGER, H.O. A scintillation camera with multichannel collimators. *J. Nuc. Med.* **5**, 515 (1964)

Baker, R.G., Scrimger, J.W. An investigation of the parameters in scintillation camera design. *P.M.B.* **12**, 51 (1967)

Barker, M.C.J., Woods, R.H.K. *et al.* Provision of anatomical markers for gamma camera pictures. *I.J.A.R.I.* **20**, 140 (1969)

Barrett, H.H. Fresnel zone plate imaging in nuclear medicine. *J. Nuc. Med.* **13**, 382 (1972)

Bliss, R.D., Weber, P.M. *et al.* An improved data recorder for scintillation cameras. *J. Nuc. Med.* **15**, 1130 (1974)

Bowley, A.R., Taylor, C.G. *et al.* A radio-isotope scanner for rectilinear, arc, transverse section and longitudinal section scanning. *Brit. J. Radiol.* **46**, 262 (1973)

*Brookeman, V.A. Computers and quality control in nuclear medicine. *Sem. Nuc Med.* **8**, 113 (1978)

*Brooks, R.A., di Chiro, G. Principles of computer assisted-tomography in radiographic and radioisotope imaging. *P.M.B.* **21**, 689 (1976)

Brown, D.W., Kirch, D.L. *et al.* Computer processing of scans using Fourier and other transformations. *J. Nuc. Med.* **12**, 287 (1971)

*Brown, D.W., Kirch, D.L. *et al.* Quantification of the radionuclide image. *Sem. Nuc. Med.* **4**, 311 (1973)

Brunning, P.F., Cuperus, P. *et al.* A simple device for the elimination of the blurring effect of respiration on organ scintigraphy. *I.J.A.R.I.* **24**, 53 (1973)

Budinger, T.F. Limitations in digital manipulation and presentation of scintillation camers imagers. *J. Nuc. Med.* **13**, 416 (1972)

Budinger, T.F., Gullberg, G.T. Three-dimensional reconstruction in nuclear medicine by iterative least-squares and Fourier techniques. *I.E.E.E. Med. Sci.* **NS21:2**, 20 (1974)

Budinger, T.F. Instrumentation trends in nuclear medicine. *Sem. Nuc. Med.* **7**, 285 (1977)

Budinger, T.G., Derenzo, S.E. *et al.* Emission computer assisted tomography with single photon and positron annihilation photon emission. *J. Computer Assisted Tomography* **1**, 13 (1977)

Budinger, T.F., Derenzo, S.E. Quantitative potentials of dynamic emission computer tomography. *J. Nuc. Med.* **19**, 309 (1978)

Burnham, C.A., Aronow, S. A hybrid positron scanner. *P.M.B.* **15**, 493 (1970)

Buzzi, K., Paul, J.M. *et al.* Improvement in figures of merit by selective summing of multiple gamma emission lines. (abstract only). *J. Nuc. Med.* **13**, 419 (1972)

Chesler, D.A., Hales, C. *et al.* Three-dimensional reconstruction of lung perfusion image with positron detection. *J. Nuc. Med.* **16**, 80 (1975)

Deland, F.H., Mauderli, W. Gating mechanism for motion-free liver and lung scintigraphy. *J. Nuc. Med.* **13**, 939 (1972)

Derenzo, S.E., Zakland, H. *et al.* Analytical study of a high-resolution positron ring detector system for transaxial reconstruction tomography. *J. Nuc. Med.* **16**, 1166 (1975)

Dowsett, D.J., Roberts, K. Long-distance transmission of analogue gamma camera signals. *J. Nuc. Med.* **15**, 896 (1974)

Dyer, G.R., McClain, W.J. *et al.* A system for recording and displaying radio-isotope scan data. *I.J.A.R.I.* **21**, 93 (1972)

ERIKSON, L., CHO, Z.H. A simple absorption correction in positron (annihilation gamma coincidence detection) transverse axial tomography. *P.M.B.* **21**, 429 (1976)

EVANS, A.L., DAVISON, M. *et al.* A gamma ray densitometer for investigation of pulmonary function. *P.M.B.* **20**, 261 (1975)

FARMELANT, M.H., DEMEESTER, G. *et al.* Initial clinical experiences with a Fresnel Zone-plate imager. *J. Nuc. Med.* **16**, 183 (1975)

FOWLER, J.F., POPOVIC, S. Single and double-headed detector systems for scanning. *I.J.A.R.I.* **20**, 619 (1969)

*FREEDMAN, G.S. Radionuclide tomography. *Sem. Nuc. Med.* **3**, 267 (1973)

HASMAN, A., GROOTEHEDDE, R.T. Gamma camera uniformity as a function of energy and countrate. *Brit. J. Radiol.* **49**, 718 (1976)

HERATH, K.B., SHARP, P.F. Effects of 'matched filter' smoothing as measured by receiver operating characteristic curve. *P.M.B.* **21**, 442 (1976)

HOFFER, P.B., HARPER, P.V. *et al.* Improved xenon images with ^{127}Xe. *J. Nuc. Med.* **14**, 172 (1973)

HOFFMAN, E.S., PHELPS, M.E. *et al.* Design and performance characteristics of a whole-body positron transaxial tomograph. *J. Nuc. Med.* **17**, 493 (1976)

HUDSON, F.R., WILLETTS, R.J. A simple dual isotope conversion for a gamma camera. *P.M.B.* **21**, 143 (1976)

HUESMAN, R.H. The effects of a finite number of projection angles and finite lateral samplings of projections on the propagation of errors in transverse section reconstruction. *P.M.B.* **22**, 511 (1977)

INIA, P. Reduction of the effects of scattered radiation. *J. Nuc. Med.* **15**, 316 (1974)

JANSON, L.G., PARKER, R.P. Pitfalls in gamma camera field uniformity correction. *Brit. J. Radiol.* **48**, 408 (1975)

JASZCZAK, R.J. Increased resolving power from scintillation cameras using electronic signal processing and a movable filter plate. *J. Nuc. Med.* **14**, 14 (1973)

KAN, K. Inexpensive EKG gate for computer processed cardiac motion study. *J. Nuc. Med.* **19**, 320 (1978)

KAY, D.B., Keyes, J.W. *et al.* Radionuclide tomographic image reconstruction using Fourier transform techniques. *J. Nuc. Med.* **15**, 981 (1974)

KEMPLAY, J.R., ASPINALL, J.A. A digital subtraction technique for dual isotope scintigraphy. *P.M.B.* **16**, 243 (1971)

KEYES, W.I. An analysis of multilevel scan display systems. *P.M.B.* **18**, 282 (1973)

KEYES, W.I., MALLARD, J.R. Observations on ratemeter-controlled colour dot scan displays. *Brit. J. Radiol.* **46**, 369 (1973)

KEYES, W.I. The fan-beam gamma camera. *P.M.B.* **20**, 489 (1975)

KORAL, K.F., ROGERS, W.L. *et al.* Digital tomographic imaging with time-modulated pseudorandom coded aperture and Anger camera. *J. Nuc. Med.* **16**, 402 (1975)

KULBERG, G.H., VAN DIJK, N. Improved resolution of the Anger scintillation camera through the use of threshold preamplifiers. *J. Nuc. Med.* **13**, 169 (1972)

LANCASTER, J.L. STOKELY, E.M. Rib erasure using nine-point smoothing. *J. Nuc. Med.* **17**, 68 (1976)

LAVOIE, L. Comparison of radiation detector materials for imaging application in nuclear medicine. *P.M.B.* **18**, 120 (1973)

LEACH, K.G., MARSHALL, R.J. A method of correcting gamma camera non-uniformity. *J. Nuc. Med.* **13**, 449 (1972)

LEEPER, R.D. POWELL, M. *et al.* Use of scaler attachment to rectilinear scanner to measure ^{99m}Tc-pertechnetate uptake by the thyroid. *J. Nuc. Med.* **15**, 1117 (1974)

LEGRAS, J., CHAU, N. A new processing algorithm for the resolution enhancement of digital scans. *P.M.B.* **22**, 1180 (1977)

LEVY, G. Comment on Fresnel zone-plate imaging in nuclear medicine. *J. Nuc. Med.* **15**, 214 (1974)

MALAMUD, H. Filter to correct for inverse-square-law non-uniformity in the pinhole collimator. *J. Nuc. Med.* **13**, 861 (1972)

METZ, C.E., BECK, R.N Quantitative effects of stationary linear image processing on noise and resolution of structure in radionuclide images. *J. Nuc. Med.* **15**, 164 (1974)

MEYER, E., CHALLIER, J.C. *et al.* Semimicroscanning device using a miniature silicon–lithium semiconductor probe. *I.J.A.R.I.* **23**, 251 (1972)

MOULD, R.F., WYLD, C. A comparison between line printer and conventional polaroid gamma camera displays. *P.M.B.* **18**, 88 (1973)

MUEHLLEHNER, G. A tomographic scintillation camera. *P.M.B.* **16**, 87 (1971)

NOHARA, N., TOMITANI, T. *et al.* Analog image processing in two dimensions by omnidirectional scanning. *J. Nuc. Med.* **15**, 884 (1974)

PHELPS, M.E., HOFFMAN, E.J. *et al.* Application of annihilation coincidence detection to transaxial reconstruction tomography. *J. Nuc. Med.* **16**, 210 (1975)

OVERTON, T.R., HESLIP, P.G. *et al.* Dual radio-isotope techniques and digital image subtraction methods in pancreas visualisation. *J. Nuc. Med.* **12**, 493 (1971)

*PHELPS, M.E. Emission computed tomography. *Sem. Nuc. Med.* **7**, 337 (1977)

*PIZER, S.M., TODD-POKROPEK, A.E. Improvement of scintigrams by computer processing. *Sem. Nuc. Med.* **8**,125 (1978)

PROFIO, A.E., CHO, Z.H. Semiconductor camera for detection of small tumours. *J. Nuc. Med.* **16**, 53 (1975)

REESE, I.C., MISHKIN, F.S. Technique for producing cardiac radionuclide motion images. *J. Nuc. Med.* **16**, 368 (1975)

ROGERS, W.L., HAN, K.S. *et al.* Application of a Fresnel zone plate to gamma ray imaging. *J. Nuc. Med.* **13**, 612 (1972)

SHARP, P.F., HERATH, K.B. Comments on the generalisation of receiver operating characteristic analysis to detection and localisation tasks. *P.M.B.* **22**, 380 (1977)

SCHMIDLIN, P. Development and comparison of computer methods for organ motion correction in scintigraphy. *P.M.B.* **20**, 465 (1975)

SHIMMINS, J.G., SUMNER, D.J. Experimental study of profile scanning optimisation. *J. Nuc. Med.* **14**, 895 (1973)

SPECTOR, J.S., BROOKEMAN, V.A. *et al.* Analysis of spatial distortion produced by the gamma camera. *J. Nuc. Med.* **13**, 307 (1972)

STARR, S.J., METZ, C.E. *et al.* Comments on the generalisation of receiver operating characteristic analysis to detection and localisation tasks. *P.M.B.* **22**, 376 (1977)

STRAUSS, M.G., SHERMAN, I.S. *et al.* Performance of a coaxial germanium gamma ray camera. *J. Nuc. Med.* **15**, 1196 (1974)

TER-POGOSSIAN, M.M., PHELPS, M.E. *et al.* A positron emission transaxial tomograph for nuclear imaging. *Radiology* **114**, 89 (1975)

*TER-POGOSSIAN, M.M. Basic principles of computed axial tomography. *Sem. Nuc. Med.* **7**, 109 (1977)

TIPTON, M.D., DOWDEY, J.E. *et al.* Coded aperture imaging using on-axis Fresnel zone plates and extended gamma ray sources. *Radiology* **112**, 155 (1974)

VILENSKY, A., TATCHER, M. *et al.* A cassette magnetic tape recording system for radio-isotope scanners. *I.J.A.R.I.* **23**, 129 (1972)

WHITEHEAD, F.R. Minimum detectable gray-scale differences in nuclear medicine images. *J. Nuc. Med.* **19**, 87 (1978)

WHITTINGHAM, T.A., MALLARD, J.R. Theoretical considerations for high-resolution radio-isotope scanning. *P.M.B.* **13**, 657 (1968)

WILLIAMS, D.L., RITCHIE, J.L. Implementation of a digital image superposition algorithm for radionuclide images. *J. Nuc. Med.* **19**, 316 (1978)

WILLIAMS, J.P. A simple marking device for use with gamma cameras. *J. Nuc. Med.* **12**, 300 (1971)

Parent–Daughter Relationships and Exponential Curve-Fitting

Berkson, J. Do radioactive decay events follow a random Poisson distribution? *I.J.A.R.I.* **26**, 543 (1975)

DUTTON, J.W.R. The determination of parent-radionuclide activity by daughter decay measured after separation of two nuclides with different radiochemical yields. *I.J.A.R.I.* **29**, 116 (1978)

ENGLAND, J.M., MILLER, R.G. Counting times in measurement of radioactivity. *I.J.A.R.I.* **20**, 1 (1969)

FRIED, J., ZIETZ, S. Curve fitting by spline and akima methods. *P.M.B.* **18**, 550 (1973)

GLASS, H.I., DE GARRETA, A.C. Quantitative analysis of exponential curve fitting for biological applications. *P.M.B.* **12**, 379 (1967)

GLASS, H.I., DE GARRETA, A.C. The quantitative limitations of exponential curve fitting. *P.M.B.* **16**, 119 (1971)

HANSSON, L., KURTEN, R. *et al.* The application of Laguerre functions for the approximation and smoothing of countrates varying with time. *I.J.A.R.I.* **26**, 347 (1975)

LEWIS, V.E. The interval distribution in parent–daughter decay. *I.J.A.R.I.* **24**, 63 (1973)

MCBETH, G.W., WINYARD, R.A. Isotope identification and radioassay by time interval analysis. *I.J.A.R.I.* **23**, 527 (1972)

PUTMAN, J.L. On counting precision and the S^2/B criterion. *I.J.A.R.I.* **20**, 205 (1969)

SIMON, W. A method of exponential separation applicable to small computers. *P.M.B.* **15**, 355 (1970)

SPENCER, R.P., HOSAIN, F. Theoretical basis for regional blood flow measurement by the use of the ratio $^{81}Rb/^{81m}Kr$. *I.J. Nuc. Med. Biol.* **3**, 55 (1976)

STEVENSON, J.K. Graphical method for the analysis of multicomponent exponential decay data. *I.J.A.R.I.* **28**, 909 (1977)

YAMAMOTO, S., BROWN, M. Radioactive growth–decay data—analysis by application of the least-squares method. *I.J.A.R.I.* **20**, 209 (1969)

Patient Organ-Dose Estimation and Measurement

BUSH, D. Measurement and control of dose equivalent to patients undergoing whole-body neutron irradiation for *in vivo* activation analysis. *P.M.B.* **17**, 32 (1972)

CLOUTIER, R.J., WATSON, E.E. *et al.* Calculating the radiation dose to an organ. *J. Nuc. Med.* **14**, 53 (1973)

COLOMBETTI, L.G. Absorbed radiation dose from radionuclide impurities in ^{99m}Tc obtained by different procedures. *I.J.A.R.I.* **25**, 455 (1974)

FINCK, R., MATTISON, S. Long-lived impurities in eluate from Mo–Tc generators and the associated absorbed dose in the patient. *I.J. Nuc. Med. Biol.* **3**, 89 (1976)

GILLESPIE, F.C., ORR, J.S. The prediction of dose due to an internal radioisotope by application of the occupancy principle. *P.M.B.* **14**, 639 (1969)

GODDARD, B.A., ACKERY, D.M. Xenon-133, Xenon-127, and Xenon-125 for lung function investigations. *J. Nuc. Med.* **16**, 780 (1975)

GUPTA, P.C., GUPTA, M.M. Gamma dose-rate at an internal point on the axis of a solid radioactive sphere. *I.J.A.R.I.* **20**, 275 (1969)

LASHMANN, W., HINZPETER, A. Measurement of the average gamma-dose rate with uniformly distributed sources. *I.J.A.R.I.* **27**, 137 (1976)

LOEVINGER, R., BERMAN, M. A formalism for calculation of absorbed dose from radionuclides. *P.M.B.* **13**, 205 (1968)

MIRD COMMITTEE PUBLICATIONS Summary of current radiation dose estimates to humans from ^{123}I, ^{124}I, ^{125}I, ^{126}I, ^{130}I, ^{131}I, and ^{132}I as sodium iodide. *J. Nuc. Med.* **16**, 857 (1975)

McEWAN, A.C. Dosimetry of radionuclides in blood. *Brit. J. Radiol.* **47**, 652 (1974)

PLATO, P., JACOBSON, A.P. *et al.* *In vivo* thyroid monitoring for ^{131}I in the environment. *I.J.A.R.I.* **27**, 547 (1976)

*SAENGER, E.L., KEREIAKES, J.G. The safe tracer dose in medical investigation. *Recent Advances in Nuclear Medicine* **3**, 139 (1971)

SWANSON, A.L., MAYRON, L.W. *et al.* Radiation dosimetry for ^{81m}Kr. *I.J. Nuc. Med. Biol.* **3**, 140 (1976)

UNNIKRISHNAN, K. Dose to the urinary bladder from radionuclides in urine. *P.M.B.* **19**, 329 (1974)

Radionuclide Decay Data: Calibration and Standardisation

BRINKMAN, G.A., LINDER, L. *et al.* The sumpeak calibration of ^{123}I, *I.J.A.R.I.* **28**, 271 (1977)

GEHRKE, R.J., HELMER, R.G. Absolute gamma-ray intensities for ^{127}Xe decay. *I.J.A.R.I.* **28**, 743 (1977)

HUDSON, F.R., GLASS, H.I. *et al.* The assay of ^{123}I. *J. Nuc. Med.* **17**, 220 (1976)

HUTCHINSON, J.M.R., Mullan, P.A. Standardisation and ground-state branching of ^{75}Se. *I.J.A.R.I.* **27**, 47 (1976)

MANN, W.B., HUTCHINSON, J.M.R. Standardisation of ^{129}I by beta–photon coincidence counting. *I.J.A.R.I.* **27**, 188 (1976)

NEIRINCKX, R.D., LAMBRECHT, R.M. *et al.* Quality control of radiopharmaceuticals XIV: Na^{123}I. *I.J.A.R.I.* **25**, 387 (1975)

RAJPUT, M.J., DUBEY, D.P. Physical characteristics of ^{99m}Tc and calculation of $\bar{E}_\beta$, Γ and Δi. *I.J.A.R.I.* **26**, 35 (1975) (See also LEBLANC , A.D., *I.J.A.R.I.* **26**, 785 (1975) and NOZZ, M.E., NAGUIRE, G.O., *I.J.A.R.I.* **26**, 785 (1975))

READ, L.R., BURNS, J.E. *et al.* Exposure-rate calibration of small radioactive sources of ^{60}Co, ^{226}Ra and ^{137}Cs. *I.J.A.R.I.* **29**, 21 (1978)

SPERNOL, A., VATIN, R. Coincidence counting of ^{129}I. *I.J.A.R.I.* **28**, 615 (1977)

Radiation-Chemical Measurements

AHMED, S.A., LAW, J. *et al.* Use of ferrous sulphate for dosimetry by post. *P.M.B.* **15**, 311 (1970)

DAY, M.J., LAW, J. Ferrous sulphate *G* values. *P.M.B.* **14**, 665 (1969)

DAY, M.J., RASOUL, M. *et al.* Measurement of ferric ion concentration in a Fricke dosimeter. *P.M.B.* **16**, 531 (1971)

FRANKENBERG, D. A ferrous sulphate dosimeter independent of photon energy in the range from 25 keV up to 50 MeV. *P.M.B.* **14**, 607 (1969)

LAW, J. Ferrous sulphate *G* values for X-rays of 'effective' energy of 48 and 25 keV. *P.M.B.* **14**, 607 (1969)

LAW, J. Methods for using the conventional ferrous sulphate dosimeter to measure doses below 100 rad. *I.J.A.R.I.* **22**, 701 (1971)

MATTHEWS, R.W., Barker, N.T. *et al.* A comparison of some aqueous chemical dosimeters for absorbed doses of less than 1000 rad. *I.J.A.R.I.* **29**, 1 (1978)

MATTHEWS, R.W. Potentiometric estimation of mega-rad doses with the Ce^{4+}/Ce^{3+} system. *I.J.A.R.I.* **23**, 179 (1872)

Radionuclide Preparations

ALFASSI, Z.B., FELDMAN, L. The preparation of carrier-free radiobromine: separation of $KBrO_3$ and KBr using an organic solvent. *I.J.A.R.I.* **27**, 125 (1976)

ALLEN, J.F., PINAJIAN, J.S. A ^{87m}Sr generator for medical application. *I.J.A.R.I.* **16**, 319 (1965)

ANDERSON, D.W., RAESIDE, D.E. *et al.* Determination of impurity activities in fission-product generator eluate. *J. Nuc. Med.* **15**, 889 (1974)

ARINO, H., KRAMER, H.H. Separation and purification of radioiodine using platinum-coated copper granules. *I.J.A.R.I.* **27**, 637 (1976)

ARINO, H., KRAMER, H.H. ^{113}Sn/^{113}In generator systems. *I.J.A.R.I.* **25**, 493 (1974)

ARINO, H., SKRABA, W.S. A new $^{68}Ge/^{68}Ga$ radio-isotope generator system. *I.J.A.R.I.* **29**, 117 (1978)

ARINO, H., KRAMER, H.H. Separation and purification of radiomolybdenum from a fission product mixture using silver coated carbon granules. *I.J.A.R.I.* **29**, 97 (1978)

BAKER, R.I. A system for routine production of ^{99}Tc by solvent extraction from ^{99}Mo. *I.J.A.R.I.* **22**, 483 (1971)

BOLMSJO, M.S., PERSSON, B.R.R. Trapping and re-use system for radioactive xenon in nuclear medicine. *P.M.B.* **23**, 77 (1978)

BOULOGNE, A.R., EVANS, A.G. Californium-252 neutron sources for medical application. *I.J.A.R.I.* **20**, 453 (1969)

CARROLL, R.G., BERKE, R.A. *et al.* A multidose ^{133}Xe generator: the disposable glass ampoule chamber. *J. Nuc. Med.* **14**, 935 (1973)

CLARK, J.C., SILVESTER, D.J. A cyclotron method for production of ^{18}F. *I.J.A.R.I.* **17**, 151 (1966)

COHEN, B.L. Production of short-lived isotopes by charged particle acceleration. *P.M.B.* **18**, 286 (1973)

COLOMBETTI, L.G., BARRALL, R.C. *et al.* Experience with purity tests of $^{113m}Sn/^{113m}In$ generators. *I.J.A.R.I.* **20**, 717 (1969)

COLOMBETTI, L.G., MAYRON, L.W. *et al.* Continuous radionuclide generation. I. Production and evaluation of a ^{81m}Kr minigenerator. *J. Nuc. Med.* **15**, 868 (1974)

CUNINGHAME, J.G., MORRIS, B. *et al.* Large-scale production of ^{123}I from a flowing liquid target using the (p,5n) reaction. *I.J.A.R.I.* **27**, 597 (1976)

DAHL, J.R., TILBURY, R.S. The use of a compact multiparticle cyclotron for production of ^{52}Fe, ^{67}Ga, ^{111}In and ^{123}I. *I.J.A.R.I.* **23**, 431 (1972)

DANIELS, R.J., GRANT, P.M. *et al.* The production, recovery and purification of ^{172}Hf for utilisation in nuclear medicine as a generator of ^{172}Lu. *I.J. Nuc. Med. Biol.* **5**, 11 (1978)

DEGLUME, C., DEUTSCH, J.P. *et al.* Note on the production of ^{123}I for radiodiagnostic purposes. *I.J.A.R.I.* **24**, 291 (1973)

EILBERT, R.F., KOEHLER, A.M. *et al.* A range modulator to produce uniform ^{38}K yield. *I.J.A.R.I.* **27**, 707 (1976)

FINCK, R., MATTISON, S. Long-lived radionuclide impurities in eluate from Mo–Tc generators and the associated absorbed dose to the patient. *I.J. Nuc. Med. Biol.* **3**, 89 (1976)

FITSCHEN, J., BECKMANN, R. *et al.* Yield and production of ^{18}F by ^{3}He irradiation of water. *I.J.A.R.I.* **28**, 781 (1977)

GODART, J., BARAT, J.L. *et al.* In-beam collection of ^{123}Xe for carrier-free ^{123}I production. *I.J.A.R.I.* **28**, 967 (1977)

GINDLER, J.E., OSELKA, M.C. *et al.* A gas target assembly for production of high purity, high specific activity ^{81}Rb. *I.J.A.R.I.* **27**, 330 (1976)

GROTH, M.W., LEWIS, G.K. Thallium-201—scintillation camera imaging considerations. *J. Nuc. Med.* **17**, 142 (1976)

HAASBROEK, F.J., STEYN, J. *et al.* Thick-target yields of ^{22}Na in Mg, ^{55}Fe in Co, ^{56}Co, ^{57}Co, and ^{58}Co in Ni, and ^{175}Hf in Ta by proton bombardment in the energy range up to 100 MeV. *I.J.A.R.I.* **28**, 533 (1977)

HARA, T., FREED, B.R. Preparation of carrier-free ^{47}Sc by chemical separation from ^{47}Ca and its distribution in tumour-bearing mice. *I.J.A.R.I.* **24**, 373 (1873)

Hara, T., Tilbury, S. *et al.* Production of ^{73}Se in cyclotron and its uptake in tumours of mice. *I.J.A.R.I.* **24**, 377 (1973)

Harper, P.V., Siemens, W.D. *et al.* Production and use of ^{125}I. *J. Nuc. Med.* **4**, 277 (1963)

Hawkins, T., Harris, R. A compact ^{133}Xe gas dispenser. *I.J.A.R.I.* **28**, 812 (1977)

Hillman, M., Bishop, W.N. *et al.* Production of ^{87}Y and a ^{87m}Sr generator. *I.J.A.R.I.* **17**, 9 (1966)

Hines, H.H., Peek, N.F. *et al.* Production and characteristics of ^{125}Xe; a new noble gas for *in vivo* studies. *J. Nuc. Med.* **16**, 143 (1975)

Homma, Y., Murakami, Y. The production of ^{125}Xe for medical use by ^{3}He bombardment of natural Te. *I.J.A.R.I.* **28**, 738 (1977)

Hondayer, A.J., Shapiro, M.M. *et al.* Cyclotron production of ^{62}Zn for medical use. *I.J. Nuc. Med. Biol.* **3**, 97 (1976)

Hondayer, A.J., Meyer, E. *et al.* Cyclotron production of ^{77}Kr for regional cerebral blood flow measurement. *I.J. Nuc. Med. Biol.* **4**, 83 (1977)

Hsieh, T.H., Fan, K.W. *et al.* Preparation of carrier-free ^{18}F. *I.J.A.R.I.* **28**, 251 (1977)

Hupf, H.B., Beaver, J.E. Cyclotron production of carrier-free ^{67}Ga. *I.J.A.R.I.* **21**, 75 (1970)

*Kaplan, E., Mayron, L.W. Evaluation of perfusion with a ^{81}Rb/^{81m}Kr generator. *Sem. Nuc. Med.* **6**, 163 (1976)

Kondo, K., Lambrecht, R.M. *et al.* Cyclotron isotopes and radiopharmaceuticals XXII: Improved targetry and radiochemistry for production of ^{123}I and ^{124}I. *I.J.A.R.I.* **28**, 765 (1977)

Kondo, K., Lambrecht, R.M. *et al.* Iodine-123 production for radiopharmaceuticals—XX. Excitation functions of the ^{124}Te(p,2n)^{123}I and ^{124}Te(p,n)^{124}I reactions and the effect of target enrichment on radionuclidic purity. *I.J.A.R.I.* **28**, 395 (1977)

Kopecký, P., Mudrová, B. ^{68}Ge–^{68}Ga generator for production of ^{68}Ga in an ionic form. *I.J.A.R.I.* **25**, 263 (1974)

Kinsley, M.T., Lebowitz, E. *et al.* The production of ^{204}Bi for medical use. *I.J. Nuc. Med. Biol.* **1**, 85 (1973)

Kosarck, L.J., Martin, J.B. *et al.* The utilisation of ^{99}Mo from 'spent' ^{99}Mo–^{99m}Tc generators. *I.J.A.R.I.* **27**, 193 (1976)

Lambrecht, R.M., Ritter, E. *et al.* Cyclotron isotopes and radiopharmaceuticals XXI: Fabrication of ^{122}Te–Au targets. *I.J.A.R.I.* **28**, 567 (1977)

Lambrecht, R.M., Wolf, A.P. *et al.* High energy alpha reactions for production of ^{123}Cs → ^{123}Xe → ^{123}I and ^{125}Xe and ^{127}Cs. *I.J.A.R.I.* **27**, 675 (1976)

Lamson, M.L., Kirschner, A.S. *et al.* $^{99m}TcO_4^-$: Carrier-free? *J. Nuc. Med.* **16**, 639 (1975)

*Laughlin, J.S., Tilbury, R.S. *et al.* The cyclotron: source of short-lived positron emitters for medicine. *Recent Advances in Nuclear Medicine* **3**, 39 (1971)

Leblanc, A.D., Johnson, P.C. The handling of ^{133}Xe in clinical studies. *P.M.B.* **16**, 105 (1971)

LEBOWITZ, E., GREENE, M.W. *et al.* On the production of ^{123}I for medical use. *I.J.A.R.I.* **22**, 489 (1971)
LEBOWITZ, E., GREENE, M.W. *et al.* Thallium-201 for medical use. *J. Nuc. Med.* **16**, 151 (1975)
*LEBOWITZ, E., RICHARDS, P. Radionuclide generator systems. *Sem. Nuc. Med.* **4**, 257 (1974)
LEVIN, V.I., KOZLOVA, M.D. *et al.* The production of carrier-free ^{111}In. *I.J.A.R.I.* **25**, 286 (1974)
LEVIN, V.I., KOZYREVA-ALEXANDROVNA, L.S. *et al.* The production of carrier-free ^{99}Mo. *I.J.A.R.I.* **28**, 601 (1977)
LEVIN, V.I., KOZLOVA, M.D. *et al.* ^{113m}In generator with modified silica-gel column. *I.J.A.R.I.* **28**, 667 (1977)
LEWIS, R.E. Production of high specific activity ^{99}Mo. *I.J.A.R.I.* **22**, 603 (1971)
LINDNER, L., SUER, T.H.G.A. *et al.* A dynamic 'loop'-target for the in-cyclotron production of ^{18}F by the $^{16}O(\alpha,d)^{18}F$ reaction on water. *I.J.A.R.I.* **24**, 124 (1973)
LINDNER, L., BIJL, J.A. *et al.* Carrier-free iodide ^{123}I generated from ^{123}Xe: kit preparation and quality control. *I.J.A.R.I.* **27**, 653 (1976)
LIUZZI, A., KEANEY, J. *et al.* Use of activated charcoal for the collection and containment of ^{133}Xe exhaled during pulmonary studies. *J. Nuc. Med.* **13**, 673 (1971)
LOBERG, M.D., PHELPS, M.E. *et al.* Preparation of pure carrier-free ^{133}Xe for rare-gas washout studies. *J. Nuc. Med.* **14**, 733 (1973)
MAEDO, S., ENOMOTO, S. Development of a high intensity ^{192}Ir radiographic source. *I.J.A.R.I.* **27**, 447 (1976)
MARCEAU, N., KRUCK, T.P.A. *et al.* The production of ^{67}Cu from natural zinc using a linear accelerator. *I.J.A.R.I.* **21**, 667 (1970)
MATTSON, S., PERSSON, B.R.R. *et al.* Preparation and quality control of high purity ^{123}I. *I.J.A.R.I.* **27**, 319 (1976)
MAYRON, L.W., KAPLAN, E. *et al.* Preparation of ^{81}Rb in high specific activity and its use in ^{81}Rb–^{81m}Kr generator. *I.J.A.R.I.* **25**, 237 (1974)
MERRIL, J.C., LAMBRECHT, R.M. *et al.* Cyclotron production of lead-203 for radiopharmaceutical applications. *I.J.A.R.I.* **24**, 701 (1973)
MCFARLAND, R.C. An improved generator of ^{131m}Xe. *I.J.A.R.I.* **25**, 567 (1974)
MOGHISSI, A.A., HUPF., H.B. A krypton-83m generator. *I.J.A.R.I.* **22**, 219 (1971)
MYERS, W.G. Radio-potassium-38 for *in vivo* studies of dynamic processes. *J. Nuc. Med.* **14**, 359 (1973)
MANTEL, J., RUSKIN, R.L. *et al.* Production of ^{79}Kr for pulmonary ventilation and blood flow studies. *I.J. Nuc. Med. Biol.* **3**, 143 (1976)
NEIRINCKX, R.D. A high-yield production method for ^{67}Ga using an electroplated natural zinc or enriched ^{66}Zn target. *I.J.A.R.I.* **27**, 1 (1976)
NEIRENCKX, R.D. Production of ^{133m}Ba for medical purposes. *I.J.A.R.I.* **28**, 323 (1977)
NEIRINCKX, R.D. Cyclotron production of ^{57}Ni and ^{55}Co and synthesis of their bleomycin complexes. *I.J.A.R.I.* **28**, 561 (1977)

NEIRENCKX, R.D. Excitation functions for the $^{60}Ni(\alpha,2n)^{62}Zn$ reaction and production of ^{62}Zn-bleomycin. *I.J.A.R.I.* **28**, 808 (1977)

NEIRINCKX, R.D. Simultaneous production of ^{67}Cu, ^{64}Cu and ^{67}Ga and labelling of bleomycin with ^{67}Cu or ^{64}Cu. *I.J.A.R.I.* **28**, 802 (1977)

O'BRIEN, H.A. Preparation of carrier-free ^{53}Fe and ^{59}Fe by reactor irradiation. *I.J.A.R.I.* **20**, 711 (1969)

OSELKA, M., GINDLER, J.E. *et al.* Non-linear behaviour of gas targets for isotope production. *I.J.A.R.I.* **28**, 804 (1977)

PHELPS, M.E., WIELAND, B.W. Production of short-lived isotopes by charged particle acceleration. *P.M.B.* **18**, 284 (1973)

PROBST, H.J., QAIM, S.M. *et al.* Excitation functions of high energy alpha particle induced nuclear reactions on aluminium and magnesium: production of ^{28}Mg. *I.J.A.R.I.* **27**, 431 (1976)

*POGGENBURG, J.K. The nuclear reactor and its products. *Sem. Nuc. Med.* **4**, 229 (1974)

QAIM, S.M., STÖCKLIN, G. *et al.* Excitation functions for the formation of neutron-deficient isotopes of bromine and krypton via high energy deuteron induced reactions on bromine: production of ^{77}Br, ^{76}Br and ^{79}Kr. *I.J.A.R.I.* **28**, 947 (1977)

RAO, D.V., GOODWIN, P.N. *et al.* ^{165}Er: an 'ideal' radionuclide for imaging with pressurised multiwire proportional gamma cameras. *J. Nuc. Med.* **15**, 1008 (1974)

ROBINSON, J.R. ^{33}P: a superior tracer for phosphorus? *I.J.A.R.I.* **20**, 531 (1969)

SCHNEIDER, R.J., GOLDBERG, C.J. Production of ^{81}Rb by the reaction $^{85}Rb(p,5n)^{81}Sr$ and decay of ^{81}Sr. *I.J.A.R.I.* **27**, 189 (1976)

SCHOLZ, K.L., SODD, V.J. *et al.* Production of thulium-167 by irradiation of Lu, Hf, Ta, and W with 590 MeV protons. *I.J.A.R.I.* **27**, 263 (1976)

SERVIAN, J.L. Production of radionuclides and labelled compounds from accelerators. (Report of an IAEA consultants meeting). *I.J.A.R.I.* **26**, 763 (1975)

SILVESTER, D.J., THAKUR, M.L. Cyclotron production of carrier-free ^{67}Ga. *I.J.A.R.I.* **21**, 630 (1970)

SKRABA, W.J., ARINO, H. *et al.* A new strontium-90/yttrium-90 radioisotope generator. *I.J.A.R.I.* **29**, 91 (1978)

SNYDER, R.E., OVERTON, T.R. A system for handling and dispensing ^{133}Xe. *J. Nuc. Med.* **14**, 56 (1973)

SODD, V.J., BLUE, J.W. Cyclotron generation of high purity ^{123}I. *J. Nuc. Med.* **9**, 349 (1968)

SODD, V.J., BLUE, J.W. *et al.* A gas-flow powder target for the cyclotron production of pure ^{123}I. *I.J.A.R.I.* **24**, 171 (1973)

STEIGMAN, J., RICHARDS, P. Chemistry of technetium-99m. *Sem. Nuc. Med.* **4**, 269 (1974)

STEYN, J., MEYER, B.R. Production of ^{67}Ga by deuteron bombardment of natural zinc. *I.J.A.R.I.* **24**, 369 (1973)

STÜHMER, W., WEGMANN, H. *et al.* Production of ^{28}Mg and ^{38}S with $(n,^{3}He)$ reaction. *I.J.A.R.I.* **28**, 629 (1977)

SULLIVAN, J.C., FRIEDMAN, A.M. *et al.* Turnover localisation with radioactive lanthanide and actinide complexes. *I.J. Nuc. Med. Biol.* **2**, 44 (1974)

SUZUKI, K., IWATA, R. A multitarget assembly in an irradiation with high-energy particles. *I.J.A.R.I.* **28**, 663 (1977)
SYME, D.B., WOOD , E. *et al.* Yield curves for cyclotron production of ^{123}I, ^{125}I, and ^{121}I(p, *x*n)Xe → (β) → I reactions. *I.J.A.R.I.* **29**, 29 (1978)
TILBURY, R.S., KRAMER, H. *et al.* Preparing and dispensing ^{133}Xe in saline. *J. Nuc. Med.* **8**, 401 (1967)
TILBURY, R.S., DAHL, J.R. *et al.* Fluorine-18 production for medical use by ^{3}He bombardment of water. *I.J.A.R.I.* **21**, 277 (1970)
TILBURY, R.S., LAUGHLIN, J.S. Cyclotron production of radioactive isotopes for medical use. *Sem. Nuc. Med.* **4**, 245 (1974)
THOMAS, C.C., SONDEL, J.A. *et al.* Production of carrier-free ^{18}F. *I.J.A.R.I.* **16**, 71 (1965)
VAN DAMME, K.J., PAUWELS, J. Minor radioactive impurities in a ^{99m}Tc eluate. *I.J.A.R.I.* **27**, 128 (1976)
VAN DEN BOSCH, R., DE GOEIJ, J.J.M. *et al.* A new approach to target chemistry for the ^{123}I production via the ^{124}Te(p,2n) reaction. *I.J.A.R.I.* **28**, 255 (1977)
VAN DER MARK, T.W., Peset, R. *et al.* Storage and retrieval of waste ^{133}Xe. *I.J.A.R.I.* **28**, 602 (1977)
VEALL, N. The handling and dispensing of ^{133}Xe. *I.J.A.R.I.* **16**, 385 (1965)
VLATKOVIC, M., PAIC, G. *et al.* Production of ^{67}Ga by deuteron irradiation of zinc. *I.J.A.R.I.* **26**, 377 (1975)
WEINREICH, R., SCHULT, O. *et al.* Production of ^{123}I via the ^{127}I(d,6n)^{123}Xe → (β^{+}E.C.)^{123}I process. *I.J.A.R.I.* **25**, 535 (1974)
*WELCH, M.S. (ed.) Radiopharmaceuticals and other compounds labelled with short-lived radionuclides. *I.J.A.R.I.* **18**, 1 (1977)
WILLIS, J.N. Specific activities of carrier-free radio-isotope preparations. *I.J.A.R.I.* **24**, 354 (1973)
WONG, S., ACHE, H.J. On the production of ^{80}Br and ^{82}Br labelled molecules via excitation labelling methods. *I.J.A.R.I.* **27**, 19 (1976)
YAGI, M., KONDO , K. Preparation of carrier-free ^{47}Sc by the ^{48}Ti(γ,p) reaction. *I.J.A.R.I.* **28**, 463 (1977)
YANO, Y., ANGER, H.O. Production and processing of ^{52}Fe for medical use. *I.J.A.R.I.* **16**, 153 (1965)
YANO, Y., CHU, P. Cyclotron-produced thulium-167 for bone and tumour scanning. *I.J. Nuc. Med. Biol.* **2**, 135 (1975)
ZATOLOKIN, B.V., KONSTANTINOV, I.O. *et al.* Thick target yields of ^{34m}Cl and ^{38}Cl produced by various charged particles on phosphorus, sulphur, and chlorine targets. *I.J.A.R.I.* **27**, 159 (1976)

Special Techniques for Nuclear Medicine

ALPERT, N.M., MCKUSICK, K.A. *et al.* Initial assessment of a simple functional image of ventilation. *J. Nuc. Med.* **17**, 88 (1976)
BEIHN, R.M., DAMRON, J.R. *et al.* Subtraction technique for the detection of subphrenic abscesses using ^{67}Ga and ^{99m}Tc. *J. Nuc. Med.* **15**, 371 (1974)
BOSNJAKOVIC, V.B., BENNETT, L.R. *et al.* Dual isotope method for diagnosis of intracardiac shunts. *J. Nuc. Med.* **14**, 514 (1973)
DAMRON, J.R., BEIHN, R.M. *et al.* Gallium–technetium subtraction scanning for the localisation of subphrenic abscess. *Radiology* **113**, 117 (1974)

DUBOVSKY, E.V., LOGIC, J.R. *et al.* Comprehensive evaluation of renal function in the transplanted kidney. *J. Nuc. Med.* **16**, 1115 (1975)

FLEMING, J.S., GODDARD, B.A. *et al.* Regional ventilation assessment by transmission scintigraphy. *Acta Radiologica* **12**, 416 (1973)

FREEDMAN, G.S., DWYER, A. *et al.* Radionuclide determination of cardiac chamber flow/volume characteristics. *J. Nuc. Med.* **17**, 84 (1976)

*GRUNFELD, J.P., SABTO, J. *et al.* Methods for measurement of renal blood flow in man. *Sem. Nuc. Med.* **4**, 39 (1974)

*HOLLENBERG, N.K., MANGEL, R. *et al.* Assessment of intra-renal perfusion with radio-xenon. *Sem. Nuc. Med.* **6**, 193 (1976)

INKLEY, S.R., MACKINTYRE, W.J. Measurement of regional area gas exchange by perfusion and clearance of ^{133}Xe from the lung. *J. Nuc. Med.* **14**, 490 (1973)

KOTRAPPA, P., RAGHUNATH, B. *et al.* Scintiphotography of lungs with dry aerosol generation and delivery system. *J. Nuc. Med.* **18**, 1082 (1977)

KRAL, M., KUBA, J. *et al.* Determination of regional cerebral blood flow using a ^{133}Xe inhalation technique. *P.M.B.* **18**, 100 (1973)

OVERTON, T.R., FRIEDENBERG, L.W. *et al.* Multidetector instrumentation and data analysis for regional pulmonary function studies using ^{133}Xe. *P.M.B.* **18**, 246 (1973)

OSTROWSKI, S.T., TOTHILL, P. Kidney depth measurements using a double isotope technique. *Brit. J. Radiol.* **48**, 291 (1975)

SECKER-WALKER, R.H., HILL, R.I. *et al.* Measurement of regional ventilation in man—a new method of quantitation. *J. Nuc. Med.* **14**, 725 (1973)

TOTHILL, P., GALT, J.M. Quantitative profile scanning for measurement of organ radioactivity. *P.M.B.* **16**, 625 (1971)

VERBIST, A., CAPON, A. *et al.* A rapid method for evaluation of regional cerebral blood flow after intra-arterial injection of ^{133}Xe. *J. Nuc. Med.* **16**, 264 (1975)

*WRIGHT, R.R., TONO, M. *et al.* Blood volume. *Sem. Nuc. Med.* **5**, 63 (1975)

YANO, Y., MACRAE, J. *et al.* Lung function studies with ^{81m}Kr and the gamma camera. *J. Nuc. Med.* **11**, 674 (1970)

Whole-Body Counting

ANDRASI, A., KOTEL, G. NaI(Tl) detector efficiency calculations for distributed sources in a human phantom. *I.J.A.R.I.* **26**, 451 (1975)

*ANDREWS, G.A., GIBBS, W.D. *et al.* Whole-body counting. *Sem. Nuc. Med.* **3**, 367 (1973)

BARNABY, C.F., JASANI, B.M. Calibration of a high sensitivity large area whole-body counter. *P.M.B.* **13**, 561 (1968)

BARNABY, C.F., SMITH, T. Calibration of a whole-body monitor suitable for use in routine clinical investigations. *P.M.B.* **16**, 97 (1971)

BODDY, K., ELLIOT, A. *et al.* A high sensitivity dual detector shadow-shield whole-body counter with an invariant response for total body. *P.M.B.* **20**, 296 (1975)

BODDY, K., WILL, G. *et al.* An evaluation of ^{57}Co-labelled vitamin B-12 in a double-tracer test of absorption using a whole-body monitor. *P.M.B.* **14**, 455 (1969)

BODDY, K., KING, P.C. *et al.* Measurement of whole-body potassium using a shadow-shield whole-body counter. *P.M.B.* **16**, 275 (1971)

BODDY, K., HOLLOWAY, I. *et al.* A simple facility for total body *in vivo* activation analysis *I.J.A.R.I.* **24**, 428 (1973)

BRAUNSFORTH, J.S., GABBE, E.E. *et al.* Performance parameters of the Hamburg 4π whole-body radioactivity detector. *P.M.B.* **22**, 1 (1977)

CHEN, N.S., ELLIS, K.J. *et al.* Application of a coincidence counting technique in a fixed geometry whole-body counter. *I.J. Nuc. Med. Biol.* **1**, 175 (1974)

*COHN, S.H., PALMER, H.E. Recent advances in whole-body counting: a review. *J. Nuc. Med. Biol.* **1**, 155 (1974)

COOK, J.D., PALMER, H.E. *et al.* Measurement of iron absorption by whole-body counting. *P.M.B.* **15**, 467 (1970)

DAVIES, I.H., JACOBS, A. The use of a whole-body counter for measuring bone marrow erythroid activity by profile scanning. *J. Nuc. Med. Biol.* **1**, 145 (1974)

DELWAIDE, P.A. A 4π plastofluor body counter for clinical use: calibration. *I.J.A.R.I.* **20**, 623 (1969)

DUDLEY, R.A., BEN HAIM, A. Comparison of techniques for whole-body counting of gamma-emitting nuclides with NaI(Tl) detectors. *I.J.A.R.I.* **13**, 181 (1968)

DYMOCK, I.W., GODFREY, B.E. *et al.* A comparison of methods for determination of iron absorption using a whole-body counter. *P.M.B.* **16**, 269 (1971)

HODGES, H.D., GIBBS, W.D. *et al.* An improved high-level whole-body counter. *J. Nuc. Med.* **15**, 610 (1974)

ELL, P.J., MYERS, M.J. A comparison of a new whole-body scanner with a large crystal scanning camera in whole-body imaging. *Eur. J. Nuc. Med.* **2**, 281 (1977)

PALMER, H.E., COOK, J.D. *et al.* A new whole-body counter for precision *in vivo* measurement of radio-iron. *P.M.B.* **15**, 457 (1970)

TOTHILL, P. The possible effects of changes in relative organ position in whole-body counting. *P.M.B.* **22**, 769 (1977)

WARNER, G.T. The use of whole-body counters for the study of iron metabolism and iron loss. *Postgraduate Medical Journal* **49**, 477 (1973)

Answers to Questions

Chapter 1

3. $y \tan(x) - x \sec(x)$
6. $y = c \sec(t)$, with c any constant
7. $\{w\} = 0.2, 0.9, 2.6, 4.8, 6.5, 6.7, 7.3, 5.0, 1.2$
9. $y = (\exp(t))/3 + 2(\exp(4t))/3$

Chapter 2

1. Mass loss is 0.030376 unified mass units per molecule, i.e. 0.75% of the initial mass.

$$\begin{aligned}\text{Energy gain} &= \Delta mc^2 \\ &= 0.75 \times 10^{-2} \times (3 \times 10^8)^2 \text{ J} \\ &= 6.75 \times 10^{14} \text{ J}\end{aligned}$$

2. Writing E_m for the K, L, . . . mesonic energy levels, M_{Ca} for the calcium atom mass and m_m for the meson mass, the Bohr equation reads

$$E_m = 2\pi^2 - \frac{M_{Ca} m_m}{M_{Ca} + m_m} \cdot \frac{e^4 z^2}{h^3} \left(\frac{1}{n^2}\right)$$

By comparison with the equation for the electron energy levels E_e, one has for any fixed value of n that

$$E_m / E_e \approx m_m / m_e \approx 300$$

Therefore mesonic energy levels have approximately 300 times the energy of the corresponding electronic levels, and the mesonic X-rays will have 300 times the energy of the electronic (i.e. normal) X-rays. The wavelength of the mesonic K X-ray will be 0.34/300 = 0.0113 nm.

5. Assume 3×10^{15} nuclei of ^{198}Au at end of bombardment. Production rate of ^{199}Au is then

$$3 \times 10^{15} \times 2.6 \times 10^{-24} \times 10^{16} = 8 \times 10^7 \text{ nuclei/s}$$

Gold foil 0.2 mm thick would attenuate slow neutron flux through it by about 10%. The flux in a large compact mass (e.g. 1 kg) would be very highly attenuated.

6. Barrier height is approximately 27 MeV. Necessary thickness of lead target = 1.2 mm.
7. The complete equation for the concentration of solute, c, eluting when just F free-column-volumes have been used is

$$c = \frac{1}{\sqrt{(2\pi W)}} \exp\{-(F - F_p)^2/2W\}$$

Writing equations for two solutes with F_p and W values $F_p(1)$ and $F_p(2)$, and $W(1)$ and $W(2)$, we find that when the eluting concentrations are equal, $c(1) = c(2)$, and

$$\begin{aligned}&F^2\{W(1) - W(2)\}\\&+F\{2W(2)F_p(1) - 2W(1)F_p(2)\}\\&-W(2)F_p(1)^2 + W(1)F_p(2)^2 - 2W(1)W(2)\ln\{W(1)/W(2)\}\\&= 0\end{aligned}$$

Inserting $F_p(1)$, $F_p(2)$, $W(1)$ and $W(2)$ values derived from $D_v(1) = 10$, $D_v(2) = 15$, $l = 0.7$, $p = 50$, we find (ignoring a negative solution) $F \approx 18$. Then by summation of $c(1)$ and $c(2)$ values we find

$$\sum_0^{18} c(1) = 0.988 \qquad \text{and} \qquad \sum_0^{18} c(2) = 0.012$$

Similar calculations for $D_v = 20$ and for $p = 100$ will show the improvement to be expected from having more divergent D_v values, or more theoretical plates in the column. In general more widely different D_v values are to be preferred in order to get good analytical separations.

Chapter 3

1. KE of alpha particle = 3.75×10^3 MeV. An approximate treatment of the relativistic alpha-particle/electron collision is as follows. Denote the alpha-particle mass by M, the electron mass before collision by m_0 and after collision by m_e. By differentiation of the relation between the alpha-particle total energy E and its mass, it is seen that a small change in E is given by

$$\Delta E = c^2 \Delta M$$

Similarly a small change in alpha-particle momentum P is found to be

$$\Delta P = E\Delta E/Pc^2$$

$$= \frac{Ec}{\sqrt{(E^2 - E_0^2)}} \Delta M$$

Now the losses in energy and momentum of the alpha-particle must equal the gains made by the electron; so

$$-\Delta m = m_e - m_0$$
$$-\Delta P = m_e \beta c$$

With the simplifying assumption that β is approximately unity, the mass loss of the alpha particle, or the gain by the electron, is then

$$-\Delta m = m_0/(1/\surd(1 - E_0^2/E^2)^{-1})$$

and with the condition that $E = 2E_0$, this amounts to 6.46 electron rest-masses. The energy transferred is then 6.46 × 0.51 = 3.3 Mev. At this energy β is in fact 0.99.

2. (a) 3 MeV. (b) 2.46 and 3.47 in units MeV/c. (c) 63.88° and 71.81°.
3. From the range-energy equation of Section 3.5, the ranges are 2.15, 0.32 and 0.0067 kg m^{-2}.
4. The fundamental formulae are

$$\Delta\lambda = 0.0242(1 - \cos\theta)$$
$$\Delta = 1.24/E$$

A suitable HP25 program is as follows:

```
01  cos x       08  ÷
02  1           09  −
03  −           10  RCL 1
04  RCL 2       11  ÷
05  ×           12  1/x
06  RCL 1       13  CHS
07  RCL 3       14  GTO 00
```

Store 12.4 in STO 1,0.0242 in STO 2, E = energy of incident gamma ray into STO 3 (in keV). Successive angles are then written into the X register. Typical results: 150° scattering for 140, 350 keV gamma rays gives scattered gamma energies of 93, 154 keV.

5. (a) 12 mm (b) 48 mm (c) 180 mm, based on assumed values for mass attenuation coefficients of 0.07, 0.017 and 0.0045 $m^2\ kg^{-1}$ respectively.
6. Ratio of unchanged 1.28 MeV and 0.51 MeV quanta is approximately 8000.
9. 9×10^{-3} g
10. 2 min

Chapter 4

1. (a) The data allow N and dN/dt to be calculated for a sample of ^{238}U, whence $t_{1/2} = 4.5 \times 10^9$ years. (b) 9000. (c) 420.
2. For N_0 initial nuclei, the number c which decay in time τ is $N_0(1 - \exp(-\lambda\tau))$. Also the decay rate at time t is $\lambda N_0 \exp(-\lambda t)$. From

the conditions given,

$$\lambda N_0 \exp(-\lambda t) = \frac{c}{\tau} = \frac{N_0}{\tau}(1 - \exp(-\lambda\tau))$$

t is clearly nearly equal to $\tau/2$, so put $t = \tau/2 - \varepsilon$, where ε is small. Then

$$\exp(-\lambda\tau/2).\exp(\lambda\varepsilon) = \frac{1}{\lambda\tau}(1 - \exp(-\lambda\tau))$$

i.e.

$$\exp(\lambda\varepsilon) = \frac{1}{\lambda\tau}[\exp(\lambda\tau/2) - \exp(-\lambda\tau/2)$$

Expanding the exponentials on the left-hand side to the second term and the others to five terms, we have

$$\varepsilon = 0.0289(\tau/t_{1/2})\tau + 0.00072(\tau/t_{1/2})^2\tau + \cdots$$

so, for $(\tau/t_{1/2})$ less than 2, the second term and later terms on the right-hand side are less than 10% of the first term, and approximately

$$\varepsilon = 0.03(\tau/t_{1/2})\tau$$

3. Reasonable estimates from a graph would be $c_0 = 9000$ and $t_{1/2} = 80$ s. A useful HP25 program to calculate $c'(t)$ from estimates of $c(0)$ and $t_{1/2}$ is as follows:

01	↑	09	↓	17	RCL 2
02	2	10	RCL 5	18	−
03	ln x	11	×	19	STO 4
04	÷	12	CHS	20	x^2
05	$1/x$	13	exp x	21	STO + 3
06	STO 5	14	RCL 1	22	RCL 4
07	R/S	15	×	23	GTO 07
08	STO 2	16	R/S		

To use the program put $c(0)$ in STO 1, zero in STO 3. Write $t_{1/2}$ in the X register. Press R/S: the display is λ. Write a value of t in the X register. Press R/S. Write the corresponding observed value of $c(t)$ in the X register: press R/S. The display is then the calculated value of $c'(t)$. Press R/S: the display is the difference between observed and calculated counts. Press R/S: the display is the square of the latter quantity. Repeat (from step 07) for all experimental values. Finally press RCL 3 to display the sum of squares.

By trial a minimum value for the sum of the squared deviations is found near $c(0) = 9100$ and $t_{1/2} = 75$ s.

The criterion for best fit is not necessarily to minimise this sum (although in many practical cases it is not worth while to use more sophisticated techniques). If the variance s for each count were known it would be theoretically better to minimise the sum of $(\text{deviation})^2/s$ over all data.

4. A useful HP25 program for the Intercept Plot is as follows:

01	2	11	R/S	21	R/S
02	ln x	12	STO 0	22	RCL 3
03	RCL 1	13	↓	23	RCL 2
04	÷	14	STO 3	24	×
05	STO 1	15	RCL 1	25	CHS
06	2	16	×	26	exp x
07	ln x	17	CHS	27	RCL 0
08	RCL 2	18	exp x	28	÷
09	÷	19	RCL 0	29	GTO 11
10	STO 2	20	÷		

To use the program put $(t_{1/2})_1$ in STO 1 and $(t_{1/2})_2$ in STO 2. Press R/S. Write a value of t in the X register. Press ENTER. Write the corresponding value of $c(t)$ in the X register.
Press R/S: the display is $y = (\exp(-\lambda_1 t))/c(t)$.
Press R/S: the display is $x = (\exp(-\lambda_2 t))/c(t)$.
Repeat (from step 11) for all data values.

From a graph of (x, y) values obtained from the given data one finds approximately 155 000 and 24 300 as the initial counting rates of the two components.

Chapter 5

1. Most line-spread functions approximate closely to a Gaussian. On this assumption the following HP25 program gives a useful approximation to the FWHM of a distribution by estimating the half-width on each side of the curve.

01	↑	11	STO + 2	21	RCL 3	31	÷
02	2	12	RCL 0	22	×	32	RCL 2
03	÷	13	GTO 06	23	.	33	+
04	STO 0	14	STO + 3	24	2	34	.
05	STO 1	15	R/S	25	3	35	5
06	R/S	16	GTO 14	26	9	36	+
07	$x \gtrless y$	17	.	27	RCL 1	37	GTO 00
08	GTO 14	18	7	28	×		
09	STO + 1	19	6	29	−		
10	1	20	1	30	RCL 0		

To use the program, clear all registers and write the value of the maximum ordinate, say y_n, in the X register. Press R/S. Write in y_{n+1}: press R/S and repeat for all succeeding values $y_{n+2}, y_{n+3}, y_{n+4} \ldots$ Finally press GTO 17, R/S when the display will be the width of the right-hand side of the distribution. Doubling this gives the FWHM if the distribution is symmetrical; otherwise the calculation can be repeated for the other side of the distribution and the results added.

The program is based on the principle that a Gaussian distribution the area under the curve between the limits of the FWHM is 76.1% of the whole. Since the calculation depends on the sums of ordinates it is rather insensitive to statistical fluctuations in individual values.

2. The hole radius must be 3.3 mm. For a ^{99m}Tc collimator, there can be 17 holes across the collimator diameter, giving a plane-source efficiency of 1.84 mm^2. For a ^{18}F collimator the wall thickness should be about 7 mm. There are then 9 holes across the diameter and the plane-source efficiency is 0.52 mm^2.
3. 0.08 Gy
4. 0.002 Gy

Chapter 6

1. $C_1 = \{k_1/(k_1 + k_2) + k_2/(k_1 + k_2) \exp[-(k_1 + k_2)t]\}C_1(0)$
   $C_2 = k_2/(k_1 + k_2)\{1 - \exp[-(k_1 + k_2)t]\}C_1(0)$
2. $R_1 = 74\ \Omega$, $C_1 = 1.7\ \mu$F, $C_2 = 0.3\ \mu$F.
3. Decay factor 0.9

$$\text{Output} = \{1, 0.9, 0.81, 0.729, \ldots\} * \{250, 600, 650, 700, 200\}$$
$$= 250, 825, 1392, 1953, 1957, 1762, 1585, \ldots$$

An elegant HP25 program to carry out the calculation is as follows:

```
01  x ≷ y
02  RCL 0
03  ×
04  +
05  GTO 00
```

To use the program put the decay factor in STO 0. Then write each number in the input sequence, followed by R/S, in the X register. The displays are the convolution required.

4. For $\nu = 0.5$, the MTF calculated for counts up to 10 mm displacement is 0.465; calculation including counts up to 15 mm gives 0.433. Further counts with values obtained by extrapolation of the given series changes the MTF only in the third place of decimals.

Index